QCD Perspectives on Hot and Dense Matter

NATO Science Series

A Series presenting the results of scientific meetings supported under the NATO Science Programme.

The Series is published by IOS Press, Amsterdam, and Kluwer Academic Publishers in conjunction with the NATO Scientific Affairs Division

Sub-Series

I. Life and Behavioural Sciences	IOS Press
II. Mathematics, Physics and Chemistry	Kluwer Academic Publishers
III. Computer and Systems Science	IOS Press
IV. Earth and Environmental Sciences	Kluwer Academic Publishers
V. Science and Technology Policy	IOS Press

The NATO Science Series continues the series of books published formerly as the NATO ASI Series.

The NATO Science Programme offers support for collaboration in civil science between scientists of countries of the Euro-Atlantic Partnership Council. The types of scientific meeting generally supported are "Advanced Study Institutes" and "Advanced Research Workshops", although other types of meeting are supported from time to time. The NATO Science Series collects together the results of these meetings. The meetings are co-organized bij scientists from NATO countries and scientists from NATO's Partner countries – countries of the CIS and Central and Eastern Europe.

Advanced Study Institutes are high-level tutorial courses offering in-depth study of latest advances in a field.
Advanced Research Workshops are expert meetings aimed at critical assessment of a field, and identification of directions for future action.

As a consequence of the restructuring of the NATO Science Programme in 1999, the NATO Science Series has been re-organised and there are currently Five Sub-series as noted above. Please consult the following web sites for information on previous volumes published in the Series, as well as details of earlier Sub-series.

http://www.nato.int/science
http://www.wkap.nl
http://www.iospress.nl
http://www.wtv-books.de/nato-pco.htm

Series II: Mathematics, Physics and Chemistry – Vol. 87

QCD Perspectives on Hot and Dense Matter

edited by

J.-P. Blaizot

Service de Physique Théorique,
CEA-Saclay, Gif-sur-Yvette, France

and

E. Iancu

Service de Physique Théorique,
CEA-Saclay, Gif-sur-Yvette, France

Kluwer Academic Publishers

Dordrecht / Boston / London

Published in cooperation with NATO Scientific Affairs Division

Proceedings of the NATO Advanced Study Institute on
QCD Perspectives on Hot and Dense Matter
Cargèse, France
August 6–18, 2001

A C.I.P. Catalogue record for this book is available from the Library of Congress.

ISBN 1-4020-1036-2

Published by Kluwer Academic Publishers,
P.O. Box 17, 3300 AA Dordrecht, The Netherlands.

Sold and distributed in North, Central and South America
by Kluwer Academic Publishers,
101 Philip Drive, Norwell, MA 02061, U.S.A.

In all other countries, sold and distributed
by Kluwer Academic Publishers,
P.O. Box 322, 3300 AH Dordrecht, The Netherlands.

Printed on acid-free paper

Printed in the Netherlands.

Contents

DOMINIQUE VAUTHERIN

Dominique Vautherin passed away on December 7, 2000. He was one of the initiators of this project and he contributed to shape the program of the school. Cargèse is a place he knew well since he had organised already two schools at the Institute for Scientific Studies: "Heavy Ion Physics" in 1984, "Hadrons and Hadronic Matter" in 1989.

Dominique Vautherin was well known in the nuclear physics community for his pioneering work on Hartree-Fock calculations using the Skyrme interaction. He had broad interests in physics, which led him to contribute to other fields: astrophysics, ultrarelativistic heavy ion collisions, development of variational methods for non abelian gauge theories and, more recently, Bose-Einstein condensation. (Details on Dominique Vautherin's carrier can be found in Ref. [1].)

All participants would have enjoyed having him around, sharing his joy at doing physics. We also believe that Dominique would have liked this school. We dedicate this book to him.

Jean-Paul Blaizot and Edmond Iancu

[1] M. Soyeur et al. Nucl. Phys. **A 690** (2001) 331.

ACKNOWLEDGEMENTS

We are indebted to Ms. Dubois-Violette, director of IESC, and to the local staff in Cargèse for their kind anf efficient help. The atmosphere at the Institute for Scientific Studies in Cargese has been ideal for promoting the interactions between participants.

Special thanks are due to Ms. Ariano and Ms. Cassegrain for their continuous assistance in the preparation and the administration of the School.

The school "QCD perspectives on hot and dense matter" has been made possible thanks to the support of several institutions:

- the Scientific Committee of NATO accepted the school as an Advanced Study Institute and provided a grant allowing also for the publication of this volume.

- the European Commission, through its "Human Potential Programme: High Level Scientific Conferences"

- the Centre National pour la Recherche Scientifique (CNRS) through the program "formation permanente".

The school was co-organized by A. Leonidov, L. McLerran and the editors of the present volume. D. Vautherin was one of the initiators of the project and contributed to the early preparation of the program.

Last but not least, we would like to thank all the participants, lecturers and students, for their enthusiasm and application. They all contributed to make this summer school a memorable experience.

J.-P. Blaizot and E. Iancu

PREFACE

Ultrarelativistic heavy-ion collisions in the newly accessible energy regimes offer the possibility to reach a deeper understanding of new phases of Quantum Chromo-Dynamics (QCD) and, more generally, of the dynamics of gauge fields under extreme and non-equilibrium conditions. The Relativistic Heavy Ion Collider (RHIC), operating at the Brookhaven National Laboratory since 1999, has already produced a wealth of new data, and brings the study of nucleus-nucleus collisions carried out over the last fifteen years at the CERN/SPS to a new energy regime. A few years from now, the Large Hadron Collider (LHC) at CERN will offer nuclear beams at still higher energies. All these experimental advances have, over the last few years, triggered numerous theoretical developments, and generated new ideas.

The compexity of the physical conditions achieved in these heavy ion experiments calls for a tight interplay between theory and experiment, and also for interdisciplinary theoretical efforts combining various methods which traditionally belong to different fields. Quantum Chromo-Dynamics stays at the heart of all these developments, and it was the main motivation for the school "QCD perspectives on hot and dense matter" to expose the students to the many facets of QCD which are involved in these complex phenomena.

Thus, *perturbative QCD* and *small-x physics* become necessary ingredients for understanding the early stages of nuclear collisions. *Nonequilibrium field theory* methods, both analytical and numerical, are required to describe the pre-thermalization stage. *QCD thermodynamics*, as studied by *weak coupling techniques* and *lattice calculations*, is relevant to describe the equilibrated Quark-Gluon Plasma. The final stages of the collision require an understanding of the mechanisms of *hadronization* and *multiparticle production*. Finally appropriate *observables* need to be identified to allow for an unambiguous characterization of the matter produced in the collisions.

The courses are organized in roughly three categories: theoretical aspects of *high-energy scattering* in QCD, the QCD *many-body problem* dealing with various aspects of the equilibrated quark-gluon plasma and the

Speaking about the final state structure, we cannot predict, say, the kaon multiplicity or the pion energy spectrum. However, one can decide to be not too picky and concentrate on global characteristics of the final states rather than on the yield of specific hadrons. Being sufficiently inclusive with respect to final hadron species, one can rely on a picture of the energy-momentum flow in hard collisions supplied by pQCD — the jet pattern.
There are well elaborated procedures for counting jets (CIS jet finding algorithms) and for quantifying the internal structure of jets (CIS jet shape variables). They allow the study of the gross features of the final states while staying away from the physics of hadronisation. Along these lines one visualizes asymptotic freedom, checks out gluon spin and colour, predicts and verifies scaling violation pattern in hard cross sections, etc. These and similar checks have constituted the basic QCD tests of the past two decades.
This epoch is over. Now the High Energy Particle physics community is trying to probe genuine confinement effects in hard processes to learn more about strong interactions. The programme is ambitious and provocative. Friendly phenomenology keeps it afloat and feeds our hopes of extracting valuable information about physics of hadronisation.

2. Bremsstrahlung gluons at work

High-energy annihilation $e^+e^- \to$ hadrons, deep inelastic lepton-hadron scattering (DIS), production in hadron-hadron collisions of massive lepton pairs, heavy quarks and their bound states, large transverse momentum jets and photons are classical examples of hard processes.
Copious production of hadrons is typical for all these processes. On the other hand, at the microscopic level, multiple quark-gluon "production" is to be expected as a result of QCD bremsstrahlung — gluon radiation accompanying abrupt creation/scattering of colour partons.
Is there a correspondence between observable hadron and calculable quark-gluon production?

2.1. SCALING VIOLATION PATTERN

An indirect evidence that gluons are there, and that they behave, can be obtained from the study of the scaling violation pattern. QCD quarks (and gluons) are not point-like particles, as the orthodox parton model once assumed. Each of them is surrounded by a proper field coat — a coherent virtual cloud consisting of gluons and "sea" $q\bar{q}$ pairs. A hard probe applied to such a dressed parton breaks coherence of the cloud. Constituents of these field fluctuations are then released as particles accompanying the hard interaction. The harder the hit, the larger an intensity of bremsstrahlung and, therefore, the fraction of the energy-momentum

of the dressed parton that the bremsstrahlung quanta typically carry away. Thus we should expect, in particular, that the probability that a "bare" core quark carries a large fraction of the energy of its dressed parent will decrease with increase of Q^2. And so it does.
The logarithmic scaling violation pattern in DIS structure functions is well established and meticulously follows the QCD prediction based on the parton evolution picture.

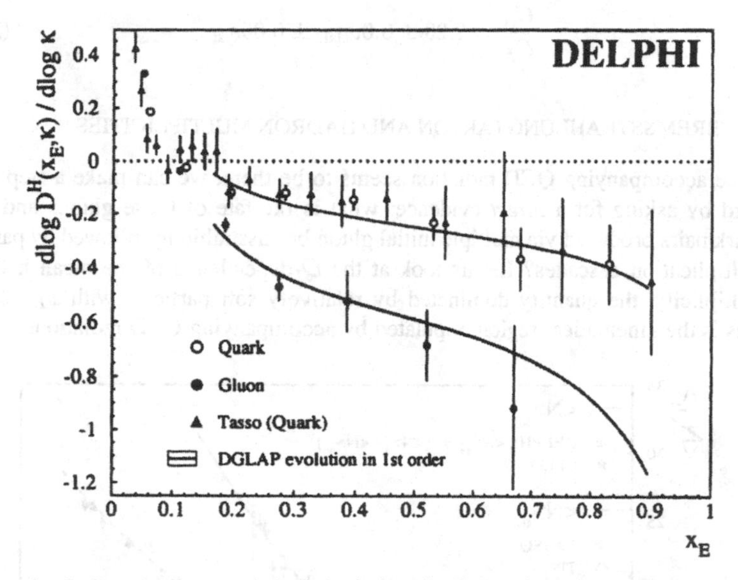

Figure 1. Scaling violation rates in inclusive hadron distributions from gluon and quark jets [1]

In DIS we look for a "bare" quark inside a target dressed one. In e^+e^- hadron annihilation at large energy $s = Q^2$ the chain of events is reversed. Here we produce instead a bare quark with energy $Q/2$, which then "dresses up". In the process of restoring its proper field-coat our parton produces (a controllable amount of) bremsstrahlung radiation which leads to formation of a hadron jet. Having done so, in the end of the day it becomes a constituent of one of the hadrons that hit the detector. Typically, this is the leading hadron. However, the fraction x_E of the initial energy $Q/2$ that is left to the leader depends on the amount of accompanying radiation and, therefore, on Q^2 (the larger, the smaller). In fact, the same rule (and the same formula) applies to the scaling violation pattern in e^+e^- fragmentation functions (time-like parton evolution) as to that in the DIS parton distributions (space-like evolution).
What makes the annihilation channel particularly interesting, is that the present day experiments are so sophisticated that they provide us with a near-to-perfect

separation between quark- and gluon-initiated jets (the latter being extracted from heavy-quark-tagged three-jet events).
In Fig. 1 a comparison is shown of the scaling violation rates in the hadron spectra from gluon and quark jets, as a function of the hardness scale κ that characterizes a given jet [1]. For large values of $x_E \sim 1$ the ratio of the logarithmic derivatives is predicted to be close to that of the gluon and quark "colour charges", $C_A/C_F = 9/4$. Experimentally, the ratio is measured to be

$$\frac{C_A}{C_F} = 2.23 \pm 0.09_{\text{stat.}} \pm 0.06_{\text{syst.}}. \tag{2.1}$$

2.2. BREMSSTRAHLUNG PARTON AND HADRON MULTIPLICITIES

Since accompanying QCD radiation seems to be there, we can make a step forward by asking for a *direct* evidence: what is the fate of those gluons and sea quark pairs produced via multiple initial gluon bremsstrahlung followed by parton multiplication cascades? Let us look at the Q-dependence of the mean hadron multiplicity, the quantity dominated by relatively soft particles with $x_E \ll 1$. This is the kinematical region populated by accompanying QCD radiation.

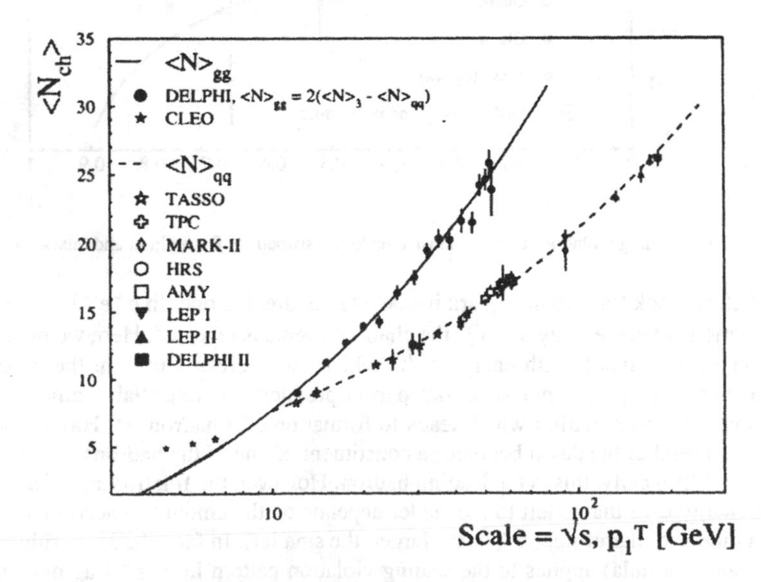

Figure 2. Charged hadron multiplicities in gluon and quark jets [1].

Fig. 2 demonstrates that the hadron multiplicity increases with the hardness of the jet proportional to the multiplicity of secondary gluons and sea quarks. The ratio of the slopes, once again, provides an independent measure of the ratio of the colour charges, which is consistent with (2.1) [1]:

$$\frac{C_A}{C_F} = 2.246 \pm 0.062_{\text{stat.}} \pm 0.008_{\text{syst.}} \pm 0.095_{\text{theo.}}. \tag{2.2}$$

2.3. INCLUSIVE HADRON DISTRIBUTION IN JETS

Since the total numbers match, it is time to ask a more delicate question about energy-momentum distribution of final hadrons versus that of the underlying parton ensemble. One should not be too picky in addressing such a question. It is clear that hadron-hadron correlations, for example, will show resonant structures about which the quark-gluon speaking pQCD can say little, if anything, at the present state of the art. Inclusive single-particle distributions, however, have a better chance to be closely related. Triggering a single hadron in the detector, and a single parton on paper, one may compare the structure of the two distributions to learn about dynamics of hadronisation.

Inclusive energy spectrum of soft bremsstrahlung partons in QCD jets has been derived in 1984 in the so-called MLLA — the Modified Leading Logarithmic Approximation ([2, 3]).

This approximation takes into account all essential ingredients of parton multiplication in the next-to-leading order. They are: parton splitting functions responsible for the energy balance in parton splitting, the running coupling $\alpha_s(k_\perp^2)$ depending on the relative transverse momentum of the two offspring and exact angular ordering. The latter is a consequence of soft gluon coherence and plays, as we shall discuss below, an essential rôle in parton dynamics. In particular, gluon coherence suppresses multiple production of very small momentum gluons. It is particles with intermediate energies that multiply most efficiently. As a result, the energy spectrum of relatively soft secondary partons in jets acquires a characteristic hump-backed shape. The position of the maximum in the logarithmic variable $\xi = -\ln x$, the width of the hump and its height increase with Q^2 in a predictable way.

The shape of the inclusive spectrum of all charged hadrons (dominated by $\pi^\pm$) exhibits the same features. This comparison, pioneered by Glen Cowan (ALEPH) and the OPAL collaboration, has later become a standard test of analytic QCD predictions. First scrutinized at LEP, the similarity of parton and hadron energy distributions has been verified at SLC and KEK e^+e^- machines, as well as at HERA and Tevatron where hadron jets originate not from bare quarks dug up from the vacuum by a highly virtual photon/Z^0 but from hard partons kicked out from initial hadron(s).

In Fig. 3 (DELPHI) the comparison is made of the all-charged hadron spectra at various annihilation energies Q with the so-called "distorted Gaussian" fit

(Fong & Webber 1989) which employs the first four moments (the mean, width, skewness and kurtosis) of the MLLA distribution around its maximum.

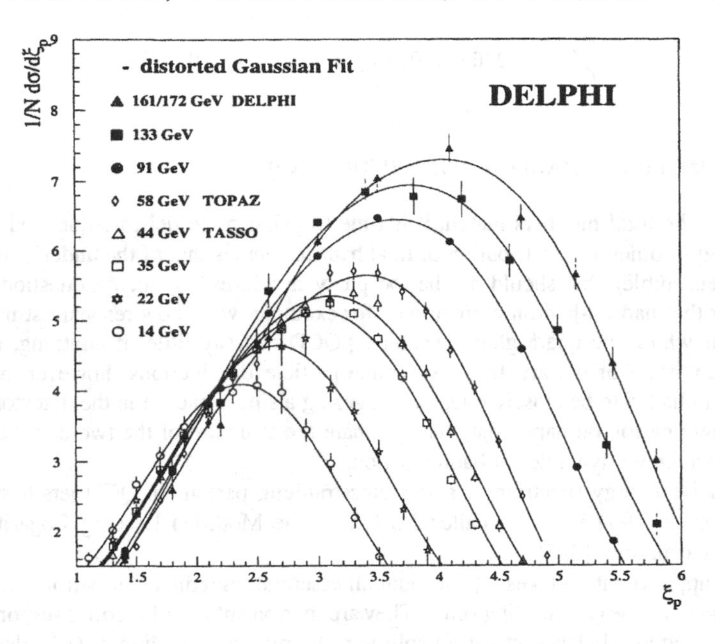

Figure 3. Inclusive energy distribution of charged hadrons in jets produced in e^+e^- annihilation

Shall we say, a (routine, interesting, wonderful) check of yet another QCD prediction? Better not. Such a close similarity offers a deep puzzle, even a worry, rather than a successful test. Indeed, after a little exercise in translating the values of the logarithmic variable $\xi = \ln(E_{\rm jet}/p)$ in Fig. 3 into GeVs you will see that the actual hadron momenta at the maxima are, for example, $p=\frac{1}{2}Q \cdot e^{-\xi_{\max}} \simeq 0.42$, 0.85 and 1.0 GeV for Q=14, 35 GeV and at LEP-1, Q=91 GeV. Is it not surprising that the pQCD spectrum is mirrored by that of the pions (which constitute 90% of all charged hadrons produced in jets) with momenta well below 1 GeV?!

For this very reason the observation of the parton-hadron similarity was initially met with a serious and well grounded skepticism: it looked more natural (and was more comfortable) to blame the finite hadron mass effects for fall-off of the spectrum at large ξ (small momenta) rather than seriously believe in applicability of the pQCD consideration down to such disturbingly small momentum scales.

This worry has been recently answered. Andrey Korytov (CDF) was the first to hear a theoretical suggestion [4] and carry out a study of the energy distribution of hadrons produced inside a restricted angular cone Θ around the jet axis. Theoretically, it is not the energy of the jet but the maximal parton transverse momentum inside it, $k_{\perp\max} \simeq E_{\rm jet}\sin\frac{\Theta}{2}$, that determines the hardness scale and thus the yield and the distribution of the accompanying radiation.

This means that by choosing a small opening angle one can study relatively small hardness scales but in a cleaner environment: due to the Lorentz boost effect, eventually all particles that form a short small-Q^2 QCD "hump" are now relativistic and concentrated at the tip of the jet.

For example, selecting hadrons inside a cone $\Theta \simeq 0.14$ around an energetic quark jet with $E_{\rm jet} \simeq 100$ GeV (LEP-2) one should see that very "dubious" $Q = 14$ GeV curve in Fig. 3 but now with the maximum boosted from 0.45 GeV into a comfortable 6 GeV range.

In the CDF Fig. 4 [5, 6, 7] a close similarity between the hadron yield and the full MLLA parton spectra can no longer be considered accidental and be attributed to non-relativistic kinematical effects.

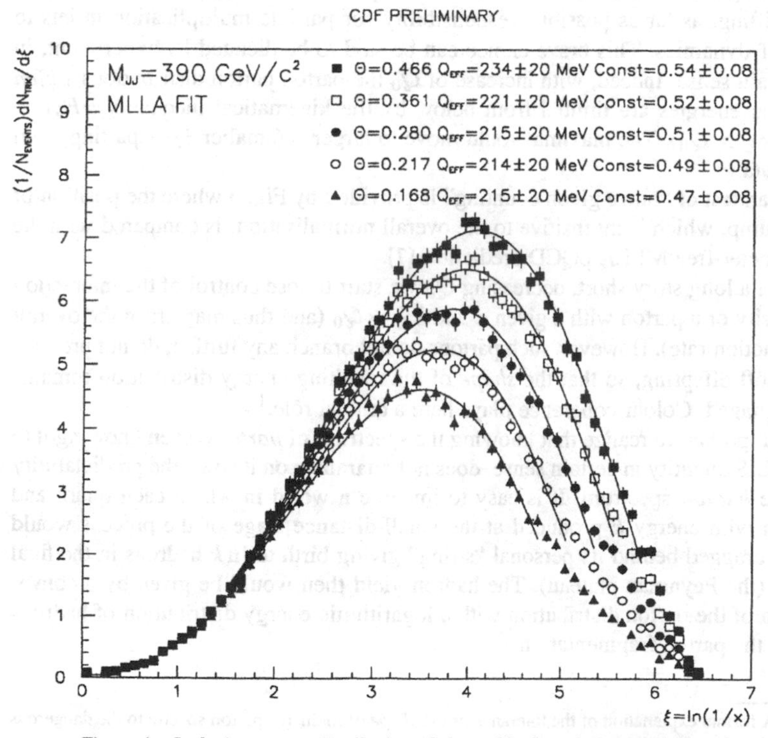

Figure 4. Inclusive energy distribution of charged hadrons in large-$p_\perp$ jets [6].

2.4. BRAVE GLUON COUNTING

Modulo $\Lambda_{\rm QCD}$, there is only one unknown in this comparison, namely, the overall normalisation of the spectrum of hadrons relative to that of partons (bremsstrahlung gluons).

Strictly speaking, there should/could have been another free parameter, the one which quantifies one's bravery in applying the pQCD dynamics. It is the minimal transverse momentum cutoff in parton cascades, $k_\perp > Q_0$. The strength of successive $1 \to 2$ parton splittings is proportional to $\alpha_s(k_\perp^2)$ and grows with $k_\perp$ decreasing. The necessity to terminate the process at some low transverse momentum scale where the PT coupling becomes large (and eventually hits the formal "Landau pole" at $k_\perp = \Lambda_{\rm QCD}$) seems imminent. Surprisingly enough, it is not.

Believe it or not, the inclusive parton energy distribution turns out to be a CIS QCD prediction. Its crazy $Q_0 = \Lambda_{\rm QCD}$ limit (the so-called "limiting spectrum") is shown by solid curves in Fig. 4.

Choosing the minimal value for the collinear parton cutoff Q_0 can be looked upon as shifting, as far as possible, responsibility for particle multiplication in jets to the PT dynamics. This brave choice can be said to be dictated by experiment, in a certain sense. Indeed, with increase of Q_0 the parton parton distributions *stiffen* (parton energies are limited from below by the kinematical inequality $xE_{\rm jet} \equiv k \geq k_\perp > Q_0$). The maxima would move to larger x (smaller ξ), departing from the data.

A clean test of "brave gluon counting" is provided by Fig. 5 where the position of the hump, which is insensitive to the overall normalisation, is compared with the parameter-free MLLA pQCD prediction [7].

To put a long story short, decreasing Q_0 we start to lose control of the interaction intensity of a parton with a given x and $k_\perp \sim Q_0$ (and thus may err in the overall production rate). However, such partons do not branch any further, do not produce any soft offspring, so that the *shape* of the resulting energy distribution remains undamaged. Colour coherence plays here a crucial rôle.[1]

It is important to realize that knowing the spectrum of *partons*, even knowing it to be a CIS quantity in certain sense, does not guarantee on its own the predictability of the *hadron* spectrum. It is easy to imagine a world in which each quark and gluon with energy k produced at the small-distance stage of the process would have dragged behind its personal "string" giving birth to $\ln k$ hadrons in the final state (the Feynman plateau). The hadron yield then would be given by a convolution of the parton distribution with a logarithmic energy distribution of hadrons from the parton fragmentation.

[1] A formal explanation of the tolerance of the *shape* of inclusive parton spectra to the dangerous small-$k_\perp$ domain will be given below in Sec. 4.4.

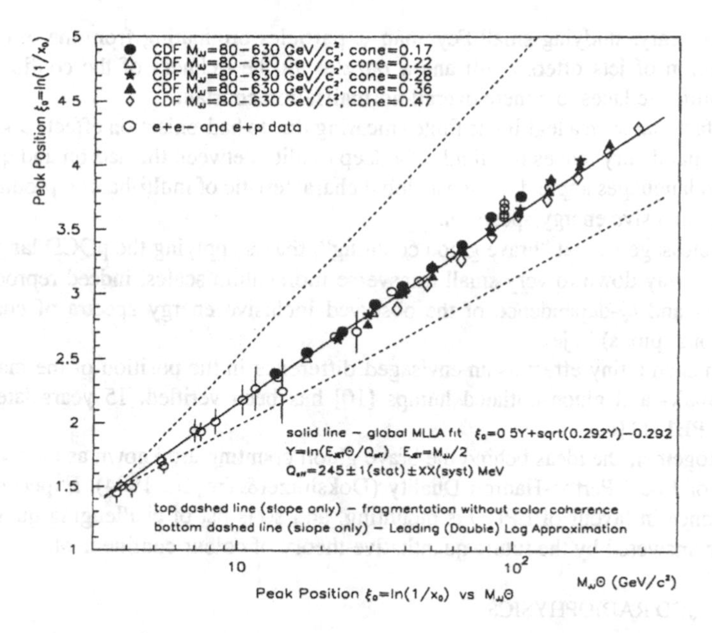

Figure 5. **The position of the maximum versus the analytic MLLA prediction (Safonov, 1999).**

If it were the case, each parton would have contributed to the yield of non-relativistic hadrons and the hadron spectra would peak at much smaller energies, $\xi_{\max} \simeq \ln Q$, in a spectacular difference with experiment.
Physically, it could be possible if the non-perturbative (NP) hadronisation physics did not respect the basic rule of the perturbative dynamics, namely, that of colour coherence.
There is nothing wrong with the idea of convoluting time-like parton production in jets with the inclusive NP parton→hadron fragmentation function, the procedure which is similar to convoluting space-like parton cascades with the NP initial parton distributions in a target proton to describe DIS structure functions.
What the nature is telling us, however, is that this NP fragmentation has a finite multiplicity and is *local* in the momentum space. Similar to its PT counterpart, the NP dynamics has a short memory: the NP conversion of partons into hadrons occurs locally in the configuration space.
In spite of a known similarity between the space- and time-like parton evolution pictures ($x \sim 1$), there is an essential difference between *small–x* physics of DIS structure functions and the jet fragmentation. In the case of the space-like evolution, in the limit of small Bjorken–x the problem becomes essentially non-perturbative and pQCD loses control of the DIS cross sections [8, 9]. On

the contrary, studying small Feynman–x particles originating from the time-like evolution of jets offers a gift and a puzzle: all the richness of the confinement dynamics reduces to a mere overall normalisation constant.
The fact that even a legitimate finite smearing due to hadronisation effects does not look mandatory makes one think of a deep duality between the hadron and quark-gluon languages applied to such a global characteristic of multi-hadron production as an inclusive energy spectrum.
The message is, that "brave gluon counting", that is applying the pQCD language all the way down to very small transverse momentum scales, indeed reproduces the x- and Q-dependence of the observed inclusive energy spectra of charged hadrons (pions) in jets.
Even such a tiny effect as an envisaged difference in the position of the maxima in quark- and gluon-initiated humps [10] has been verified, 15 years later, by DELPHI [11].
Put together, the ideas behind the brave gluon counting are known as the hypothesis of Local Parton-Hadron Duality (Dokshitzer&Troyan, 1984). Experimental evidence in favour of LPHD is mounting, and so is list of challenging questions to be answered by the future quantitative theory of colour confinement.

2.5. QCD RADIOPHYSICS

Even more striking is *miraculously* successful rôle of gluons in predicting the pattern of hadron multiplicity flows in the inter-jet regions — realm of various *string/drag* effects.This is another class of multi-hadron production phenomena speaking in favour of LPHD. It deals with particle flows in the angular regions *between jets* in various multi-jet configurations. These particles do not belong to any particular jet, and their production, at the pQCD level, is governed by *coherent* soft gluon radiation off the multi-jet system as a whole. The ratios of particle (gluon) flows in different inter-jet valleys are given by parameter-free pQCD predictions and reveal the so-called "string" or "drag" effects. For a given kinematical jet configuration such ratios depend only on the number of colours (N_c).
It isn't strange at all that with *gluons* one can get, e.g., $1+1 = 2$ while $1+1+9/4 = 7/16$, which is a simple *radiophysics* of composite antennas, or *quantum mechanics* of conserved colour charges.
This particular example of "quantum arithmetics" has to do with comparison of hadron flows in the inter-quark valleys in $q\bar{q}\gamma$ and $q\bar{q}g$ (3-jet) events. The first equation describes the density of soft gluon radiation produced by two quarks in a $q\bar{q}\gamma$ event, with 1 standing for the colour quark charge.
Replacing the colour-blind photon by a gluon one gets an additional emitter with the relative strength 9/4, as shown in the l.h.s. of the second equation. The resulting soft gluon yield in the $q\bar{q}$ direction, however, *decreases* substantially as a result of destructive interference between three elements of a composite colour

antenna. In Fig. 6 the OPAL measurements are compared with the parameter-free theoretical prediction [12].

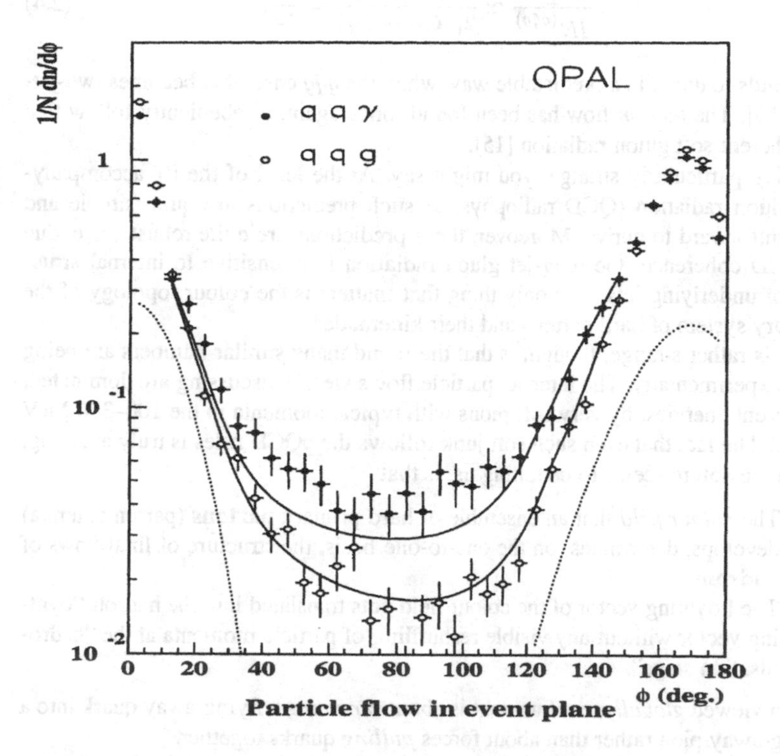

Figure 6. Comparison of particle flows in the $q\bar{q}$ valley in $q\bar{q}\gamma$ and $q\bar{q}g$ 3-jet events versus a parameter-free analytic prediction based on the soft gluon radiation pattern. [13]

Another example is the ratio of the multiplicity flow between a quark (antiquark) and a gluon to that in the $q\bar{q}$ valley in symmetric ("Mercedes") three-jet $q\bar{q}g$ e^+e^- annihilation events:

$$\frac{dN_{qg}^{(q\bar{q}g)}}{dN_{q\bar{q}}^{(q\bar{q}g)}} \simeq \frac{5N_c^2 - 1}{2N_c^2 - 4} = \frac{22}{7}. \tag{2.3}$$

Emitting an energetic gluon off the initial quark pair depletes accompanying radiation in the backward direction: colour is *dragged* out of the $q\bar{q}$ valley. This destructive interference effect is so strong that the resulting multiplicity flow falls below that in the least favourable direction *transversal* to the three-jet event plane.

For symmetric "Mercedes" events the expected ratio is

$$\frac{dN_{\perp}^{(q\bar{q}\gamma)}}{dN_{q\bar{q}}^{(q\bar{q}g)}} \simeq \frac{N_C + 2C_F}{2(4C_F - N_c)} = \frac{17}{14}, \tag{2.4}$$

and tends to unity, in a predictable way, when the $q\bar{q}g$ ensemble becomes two-jet-like [14]. The *hadron* flow has been found, once again, to obediently follow that of coherent soft gluon radiation [15].
Nothing particularly strange, you might say. At the level of the PT accompanying gluon radiation (QCD radiophysics) such predictions are quite simple and straightforward to derive. Moreover, these predictions are quite robust since, due to QCD coherence, the inter-jet gluon radiation is insensitive to internal structure of underlying jets. The only thing that matters is the colour topology of the primary system of hard partons and their kinematics.
What **is** rather strange, though, is that these and many similar numbers are being seen experimentally. The inter-jet particle flows we are discussing are dominated, at present energies, by very soft pions with typical momenta in the 100–300 MeV range! The fact that even such soft junk follows the pQCD rules is truly amazing. What the nature seems to be telling us, is that

- The *colour field* that an ensemble of hard primary **partons** (parton antenna) develops, determines, on the one-to-one basis, the structure of final flows of **hadrons.**
- The Poynting vector of the colour field gets translated into the hadron Poynting vector without any visible reshuffling of particle momenta at the "hadronisation stage".

When viewed *globally*, confinement is about *renaming* a flying-away quark into a flying-away pion rather than about forces *pulling* quarks together.

3. Basics of parton multiplication

In this lecture we shall recall the basic properties of accompanying radiation. The gluon bremsstrahlung (off quarks and gluons) is not much different from the photon emission off electric charges. So, we shall start from the electromagnetic radiation and turn to gluons later.

3.1. PHOTON BREMSSTRAHLUNG

Let us consider photon bremsstrahlung induced by a charged particle (electron) which scatters off an external field (e.g., a static electromagnetic field). The derivation is included in every textbook on QED, so we confine ourselves to the essential aspects.

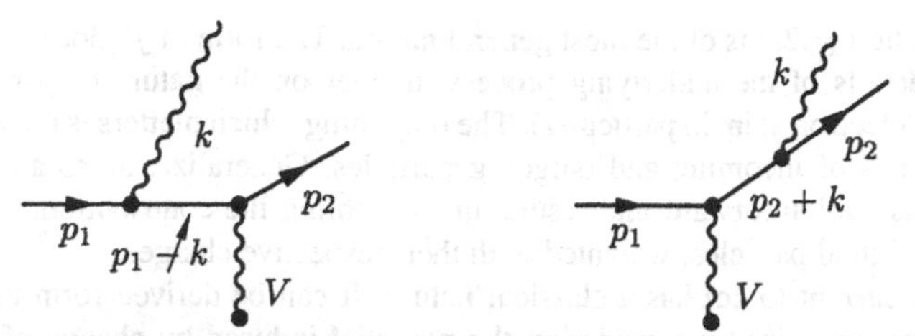

Figure 7. **Photon Bremsstrahlung diagrams for scattering off an external field.**

The lowest order Feynman diagrams for photon radiation are depicted in Fig. 7, where p_1, p_2 are the momenta of the incoming and outgoing electron respectively and k represents the momentum of the emitted photon. The corresponding amplitudes, according to the Feynman rules, are given in momentum space by

$$M_{\mathrm{i}}^{\mu} = e\,\bar{u}(p_2, s_2)\, V(p_2 + k - p_1)\, \frac{m + \not{p}_1 - \not{k}}{m^2 - (p_1 - k)^2}\, \gamma^{\mu}\, u(p_1, s_1), \tag{3.1a}$$

$$M_{\mathrm{f}}^{\mu} = e\,\bar{u}(p_2, s_2)\, \gamma^{\mu}\, \frac{m + \not{p}_2 + \not{k}}{m^2 - (p_2 + k)^2}\, V(p_2 + k - p_1)\; u(p_1, s_1). \tag{3.1b}$$

Here V stands for the basic interaction amplitude which may depend in general on the momentum transfer (for the case of scattering off the static e.m. field, $V = \gamma^0$). First we apply the soft-photon approximation, $\omega \ll p_1^0, p_2^0$, to neglect $\not{k}$ terms in the numerators. To deal with the remaining matrix structure in the numerators of (3.1) we use the identity $\not{p}\,\gamma^{\mu} = -\gamma^{\mu}\not{p} + 2\,p^{\mu}$ and the Dirac equation for the on-mass-shell electrons,

$$(m + \not{p}_1)\, \gamma^{\mu}\, u(p_1) = (2p_1^{\mu} + [\,(m - \not{p}_1\,])\, u(p_1) = 2p_1^{\mu}\, u(p_1)\,,$$
$$\bar{u}(p_2)\, \gamma^{\nu}\, (m + \not{p}_2) = \bar{u}(p_2)\, ([\,(m - \not{p}_2\,] + 2p_1^{\nu}) = 2p_2^{\nu}\, \bar{u}(p_2)\,.$$

Denominators for real electrons ($p_i^2 = m^2$) and the photon ($k^2 = 0$) become $m^2 - (p_1 - k)^2 = 2(p_1 k)$ and $m^2 - (p_2 + k)^2 = -2(p_2 k)$, so that for the total amplitude we obtain the factorized expression

$$M^{\mu} = e\, j^{\mu}\, \times\, M_{\mathrm{el}}\,. \tag{3.2a}$$

Here M_{el} is the Born matrix element for non-radiative (elastic) scattering,

$$M_{\mathrm{el}} = \bar{u}(p_2, s_2)\, V(p_2 - p_1)\, u(p_1, s_1) \tag{3.2b}$$

(in which the photon recoil effect has been neglected, $q = p_2 + k - p_1 \simeq p_2 - p_1$), and j^{μ} is the *soft accompanying radiation current*

$$j^{\mu}(k) = \frac{p_1^{\mu}}{(p_1 k)} - \frac{p_2^{\mu}}{(p_2 k)}\,. \tag{3.2c}$$

Factorisation (3.2a) is of the most general nature. The form of j^μ does not depend on the details of the underlying process, neither on the nature of participating charges (electron spin, in particular). The only thing which matters is the momenta and charges of incoming and outgoing particles. Generalization to an arbitrary process is straightforward and results in assembling the contributions due to all initial and final particles, weighted with their respective charges.

The soft current (3.2c) has a classical nature. It can be derived form the classical electrodynamics by considering the potential induced by change of the e.m. current due to scattering.

3.2. SOFT RADIATION CROSS SECTION

To calculate the radiation probability we square the amplitude projected onto a photon polarization state ε^λ_μ, sum over λ and supply the photon phase space factor to write down

$$dW = e^2 \sum_{\lambda=1,2} \left|\varepsilon^\lambda_\mu j^\mu\right|^2 \frac{\omega^2\, d\omega\, d\Omega_\gamma}{2\,\omega\,(2\pi)^3}\, dW_{\rm el}\ . \tag{3.3}$$

The sum runs over two physical polarization states of the real photon, described by normalized polarization vectors orthogonal to its momentum:

$$\epsilon^\mu_\lambda(k)\cdot\epsilon^*_{\mu,\lambda'}(k) = -\delta_{\lambda\lambda'}\,,\quad \epsilon^\mu_\lambda(k)\cdot k_\mu = 0\,;\qquad \lambda,\lambda'=1,2\,.$$

Within these conditions the polarization vectors may be chosen differently. Due to the gauge invariance such an uncertainty does not affect physical observables. Indeed, the polarization tensor may be represented as

$$\sum_{\lambda=1,2} \epsilon^\mu_\lambda \epsilon^{*\nu}_\lambda \;=\; -g^{\mu\nu} \;+\; \text{tensor proportional to } k^\mu \text{ and/or } k^\nu\,. \tag{3.4}$$

The latter, however, can be dropped since the classical current (3.2c) is explicitly conserving, $(j^\mu k_\mu)=0$. Therefore one may enjoy the gauge invariance and employ an arbitrary gauge, instead of using the physical polarizations, to calculate accompanying photon production.

The Feynman gauge being the simplest choice, $\sum_{\lambda=1,2}\epsilon^\mu_\lambda\epsilon^{*\nu}_\lambda \Longrightarrow -g^{\mu\nu}$, we get

$$\begin{aligned} dN \equiv \frac{dW}{dW_{\rm el}} &= -\frac{\alpha}{4\pi^2}\,(j^\mu)^2\,\omega\, d\omega\, d\Omega_\gamma \\ &\simeq \frac{\alpha}{\pi}\frac{d\omega}{\omega}\frac{d\Omega_\gamma}{2\pi}\,\frac{1-\cos\Theta_s}{(1-\cos\Theta_1)(1-\cos\Theta_2)}\,. \end{aligned} \tag{3.5}$$

The latter expression corresponds to the relativistic approximation $1-v_1, 1-v_2 \ll 1$:

$$-(j^\mu)^2 = \frac{2(p_1p_2)}{(p_1k)(p_2k)} + \mathcal{O}\left(\frac{m^2}{p_0^2}\right) \simeq \frac{2}{\omega^2}\,\frac{(1-\vec{n}_1\cdot\vec{n}_2)}{(1-\vec{n}_1\cdot\vec{n})(1-\vec{n}_2\cdot\vec{n})}\,;$$

it disregards the contribution of *very* small emission angles $\Theta_i^2 \lesssim (1 - v_i^2) = m^2/p_{0i}^2 \ll 1$, where the soft radiation vanishes (the so-called "Dead Cone" region).
If the photon is emitted at a small angle with respect to, say, the incoming particle, i.e. $\Theta_1 \ll \Theta_2 \simeq \Theta_s$, the radiation spectrum (3.5) simplifies to

$$dN \simeq \frac{\alpha}{\pi} \frac{\sin\Theta_1 \, d\Theta_1}{(1 - \cos\Theta_1)} \frac{d\omega}{\omega} \simeq \frac{\alpha}{\pi} \frac{d\Theta_1^2}{\Theta_1^2} \frac{d\omega}{\omega} .$$

Two bremsstrahlung cones appear, centred around incoming and outgoing electron momenta. Inside these cones the radiation has a *double-logarithmic* structure, exhibiting both the *soft* ($d\omega/\omega$) and *collinear* ($d\Theta^2/\Theta^2$) enhancements.

3.2.1. *Low-Barnett-Kroll wisdom*

Soft factorisation (3.2a) is an essence of the celebrated soft bremsstrahlung theorem, formulated by Low in 1956 for the case of scalar charged particles and later generalized by Barnett and Kroll to charged fermions. The very classical nature of soft radiation makes it universal with respect to intrinsic quantum properties of participating objects and the nature of the underlying scattering process: it is only the classical movement of electromagnetic charges that matters.
It is interesting that according to the LBK theorem both the leading $d\omega/\omega$ and the first subleading, $\propto d\omega$, pieces of the soft photon spectrum prove to be "classical". For the sake of simplicity we shall leave aside the angular structure of the accompanying photon emission and concentrate on the energy dependence. Then, the relation between the basic cross section $\sigma^{(0)}$ and that with one additional photon with energy ω can be represented symbolically as

$$d\sigma^{(1)}(p_i, \omega) \propto \frac{\alpha}{\pi} \frac{d\omega}{\omega} \left[\left(1 - \frac{\omega}{E}\right) \cdot \sigma^{(0)}(p_i) + \left(\frac{\omega}{E}\right)^2 \cdot \tilde{\sigma}(p_i, \omega) \right] . \qquad (3.6)$$

The first term in the right-hand side is proportional to the non-radiative cross section $\sigma^{(0)}$. The second term involves the new ω-dependent cross section $\tilde{\sigma}$ which is finite at $\omega = 0$, so that this contribution is suppressed for small photon energies as $(\omega/E)^2$.
This general structure has important consequences, the most serious of which can be formulated, in a dramatic fashion, as

3.2.2. *Soft Photons don't carry quantum numbers*

We are inclined to think that the photon has definite quantum numbers (negative C-parity, in particular). Imagine that the basic process is forbidden, say, by C-parity conservation. Why not to take off the veto by adding a photon to the system? Surely enough it can be done. There is, however, a price to pay: the selection rules cannot be overcome by *soft* radiation. Since the classical part of the radiative cross section in (3.6) is explicitly proportional to the non-radiative cross section $\sigma^{(0)} = 0$, only *energetic* photons (described by the $\tilde{\sigma}$ term) could do the job. The energy distribution

$$|M|^2 \cdot \frac{d^3k}{\omega} \propto \omega d\omega$$

is typical for a quantum particle, where the production matrix element M is finite in the $\omega \to 0$ limit, $M = \mathcal{O}(1)$. An enhanced radiation matrix element, $M \propto \omega^{-1}$ characterizes a classical field rather than a quantum object.
So, the price one has to pay to overrule the quantum-number veto by emitting a soft photon with $\omega \ll E$ is the suppression factor $(\omega/E)^2 \ll 1$. We conclude that the photons that are capable of changing the quantum numbers of the system (be it parity, C-parity or angular momentum) cannot be *soft*. Neither can they be *collinear*, by the way, as it follows from the

3.2.3. *Gribov Bremsstrahlung theorem*
This powerful generalisation of the Low theorem states that a simple factorisation holds at the level of the *matrix element*, provided the photon transverse momentum with respect to the radiating charged particle is small compared to the momentum transfers characterising the underlying scattering process:

$$M^{(1)} \propto \frac{(\vec{k}_\perp \cdot \vec{e})}{k_\perp^2} \cdot M^{(0)} + \tilde{M}. \tag{3.7}$$

Here again $\tilde{M} = \text{const}$ in the $k_\perp \to 0$ limit. This factorisation holds for *hard* photons ($\omega \sim E$) as well as for soft ones.
Both the Low-Barnett-Kroll and the Gribov theorems hold in QCD as well. In particular, it is the Gribov collinear factorisation that leads to the probabilistic evolution picture describing collinear QCD parton multiplication which we shall briefly discuss in the next lecture.
In the QCD context, our statement that "soft photons don't carry quantum numbers" should be strengthened to even more provocative (but true)

3.2.4. *Soft Gluons don't carry away no colour*
Don't rush to protest. Just think it over. In more respectable terms this title can be abbreviated as the NSFL (no-soft-free-lunch) theorem.
Imagine we want to produce a heavy quark $Q\overline{Q}$ bound state (onium) in a hadron-hadron collision. The C-even (χ_Q) mesons can be produced by fusing two quasi-real gluons (with opposite colours) from the QCD parton clouds of the colliding hadrons:

$$(g+g)_{(1)} \to Q + \overline{Q} \to \chi_Q \,. \tag{3.8}$$

In particular, radiative decays of such χ_c mesons are responsible for about 40% of the J/ψ yield. How about the remaining 60% ? To directly create a J/ψ (or ψ' —

$^3S_1$ C-odd $c\bar{c}$ states) two gluons isn't enough. A C-odd meson can decay into, or couple to, *three* photons (like para-positronium does), a photon plus two gluons, or *three* gluons (in a colour-symmetric d_{abc} state).
So, we need one more gluon to attach, for example, in the final state:

$$(g+g)_{(8)} \to \left(Q+\overline{Q}\right)_{(8)} \to J/\psi + g\,. \tag{3.9}$$

To pick up an initial gg pair in a colour octet state is easier than in the singlet as in (3.8). This, however, does not help to avoid the trouble: the perturbative cross section turns out to be too small to meet the need. It underestimates the Tevatron $p\bar{p}$ data on direct J/ψ and ψ' production by a large factor (up to 50, at large $p_\perp$). That very same effect that makes the J/ψ so narrow a meson with the small hadronic decay width $\Gamma_{J/\psi}/M \propto \alpha_s^3(M)$, suppresses its perturbative production cross section (3.9) as well.
Since the perturbative approach apparently fails, it seemed natural to blame the non-perturbative physics. Why not to perturbatively form a colour-octet "J/ψ" and then to get rid of colour in a smooth (free of charge) non-perturbative way? To *evaporate* colour does not look problematic: on the one hand, the soft glue distribution is $d\omega/\omega = \mathcal{O}(1)$, on the other hand, the coupling α_s/π in the NP domain may be of the order of unity as well. So why not?
The LBK theorem tells us that either the radiation is soft-enhanced, $\propto d\omega/\omega = \mathcal{O}(1)$, and *classical*, or hard, $\propto \omega d\omega$ and capable of changing the quantum state of the system. Therefore, to rightfully participate in the J/ψ formation as a quantum field, a NP gluon with $\omega \sim \Lambda_{\rm QCD}$ would have to bring in the suppression factor

$$\left(\frac{\Lambda_{\rm QCD}}{M_c}\right)^2 \ll 1\,.$$

The *language* of the LBK is perturbative, 'tis true. The question is, and a serious one indeed, whether the NP phenomena respect the basic dynamical features that its PT counterpart does? Or shall we rather forget about quantum mechanics, colour conservation, etc. and accept an "anything goes" motto in the NP domain? To avoid our discussion turning theological, we better address another verifiable issue namely, photoproduction of J/ψ at HERA. Here we have instead of (3.9) the fusion process of a real (photoproduction) or virtual (electroproduction) photon with a quasi-real space-like gluon from the parton cloud of the target proton:

$$\gamma^{(*)} + g \to \left(Q+\overline{Q}\right)_{(8)} \to J/\psi + g\,. \tag{3.10}$$

If the final-state gluon were soft NP junk, the J/ψ meson would have carried the whole photon momentum and its distribution in Feynman z would peak at $z = 1$ as $(1-z)^{-1}$. The HERA experiments have found instead a flatish (if not vanishing) z-spectrum at large z. The NSFL theorem seems to be up and running.

By the way, the conventional PT treatment of the photoproduction (3.10) is reportedly doing well. So, what is wrong with the hadroproduction then? Strictly speaking, the problem is still open. An alternative to (3.9) would be to look for the third (hard or hard*ish*) gluon in the initial state[2].
The NSFL QCD discourse has taken us quite far from the mainstream of the introductory lecture. Let us return to the basic properties of QED bremsstrahlung and make a comparative study of

3.3. INDEPENDENT AND COHERENT RADIATION

In the Feynman gauge, the accompanying radiation factor dN in (3.5) is dominated by the *interference* between the two emitters:

$$dN \propto -\left[\frac{p_1^\mu}{(p_1k)} - \frac{p_2^\mu}{(p_2k)}\right]^2 \approx \frac{2(p_1p_2)}{(p_1k)(p_2k)}\,.$$

Therefore it does not provide a satisfactory answer to the question, which part of radiation is due to the initial charge and which is due to the final one?
There is a way, however, to give a reasonable answer to this question. To do that one has to sacrifice simplicity of the Feynman-gauge calculation and recall the original expression (3.3) for the cross section in terms of physical photon polarizations. It is natural to choose the so-called *radiative* (temporal) gauge based on the 3-vector potential $\vec{A}$, with the scalar component set to zero, $A_0 \equiv 0$. Our photon is then described by (real) 3-vectors orthogonal to one another and to its 3-momentum:

$$(\vec{\epsilon}_\lambda \cdot \vec{\epsilon}_{\lambda'}) = \delta_{\lambda\lambda'}\,, \qquad (\vec{\epsilon}_\lambda \cdot \vec{k}\,) = 0\,. \tag{3.11}$$

This explicitly leaves us with *two* physical polarization states. Summing over polarizations obviously results in

$$dN \propto \sum_{\lambda=1,2} \left|\vec{j}(k)\cdot\vec{e}_\lambda\right|^2 = \sum_{\alpha,\beta=1...3} \vec{j}^{\,\alpha}(k)\cdot\left[\delta_{\alpha\beta} - \vec{n}_\alpha\vec{n}_\beta\right]\cdot\vec{j}^{\,\beta}(k)\,, \tag{3.12}$$

with α, β the 3-dimensional indices. We now substitute the soft current (3.2c) in the 3-vector form, $p_i^\mu \to \vec{v}_i p_{0i}$, and make use of the relations

$$(\vec{v}_i)_\alpha\left[\delta_{\alpha\beta} - \frac{k_\alpha\, k_\beta}{\vec{k}^2}\right](\vec{v}_i)_\beta = v_i^2 \sin^2\Theta_i\,, \tag{3.13a}$$

$$(\vec{v}_1)_\alpha\left[\delta_{\alpha\beta} - \frac{k_\alpha\, k_\beta}{\vec{k}^2}\right](\vec{v}_2)_\beta = v_1v_2(\cos\Theta_{12} - \cos\Theta_1\cos\Theta_2)\,, \tag{3.13b}$$

[2] an interesting, reliable and predictive model for production of onia in the gluon field of colliding hadrons is being developed by Paul Hoyer and collaborators [16]

to finally arrive at

$$dN = \frac{\alpha}{\pi}\left\{\mathcal{R}_1 + \mathcal{R}_2 - 2\mathcal{J}\right\}\cdot\frac{d\omega}{\omega}\frac{d\Omega}{4\pi}. \tag{3.14a}$$

Here

$$\mathcal{R}_i = \frac{v_i^2\sin^2\Theta_i}{(1-v_i\cos\Theta_i)^2}, \quad i=1,2, \tag{3.14b}$$

$$\mathcal{J} \equiv \frac{v_1v_2(\cos\Theta_{12}-\cos\Theta_1\cos\Theta_2)}{(1-v_1\cos\Theta_1)(1-v_2\cos\Theta_2)}. \tag{3.14c}$$

The contributions $\mathcal{R}_{1,2}$ can be looked upon as being due to *independent radiation* off initial and final charges, while the $\mathcal{J}$-term accounts for *interference* between them. The independent and interference contribution, taken together, describe the *coherent* emission. It is straightforward to verify that (3.14) is identical to the Feynman-gauge result (3.5):

$$\mathcal{R}_{\text{coher.}} \equiv \mathcal{R}_{\text{indep}} - 2\mathcal{J} = -\omega^2(j^\mu)^2, \qquad \mathcal{R}_{\text{indep.}} \equiv \mathcal{R}_1 + \mathcal{R}_2. \tag{3.15}$$

3.3.1. *The rôle of interference: strict angular ordering*

In the relativistic limit we have

$$\mathcal{R}_1 \simeq \frac{\sin^2\Theta_1}{(1-\cos\Theta_1)^2} = \frac{2}{a_1} - 1, \tag{3.16a}$$

$$\mathcal{J} \simeq \frac{\cos\Theta_{12}-\cos\Theta_1\cos\Theta_2}{(1-\cos\Theta_1)(1-\cos\Theta_2)} = \frac{a_1+a_2-a_{12}}{a_1a_2} - 1 \tag{3.16b}$$

where we introduced a convenient notation

$$a_1 = 1 - \vec{n}\vec{n}_1 = 1-\cos\Theta_1, \quad a_2 = 1-\cos\Theta_2,$$
$$a_{12} = 1-\vec{n}_1\vec{n}_2 = 1-\cos\Theta_s.$$

The variables a are small when the angles are small: $a \simeq \frac{1}{2}\Theta^2$.
The independent radiation has a typical logarithmic behaviour up to large angles:

$$dN_1 \propto \mathcal{R}_1\sin\Theta d\Theta \propto \frac{da_1}{a_1}, \qquad a_1 \lesssim 1.$$

However, the interference effectively cuts off the radiation at angles exceeding the scattering angle:

$$dN \propto \mathcal{R}_{\text{coher.}}\sin\Theta d\Theta = 2a_{12}\frac{da}{a_1a_2} \propto \frac{da}{a^2} \propto \frac{d\Theta^2}{\Theta^4}, \qquad a \equiv a_1 \simeq a_2 \gg a_{12}.$$

To quantify this coherent effect, let us combine an independent contribution with a half of the interference one to define

$$V_1 = \mathcal{R}_1 - \mathcal{J} = \frac{2}{a_1} - \frac{a_1 + a_2 - a_{12}}{a_1 a_2} = \frac{a_{12} + a_2 - a_1}{a_1 a_2}, \qquad V_2 = \mathcal{R}_2 - \mathcal{J} = \frac{2}{a_2} - \frac{a_1 + a_2 - a_{12}}{a_1 a_2} = \frac{a_{12} + a_1 - a_2}{a_1 a_2}; \tag{3.17a}$$

$$\mathcal{R}_{\text{coher}} = V_1 + V_2 . \tag{3.17b}$$

The emission probability V_i can be still considered as "belonging" to the charge $\#i$ (V_1 is singular when $a_1 \to 0$, and vice versa). At the same time these are no longer *independent* probabilities, since V_1 explicitly depends on the direction of the partner-charge #2; *conditional* probabilities, so to say.
It is straightforward to verify the following remarkable property of the "conditional" distributions V: after *averaging* over the *azimuthal angle* of the radiated quantum, $\vec{n}$, with respect to the direction of the parent charge, $\vec{n}_1$, the probability $V_1(\vec{n}, \vec{n}_1; \vec{n}_2)$ *vanishes* outside the Θ_s-cone. Namely

$$\langle V_1 \rangle_{\text{azimuth}} \equiv \int_0^{2\pi} \frac{d\phi_{n,n_1}}{2\pi} V_1(\vec{n}, \vec{n}_1; \vec{n}_2) = \frac{2}{a_1} \vartheta\left(a_{12} - a_1\right) . \tag{3.18}$$

It is only a_2 that changes under the integral (3.18), while a_1, and obviously a_{12}, stay fixed. The result follows from the angular integral

$$\int_0^{2\pi} \frac{d\phi_{n,n_1}}{2\pi} \frac{1}{a_2} = \frac{1}{|\cos\Theta_1 - \cos\Theta_s|} = \frac{1}{|\, a_{12} - a_1|} .$$

Naturally, a similar expression for V_2 emerges after the averaging over the azimuth around $\vec{n}_2$ is performed.
We conclude that as long as the *total* (angular-integrated) emission probability is concerned, the result can be expresses as a sum of two independent bremsstrahlung cones centred around $\vec{n}_1$ and $\vec{n}_2$, both having the finite opening half-angle Θ_s.
This nice property is known as a "strict angular ordering". It is an essential part of the so-called Modified Leading Log Approximation (MLLA), which describes the internal structure of parton jets with a single-logarithmic accuracy.

3.3.2. *Angular ordering on the back of envelope*

What is the reason for radiation at angles exceeding the scattering angle to be suppressed? Let us try our physical intuition and consider semi-classically how the radiation process really develops.
A physical electron is a charge surrounded by its proper Coulomb field. In quantum language the Lorentz-contracted Coulomb-disk attached to a relativistic particle may be treated as consisting of photons virtually emitted and, in due time,

re-absorbed by the core charge. Such virtual emission and absorption processes form a coherent state which we call a physical electron ("dressed" particle).
This coherence is partially destroyed when the charge experiences an impact. As a result, a part of intrinsic field fluctuations gets released in the form of real photon radiation: the bremsstrahlung cone in the direction of the initial momentum develops. On the other hand, the deflected charge now leaves the interaction region as a "half-dressed" object with its proper field-coat lacking some field components (eventually those that were lost at the first stage). In the process of regenerating the new Coulomb-disk adjusted to the final-momentum direction, an extra radiation takes place giving rise to the second bremsstrahlung cone.
Now we need to be more specific to find out which momentum components of the electromagnetic coat do actually take leave.
A typical time interval between emission and re-absorption of the photon k by the initial electron p_1 may be estimated as the Lorentz-dilated lifetime of the virtual intermediate electron state $(p_1 - k)$ (see the left graph in Fig. 7,

$$t_{\text{fluct}} \sim \frac{E_1}{|m^2 - (p_1 - k)^2|} = \frac{E_1}{2p_1 k} \sim \frac{1}{\omega\Theta^2} = \frac{\omega}{k_\perp^2} \cdot \tag{3.19}$$

Here we restricted ourselves, for simplicity, to small radiation angles, $k_\perp \approx \omega\Theta \ll k_{||} \approx \omega$. The fluctuation time (3.19) may become macroscopically large for small photon energies ω and enters as a characteristic parameter in a number of QED processes. As an example, let us mention the so called Landau-Pomeranchuk effect — suppression of soft radiation off a charge that experiences multiple scattering propagating through a medium. Quanta with too large a wavelength get not enough time to be properly formed before successive scattering occurs, so that the resulting bremsstrahlung spectrum behaves as $dN \propto d\omega/\sqrt{\omega}$ instead of the standard logarithmic $d\omega/\omega$ distribution.
The characteristic time scale (3.19) responsible for this and many other radiative phenomena is often referred to as the *formation time*.
Now imagine that within this interval the core charge was kicked by some external interaction and has changed direction by some Θ_s. Whether the photon will be re-absorbed or not depends on the position of the scattered charge with respect to the point where the photon was expecting to meet it "at the end of the day". That is, we need to compare the spatial displacement of the core charge $\Delta\vec{r}$ with the characteristic size of the photon field, $\lambda_{||} \sim \omega^{-1}$, $\lambda_\perp \sim {k_\perp}^{-1}$:

$$\begin{aligned} \Delta r_{||} &\sim \left| v_{2||} - v_{1||} \right| \cdot t_{\text{fluct}} \sim \Theta_s^2 \cdot \frac{1}{\omega\Theta^2} = \left(\frac{\Theta_s}{\Theta}\right)^2 \lambda_{||} \Leftrightarrow \lambda_{||}\,; \\ \Delta r_\perp &\sim \quad c\Theta_s \cdot t_{\text{fluct}} \sim \Theta_s \cdot \frac{1}{\omega\Theta^2} = \left(\frac{\Theta_s}{\Theta}\right) \lambda_\perp \Leftrightarrow \lambda_\perp\,. \end{aligned} \tag{3.20}$$

For large scattering angles, $\Theta_s \sim 1$, the charge displacement exceeds the photon wavelength for arbitrary Θ, so that the two full-size bremsstrahlung cones are

present. For numerically small $\Theta_s \ll 1$, however, it is only photons with $\Theta \lesssim \Theta_s$ that can notice the charge being displaced and thus the coherence of the state being disturbed. Therefore only the radiation at angles smaller than the scattering angle actually emerges. The other field components have too large a wavelength and are easily re-absorbed *as if* there were no scattering at all.
So what counts is a change in the current, which is sharp enough to be noticed by the "to-be-emitted" quantum within the characteristic formation/field-fluctuation time (3.19) of the latter.
Radiation at large angles has too short a formation time to become aware of the acceleration of the charge. No scattering — no radiation.
The same argument applies to the dual process of production of two opposite charges (decay of a neutral object, vacuum pair production, etc.). The only difference is that now one has to take for $\Delta\vec{r}$ not a displacement between the initial and the final charges, but the actual distance between the produced particles (spatial size of a dipole), to be compared with the radiation wavelength.

3.4. QCD SCATTERING AND CROSS-CHANNEL RADIATION

Both the qualitative arguments of the previous sections and the quantitative analysis of the two-particle antenna pattern apply to the QCD process of gluon emission in the course of quark scattering. So two gluon-bremsstrahlung cones with the opening angles restricted by the scattering angle Θ_s would be expected to appear. There is an important subtlety, however. In the QED case it was deflection of an electron that changed the e.m. current and caused photon radiation. In QCD there is another option, namely to "repaint" the quark. Rotation of the *colour state* would affect the colour current as well and, therefore, must lead to gluon radiation irrespectively of whether the quark-momentum direction has changed or not.
This is what happens when a quark scatters off a *colour* field. To be specific, one may consider as an example two channels of Higgs production in hadron-hadron collisions.

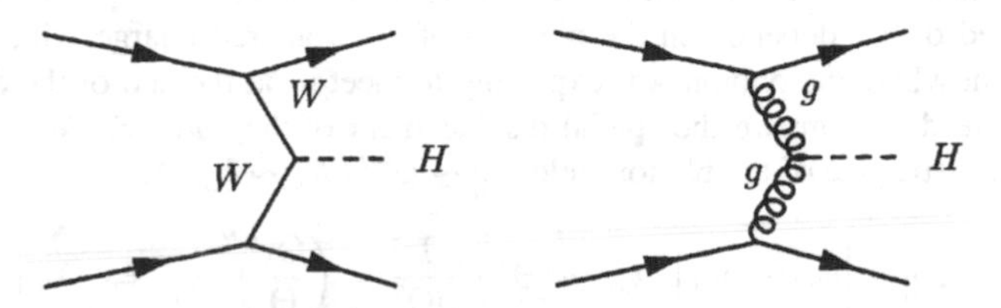

Figure 8. WW and gluon-gluon fusion graphs for Higgs production

At very high energies two mechanisms of Higgs production become competitive: $W^+W^- \to H$ and the gluon-gluon fusion $gg \to H$ (see Fig. 8).

Since the typical momentum transfer is large, of the order of the Higgs mass, $(-t) \sim M_H^2$, Higgs production is a *hard* process. Colliding quarks experience hard scattering with characteristic scattering angles $\Theta_s^2 \simeq |t|/s \sim M_H^2/s$. As far as the accompanying gluon radiation is concerned, the two subprocesses differ with respect to the nature of the "external field", which is *colorless* for the W-exchange and *colorful* for the gluon fusion.

The gluon bremsstrahlung amplitudes for the second case are shown in Fig. 9. In principle, a graph with the gluon-gluon interaction vertex should also be considered. However, in the limit $k_\perp \ll q_\perp$, with $\vec{q}_\perp \approx \vec{p}_{2\perp} - \vec{p}_{1\perp}$ the momentum transfer in the scattering process, emission off the external lines dominates (the "soft insertion rules").

The accompanying soft radiation current j^μ factors out from the Feynman amplitudes of Fig. 9, the only difference with the Abelian current (3.2c) being the order of the colour generators:

$$j^\mu = \left[t^b\, t^a \left(\frac{p_1^\mu}{(p_1 k)} \right) - t^a\, t^b \left(\frac{p_2^\mu}{(p_2 k)} \right) \right] . \tag{3.21}$$

Introducing the abbreviation $A_i = \frac{p_i^\mu}{(p_i k)}$, we apply the standard decomposition of the product of two triplet colour generators,

$$t^a t^b = \frac{1}{2N_c} \delta_{ab} + \tfrac{1}{2} \left(d_{abc} + \mathrm{i} f_{abc} \right) t^c ,$$

to rewrite (3.21) as

$$j^\mu = \tfrac{1}{2}(A_1 - A_2) \left(\frac{1}{N} \delta^{ab} + d^{abc} t^c \right) - \tfrac{1}{2}(A_1 + A_2)\, \mathrm{i} f^{abc} t^c .$$

To find the emission *probability* we need to construct the product of the currents and sum over colours. Three colour structures do not "interfere", so it suffices to

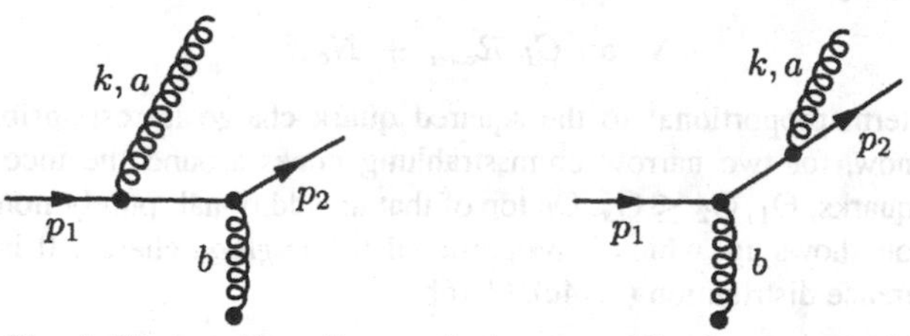

Figure 9. Gluonic Bremsstrahlung diagrams for $k_\perp \ll q_\perp$. The characters a and b denote the colours of the radiated and exchanged gluons.

evaluate the squares of the singlet, $\mathbf{8}_s$ and $\mathbf{8}_a$ structures:

$$\sum_{a,b}\left(\frac{1}{2N_c}\delta_{ab}\right)^2 = \left(\frac{1}{2N}\right)^2 (N_c^2-1) = \frac{1}{2N_c}\cdot C_F\,;$$

$$\sum_{a,b}\left(\tfrac{1}{2}\,d_{abc}\,t^c\right)^2 = \frac{1}{4}\,\frac{N_c^2-4}{N_c}\,(t^c)^2 = \frac{N_c^2-4}{4N_c}\cdot C_F\,;$$

$$\sum_{a,b}\left(\tfrac{1}{2}\,if_{abc}\,t^c\right)^2 = \frac{1}{4}\,N_c\,(t^c)^2 = \frac{N_c}{4}\cdot C_F\,.$$

The common factor $C_F = \left(t^b\right)^2$ belongs to the Born (non-radiative) cross section, so that the radiation spectrum takes the form

$$dN \propto \frac{1}{C_F}\sum_{\text{colour}} j^\mu\cdot(j_\mu)^* = \left(\frac{1}{2N_c}+\frac{N_c^2-4}{4N_c}\right)(A_1-A_2)\cdot(A_1-A_2) + \frac{N_c}{4}\,(A_1+A_2)\cdot(A_1+A_2)\,.$$

A simple algebra leads to

$$dN \propto C_F\,(A_1-A_2)\cdot(A_1-A_2) \,+\, N_c\,A_1\cdot A_2\,. \tag{3.22}$$

Dots here symbolize the sum over gluon polarization states. To calculate the cross section some care should be exercised: the current (3.21) *does not conserve* because of non-commuting colour matrices. We would need to include gluon radiation from the exchange-gluon line *and* from the source, to be in a position to use an arbitrary gauge (e.g. the Feynman gauge) for the emitted gluon. The physical polarization technique (3.11) simplifies our task. To obtain the true accompanying radiation pattern (in the $k_\perp \ll q_\perp$ region) it suffices to use the projectors (3.13) for the dots in (3.22). In particular,

$$A_1\cdot A_2 \equiv \sum_{\lambda=1,2}\left(A_1 e^{(\lambda)}\right)\left(A_2 e^{(\lambda)}\right)^* = \mathcal{J} \qquad \{\neq -(A_1\,A_2)\ \textit{sic!}\}\,.$$

Accompanying radiation intensity finally takes the form

$$dN \;\propto\; C_F\,\mathcal{R}_{\text{coher}} \,+\, N_c\,\mathcal{J}\,. \tag{3.23}$$

The first term proportional to the squared quark charge is responsible, as we already know, for two narrow bremsstrahlung cones around the incoming and outgoing quarks, $\Theta_1, \Theta_2 \le \Theta_s$. On top of that an additional, purely non-Abelian, contribution shows up, which is proportional to the *gluon* charge. It is given by the interference distribution (3.14c), (3.16b),

$$\mathcal{J} = \frac{a_1+a_2-a_{12}}{a_1 a_2} - 1\,,$$

which remains *non-singular* in the forward regions $\Theta_1 \ll \Theta_s$ and $\Theta_2 \ll \Theta_s$. At the same time, it populates large emission angles $\Theta = \Theta_1 \approx \Theta_2 \gg \Theta_s$ where

$$dN \propto d\Omega\, \mathcal{J} \propto \sin\Theta d\Theta \left(\frac{2}{a} - 1\right) \sim \frac{d\Theta^2}{\Theta^2}\,. \tag{3.24}$$

Indeed, evaluating the azimuthal average, say, around the *incoming* quark direction we obtain

$$\int \frac{d\phi_1}{2\pi}\, \mathcal{J} = \frac{1}{a_1}\left(1 + \frac{a_1 - a_{12}}{|a_1 - a_{12}|}\right) - 1 = \frac{2}{a_1}\,\vartheta(\Theta_1 - \Theta_s) - 1\,.$$

Thus we conclude that the third complementary bremsstrahlung cone emerges. It basically corresponds to radiation at angles *larger* than the scattering angle and its intensity is proportional to the colour charge of the t-channel exchange.
We could have guessed without actually performing the calculation that at large angles the gluon radiation is related to the *gluon* colour charge. As far as large emission angles $\Theta \gg \Theta_s$ are concerned, one may identify the directions of initial and final particles to simplify the total radiation amplitude as

$$j^\mu = T^b T^a \cdot \frac{p_1^\mu}{p_1 k} - T^a T^b \cdot \frac{p_2^\mu}{p_2 k} \approx \left(T^b T^a - T^a T^b\right) \cdot \frac{p^\mu}{pk}\,.$$

Recalling the general commutation relation for the $SU(N_c)$ generators,

$$\left[T^a(R)\,,\, T^b(R)\right] = \mathrm{i} \sum_c f_{abc}\, T^c(R)\,, \tag{3.25}$$

we immediately obtain the factor $N_c \propto (\mathrm{i} f_{abc})^2$ as the proper colour charge. Since (3.25) holds for arbitrary colour representation R, we see that the accompanying gluon radiation at large angles $\Theta > \Theta_s$ does not depend on the nature of the projectile.
The bremsstrahlung gluons we are discussing transform, in the end of the day, into observable final hadrons. We are ready now to derive an interesting physical prediction from our QCD soft radiation exercise.
Translating the emission angle into (pseudo)rapidity $\eta = \ln \Theta^{-1}$, the logarithmic angular distribution (3.24) converts into the rapidity plateau. We conclude that in the case of the gluon fusion mechanism, the second in Fig. 8, the hadronic accompaniment should form a practically uniform rapidity plateau. Indeed, the hadron density in the centre (small η, large c.m.s. angles) is proportional to the gluon colour charge N_c, while in the "fragmentation regions" ($\eta_{\max} > |\eta| > \ln \Theta_s^{-1}$, or $\Theta < \Theta_s$) the two quark-generated bremsstrahlung cones give, roughly speaking, the density $\sim 2 \times C_F \approx N_c$.
At the same time, the WW-fusion events (the first graph in Fig. 8) should have an essentially different final state structure. Here we have a colorless exchange, and

the QED-type angular ordering, $\Theta < \Theta_s$, restricts the hadronic accompaniment to the two projectile fragmentation humps as broad as $\Delta\eta = \eta_{\max} - \ln\Theta_s \simeq \ln M_H$, while the central rapidity region should be devoid of hadrons. The "rapidity gap" is expected which spans over $|\eta| < \ln(\sqrt{s}/M_H)$.

3.5. CONSERVATION OF COLOUR CURRENT

In physical terms universality of the generator algebra is intimately related with *conservation of colour.* To illustrate this point let us consider production of a quark-gluon pair in some hard process and address the question of how this system radiates. Let p and k be the momenta of the quark and the gluon, with b the octet colour index of the latter. For the sake of simplicity we concentrate on *soft* accompanying radiation, which determines the bulk of particle multiplicity inside jets, the structure of the hadronic plateau, etc. As far as emission of a soft gluon with momentum $\ell \ll k, p$ is concerned, the so-called "soft insertion rules" apply, which tell us that the Feynman diagrams dominate where ℓ is radiated off the external (real) partons — the final quark line p and the gluon k. The corresponding Feynman amplitudes are shown in Fig. 10.

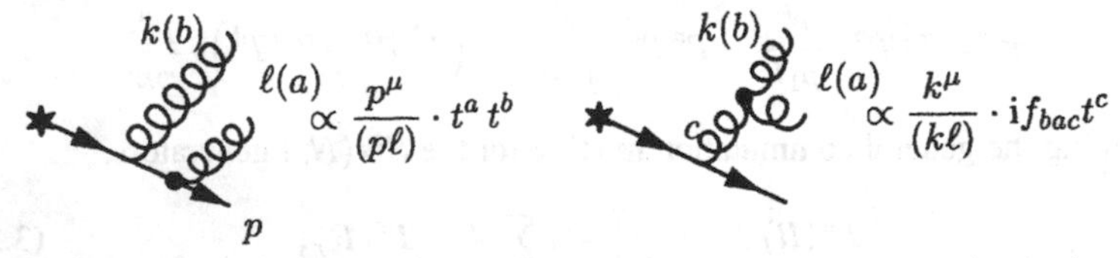

Figure 10. Feynman diagrams for radiation of the soft gluon with momentum ℓ and colour a off the qg system.

Do two emission amplitudes interfere with each other? It depends on the direction of the radiated gluon $\vec{\ell}$.

In the first place, there are two bremsstrahlung cones centred around the directions of $\vec{p}$ and $\vec{k}$:

$$\text{quark cone:} \quad \Theta_{\vec{\ell}} \equiv \Theta_{\vec{\ell},\vec{p}} \ll \Theta \approx \Theta_{\vec{\ell},\vec{k}}\,,$$

$$\text{gluon cone:} \quad \Theta_{\vec{\ell}} \equiv \Theta_{\vec{\ell},\vec{k}} \ll \Theta \approx \Theta_{\vec{\ell},\vec{p}}\,,$$

with Θ the angle between $\vec{p}$ and $\vec{k}$ — the aperture of the qg fork. In these regions one of the two amplitudes of Fig. 10 is much larger than the other, and the interference is negligible: the gluon ℓ is radiated independently and participates in the formation of the quark and gluon sub-jets.

If Θ is sufficiently large and the gluon k sufficiently energetic (relatively hard, $k \sim p$), these two sub-jets can be distinguished in the final state. The particle density in q and g jets should be remarkably different. It should be proportional (at least asymptotically) to the probability of soft gluon radiation which, in turn, is proportional to the "squared colour charge" of a the jet-generating parton, quark or gluon:

$$\left(\ell\frac{dn}{d\ell}\right)^{g}_{\Theta_{\vec\ell}<\Theta} : \left(\ell\frac{dn}{d\ell}\right)^{q}_{\Theta_{\vec\ell}<\Theta} = N_c : \frac{N_c^2-1}{2N_c} = 3 : \frac{4}{3} = \frac{9}{4}\,.$$

Multi-jet configurations are comparatively rare: emission of an additional hard gluon $k \sim p$ at large angles $\Theta \sim 1$ constitutes a fraction $\alpha_s/\pi \lesssim 10\%$ of all events. Typically k would prefer to belong to the quark bremsstrahlung cone itself, that is to have $\Theta \ll 1$. In such circumstances the question arises about the structure of the accompanying radiation at comparatively *large* angles

$$\Theta_{\vec\ell} \equiv \Theta_{\vec\ell,\vec p} \simeq \Theta_{\vec\ell,\vec k} \gg \Theta\,. \tag{3.26}$$

If the quark and the gluon were acting as independent emitters, we would expect the particle density to increase correspondingly and to overshoot the standard quark jet density by the factor

$$\left(\ell\frac{dn}{d\ell}\right)^{g+q}_{\Theta_{\vec\ell}>\Theta} : \left(\ell\frac{dn}{d\ell}\right)^{q}_{\Theta_{\vec\ell}>\Theta} = N_c : \frac{N_c^2-1}{2N_c} + 1 = \frac{13}{4}\,. \tag{3.27}$$

However, in this angular region our amplitudes start to interfere significantly, so that the radiation off the qg pair is no longer given by the *sum of probabilities* $q \to g(\ell)$ plus $g \to g(\ell)$. We have to square the *sum of amplitudes* instead.
This can be easily done by observing that in the large-angle kinematics (3.26) the angle Θ between $\vec p$ and $\vec k$ can be neglected, so that the accompanying soft radiation factors in Fig. 10 become indistinguishable,

$$\frac{p^\mu}{(p\ell)} \simeq \frac{k^\mu}{(k\ell)}\,.$$

Thus the Lorentz structure of the amplitudes becomes the same and it suffices to sum the colour factors:

$$t^a\,t^b \,+\, \mathrm{i}f_{bac}t^c \,=\, t^a\,t^b + \left[t^b\,,\,t^a\right] \,\equiv\, t^b\,t^a\,. \tag{3.28}$$

We conclude that the coherent sum of two amplitudes of Fig. 10 results in radiation at large angles *as if* off the initial quark, as shown in Fig. 11.
This means that the naive probabilistic expectation of enhanced density (3.27) fails and the particle yield is equal to that for the quark-initiated jet instead: $13/4 \to 1$.
It actually does not matter whether the gluon k was present at all, or whether there was instead a whole bunch of partons with small relative angles between them. Soft gluon radiation at large angles is sensitive only to the *total* colour charge of the final parton system, which equals the colour charge of the initial parton.

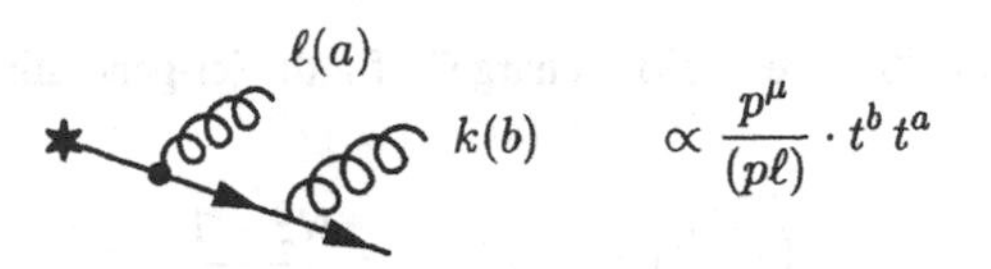

Figure 11. Soft radiation at large angles is determined by the total colour charge

This physically transparent statement holds not only for the quark as in Figs. 10, 11 but for an arbitrary object R (gluon, diquark, ..., you name it) as an initial object. In this case the matrices $t = T(3)$ should be replaced by the generators $T(R)$ corresponding to the colour representation R, and (3.28) holds due to the universality of the generator algebra (3.25).

4. Coherence in QCD parton cascades

Allowing small relative angles between partons in a process with a large hardness Q^2 results in a logarithmic enhancement of the emission probability:

$$\alpha_s \Longrightarrow \alpha_s \frac{d\Theta^2}{\Theta^2} \to \alpha_s \log Q^2 . \tag{4.1a}$$

As a result, the total probability of one parton (E) turning into two ($E_1 \sim E_2 \sim \frac{1}{2}E$) may become of order 1, in spite of the smallness of the characteristic coupling, $\alpha_s(Q^2) \propto 1/\log Q^2$. A typical example of such a "collinear" enhancement — the splitting process $g \to q\bar{q}$.
Moreover, when we consider the *gluon* offspring, another — "soft" — enhancement enters the game, which is due to the fact that the gluon bremsstrahlung tends to populate the region of *relatively* small energies ($E \simeq E_1 \gg E_2 \equiv \omega$):

$$\alpha_s \Longrightarrow \alpha_s \frac{d\omega}{\omega} \frac{d\Theta^2}{\Theta^2} \to \alpha_s \log^2 Q^2 . \tag{4.1b}$$

Thus the true perturbative "expansion parameter" responsible for parton multiplication via $q \to qg$ and $g \to gg$ may actually become *much larger* that 1!
In such circumstances we cannot trust the perturbative expansion in $\alpha_s \ll 1$ unless the logarithmically enhanced contributions (4.1) are taken full care of in all orders. Fortunately, in spite of the complexity of high order Feynman diagrams, such a programme can be carried out. The very fact that the all-order logarithmic asymptotes can be written down in a closed form and, more than that, that they *a posteriori* prove to be quite simple, follows from the **classical nature** of

- *soft* enhancement of bremsstrahlung amplitudes ("infrared" singularities) and
- *collinear* enhancement of basic $1 \to 2$ parton splitting amplitudes (or "mass" singularities).

As a result, the leading logarithmic asymptotes can be found without performing laborious calculations. It suffices to invoke an intuitively clear picture of parton cascades described in probabilistic fashion in terms of sequential independent elementary parton branchings *ordered in fluctuation times.*

4.1. PUZZLE OF DIS AND PARTONS

Let us invoke the deep inelastic lepton-hadron scattering — a classical example of a hard process and the standard QCD laboratory for carrying out the PT-resummation programme.
Here, the momentum q with a large space-like virtuality $Q^2 = |q^2|$ is transferred from an incident electron (muon, neutrino) to the target proton, which then breaks up into the final multi-parton → multi-hadron system. Introducing an invariant energy s between the exchange photon (Z^0, $W^\pm$) and the proton with 4-momentum P ($P^2 = M_p^2$), one writes the invariant mass of the produced hadron system which measures *inelasticity* of the process as

$$W^2 \equiv (q+P)^2 - M_p^2 = q^2 + 2(Pq) = s(1-x),\; s = 2(Pq),\; x \equiv \frac{Q^2}{2(Pq)} \le 1,$$

with x the Bjorken variable. The cross section of the process depends on two variables: Q^2 and x. For the case of *elastic* lepton-proton scattering one has $x=1$ and it is natural to write the cross section as

$$\frac{d\sigma_{el}}{dQ^2\,[dx]} = \frac{d\sigma_{\text{Ruth}}}{dQ^2} \cdot f_{el}^2(Q^2) \cdot [\delta(1-x)]\,. \tag{4.2a}$$

Here $\sigma_{\text{Ruth}} \propto \alpha^2/Q^4$ is the standard Rutherford cross section for e.m. scattering off a point charge and f_{el} stands for the elastic proton form factor.
For inclusive *inelastic* cross section one can write an analogous expression by introducing an inelastic proton "form factor" which now depends on both the momentum transfer Q^2 and the inelasticity parameter x:

$$\frac{d\sigma_{in}}{dQ^2\,dx} = \frac{d\sigma_{\text{Ruth}}}{dQ^2} \cdot f_{in}^2(x, Q^2)\,. \tag{4.2b}$$

What kind of Q^2-behaviour of the form factors (4.2) could one expect in the Bjorken limit $Q^2 \to \infty$? Quantum mechanics tells us how the Q^2-behaviour of the electromagnetic form factor can be related to the charge distribution inside a proton:

$$f_{el}(Q^2) = \int d^3r\, \rho(\vec{r})\, \exp\left\{i\vec{Q}\vec{r}\right\}. \tag{4.3}$$

For a point charge $\rho(\vec{r}) = \delta^3(\vec{r})$, it is obvious that $f \equiv 1$. On the contrary, for a smooth charge distribution $f(Q^2)$ falls with increasing Q^2, the faster the smoother

ρ is. Experimentally, the elastic *e-p* cross section does decrease with Q^2 *much faster* that the Rutherford one ($f_{el}(Q^2)$ decays as a large power of Q^2). Does this imply that $\rho(\vec{r})$ is indeed regular so that there is no well-localized — point-charge inside a proton? If it were the case, the *inelastic* form factor would decay as well in the Bjorken limit: a tiny photon with the characteristic size $\sim 1/Q \to 0$ would penetrate through a "smooth" proton like a knife through butter, inducing neither elastic nor inelastic interactions.

However, as was first observed at SLAC in the late sixties, for a fixed x, f_{in}^2 stays practically constant with Q^2, that is, the inelastic cross section (with a given inelasticity) is similar to the Rutherford cross section (Bjorken scaling). It looks *as if* there was a point-like scattering in the guts of it, but in a rather strange way: it results in inelastic break-up dominating over the elastic channel. Quite a paradoxical picture emerged; Feynman-Bjorken partons came to the rescue.

Imagine that it is not the proton itself that is a point-charge-bearer, but some other guys (quark-partons) inside it. If those constituents were *tightly* bound to each other, the elastic channel would be bigger than, or comparable with, the inelastic one: an excitation of the parton that takes an impact would be transferred, with the help of rigid links between partons, to the proton as a whole, leading to the elastic scattering or to the formation of a quasi-elastic finite-mass system ($N\pi$, $\Delta\pi$ or so), $1-x \ll 1$.

To match the experimental pattern $f_{el}^2(Q^2) \ll f_{in}^2(Q^2)$ one has instead to view the parton ensemble as a *loosely* bound system of quasi-free particles. Only under these circumstances does knocking off one of the partons inevitably lead to deep inelastic breakup, with a negligible chance of reshuffling the excitation among partons.

The parton model, forged to explain the DIS phenomenon, was intrinsically paradoxical by itself. In sixties and seventies, there was no other way of discussing particle interactions but in the field-theoretical framework, where it remains nowadays. But all reliable (renormalisable, 4-dimensional) quantum field theories (QFTs) known by then had one feature in common: an effective interaction strength (the running coupling $g^2(Q^2)$) *increasing* with the scale of the hard process Q^2. Actually, this feature was widely believed to be a general law of nature, and for a good reason [3]. At the same time, it would be preferable to have it the other way around so as to be in accord with the parton model, which needs parton-parton interaction to *weaken* at small distances (large Q^2).

Only with the advent of non-Abelian QFTs (and QCD among them) exhibiting an anti-intuitive asymptotic-freedom behaviour of the coupling, the concept of partons was to become more than a mere phenomenological model.

---

[3] relation between screening and unitarity

4.2. QCD PARTON PICTURE

Typical QCD graphs for DIS amplitudes are shown in Fig. 12.

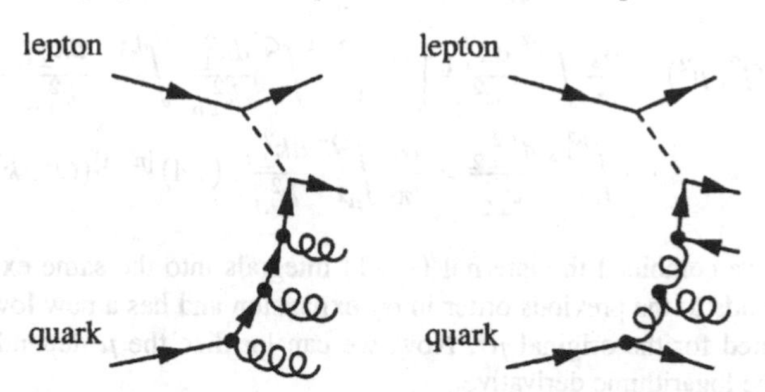

Figure 12. Valence (left) and Bethe-Heitler mechanism (right) of DIS.

For moderate x-values (say, $x > 0.1$) the process is dominated by lepton scattering off a valence quark in the proton. The scattering cross section has a standard energy behaviour $\sigma \propto x^{-2(J_{\rm ex}-1)}$, where $J_{\rm ex}$ is the spin of the exchanged particle in the t-channel. It is the quark with $J_{\rm ex} = \frac{1}{2}$ in the left picture of Fig. 12, so that the valence contribution to the cross section decreases at small x as $\sigma \propto x$.
For high-energy scattering, $x \ll 1$, the Bethe-Heitler mechanism takes over which corresponds to the t-channel *gluon* exchange: $J_{\rm ex} = 1$, $\sigma \propto x^0$ =const (modulo logarithms).
In the Leading Logarithmic Approximation (LLA) one insists on picking up, for each new parton taken into consideration, a logarithmic enhancement factor $\alpha_s \to \alpha_s \log Q^2$. In this approximation the scattering *probability* can be simply obtained by convoluting elementary probabilities of independent $1 \to 2$ parton splittings.
To cut a long story short, the appearance of the log-enhanced contributions in (4.1a) is due to the following structure

$$\frac{1}{n!}\left[\frac{\alpha_s}{\pi}\ln\frac{Q^2}{\mu^2}\right]^n = \left[\frac{\alpha_s}{\pi}\right]^n \int^{Q^2}\frac{dk_{\perp n}^2}{k_{\perp n}^2}\int^{k_{\perp n}^2}\frac{dk_{\perp n-1}^2}{k_{\perp n-1}^2}\cdots$$

$$\cdots\int^{k_{\perp 3}^2}\frac{dk_{\perp 2}^2}{k_{\perp 2}^2}\int_{\mu^2}^{k_{\perp 2}^2}\frac{dk_{\perp 1}^2}{k_{\perp 1}^2},$$

with $k_{\perp i}^2$ the squared transverse momenta of produced partons.
To contribute to the LLA, the transverse momenta of produced partons should be strongly ordered, increasing up the "ladder": $k_{1\perp}^2 \ll \ldots \ll k_{n\perp}^2 \ll Q^2$. (At the level of Feynman *amplitudes* the ladder diagrams dominate, provided a special physical gauge is chosen for gluon fields.) This expression diverges in the zero-quark-mass limit, $\mu \to 0$. Well, when you see a nasty thing happen beyond your reach, you can do no better than make use of it. This "mass singularity", according

to (4.4), occurs in the lower limit of the $k_\perp$ integration of the very first (and only!) parton branch. Let us drag this misbehaving integral to the left by rewriting (4.4) as

$$(4.4)^{[n]}(Q^2, \mu^2) = \frac{\alpha_s}{\pi}\int_{\mu^2}^{Q^2}\frac{dk_{\perp 1}^2}{k_{\perp 1}^2}\left[\frac{\alpha_s}{\pi}\right]^{n-1}\int^{Q^2}\frac{dk_{\perp n}^2}{k_{\perp n}^2}\int^{k_{\perp n}^2}\frac{dk_{\perp n-1}^2}{k_{\perp n-1}^2}\cdots$$
$$\cdots\int_{k_{\perp 1}^2}^{k_{\perp 3}^2}\frac{dk_{\perp 2}^2}{k_{\perp 2}^2} = \frac{\alpha_s}{\pi}\int_{\mu^2}^{Q^2}\frac{dk_{\perp 1}^2}{k_{\perp 1}^2}\,(4.4)^{[n-1]}(Q^2, k_{\perp 1}^2),$$

where we have combined the internal $(n-1)$ integrals into the same expression that corresponds to the previous order in α_s-expansion and has a new lower limit $k_{\perp 1}^2$ substituted for the original μ^2. Now, we can localize the μ-dependence by evaluating the logarithmic derivative:

$$\mu^2\frac{\partial}{\partial\mu^2}\,(4.4)^{[n]} \;=\; -\frac{\alpha_s}{\pi}\cdot(4.4)^{[n-1]}\,.$$

This equation relates the nth order of the PT expansion to the previous one. To put this symbolic relation at work one first has to recall the satellite x-dependence. By extracting the first step one may look upon the rest as DIS off a new "target" — the parton with transverse momentum $k_{\perp 1}^2$ and a finite fraction z of the initial longitudinal momentum P. As a result, there appears an additional integration with the probability of the first splitting, $\phi(z)$, and the differential equation for the resummed $F^{(\mathrm{LLA})}$ takes the form

$$\mu^2\frac{\partial}{\partial\mu^2}\,F\left(x,\,Q^2,\mu^2\right) \;=\; -\int_x^1\frac{dz}{z}\,\phi(z)\,\frac{\alpha_s}{\pi}\cdot F\left(\frac{x}{z},\,Q^2,\mu^2\right). \tag{4.4a}$$

Since a logarithm (like a stick) has two ends, differentiation over the overall hardness scale Q^2 would do the same job, the result being the *evolution equation* in a familiar form:

$$Q^2\frac{\partial}{\partial Q^2}F(x,Q^2) = \frac{\alpha_s}{\pi}\phi(x)\otimes F(x,Q^2)\,, \tag{4.4b}$$

where the symbol $\otimes$ stands for convolution in the x-space.

4.3. LLA PARTON EVOLUTION

4.3.1. *Apparent and hidden symmetries of the QCD evolution*

Since (4.4b) reminds of a Schrödinger (diffusion) equation (with $dt = \frac{\alpha_s}{\pi}dQ^2/Q^2$ as the "evolution time" differential), we can refer to the kernels $\phi(x)$ as the matrix elements of the "evolution Hamiltonian" in the (F, G) space, where F marks a spin-$\frac{1}{2}$ quark (fermion) and G a vector gluon.

To discuss the relations between kernels it is convenient to extract colour factors

$$t^a_{ij}\, t^a_{jk} = C_F\, \delta_{ik}\,, \quad t^a_{ik}\, t^b_{ki} = T_R\, \delta^{ab} = \tfrac{1}{2}\delta^{ab}\,, \quad f_{ade}\, f_{bde} = N_c\, \delta_{ab}\,;$$

$$\begin{array}{ll} \Phi^F_F(z) = C_F \cdot V^F_F(z) & \Phi^G_F(z) = C_F \cdot V^G_F(z), \\ \Phi^F_G(z) = T_R \cdot V^F_G(z) & \Phi^G_G(z) = N_c \cdot V^G_G(z). \end{array} \tag{4.5}$$

The splitting functions then take the form [18, 19]:

$$V^F_F(z) = 2\,\frac{1+z^2}{1-z} \tag{4.6a}$$

$$V^G_F(z) = 2\,\frac{1+(1-z)^2}{z} \tag{4.6b}$$

$$V^F_G(z) = 2\,\left[z^2+(1-z)^2\right] \tag{4.6c}$$

$$V^G_G(z) = 4\left[z(1-z) + \frac{1-z}{z} + \frac{z}{1-z}\right]. \tag{4.6d}$$

The most important symmetry properties of the LLA parton splitting functions are:

Parton Exchange results in an obvious relation between probabilities to find decay products with complementary momenta fractions:

$$V^{BC}_A(z) = V^{CB}_A(1-z)\,. \tag{4.7}$$

A Crossing Relation [17] emerges when one links together two splitting processes corresponding to opposite evolution "time" sequences:

$$V^B_A\left(\frac{1}{z}\right) = (-1)^{2s_A+2s_B-1}\,\frac{1}{z}\; V^A_B(z) \tag{4.8}$$

with s_A the spin of the particle A.

The Super-Symmetry Relation [19] exploits the existence of the super-symmetric field theory closely related to real QCD:

$$V_F^F(z) + V_F^G(z) = V_G^F(z) + V_G^G(z) \,. \tag{4.9}$$

Conformal Invariance [20] of the leading twist approximation leads to a number of relations between splitting functions, the simplest of which reads

$$\left(z\frac{d}{dz} - 2\right) V_G^F(z) = \left(z\frac{d}{dz} + 1\right) V_F^G(z) \,. \tag{4.10}$$

As we see, these relations leave not much freedom for splitting functions. In fact, one could borrow V_F^F from a QED text book and reconstruct successively V_F^G (with use of (4.7)), V_G^F (4.8), and then even the gluon selfinteraction V_G^G from (4.9).

The general character of the symmetry properties makes them practically useful when studying next-to-leading corrections to anomalous dimensions and coefficient functions where one faces technically difficult calculations. For example, the super-symmetric QCD analog had been used to choose between two contradictory calculations for the two-loop anomalous dimensions see, e.g. [21]. We illustrate the idea by another example of the next-to-leading result, e.g. the ratio of parton multiplicities in gluon and quark jets which reads [22]

$$\begin{aligned} R\frac{\mathcal{N}_g}{\mathcal{N}_q} &= 1 - \frac{\gamma_0}{6}\left\{1 + T(1-2R)\right\} \\ &\quad + \frac{1}{8}\left(\frac{\gamma_0}{6}\right)^2 \left\{-25(58R-19)T + (6-4R-16R^2)T^2\right\} \end{aligned} \tag{4.11}$$

with $\gamma_0 = \sqrt{2N_c\alpha_s/\pi}$ the characteristic PT expansion parameter. R and T in (4.11) are the following ratios of the colour factors:

$$R \equiv \frac{C_F}{N_c}\,, \quad T \equiv \frac{2n_f T_R}{N_c}\,.$$

In *susy-QCD* "quark" and "gluon" belong to the same (adjoint) representation of the color group, so that all color factors (4.5) become equal: $C_F = N_c = T_R$. Moreover, the "quark" here is the Majorana fermion (identical to the "antiquark"), so that the total number of $q\bar{q}$ flavor states becomes $2n_f = 1$. Then, as one can easily check, all corrections to the multiplicity ratio $\mathcal{N}_g/\mathcal{N}_q = 1$ in (4.11) vanish at $R = T = 1$.

4.3.2. *Space-like parton evolution*

The decay phase space for the *space-like* evolution determining the DIS structure functions is

$$dw^{A\to B+C} = \frac{dk_\perp^2}{k_\perp^2}\,\frac{\alpha_s(k_\perp^2)}{2\pi}\,\frac{dz}{z}\,\Phi_A^{BC}(z) \tag{4.12a}$$

with z the longitudinal momentum fraction carried by the offspring parton B.

In the DIS environment the initial parton A with a negative (space-like) virtuality decays into $B[z]$ with the large space-like virtual momentum $|k_B^2| \gg |k_A^2|$ and a positive virtuality (time-like) $C[1-z]$. The parton C generates a subjet of secondary partons ($\to$ hadrons) in the final state. As long as the process is inclusive, that is that no details of the final-state structure are measured, integration over the subjet mass is due, dominated in LLA by the region $k_C^2 \ll |k_B^2|$. The latter condition makes C look *quasi-real*, compared with the hard scale of $|k_B^2|$. The same is true for the initial parton A.

Splitting can be viewed as a large momentum-transfer process of scattering (turn-over) of a "real" target parton A into a "real" C on the external field mediated by high-virtuality B. At the next step of evolution it is B's turn to play a role of a next target $B \equiv A'$, "real" with respect to yet deeper probe $|k_B'^2| \gg |k_B^2|$, and so on.

Successive parton decays with step-by-step *increasing* space-like virtualities (transverse momenta) constitute the picture of parton wave-function fluctuations inside the proton. The sequence proceeds until the overall hardness scale Q^2 is reached.

4.3.3. *Time-like parton cascades*

A similar picture emerges for the time-like branching processes determining the internal structure of jets produced, for example, in e^+e^- annihilation. Here the flow of *hardness* is opposite to that in DIS: evolution starts from a highly virtual quark with positive virtuality, originating from the e.m. vertex, while time-like virtualities of its products ("quasi-real" with respect to predecessors; "high-virtuality" with respect to offspring) degrade.

It is important to notice, however, that the flow of *energy* (longitudinal momentum) is governed, in the LLA, by the same functions Φ_A^{BC}: it does not matter that now, in contrast to space-like evolution, A is the "virtual" one and B and C are "real".

In the time-like case the longitudinal phase space is symmetric in offspring parton energy fractions, and the differential decay probability reads

$$dw^{A\to B+C} = \frac{d\Theta^2}{\Theta^2}\,\frac{\alpha_s(k_\perp^2)}{2\pi}\,dz\,\Phi_A^{BC}(z)\,. \tag{4.12b}$$

4.3.4. *Fluctuation Time and Evolution Times: Coherence*

An attentive reader has noticed that we wrote the phase space differently: for the space-like case (4.12a) in terms of transverse momentum $k_\perp$ and for the time-like evolution (4.12b) via the decay angle Θ. Logarithmic differentials by themselves are identical, since $k_\perp^2$ and Θ^2 are proportional for fixed z. We have made this distinction to stress an important difference between a probabilistic interpretation of DIS and the e^+e^- evolution: the different *evolution times*.

An appearance of the angle as a proper evolution parameter in (4.12b) is readily understood: it is a consequence of the Angular Ordering. In the DIS case, (4.12a) we aim at describing, in probabilistic terms, the inclusive cross section, a quantity that could not care less about the finite-state structure and that of soft accompaniment in particular.

To be honest, within the LLA framework it does not make much sense to argue which of evolution parameters $\ln k_{\perp i}^2$, $\ln \Theta_i^2$ or $\ln |k_i^2|$ (with k^2 the total parton virtuality) does a better job: these choices differ by *subleading* terms formally negligible from the LLA point of view. A mismatch is of the order of

$$\sim \alpha_s \ln^2 x\,. \tag{4.13}$$

It becomes significant, however, and should be "resummed" when numerically small values of the Bjorken x are concerned.[4]

The $k_\perp$-ordering proves to be the one that takes care of potentially disturbing corrections (4.13) in all orders, and in this sense becomes a preferable choice for constructing the probabilistic scheme for single-inclusive parton distributions in DIS.

It is instructive to see how this comes about.

Let us introduce two light-like vectors p_1^μ, p_2^μ and write down the Sudakov (light-cone) decomposition of momenta:

$$k^\mu = \beta p_1^\mu + \alpha p_2^\mu + k_\perp{}^\mu\,. \tag{4.14}$$

For $k_1^\mu + k_2^\mu + k_3^\mu = 0$ the general relation holds:

$$\frac{k_1^2}{\beta_1} + \frac{k_2^2}{\beta_2} + \frac{k_3^2}{\beta_3} = \frac{\beta_1\beta_2}{\beta_3}\left(\frac{\vec{k}_{1\perp}}{\beta_1} - \frac{\vec{k}_{2\perp}}{\beta_2}\right)^2. \tag{4.15}$$

Each parton cell in Fig. 12 involves a space-like parton A decaying into $B[z] + C[1-z]$. Relation (4.15) applied to our basic space-like splitting $A \to B[z]C[1-z]$ gives

$$\frac{-k_B^2}{z} = \frac{-k_A^2}{1} + \frac{k_C^2}{1-z} + \frac{k_\perp^2}{z(1-z)}\,, \tag{4.16}$$

[4] The word "numerically" stands here as a warning for not confusing this kinematical region with a "parametrically" small x, such that $\alpha_s \ln 1/x \sim 1$ — the Regge region — where essentially different physics comes into play.

with z the longitudinal momentum fraction — the ratio of the Sudakov light-cone variables β. (We have chosen the direction of p_1 such that $\vec{k}_{A\perp} = 0$, so that $\vec{k}_{B\perp} = -\vec{k}_{C\perp} \equiv \vec{k}_\perp$ is the relative transverse momentum in the splitting.)
Since the 4-momenta of A and B are space-like, all terms in (4.16) are positive. B being an intermediate virtual state, k_B^2 enters in the Feynman denominators in the matrix element. The collinear-log contribution arises upon integration over $k_\perp^2$, over the region where the last term dominates in the r.h.s. of (4.16), that is from the region

$$\frac{|k_B^2|}{z} \simeq \frac{k_\perp^2}{z(1-z)} \gg \left|k_A^2\right|, \frac{k_C^2}{1-z}\,. \tag{4.17}$$

The physical origin of this strong inequality becomes transparent in terms of lifetimes of virtual states

$$\tau_i = \frac{k_i^0}{|k_i^2|}, \tag{4.18}$$

namely $\tau_B \ll \tau_A\,,\ \tau_C$. This shows that the LLA contributions originate from the sequence of branchings well separated in the *fluctuation time* (4.18). Invoking the local-scattering analogy (recall $A \to C$ on the "external field" B), we can say that the classical picture naturally implies "fast scattering": probing time τ_B much smaller than the lifetime(s) of the "target" before (τ_A) and after the scattering occurs (τ_C).
Assembling a "ladder" of successive parton splittings $i = 1, 2\ldots, n$, and tracing the space-like parton state the nth-order LLA contribution $\sim [\alpha_s \ln(Q^2/\mu^2)]^n$ is expected to come from time-ordered kinematics

$$\frac{P}{\mu^2} \gg \tau_1 \gg \tau_2 \gg \ \ldots\ \gg \tau_n \gg \frac{xP}{Q^2}\,.$$

Let us now dig into the $k_\perp^2$ against Θ^2 problem. Equation (4.17) relates virtuality and transverse momentum of the "t-channel" parton after the ith splitting with the relative longitudinal momentum (β-) fraction z_i:

$$\left|k_i^2\right| \simeq \frac{k_{i\perp}^2}{1-z_i} \approx k_{i\perp}^2\,. \tag{4.19}$$

The latter approximation is made by remembering that, because of cancellation between real and virtual contributions in inclusive parton distributions (DIS structure functions), the soft s-channel radiation $1 - z_i \ll 1$ does not matter (as long as we stay away from the quasi-elastic kinematics, $1 - x \ll 1$, where it does).
The two-dimensional emission angle (the angle between C and A) can be written and estimated as (cf. (4.15))

$$\vec{\Theta}_i = \frac{\vec{k}_{C\perp}}{\beta_C P} - \frac{\vec{k}_{A\perp}}{\beta_A P} = \frac{\vec{k}_{i-1\perp} - \vec{k}_{i\perp}}{\beta_{i-1}(1-z_i)P} - \frac{\vec{k}_{i-1\perp}}{\beta_{i-1}P} \approx -\frac{\vec{k}_{i\perp}}{\beta_{i-1}P}\,. \tag{4.20}$$

We are now in a position to compare different orderings. It is straightforward to get

$$\text{Fluctuation times} \implies k_{i\perp}^2 \gg z_i \cdot k_{i-1\perp}^2\,, \tag{4.21a}$$

$$\text{Emission angles} \implies k_{i\perp}^2 \gg z_{i-1}^2 \cdot k_{i-1\perp}^2\,, \tag{4.21b}$$

to be compared with

$$\text{Transverse momenta} \implies k_{i\perp}^2 \gg k_{i-1\perp}^2\,. \tag{4.21c}$$

For $x = z_1 \cdot z_2 \cdot \ldots \cdot z_n \sim 1$ prescriptions (4.21) are essentially equivalent since each decay fraction stays finite, $z_i \sim 1$, and may be neglected in the logarithmic integral over $k_{i\perp}^2$.

For small x, however, the Bethe-Heitler mechanism of DIS off sea quarks dominates (see Fig. 12b). The multi-gluon "ladders" provide longitudinally enhanced contributions $\propto \ln z_i$, which combine with the $\ln z_i$ factors from the collinear-integration phase space to produce, at the end of the day, the "DL" mismatch $[\alpha_s \ln^2 x]^{n-2}$ between the different options (4.21).

So, which one of the possible orderings (4.21) is correct? The first two prescriptions are more liberal than the last one: they both allow for *disordered* transverse momentum configurations. For example, the fluctuation-time ordering (4.21a) embodies the region

$$z_i \cdot k_{i-1\perp}^2 \ll k_{i\perp}^2 \ll k_{i-1\perp}^2\,, \tag{4.22}$$

which may be quite broad for $z_i \ll 1$ and where the $k_\perp$-ordering is violated. The truth is, this region does not contribute to the answer, so that the $k_\perp$-ordering (4.21c) proves to be the correct one. The reason is quantum mechanical coherence.

4.3.5. *Vanishing of the forward inelastic diffraction*

Consider a two-step process shown by the first graph in Fig. 13 Let the second decay be *soft* in the t-channel direction, that is $z_2 = \beta_2/\beta_1 \ll 1$. (The first one can be either soft, $z_1 = \beta_1/\beta_0 \ll 1$ or hard, $z_1 \sim 1$.) In the kinematical region (4.22),

$$z_2 \cdot k_{1\perp}^2 \ll k_{2\perp}^2 \ll k_{1\perp}^2\,, \tag{4.23}$$

the *time-ordering* is still intact, which means that the momentum k_2 is transferred fast as compared with the lifetime of the first fluctuation $P \to P' + k_1$.

Since β_2 is small, the process can be viewed as inelastic relativistic scattering $P \to P' + k_1$ in the external gluon field. The transverse size of the field is $\rho_\perp \sim k_{2\perp}^{-1}$. The characteristic size of the fluctuation $P' + k_1$, according to (4.23), is smaller: $\Delta r_{01\perp} \sim k_{1\perp}^{-1} \ll \rho_\perp$. We thus have a *compact* state propagating through the field that is smooth at distances of the order of the size of the system. In such

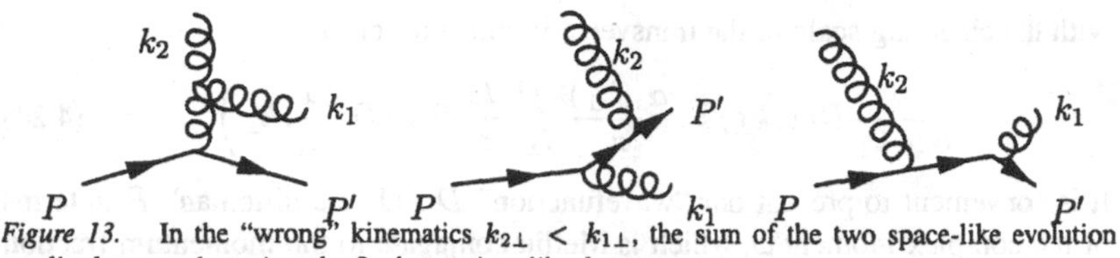

Figure 13. In the "wrong" kinematics $k_{2\perp} < k_{1\perp}$, the sum of the two space-like evolution amplitudes cancels against the final state time-like decay

circumstances the field cannot resolve the internal structure of the fluctuation. Components of the fluctuation, partons P' and k_1 in the first two graphs of Fig. 13, scatter *coherently* with the total amplitude identical and opposite in sign to that for scattering of the initial state P (the last graph): inelastic breakup does not occur. This general physical phenomenon is due to Gribov, who has proved that the diffractive deuteron disintegration process vanishes in the forward kinematics. He then used this observation to argue in favour of the so-called weak-Reggeon-coupling regime based on the vanishing of inelastic processes in the $k_\perp \to 0$ limit. In our context the cancellation between the amplitudes of Fig. 13 in the region (4.23) is a direct consequence of the conservation of the colour current.

4.4. HUMPBACKED PLATEAU AND LPHD

QCD coherence is crucial for treating particle multiplication **inside** jets, as well as for hadron flows **in-between** jets.
Here we are going to derive together the QCD "prediction" of the inclusive energy spectrum of relatively soft particles from QCD jets. I put the word *prediction* in quotation marks on purpose. This is a good example to illustrate the problem of filling the gap between the QCD formulae, talking quarks and gluons, and phenomena dealing, obviously, with hadrons.
Let me first make a statement:

> It is QCD coherence that allows the prediction of the inclusive soft particle yield in jets practically from the "first principles".

4.4.1. *Solving the DIS evolution*

You have all the reasons to feel suspicious about this. Indeed, we have stressed above the similarity between the dynamics of the evolution of space-like (DIS structure functions) and time-like systems (jets). On the other hand, you are definitely aware of the fact that the DIS structure functions cannot be calculated perturbatively. There are input parton distributions for the target proton, which have to be plugged in as an initial condition for the evolution at some finite hardness scale $Q_0 = \mathcal{O}(1\,\text{GeV})$. These initial distributions cannot be calculated "from first principles" nowadays but are subject to fitting. What pQCD controls then, is the scaling violation pattern. Namely, it tells us how the parton densities change with the changing scale of the transverse-momentum probe:

$$\frac{\partial}{\partial \ln k_\perp} D(x, k_\perp) = \frac{\alpha_s(k_\perp)}{\pi} \int_x^1 \frac{dz}{z} P(z)\, D\left(\frac{x}{z}, k_\perp\right). \tag{4.24}$$

It is convenient to present our "wavefunction" D and "Hamiltonian" P in terms of the complex moment ω, which is Mellin conjugate to the momentum fraction x:

$$D_\omega = \int_0^1 dx\, x^\omega \cdot D(x), \quad D(x) = x^{-1} \int_{(\Gamma)} \frac{d\omega}{2\pi i} x^{-\omega} \cdot D_\omega ; \tag{4.25a}$$

$$P_\omega = \int_0^1 dz\, z^\omega \cdot P(z), \quad P(z) = z^{-1} \int_{(\Gamma)} \frac{d\omega}{2\pi i} z^{-\omega} \cdot P_\omega , \tag{4.25b}$$

where the contour Γ runs parallel to the imaginary axis, to the right from singularities of D_ω (P_ω). It is like trading the coordinate ($\ln x$) for the momentum (ω) in a Schrödinger equation.
Substituting (4.25) into (4.24) we see that the evolution equation becomes algebraic and describes propagation in "time" $dt = \frac{\alpha_s}{\pi} d \ln k_\perp$ of a free quantum mechanical "particle" with momentum ω and the dispersion law $E(\omega) = P_\omega$:

$$\hat{d}\, D_\omega(k_\perp) = \frac{\alpha_s(k_\perp)}{\pi} \cdot P_\omega\, D_\omega(k_\perp) ; \qquad \hat{d} \equiv \frac{\partial}{\partial \ln k_\perp} . \tag{4.26}$$

To continue the analogy, our wavefunction D is in fact a multi-component object. It embodies the distributions of valence quarks, gluons and secondary sea quarks which evolve and mix according the 2×2 matrix LLA "Hamiltonian" Φ_A^B.
At small x, however, the picture simplifies. Here the valence distribution is negligible, $\mathcal{O}(x)$, while the gluon and sea quark components form a system of two coupled oscillators which is easy to diagonalise. What matters is one of the two energy eigenvalues (one of the two branches of the dispersion rule) that is *singular* at $\omega = 0$. The problem becomes essentially one-dimensional. Sea quarks are driven by the gluon distribution while the latter is dominated by gluon cascades. Correspondingly, the leading energy branch is determined by gluon-gluon splitting $g \to gg$, with a subleading correction coming from the $g \to q(\bar{q}) \to g$ transitions,

$$P_\omega = \frac{2N_c}{\omega} - a + \mathcal{O}(\omega) , \quad a = \frac{11 N_c}{6} + \frac{n_f}{3N_c^2} . \tag{4.27}$$

The solution of (4.26) is straightforward:

$$D_\omega(k_\perp) = D_\omega(Q_0) \cdot \exp\left\{ \int_{Q_0}^{k_\perp} \frac{dk}{k} \gamma_\omega(\alpha_s(k)) \right\} , \tag{4.28a}$$

$$\gamma_\omega(\alpha_s) = \frac{\alpha_s}{\pi} P_\omega . \tag{4.28b}$$

The structure (4.28a) is of the most general nature. It follows from *renormalisability* of the theory, and does not rely on the LLA which we used to derive it. The function $\gamma(\alpha_s)$ is known as the "anomalous dimension".[5] It can be perfected by including higher orders of the PT expansion. Actually, modern analyses of scaling violation are based on the improved next-to-LLA (two-loop) anomalous dimension, which includes α_s^2 corrections to the LLA expression (4.28b).

The structure (4.28a) of the x-moments of parton distributions (DIS structure functions) gives an example of a clever separation of PT and NP effects; in this particular case — in the form of two factors. It is the ω-dependence of the input function $D_\omega(Q_0)$ ("initial parton distributions") that limits predictability of the Bjorken-x dependence of DIS cross sections.

So, how comes then that in the time-like channel the PT answer turns out to be more robust?

4.4.2. *Coherent hump in $e^+e^- \to h(x) + \ldots$*

We are ready to discuss the time-like case, with $D_j^h(x, Q)$ now the inclusive distribution of particles h with the energy fraction (Feynman-x) $x \ll 1$ from a jet (parton j) produced at a large hardness scale Q.

Here the general structure (4.28a) still holds. We need, however, to revisit the expression (4.28b) for the anomalous dimension because, as we have learned, the proper evolution time is now different from the case of DIS.

In the time-like jet evolution, due to Angular Ordering, the evolution equation becomes non-local in $k_\perp$ space:

$$\frac{\partial}{\partial \ln k_\perp} D(x, k_\perp) = \frac{\alpha_s(k_\perp)}{\pi} \int_x^1 \frac{dz}{z} P(z)\, D\left(\frac{x}{z}, z \cdot k_\perp\right). \tag{4.29}$$

Indeed, successive parton splittings are ordered according to

$$\theta = \frac{k_\perp}{k_\parallel} > \theta' = \frac{k'_\perp}{k'_\parallel}.$$

Differentiating $D(k_\perp)$ over the scale of the "probe", $k_\perp$, results then in the substitution

$$k'_\perp = \frac{k'_\parallel}{k_\parallel} \cdot k_\perp \equiv z \cdot k_\perp$$

in the argument of the distribution of the next generation $D(k'_\perp)$.

[5] The name is a relict of those good old days when particle and solid state physicists used to have common theory seminars. If the coupling α_s were constant (had a "fixed point"), then (4.28a) would produce the function with a non-integer (non-canonical) dimension $D(Q) \propto Q^\gamma$ (analogy — critical indices of thermodynamical functions near the phase transition point).

The evolution equation (4.29) can be elegantly cracked using the Taylor-expansion trick,

$$D(z \cdot k_\perp) = \exp\left\{\ln z \frac{\partial}{\partial \ln k_\perp}\right\} D(k_\perp) \;=\; z^{\frac{\partial}{\partial \ln k_\perp}} \cdot D(k_\perp). \tag{4.30}$$

Turning as before to moment space (4.25), we observe that the solution comes out similar to that for DIS, (4.28), but for one detail. The exponent $\hat{d}$ of the additional z-factor in (4.30) combines with the Mellin moment ω to make the argument of the splitting function P a *differential operator* rather than a complex number:

$$\hat{d} \cdot D_\omega = \frac{\alpha_s}{\pi} P_{\omega+\hat{d}} \cdot D_\omega \,. \tag{4.31}$$

This leads to the differential equation

$$\left(P^{-1}_{\omega+\hat{d}}\, \hat{d} \;-\; \frac{\alpha_s}{\pi} - \left[P^{-1}_{\omega+\hat{d}}, \frac{\alpha_s}{\pi} \right] P_{\omega+\hat{d}} \right) \cdot D = 0\,. \tag{4.32}$$

Recall that, since we are interested in the small-x region, the essential moments are small, $\omega \ll 1$.
For the sake of illustration, let us keep only the most singular piece in the "dispersion law" (4.27) and neglect the commutator term in (4.32) generating a subleading correction $\propto \hat{d}\alpha_s \sim \alpha_s^2$. In this approximation (DLA),

$$P_\omega \simeq \frac{2N_c}{\omega}\,, \tag{4.33}$$

(4.32) immediately gives a quadratic equation for the anomalous dimension,[6]

$$(\omega + \gamma_\omega)\gamma_\omega \;-\; \frac{2N_c\alpha_s}{\pi} + \mathcal{O}\left(\frac{\alpha_s^2}{\omega}\right) \;=\; 0\,. \tag{4.34}$$

The leading anomalous dimension following from (4.34) is

$$\gamma_\omega = \frac{\omega}{2}\left(-1 + \sqrt{1 + \frac{8N_c\alpha_s}{\pi\,\omega^2}} \right). \tag{4.35}$$

When expanded to first order in α_s, it coincides with that for the space-like evolution, $\gamma_\omega \simeq \alpha_s/\pi \cdot P_\omega$, with P given in (4.33). Such an expansion, however, fails when characteristic $\omega \sim 1/|\ln x|$ becomes as small as $\sqrt{\alpha_s}$, that is when

$$\frac{8N_c\alpha_s}{\pi} \ln^2 x \gtrsim 1\,.$$

[6] It suffices to use the next-to-leading approximation to the splitting function (4.27) and to keep the subleading correction coming from differentiation of the running coupling in (4.32) to get the more accurate MLLA anomalous dimension γ_ω.

This inequality is an elaboration of the heuristic estimate (4.13).
Now what remains to be done is to substitute our new weird anomalous dimension into (4.28a) and perform the inverse Mellin transform to find $D(x)$. If there were no QCD parton cascading, we would expect the particle *density* $xD(x)$ to be constant (Feynman plateau). It is straightforward to derive that plugging in the DLA anomalous dimension (4.35) results in the plateau density increasing with Q and with a maximum (hump) "midway" between the smallest and the highest parton energies, namely, at $x_{\max} \simeq \sqrt{Q_0/Q}$. The subleading MLLA effects shift the hump to smaller parton energies,

$$\ln \frac{1}{x_{\max}} = \ln \frac{Q}{Q_0}\left(\frac{1}{2} + c \cdot \sqrt{\alpha_s} + \ldots\right) \simeq 0.65 \ln \frac{Q}{Q_0},$$

with c a known analytically calculated number. Moreover, defying naive probabilistic intuition, the softest particles do not multiply at all. The density of particles (partons) with $x \sim Q_0/Q$ stays constant while that of their more energetic companions increases with the hardness of the process Q.
This is a powerful legitimate consequence of pQCD coherence. We turn now to another, no less powerful though less legitimate, consequence.

4.4.3. *Coherent damping of the Landau singularity*
The time-like DLA anomalous dimension (4.35), as well as its MLLA improved version, has a curious property. Namely, in sharp contrast with DIS, it allows the momentum integral in (4.28) to be extended to very small scales. Even integrating down to $Q_0 = \Lambda$, the position of the "Landau pole" in the coupling, one gets a finite answer for the distribution (the so-called *limiting spectrum*), simply because the $\sqrt{\alpha_s(k)}$ singularity happens to be integrable!
It would have been poor taste to trust this formal integrability, since the very PT approach to the problem (selection of dominant contributions, parton evolution picture, etc.) relied on α_s being a numerically small parameter. However, the important thing is that, due to time-like coherence effects, the (still perturbative but "smallish") scales, where $\alpha_s(k) \gg \omega^2$, contribute to γ basically in a ω-*independent* way, $\gamma + \omega/2 \propto \sqrt{\alpha_s(k)} \neq f(\omega)$. This means that "smallish" momentum scales k affect only an overall *normalization* without affecting the *shape* of the x-distribution.
Since this is the rôle of the "smallish" scales, it is natural to expect the same for the truly small — non-perturbative — scales where the partons transform into the final hadrons. This hypothesis (LPHD) reduces, mathematically, to the statement (guess) that the NP factor in (4.28a) has a finite $\omega \to 0$ limit:

$$D_\omega^{(h)}(Q_0) \;\to\; K^h = \text{const}, \quad \omega \to 0\,.$$

Thus, according to LPHD, the x-shape of the so-called "limiting" parton spectrum which is obtained by formally setting $Q_0 = \Lambda$ in the evolution equations, should be mathematically similar to that of the inclusive distribution of *hadrons* (h). Another essential property is that the "conversion coefficient" K^h should be a true constant independent of the hardness of the process producing the jet under consideration.

References

1. Abreu, P. et al., DELPHI Collab., Phys.Lett. B 449 (1999) 383; Eur.Phys.J. C 13 (2000) 573.
2. Dokshitzer, Yu.L. and S.I. Troyan, "Asymptotic Freedom and Local Parton-Hadron Duality in QCD Jet Physics". In *Proceedings of 19th Leningrad Winter School* (in Russian), vol. 1, p. 144, Leningrad 1984;
Ya.I. Azimov et al., Z.Phys. C 27 (1985) 65;
Yu.L. Dokshitzer, V.A. Khoze, A.H. Mueller and S.I. Troyan, *Basics of Perturbative QCD* (ed. J. Tran Thanh Van) Gif-sur-Yvette, Editions Frontiéres, 1991.
3. Mueller, A.H., Nucl.Phys. B 213 (1983) 85. Erratum quoted *ibid.*, B 241 (1984) 141.
4. Dokshitzer, Yu.L., V.A. Khoze, A.H. Mueller and S.I. Troyan, Rev.Mod.Phys. 60 (1988) 373.
5. Korytov, A., 1996, private communication.
6. Goulianos, D., CDF Collab., In *Proceedings 32nd Recontres de Moriond*, Les Arcs, France, March 22, 1997.
7. Safonov, A.N., CDF Collab., In *Proceedings International Euroconference in Quantum Chromodynamics*, Montpellier, France, July 7, 1999.
8. Mueller, A.H., Phys.Lett. B 396 (1997) 251.
9. Camici, G. and M. Ciafaloni, Phys.Lett. B 395 (1997) 118.
10. Fong, C.P. and B.R. Webber, Nucl.Phys. B 355 (1991) 54.
11. Hamacher, K., O. Klapp, P. Langefeld and M. Siebel, DELPHI Collab., submitted to EPS-HEP Conference, Tampere, Finland, June 15, 1999; contribution 1-571.
12. Azimov, Ya.I., Yu.L. Dokshitzer, V.A. Khoze and S.I. Troyan, Phys.Lett. 165 B (1985) 147.
13. Akers, R. et al., OPAL Collab., Z.Phys. C 68 (1995) 531.
14. Khoze, V.A., S. Lupia and W. Ochs, Phys.Lett. B 394 (1997) 179.
15. DELPHI Collab., K. Hamacher et al., submitted to EPS-HEP Conference, Tampere, Finland, June 15, 1999; contribution 1-145.
16. Hoyer, P., N. Marchal and S. Peigne, Phys.Rev. D 62 (2000) 114001
17. Bukhvostov, A.P., L.N. Lipatov and N.P. Popov, Sov.J.Nucl.Phys. 20 (1975) 287.
18. Altarelli, G. and G. Parisi, Nucl.Phys. B 126 (1977) 298.
19. Dokshitzer, Yu.L., Sov.Phys.JETP 46 (1977) 641.
20. Bukhvostov, A.P., G.V. Frolov, L.N. Lipatov and E.A. Kuraev. Nucl.Phys. B 258 (1985) 601.
21. Altarelli, G., Phys.Rep. 81 (1982) 1.
22. Gaffney, J.B. and A.H. Mueller, Nucl.Phys. B 250 (1985) 109.

PARTON SATURATION: AN OVERVIEW *

A.H. MUELLER[†]
Department of Physics, Columbia University
New York, New York 10027

Abstract. The idea of partons and the utility of using light-cone gauge in QCD are introduced. Saturation of quark and gluon distributions are discussed using simple models and in a more general context. The Golec-Biernat Wüsthoff model and some simple phenomenology are described. A simple, but realistic, equation for unitarity, the Kovchegov equation, is discussed, and an elementary derivation of the JIMWLK equation is given.

1. Introduction

These lectures are meant to be an introduction, and an overview, of parton saturation in QCD. Parton saturation is the idea that the occupation numbers of small-x quarks and gluons cannot become arbitrarily large in the light-cone wavefunction of a hadron or nucleus. Parton saturation is an idea which is becoming well established theoretically and has important applications in small-x physics in high-energy lepton-hadron collisions and in the early stages of high-energy heavy ion collisions. The current experimental situation is unclear. Although saturation based models have had considerable phenomenological success in explaining data at HERA and at RHIC more complete and decisive tests are necessary before it can be concluded that parton saturation has been seen. These lectures begin very simply by describing, through an example, some important features of light-cone perturbation theory in QCD, and they end by describing some rather sophisticated equations which govern light-cone wavefunctions when parton densities are very large.

* Lectures given at the Cargese Summer School, August 6-18, 2001

† This work is supported in part by the Department of Energy

J.-P. Blaizot and E. Iancu (eds.), QCD Perspectives on Hot and Dense Matter, 45–72.

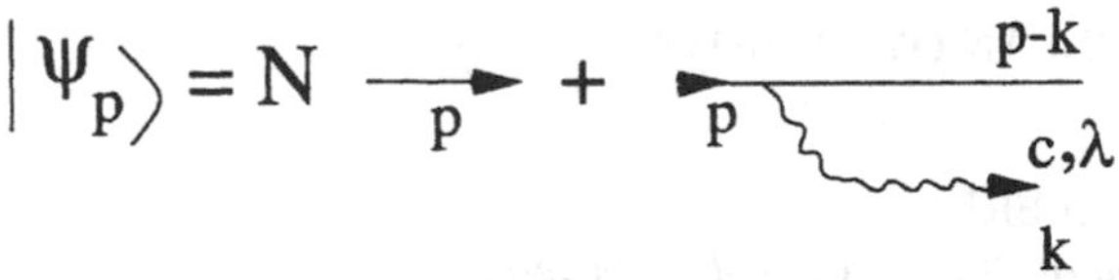

$$|\Psi_p\rangle = N \xrightarrow[p]{} + \ldots$$

Figure 1.

2. States in QCD Perturbation Theory

In light-cone perturbation theory states of QCD are described in terms of the numbers and distributions in momentum of quarks and gluons. The essential features of light-cone perturbation theory that will be needed in these lectures can be illustrated by considering the wavefunction of a quark through lowest order in the QCD coupling g. One can write

$$|\psi_p >= N|p> + \sum_{\lambda=\pm} \sum_{c=1}^{N_c^2-1} \int d^3k \psi_\lambda^c(k)|p-k; k(\lambda, c)> \tag{2.1}$$

where $|p>$ is a free quark state with momentum p, $|\psi_p>$ is a dressed quark state, and $|p-k; k(\lambda, c)>$ is a state of a quark, of momentum $p-k$, and a gluon of momentum k, helicity λ and color c. We suppress quark color indices which will appear in matrix form in what follows. We label states by momenta $p_+ = \frac{1}{\sqrt{2}}(p_0+p_3), p_1, p_2$, and $d^3k = dk_+ d^2k = dk_+ dk_1 dk_2$. Recall that in light-cone quantization momenta P_+ and $\underline{P} = (P_1, P_2)$ are kinematic while P_- plays the role of a Hamiltonian and generates evolution in the "time" variable $x_+ = \frac{1}{\sqrt{2}}(x_0+x_3)$. For an on-shell zero mass particle $p_- = \underline{p}^2/2p_+$. N in (1) is a normalization factor. Eq.(1) is illustrated in Fig.1. ψ_λ^c is determined from light-cone perturbation theory to be

$$\delta^3(p-p')\psi_\lambda^c(k) = \frac{< p'-k; k(\lambda, c)|H_I|p>}{(p'-k)_- + k_- - p_-} \tag{2.2}$$

with

$$H_I = g\int d^3x \bar{q}(x)\gamma_\mu(\frac{\lambda^c}{2})q(x)A_\mu^c(x) \tag{2.3}$$

where the gluon field is

$$A_\mu^c(x) = \sum_{\lambda=\pm}\int \frac{d^3k}{\sqrt{(2\pi)^3 2k_+}}[\epsilon_\mu^\lambda(k)a_\lambda^c(k)e^{i\underline{k}\cdot\underline{x}-ik_+x_--ik_-x_+} + h.c.] \tag{2.4}$$

with

$$[a_{\lambda'}^{c'}(k'), a_\lambda^{\dagger c}(k)] = \delta_{\lambda'\lambda}\delta_{c'c}\delta^3(k'-k). \tag{2.5}$$

In (2) and (5) $\delta^3(p) = \delta(p_+)\delta^2(\underline{p})$ while $d^3x = dx_- d^2\underline{x}$ in (3). It is useful to imagine the calculation being done in a frame where p_+ is large and $\underline{p} = 0$ in which case

$$k_- = \frac{\underline{k}^2}{2k_+}, \qquad (p-k)_- = \frac{\underline{k}^2}{2(p-k)_+}. \tag{2.6}$$

In the soft gluon approximation $k_+/p_+ << 1$ and thus $k_- >> (p-k)_-$ so that only k_- need be kept in the denominator in (2). In addition, in light-cone gauge, $A_+ = 0$, the polarization vectors can be written as

$$\epsilon_\mu^\lambda(k) = (\epsilon_+^\lambda, \epsilon_-^\lambda, \underline{\epsilon}^\lambda) = (0, \frac{\underline{\epsilon}^\lambda \cdot \underline{k}}{k_+}, \underline{\epsilon}^\lambda) \tag{2.7}$$

and, because of the $1/k_+$ term, only ϵ_-^λ need be kept in (2) in the soft gluon approximation. Using (3)-(7) in (2) one finds

$$\psi_\lambda^c(k) = (\frac{\lambda^c}{2})2g\frac{(\underline{\epsilon}^\lambda)^* \cdot \underline{k}}{\underline{k}^2}\frac{1}{\sqrt{(2\pi)^3 2k_+}} \tag{2.8}$$

Problem 1(E): Using the formula $\bar{U}(p-k)\gamma_\mu \bar{U}(p) \approx 2p_+ g_{\mu -}$ for high momentum Dirac spinors derive (8).

3. Partons

Define the gluon distribution of a state $|S(p)>$ by

$$xG_S(x,Q^2) = \sum_{\lambda,c}\int d^3k x\delta(x - k_+/p_+)\Theta(Q^2 - \underline{k}^2) < S(p)|a_\lambda^{c\dagger}(k)a_\lambda^c(k)|S(p)> . \tag{3.9}$$

The meaning of xG_S is clear. $xG_S(x,Q^2)dx$ is the number of gluons, having longitudinal momentum between xp_+ and $(x+dx)p_+$, localized in transverse coordinate space to a region $\Delta x_\perp \sim 1/Q$, in the state $|S(p)>$. For a quark, at order g^2, one finds from (1)

$$xG_q(x,Q^2) = \sum_{\lambda,c}\int d^3k x\delta(x - k_+/p_+)\Theta(Q^2 - \underline{k}^2)\psi_\lambda^{c\dagger}(k)\psi_\lambda^c(k). \tag{3.10}$$

Using (8) in (10) one finds

$$xG_q(x,Q^2) = \sum_c \frac{\lambda^c}{2}\frac{\lambda^c}{2}\int \frac{4g^2}{(2\pi)^3}\frac{d^2k}{\underline{k}^2}\frac{dk_+}{2k_+}x\delta(x-k_+/p_+)\Theta(Q^2-\underline{k}^2). \quad (3.11)$$

Using $\Sigma_c \frac{\lambda^c}{2}\frac{\lambda^c}{2} = C_F = \frac{N_c^2-1}{2N_c}$ and introducing an infrared cutoff, μ, for the transverse momentum integral in (11) gives

$$xG_q(x,Q^2) = \frac{\alpha C_F}{\pi}\ell n(Q^2/\mu^2). \quad (3.12)$$

We note that if $xG(x,Q^2) = 3xG_q(x,Q^2)$ is taken one obtains a result for the proton which is not unreasonable phenomenologically for $x \sim 10^{-2} - 10^{-1}$ and moderate Q^2 if μ is taken to be 100 MeV.

4. Classical Fields

One can associate a classical field with gluons in the quark.

$$A_i^{(c\ell)}(x) = \int d^3p' < \psi_{p'}|A_i^c(x)|\psi_p > . \quad (4.13)$$

Using (1) and (8) in (13) one finds

$$A_i^{c(c\ell)}(x) = \int \frac{d^3k}{(2\pi)^3}e^{-ik\cdot x}(\frac{\lambda^c}{2})\frac{gk_i}{\underline{k}^2k_+} \quad (4.14)$$

or

$$A_i^{c(c\ell)}(x) = \int \frac{d^3k}{(2\pi)^3}e^{-ik\cdot x}A_i^{c(c\ell)}(k) \quad (4.15)$$

with

$$A_i^{c(c\ell)}(k) = \frac{\lambda^c}{2}\frac{gk_i}{k_+}. \quad (4.16)$$

In (14) the k_+ integration goes from $-\infty$ to $+\infty$. The region $k_+ > 0$ comes from $|\psi_p >$ consisting of a bare quark and a gluon of momentum k while the region $k_+ < 0$ comes from $|\psi_{p'} >$ consisting of a bare quark and a gluon.

Problem 2(E): Take $\frac{1}{k_+} = \frac{1}{k_+ - i\epsilon}$ in (16) and show that

$$A_i^{c(c\ell)}(x) = -g(\frac{\lambda^2}{2})\frac{x_i}{2\pi\underline{x}^2}\Theta(-x_-) \text{ and } F_{+i}^{c(c\ell)} = \frac{\partial}{\partial x_-}A_i^{c(c\ell)} = g(\frac{\lambda^c}{2})\frac{x_i}{2\pi\underline{x}^2}\delta(x_-).$$

Problem 3(E): Show that G_q, as given in (10), can also be written as

$$xG_q(x,Q^2) = \int \frac{d^3k}{(2\pi)^3}\delta(x_-k_+/p_+)\Theta(Q^2-\underline{k}^2)\sum_{i,c}[A_i^{c(c\ell)}(k)]^2 2k_+.$$

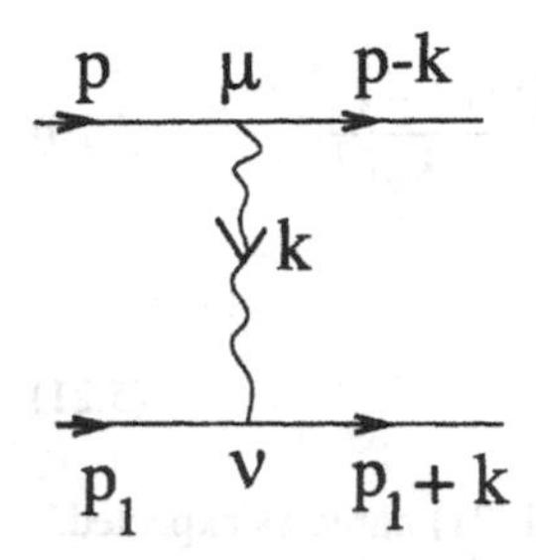

Figure 2.

5. Why Light-Cone Gauge is Special

In order to understand why light-cone gauge plays a special role in describing high-energy hadronic states a simple calculation of the near forward high-energy elastic amplitude for electron-electron scattering in QED is useful. The graph to be calculated is shown in Fig.2 and we imagine the calculation being done in the center of mass frame with p being a right-mover (p_+large) and p_1 being a left-mover (p_{1-} large). First we shall do the calculation in covariant gauge and afterwards in light-cone gauge.
In covariant gauge the photon propagator, the k-line in Fig.2, is $D_{\mu\nu} = \frac{-i}{k^2} g_{\mu\nu}$ and the dominant term comes from taking $g_{\mu\nu} \to g_{-+} = 1$ giving

$$T = ig\frac{\bar{U}(p-k)\gamma_+ U(p)}{2p_+}\ ig\ \frac{\bar{U}(p_1+k)\gamma_- U(p_1)}{2p_{1-}}\ \frac{-i}{k^2}. \tag{5.17}$$

Using $\bar{U}(p-k)\gamma_+ U(p) \approx 2p_+$ and $\bar{U}(p_1+k)\gamma_- U(p_1) \approx 2p_{1-}$ one finds

$$T = \frac{-ig^2}{\underline{k}^2} \tag{5.18}$$

where we have used $k^2 = 2k_+k_- - \underline{k}^2 \approx -\underline{k}^2$ since both k_+ and k_- are required to be small from the mass shell conditions $(p-k)^2 = (p_1+k)^2 = 0$ for the zero mass electrons.
Now suppose we do the calculation in light-cone gauge $A_+ = 0$ where the propagator is

$$D_{\mu\nu}(k) = \frac{-i}{k^2}[g_{\mu\nu} - \frac{\eta_\mu k_\nu + \eta_\nu k_\mu}{\eta \cdot k}] \tag{5.19}$$

where $\eta \cdot V = V_+$ for any vector V_μ. Now the dominant term comes from taking D_{-i} giving

$$T = iq\frac{\bar{U}(p-k)\gamma_+ U(p)}{2p_+}\, ig\, \frac{\bar{U}(p_1+k)\underline{\gamma}\cdot\underline{k}\, U(p_1)}{2p_{1-}}\, \frac{-i}{k_+\underline{k}^2}. \tag{5.20}$$

Using $\bar{U}(p_1+k)\underline{\gamma}\cdot\underline{k}\, U(p_1) = \underline{k}^2$ one finds

$$T = \frac{-ig^2}{2p_{1-}k_+}. \tag{5.21}$$

Now $(p_1+k)^2 = 0$ gives $2p_{1-}k_+ \approx \underline{k}^2$ so that (18) and (21) agree as expected. The result (18) comes about in a natural way: $\frac{\bar{U}(p-k)\gamma_+ U(p)}{2p_+}$ and $\frac{\bar{U}(p_1+k)\gamma_- U(p_1)}{2p_{1-}}$ are the classical currents, equal to 1, of particles moving along the light-cone, while the $1/\underline{k}^2$ factor is just the (instantaneous) potential between the charges. However, when one uses $A_+ = 0$ light-cone gauge the dominant part of the current for left moving particles is forbidden and one must keep the small transverse current $\underline{k}/2p_{1-}$. However, the smallness of the current is compensated by the factor $1/k_+$ in the light-cone gauge propagator. In coordinate space the $1/k_+$ comes about from the potential acting over distances $x_- \approx 1/k_+ = \frac{2p_{1-}}{\underline{k}^2}$ so that the potential is very non-local and non-causal. By choice of the $i\epsilon$ prescription[1, 2] for $1/k_+$ one can put these non-causal interactions completely before the scattering (initial state) or completely after the scattering (final state). In Problem 2 the potential A_i exhibits this non-causal behavior with our choice of $i\epsilon$ placing the long time behavior in the initial state, the $\Theta(-x_-)$ term.

6. High Momentum Particles and Wilson Lines[3-5]

Suppose a quark of momentum p is a high-energy right mover, that is $p_+ >> p_-, \underline{p}$. Then so long as one does not choose to work in $A_- = 0$ light-cone gauge the dominant coupling of gauge fields to p are classical (eikonal) when p passes some QCD hadron or source. To be specific suppose p scatters on a hadron elastically and with a small momentum transfer. We may view the interactions as shown in Fig.3 for a three-gluon exchange term. Of course to get the complete scattering one has to sum over all numbers of gluon exchanges. Call $S(p_+, \underline{b})$ the S-matrix for scattering of the right moving quark on the target. (We imagine that the target hadron has large gluon fields making it necessary to find a formula which includes all gluon interactions.) Although the right moving quark has sufficient momentum so that it moves close to the light-cone we do assume that the momentum is not so large that higher gluonic components of the quark wavefunction need be considered. Since the probability of extra transverse gluons being present in the wavefunction is in general proportional to αy, with y being the quark rapidity, we suppose $\alpha y << 1$.

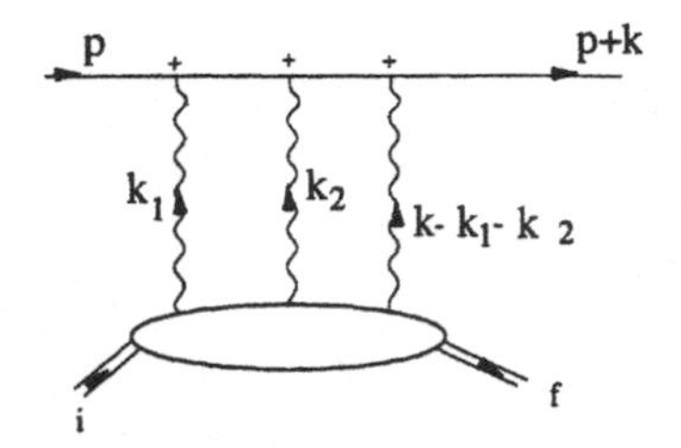

Figure 3.

The graph shown in Fig.3 can be written as

$$S(p_+, \underline{b}) = \int \{ \frac{1}{\sqrt{2(p+k)_+}} \bar{U}(p+k) igT^a \gamma_+ \frac{i}{\gamma \cdot (p+k_1+k_2)} igT^b \gamma_+ \frac{i}{\gamma \cdot (p+k_1)}$$

$$\cdot igT^c \gamma_+ U(p) \frac{1}{\sqrt{(2p_+)}} \} \frac{d^2k dk_+}{(2\pi)^3} e^{i\underline{k}\cdot\underline{b}} \frac{d^4k_1 d^4k_2}{(2\pi)^8} M_{abc}(k_1, k_2) \tag{6.22}$$

where

$$M_{abc}(k_1, k_2) = \int d^4x_1 d^4x_2 d^4x e^{i(k-k_1-k_2)\cdot x + ik_2\cdot x_2 + ik_1\cdot x_1}$$

$$\cdot < f|TA^a_+(x) A^b_+(x_2) A^c_+(x_1)|i > . \tag{6.23}$$

It is straightforward to evaluate the { }-term in (22) and one gets

$$\{ \} = igT^a igT^b igT^c \frac{i}{k_{1-} + i\epsilon} \frac{i}{(k_1+k_2)_- + i\epsilon}. \tag{6.24}$$

Now do the $d^2k_1 dk_{1+}$ and $d^2k_2 dk_{2+}$ integrals followed by $d^2x_1 dx_{1-}$ and $d^2x_2 dx_{2-}$. This sets $\underline{x}_1 = \underline{x}_2 = \underline{x}$ and $x_{1-} = x_{2-} = x_-$. The only non-zero term in the time-ordered product is proportional to $\Theta(x_- - x_{2-})\Theta(x_{2-} - x_{1-})$ and this factor along with the exponentials in x_{1-}, x_{2-} and x_- allow the dk_{1-} and dk_{2-} integrals to be done over the poles in (24). We get finally

$$S_{fi}(p_+, \underline{b}) = (ig)^3 < f| \int_{-\infty}^{\infty} dx_+ A_-(\underline{b}, x_+) \int_{-\infty}^{x_+} dx_{1+} A_-(\underline{b}, x_{2+})$$

$$\cdot \int_{-\infty}^{x_{2+}} dx_{1+} A_-(\underline{b}, x_{1+})|i > \tag{6.25}$$

where $A_\mu = T^a A^a_\mu$ and we have suppressed the variable $x_- = 0$ in the $A's$ in (25).

The general term is now apparent, and one has in the general case

$$S_{fi}(p_+, \underline{b}) =< f|Pe^{ig\int_{-\infty}^{\infty} dx_+ A_-(\underline{b}, x_+)}|i > \tag{6.26}$$

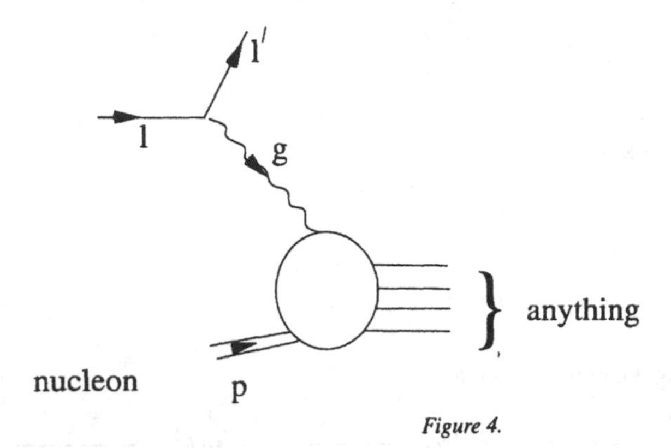

Figure 4.

where P denote an x_+- ordering of the matrices A where $A's$ having larger values of x_+ come to the left of those having smaller values.

Problem 4(M): Show that the time orderings different from $\Theta(x_+ - x_{2+})\Theta(x_{2+} - x_{1+})$ give no contribution to (22).
Finally, a word of caution in using (26). In general there are singularities present when two values of x_+, in adjoining $A's$, become equal, although in many simple models the x_+- integrations are regular. If singularities in the x_+ integrations arise it is generally possible to extract the leading logarithmic contributions to the scattering amplitude by carefully examining the singularities[5]. A detailed discussion of this is, however, far beyond the scope of these lectures.

7. Dual Descriptions of Deep Inelastic Scattering; Bjorken and Dipole Frames

Particular insight into the dynamics of a process often occurs by choosing a particular frame and an appropriate gauge. Indeed the physical picture of a process may change dramatically in different frames and in different gauges. In a frame where the parton picture of a hadron is manifest saturation shows up as a limit on the occupation number of quarks and gluons, however, in a different (dual) frame saturation appears as the unitarity limit for scattering of a quark or of a gluon dipole at high-energy[6].
To see all this a bit more clearly consider inelastic lepton-nucleon scattering as illustrated in Fig.4 where a lepton emits a virtual photon which then scatters on a nucleon. We suppose $Q^2 = -q^2$ is large. The two structure functions, F_1 and F_2, which describe the cross section can depend on the invariants Q^2 and $x = \frac{Q^2}{2P\cdot q}$. To make the parton picture manifest we choose $A_+ = 0$ light-cone gauge along with the frame

$$P = (P + \frac{M^2}{2P}, 0, 0, P)$$

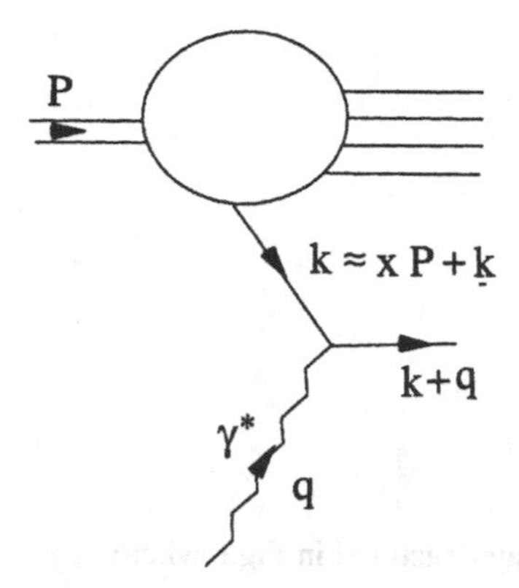

Figure 5.

and

$$q = (q_0, \underline{q}, q_z = 0),$$

and where $P \to \infty$. This is the Bjorken frame. We note that $q_0 = \frac{P\cdot q}{P}$ goes to zero as $P \to \infty$ so that the virtual photon momentum is mainly transverse to the nucleon direction. This last fact means that the virtual photon is a good analyzer of transverse structure since it is absorbed over a transverse distance $\Delta x_\perp \sim 1/Q$. Since $\Delta x_\perp$ is very small at large Q the virtual photon is absorbed by, and measures, individual quarks.

Problem 5(H): Use the uncertainty principle to show that the time, $\Delta\tau$, over which $\gamma^*(q)$ is absorbed by a quark is $\Delta\tau \approx 2xP/Q^2$. You may assume that $\underline{k}^2/Q^2 << 1$ where the process of absorption of the photon is illustrated in Fig.5. The result of problem 5 and of the result that $\Delta x_\perp \sim 1/Q$ motivates the formula

$$F_2(x, Q^2) = \sum_f e_f^2[xq_f(x, Q^2) + x\bar{q}_f(x, Q^2)] \tag{7.27}$$

which says that the structure function F_2 is proportional to the charge squared of the quark, having flavor f, absorbing the photon and proportional to the number of quarks having longitudinal momentum fraction x and localized in transverse coordinate space to a size $1/Q$. F_2 is given in terms of the longitudinal and transverse virtual photon cross sections on the proton as

$$F_2 = \frac{1}{4\pi^2\alpha_{em}} Q^2[\sigma_T + \sigma_L]. \tag{7.28}$$

Eq.(27) is the QCD improved parton model. In more technical terms Q^2 is a renormalization point which in the parton picture is a cutoff of the type given in (9) for gluon distributions and here occurring for quark distributions.

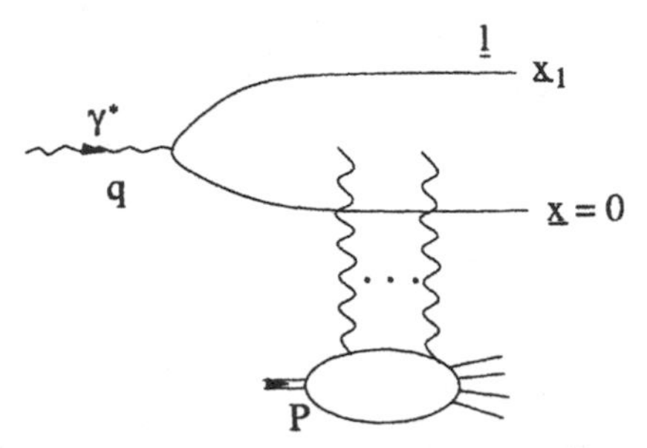

Figure 6.

Now consider the same process in the dipole frame pictured in Fig.6 where

$$P = (P + \frac{M^2}{2P}, 0, 0, P)$$

$$q = (\sqrt{q^2 - Q^2}, 0, 0, -q)$$

and where $q/Q >> 1$ but where q is fixed as x becomes small so that most of the energy in a small-x process, and it is only for small-x scattering that the dipole frame is useful, still is carried by the proton. Now we suppose that a gauge different from $A_+ = 0$ is being used, for example a covariant gauge or the gauge $A_- = 0$. In the dipole frame the process looks like $\gamma^* \to$ quark $-$ $\overline{\text{quark}}$ followed by the scattering of the quark-antiquark dipole on the nucleon[7, 8, 9]. The splitting of γ^* into the quark-antiquark pair is given by lowest order perturbation theory while all the dynamics is in the dipole-nucleon scattering. In this frame the partonic structure of the nucleon is no longer manifest, and the virtual photon no longer acts as a probe of the nucleon.
Equating these two pictures and fixing the transverse momentum of the leading quark or antiquark one has[6]

$$e_f^2 \frac{d(xq_f + x\bar{q}_f)}{d^2b d^2\underline{\ell}} = \frac{Q^2}{4\pi^2\alpha_{em}} \int \frac{d^2x_1 d^2x_2}{4\pi^2} \int_0^1 dz \frac{1}{2} \sum_\lambda \psi^{f*}_{T\lambda}(\underline{x}_2, z, Q) \psi^f_{T\lambda}(\underline{x}_1, z, Q)$$
$$\cdot e^{-i\underline{\ell}\cdot(\underline{x}_1 - \underline{x}_1)} [S^\dagger(\underline{x}_2) S(\underline{x}_1) - S^\dagger(\underline{x}_2) - S(\underline{x}_1) + 1], \qquad (7.29)$$

where the γ^* wavefunction is

$$\psi^f_{T\lambda}(\underline{x}, z, Q) = \left\{ \frac{\alpha_{em} N_c}{2\pi^2} z(1-z)[z^2 + (1-z)^2] Q^2 \right\}^{1/2}$$
$$\times \; e_f K_1(\sqrt{Q^2 \underline{x}^2 z(1-z)}) \frac{\underline{\epsilon}^\lambda \cdot \underline{x}}{|\underline{x}|}. \qquad (7.30)$$

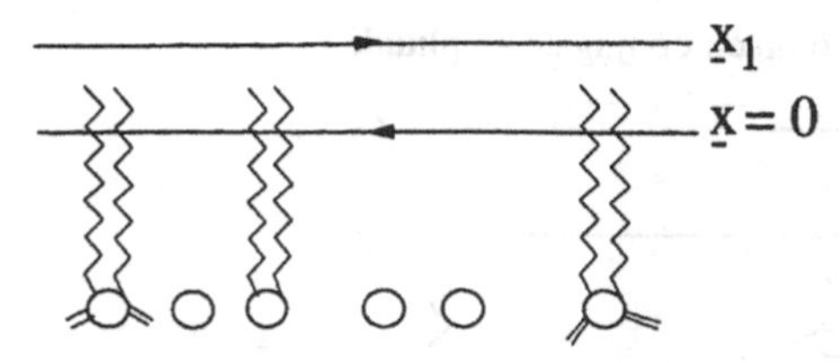

Figure 7.

We have written (29) for a fixed impact parameter, $\underline{b}$, which is a (suppressed) variable in S. To make the identification exhibited in (29) requires that the struck quark shown in Fig.5 not have final state interactions, and this requires the special choice of $i\epsilon's$ in the light-cone gauge, $A_+ = 0$, used to identify the quark distributions. With this choice of $i\epsilon's$ the lefthand side of (29) refers to the density of quarks in the nucleon wavefunction while the righthand side of (29) refers to the transverse momentum spectrum of quark jets produced. This identification is only possible when final state interactions are absent. Finally, it is not hard to see the origin of the final factor on the righthand side of (29). This is just $[S^\dagger(\underline{x}_2) - 1][S(\underline{x}_1) - 1]$, the product of the $T-$ matrices for the dipole $\underline{x}_1$ in the amplitude and the dipole $\underline{x}_2$ in the complex conjugate amplitude.

8. Quark Distributions in a large Nucleus; Quark Saturation at One-Loop

In general it is very difficult to evaluate the $S-$ matrices appearing in (29). However, if the target is a large nucleus one can define an interesting model, if not a realistic calculation for real nuclei, by limiting the dipole nucleon interaction to one and two gluon exchanges. That is, we suppose the dipole-nucleon interaction is weak. Despite this assumed weakness of interaction the dipole nucleus scattering can be very strong if the nucleus is large enough. We begin with the term $S(\underline{x}_1)$ in (29), an elastic scattering of the dipole on the nucleus in the amplitude with no interaction at all in the complex conjugate amplitude. This term is illustrated in Fig.7.

Now $|S(\underline{x}_1)|^2$ is the probability that the dipole does not have an inelastic interaction as it passes through the nucleus. We can write

$$|S(\underline{x}_1)|^2 = e^{-L/\lambda} \tag{8.31}$$

where $L = 2\sqrt{R^2 - \underline{b}^2}$ is the length of nuclear matter that the dipole traverses at impact parameter $\underline{b}$ for a uniform spherical nucleus of radius R, and λ is the mean free path for inelastic dipole-nucleon interactions. Using

$$\lambda = [\rho\sigma]^{-1} \tag{8.32}$$

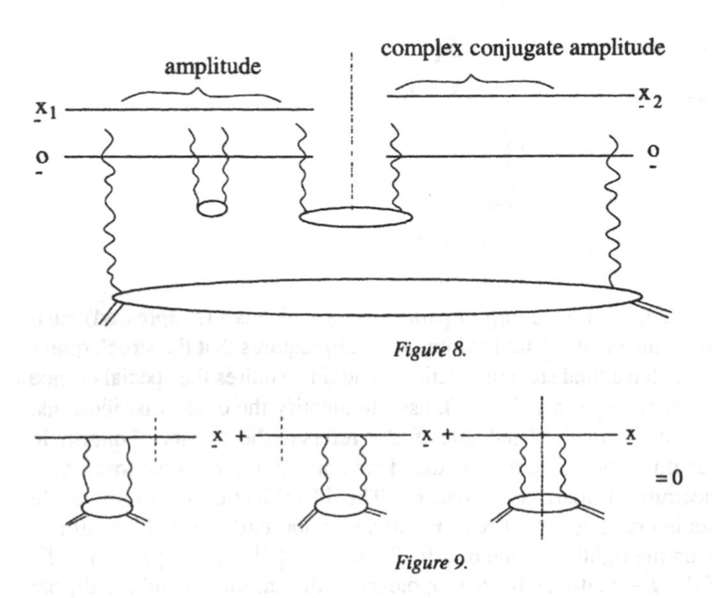

Figure 8.

Figure 9.

with ρ the nuclear density one has

$$S(\underline{x}_1, \underline{b}) = e^{-2\sqrt{R^2-b^2}\rho\sigma(\underline{x}_1)/2} \tag{8.33}$$

if we suppose S is purely real. Detailed calculation gives[10]

$$\sigma(\underline{x}) = \frac{\pi^2\alpha}{N_c} xG(x, 1/\underline{x}^2)\underline{x}^2. \tag{8.34}$$

Problem 6(H*): Check (34) for scattering of a dipole on a bare quark.
Thus we know how to calculate the last three terms in [ ] in (29). What about the $S^\dagger S$ term? Graphically this term is illustrated in Fig.8 for some typical elastic and inelastic interactions of the dipoles with the nucleons in the nucleus. Let us focus on the final interaction, the one closest to the cut (the vertical line) in Fig.8. One can check that interactions with the $\underline{x} = \underline{0}$ line cancel between interactions in the amplitude and the complex conjugate amplitude.

Problem 7(H): Verify, using $S^\dagger S = 1$, that the three interactions shown in Fig.9 cancel. You may assume that T is purely imaginary ($S = 1 - iT$) although this is not necessary for the result.
Problem 8(H): Again, using $S^\dagger S = 1$, show that the two interactions shown in Fig.10 cancel. Assume T is purely imaginary.
The results of problems 7 and 8 establish that one need not consider interactions with the $\underline{x} = 0$ line in evaluating the $S^\dagger(\underline{x}_2)S(\underline{x}_1)$ term in (29). Thus, we are left

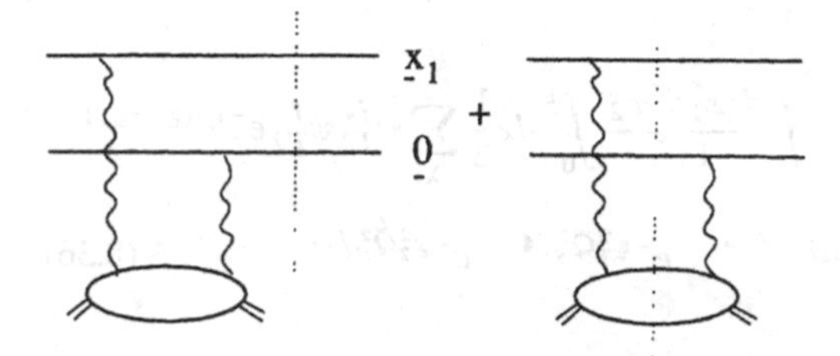

Figure 10.

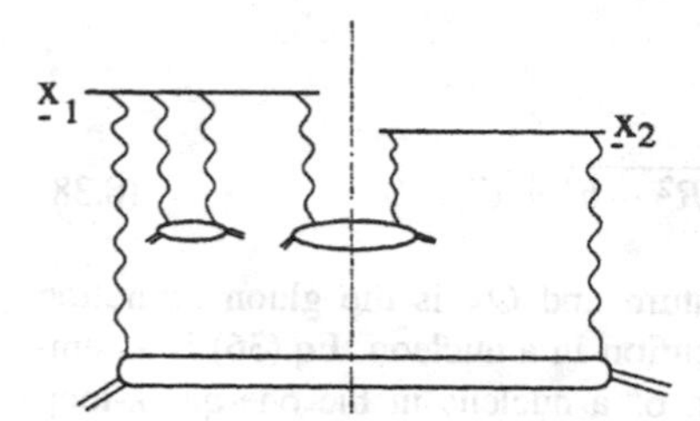

Figure 11.

only with interactions on the $\underline{x}_1$−line in the amplitude andwith the $\underline{x}_2$-line in the complex conjugate amplitude. The result of problem 8 allows one to transfer the interactions with the $\underline{x}_2$-line in the complex conjugate amplitude to interactions with a line placed at $\underline{x}_2$ in the amplitude[11], although this does require that S be real. This is illustrated in the equality between the terms in Fig.11 and Fig.12 and reads

$$S^{\dagger}(\underline{x}_2)S(\underline{x}_1) = S(\underline{x}_1 - \underline{x}_2). \tag{8.35}$$

Eq.(35) is a beautiful result, but unfortunately it likely has a limited validity. It appears to require a maximum of two gluon lines exchanged between any given nucleon and the dipole lines as well as the requirement of a purely real S-matrix for elastic scattering. Thus using (29), (33) and (35) we arrive at[6]

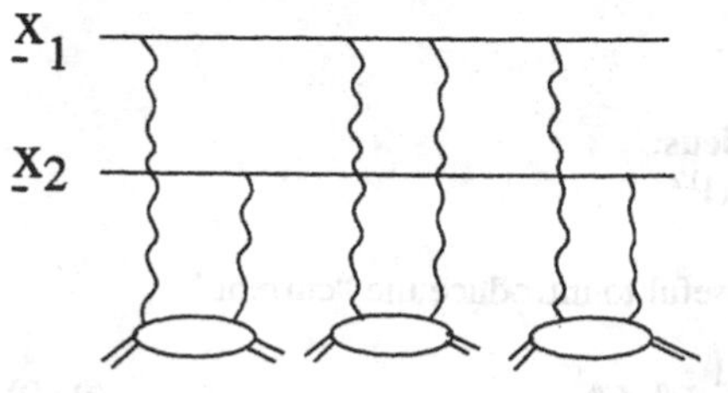

Figure 12.

$$e_f^2 \frac{d(xq_f + x\bar{q}_f)_A}{d^2\ell d^2b} = \frac{Q^2}{4\pi^2\alpha_{em}} \int \frac{d^2x_1 d^2x_2}{4\pi^2} \int_0^1 dz \frac{1}{2} \sum_\lambda \psi_{T\lambda}^{f*} \psi_{T\lambda}^f e^{-i\underline{\ell}\cdot(\underline{x}_1 - \underline{x}_2)}.$$

$$\cdot [1 + e^{-(\underline{x}_1 - \underline{x}_2)^2 \bar{Q}_S^2/4} - e^{-\underline{x}_1^2 \bar{Q}_S^2/4} - e^{-\underline{x}_2^2 \bar{Q}_S^2/4}] \tag{8.36}$$

with

$$\bar{Q}_S^2 = \frac{C_F}{N_c} Q_S^2 \tag{8.37}$$

and

$$Q_S^2 = \frac{8\pi^2 \alpha N_c}{N_c^2 - 1} \sqrt{R^2 - b^2}\, \rho x G, \tag{8.38}$$

where $\bar{Q}_S$ is the quark saturation momentum and Q_S is the gluon saturation momentum. In (38) xG is the gluon distribution in a nucleon. Eq.(36) is a complete solution to the sea quark distribution of a nucleus in the one-quark-loop approximation.

Problem 9(M-H): Use (30) and (36) to show that[6]

$$\frac{dx(q_f + \bar{q}_f)A}{d^2b d^2\ell} = \frac{N_c}{2\pi^4} \tag{8.39}$$

when $\underline{\ell}^2/\bar{Q}_S^2 << 1$.

Eq.(39) gives meaning to the idea of saturation as a maximum occupation number for, in this case, quarks. As the nucleus gets larger and larger, that is as R grows, $\bar{Q}_S^2$ grows and so there are more and more sea quarks in the nuclear wavefunction, nevertheless, the 2-dimensional occupation number hits a constant upper bound for momenta below the saturation momentum.

Problem 10(M-H): Show that $1/2$ of (39) comes from the first term in [ ] in (36) and that $1/2$ comes from the second term with the third and fourth term being small. The second term can be viewed as due to inelastic reactions while the first term is the elastic shadow of these inelastic reactions of a dipole passing over the nucleus.

9. Gluon Saturation in a Large Nucleus; the McLerran Venugopalan Model[12]

To directly probe gluon densities it is useful to introduce the "current"

$$j = -\frac{1}{4} F_{\mu\nu}^a F_{\mu\nu}^a. \tag{9.40}$$

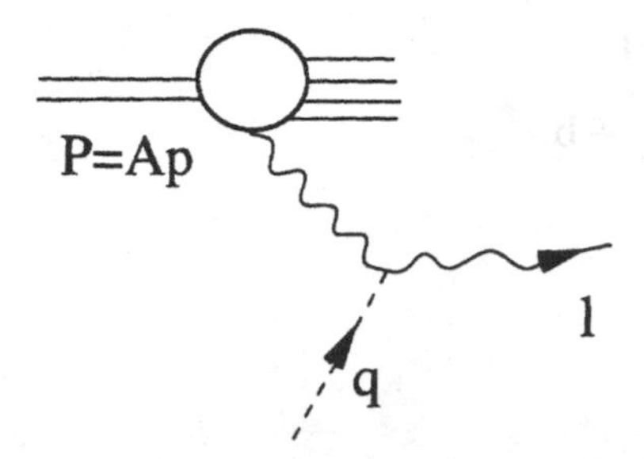

Figure 13.

We shall then calculate the process $j + A \to$ gluon ($\underline{\ell}$)+ anything. I shall interpret the process in a slightly modified Bjorken frame, and in light-cone gauge, while we shall do the calculation in a covariant gauge and in the rest system of the nucleus[2].

For the interpretation we take the momenta of a nucleon in the nucleus and the current to be

$$p = (p + \frac{M^2}{2p}, 0, 0, p)$$

$$q = (0, 0, 0, -2xp).$$

For small x p is large since $p = \frac{Q}{2x}$ and this frame is much like an infinite momentum frame. If we choose an appropriate light-cone gauge, one that eliminates final state interactions, then the transverse momentum and the x distribution of gluons in the nuclear wavefunction is the same as the distribution of produced gluon jets labeled by ℓ in Fig.13.

In order to do the calculation of the spectrum of produced gluon jets we carry out a multiple scattering calculation, a term of which is illustrated in Fig.14[2]. The calculation is simplest to do in covariant gauge. Then

$$\frac{dxG_A}{d^2b d^2\ell} = \int_0^{2\sqrt{R^2-b^2}} dz \rho x G(x, 1/\underline{x}^2) e^{-\frac{z}{L}\underline{x}^2 Q_S^2/4} e^{-i\underline{\ell}\cdot\underline{x}} \frac{d^2x}{4\pi^2} \qquad (9.41)$$

where xG is the gluon distribution for a single nucleon, and we make use of the coordinate space interpretation where it corresponds to the gluon in the complex conjugate amplitude being separated from that in the amplitude by $\Delta x_\perp \sim 1/Q$. We note that the momentum space distribution of gluon produced off a single nucleon, the unintegrated gluon distribution, is given by

$$\frac{dxG}{d^2\ell} = \int \frac{d^2x}{4\pi^2} xG(x, 1/\underline{x}^2) e^{-i\underline{\ell}\cdot\underline{x}} \qquad (9.42)$$

as can be easily verified by integrating the lefthand side of (42) over $d^2\underline{\ell}\Theta(Q^2 - \underline{\ell}^2)$. Carrying out the $z-$integration in (4) one gets[13]

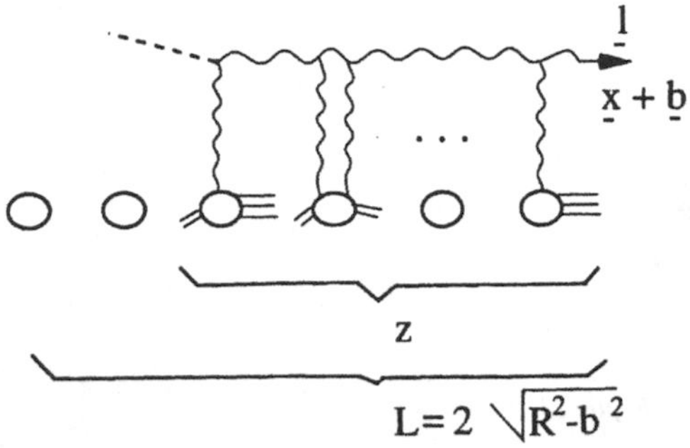

Figure 14.

$$\frac{dxG_A}{d^2bd^2\ell} = \frac{N_c^2-1}{4\pi^4\alpha N_c}\int d^2x \frac{1-e^{-\underline{x}^2Q_S^2/4}}{\underline{x}^2}e^{-i\underline{\ell}\cdot\underline{x}} \tag{9.43}$$

where Q_S^2 is given in (38). We also note that the $e^{-\frac{z}{L}\underline{x}^2Q_S^2/4}$ factor in (41) is just $S^\dagger(\underline{b})S(\underline{b}+\underline{x}) = S(\underline{x})$ for a gluon dipole to pass over a length z of nuclear material. Thus the derivation of (41) closely resembles that leading to (35) but now for gluons rather than quarks. Finally, when $\underline{\ell}^2/Q_S^2 << 1$ we find from (43)

$$\frac{dxG_A}{d^2bd^2\ell} \underset{\ell^2/Q_s^2<<1}{\longrightarrow} \frac{N_c^2-1}{4\pi^3\alpha N_c}\ell n(Q_S^2/\underline{\ell}^2) \tag{9.44}$$

while for $\underline{\ell}^2/Q_S^2 >> 1$ Eq.(43) gives the nuclear gluon distribution as simply a factor A times the nucleon gluon distribution. Eq.(44) shows that saturation is somewhat more complicated for gluons. The factor of N_c^2-1 counting the number of species of gluons is expected as is the factor of αN_c in the denominator. What is a little surprising is the log factor for which there is not yet a good intuitive understanding. Whether this log is a general factor or a peculiarity of the present model is not known for sure[6, 14].

10. The Golec-Biernat Wüsthoff Model[15]

We turn for a while to some phenomenology to see whether there is evidence for saturation of parton densities in deep inelastic lepton-proton scattering. So far the best way that has been found to approach this problem is through a simple model of deep inelastic and diffractive scattering inspired by the idea of saturation. We can motivate this discussion by going back to (36) and, supposing that such a picture might apply to a proton as well as a large nucleus, summing over f and integrating over $\underline{\ell}$ and $\underline{b}$ obtain

$$F_2(x,Q^2) = \frac{Q^2}{4\pi^2\alpha_{em}} \int d^2x \int_0^1 dz \sum_\lambda |\psi^f_{T\lambda}(\underline{x},z,Q)|^2 \int d^2b[1 - e^{-\underline{x}^2\bar{Q}_S^2/4}]. \tag{10.45}$$

Now our $\bar{Q}_S^2$ naturally depends on the impact parameter b, as indicated in (37) and (38) for a nuclear target, however, as an approximation we suppose

$$\int d^2b[1 - e^{-\underline{x}^2\bar{Q}_S^2/4}] = \sigma_0(1 - e^{-\underline{x}^2/4R_0^2}) \tag{10.46}$$

where R_0 will now be taken to depend only on x. Thus

$$F_2(x,Q^2) = \frac{Q^2}{4\pi^2\alpha_{em}} \int d^2x \int_0^1 dz \sum_\lambda |\psi^f_{T\lambda}(\underline{x},z,Q)|^2 (1 - e^{-\underline{x}^2/4R_0^2})\sigma_0 \tag{10.47}$$

which is the formula used by Golec-Biernat and Wüsthoff[15]. In addition it is then natural to take the diffractive cross section to be given by the shadow of the inelastic collisions in which case one replaces $\sigma_0(1 - e^{\underline{x}^2/4R_0^2}) = \sigma_0(1 - S)$ by $\frac{1}{2}\sigma_0(1 - S)^2$ giving

$$F_2^D(x,Q^2) = \frac{Q^2}{4\pi^2\alpha_{em}} \int d^2x \int_0^1 dz \sum_\lambda |\psi^f_{T\lambda\lambda}(\underline{x},z,Q)|^2 \frac{1}{2}\sigma_0(1 - e^{-\underline{x}^2/4R_0^2})^2. \tag{10.48}$$

Eq.(47) represents a total cross section, (48) represents an "elastic" cross section while the inelastic contribution would have a factor of $\frac{1}{2}\sigma_0(1 - e^{-\underline{x}^2/2R_0^2})$ replacing the last factors in (47) and (48).

Golec-Biernat and Wüsthoff include a quark-antiquark-gluon scattering term in addition to the quark-antiquark dipole term given by (48) so that larger mass diffractive states can also be described. For our purposes this is a detail which in any case introduces no new parameters. The model has three parameters

$$\sigma_0 = 23mb, \quad R_0^{-2} =< \bar{Q}_S^2 >= (\frac{x_0}{x})^\lambda GeV^2 \tag{10.49}$$

where $\lambda = 0.3$ and $x_0 = 3\text{x}10^{-4}$. With these three parameters a good fit to low and moderate Q^2 and low x F_2 and F_2^D data is obtained. In fact the fit is surprisingly good over a range of Q^2 which is remarkably large given that there is no QCD evolution present in the model. We shall have to wait for further tests and refinements to be sure that the fits are meaningful, but we may have the first bit of evidence for saturation effects. The fact that $< \bar{Q}_S^2 >$ is in the $1GeV^2$ region is reasonable.

11. Measuring Dipole Cross Sections

Refer back to (47). It is easy to check that

$$F_2 = c\int_{1/Q^2}^{1/Q_S^2} \frac{d\underline{x}^2}{\underline{x}^2} \tag{11.50}$$

when $Q^2/Q_S^2 >> 1$. Thus, although we may view F_2 as being given by a dipole cross section the size of the dipole is not well determined by Q^2 but, rather, varies between $1/Q$ and $1/Q_S$. Thus currently deep inelastic structure functions are not well suited for determining dipole cross sections. The situation should improve considerably when the longitudinal structure function is measured, for in this case the dipole size will be fixed to be of size $1/Q$ and will give a direct measure of the dipole cross section.

Problem 11(E): Use 30 and (45) to derive (50). What is c? ,
At present the best place to measure dipole cross sections, and thus to see how close or how far one is from finding unitarity limits, appears to be in the production of longitudinally polarized vector mesons[16].
The cross section for γ_L^*+ proton $\rightarrow \rho_L$+ proton can be written as

$$\frac{d\sigma^{\gamma_L^*\rightarrow\rho}}{dt} = \frac{1}{4\pi}|\int d^2x d^2b \int_0^1 dz \psi_\rho^*(\underline{x},z)(1-S(\underline{x},\underline{b}))\psi_{\gamma^*}(\underline{x},z,Q)^{i\underline{b}\cdot\underline{\Delta}}|^2 \tag{11.51}$$

where $t = -\underline{\Delta}^2$ is the momentum transfer. The process is pictured in Fig.15 and can be viewed in three steps. (i) The virtual photon breaks up into a quark-antiquark dipole of size $\underline{x}$ with z being the longitudinal momentum fraction of the γ^* carried by either the quark or antiquark.(ii) The dipole scatters elastically on the proton with scattering amplitude $1 - S(\underline{x}, \underline{b})$ where $\underline{x}$ is the dipole size and $\underline{b}$ the impact parameter of the scattering. (iii) The quark-antiquark pair then become a ρ long after they have passed the proton. This is the sequence of steps given in (51) where one also integrates over all possible dipole sizes and where the integration over impact parameters gives a definite transverse momentum.
Now suppose S is purely real, and define $N(Q) = (\psi_\rho, \psi_{\gamma^*})$. Then one can take the square root of both ides of (51), and after taking the inverse Fourier transform one finds

$$< S(x, r_0, b) >= 1 - \frac{1}{2N\pi^{3/2}} \int d^2\Delta e^{-i\underline{b}\cdot\underline{\Delta}} \sqrt{\frac{d\sigma}{dt}}. \tag{11.52}$$

In (52) $< S >= (\psi_\rho, S\psi_{\gamma^*})$ while x denotes the Bjorken-x value of the scattering and $r_0(Q)$ denotes the typical value of the dipole size contributing to (51). For (52) to have meaning it is important that the range of dipole sizes contributing to (51) not be too large, and this appears to be the case for longitudinal ρ production. This

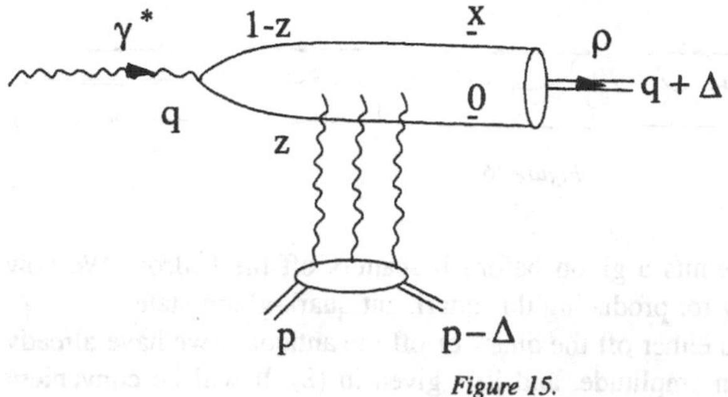

Figure 15.

analysis is modeled on the classic analysis of Amaldi and Shubert[17] for proton-proton elastic scattering. Thus one can estimate the $S-$matrix for a dipole of size r_0, determined by Q, scattering on a proton at a given impact parameter if the data are good enough to carry out the integral in (52). The analysis is not model independent as one needs to take a wavefunction for the ρ[18, 19, 20]in order to evaluate N. However, N does not appear to be very sensitive to the choice of wavefunction. Also the data are not good enough at larger values of t to accurately determine S below $b \approx 0.3.fm$. One finds, for example, at $Q^2 = 3.5GeV$ where $r_0 \approx 1/5fm$ and for $x \approx 10^{-3}$ that: (i) $S(b \approx 0) \approx 0.5 - 0.7$; (ii) the probability of an inelastic collision $= 1 - S^2(b)$ is considerable at small values of b indicating a reasonable amount of blackness at central impact parameters; (iii) $\sigma_{tot}^{q\bar{q}-Proton} \approx 10mb$; (iv) $\bar{Q}_S$ is consistent with that found in the Golec-Biernat Wüsthoff model. This adds, perhaps, another piece of evidence that saturation is approached in the HERA regime for moderate values of Q^2.

12. A Simple Equation for Unitarity; the Kovchegov equation[21]

Consider a (not too high momentum) dipole scattering on a high-energy hadron. We suppose the quark-antiquark dipole is left moving while the hadron is right moving. Further we suppose that the rapidity, y, of the dipole is such that $\alpha y << 1$ so that one need not consider radiative corrections to the wavefunction of the dipole to evaluate the scattering amplitude. We wish to study the dependence of the elastic scattering amplitude as one changes the relative rapidity of the dipole and the hadron by an amount dY when the relative rapidity is Y. Clearly one can view the change dY either as increasing the momentum of the hadron and thus allowing its wavefunction to evolve further or as increasing the momentum of the dipole. The latter is easier to deal with since the dipole is a simple object. When the rapidity of the dipole is increased there is a small probability, proportional

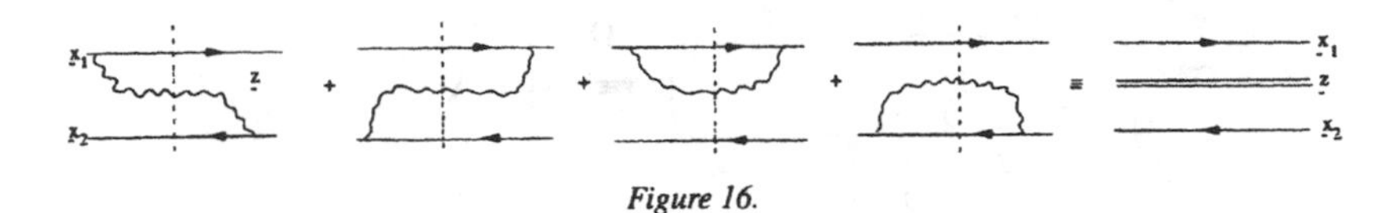

Figure 16.

to dY, that the dipole emits a gluon before it scatters off the hadron. We now calculate the probability for producing this quark-antiquark-gluon state.
Since a gluon is emitted either off the quark or off the antiquark we have already done the basic emission amplitude, and it is given in (8). It will be convenient to work in a basis where transverse coordinate are used rather than transverse momenta so one must take the Fourier transform of (8). Thus the amplitude for a quark having transverse coordinate $\underline{x}_1$ to emit a gluon having transverse coordinate $\underline{z}$, longitudinal momentum k_- (Recall that in Sec.2 we were dealing with right movers while here we are concerned with left movers.) and polarization λ is

$$\psi_\lambda^c(\underline{z}-\underline{x}_1) = \int \frac{d^2k}{(2\pi)^2}\, e^{i(\underline{z}-\underline{x})\cdot\underline{k}}\psi_\lambda^c(k) \tag{12.53}$$

with $\psi_\lambda^c(k)$ given by (8) with the replacement $k_+ \to k_-$. We are now using $A_- = 0$ light-cone gauge to evaluate the left moving quark-antiquark-gluon state. Using

$$\int \frac{d^2k}{2\pi} e^{i\underline{x}\cdot\underline{k}} \frac{\underline{k}}{\underline{k}^2} = i\frac{\underline{x}}{\underline{x}^2} \tag{12.54}$$

one gets

$$\psi_\lambda^c(\underline{z}-\underline{x}_1) = (\frac{\lambda^c}{2})2ig\frac{(\underline{z}-\underline{x}_1)\cdot\underline{\epsilon}^{\lambda^*}}{(\underline{z}-\underline{x}_1)^2}\frac{1}{\sqrt{(2\pi)^3 2k_-}}. \tag{12.55}$$

To calculate the probability of a quark-antiquark-gluon state one adds the graphs in Fig.16 to get[22]

$$dP_r = \sum_c (\frac{\lambda^c}{2}\frac{\lambda^2}{2})4g^2\frac{d^2z dk_-}{(2\pi)^3 2k_-}[-2\frac{(\underline{z}-\underline{x}_1)\cdot(\underline{z}-\underline{x}_2)}{(\underline{z}-\underline{x}_1)^2(\underline{z}-\underline{x}_2)^2}+\frac{1}{(\underline{z}-\underline{x}_1)^2}+\frac{1}{(\underline{z}-\underline{x}_2)^2}] \tag{12.56}$$

or

$$dP_r = \frac{\alpha N_c}{2\pi^2}d^2z dY\frac{(\underline{x}_1-\underline{x}_2)^2}{(\underline{x}_1-\underline{z})^2(\underline{x}_2-\underline{z})^2} \tag{12.57}$$

where we have set $dk_-/k_- = dY$ and $C_F = \frac{N_c}{2}$ in the large N_c limit where the Kovchegov equation will be valid.

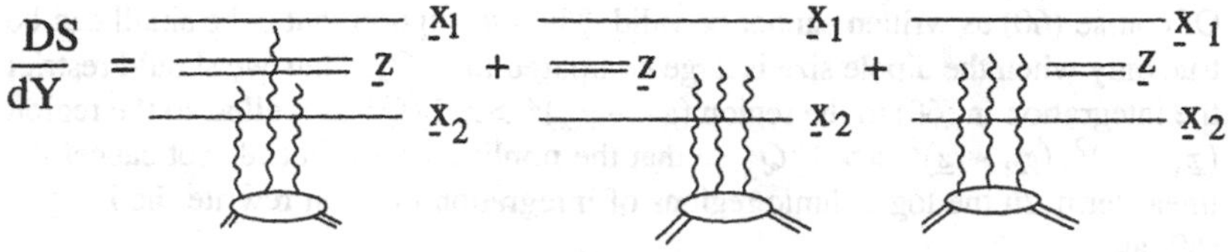

Figure 17.

Then the $S-$matrix for the quark-antiquark-gluon state to elastically scatter on the hadron multiplied by dP_τ in (57) gives the change in the $S-$matrix, dS, for dipole-hadron scattering. The result is the Kovchegov equation[21]

$$\frac{dS(\underline{x}_1-\underline{x}_2,Y)}{dY} = \frac{\alpha N_c}{2\pi^2}\int d^2z \frac{(\underline{x}_1-\underline{x}_2)^2}{(\underline{x}_1-\underline{z})^2(\underline{x}_2-\underline{z})^2}$$

$$[S(\underline{x}_1-\underline{z},Y)S(\underline{z}-\underline{x}_2,Y) - S(\underline{x}_1-\underline{x}_2,Y)], \qquad (12.58)$$

and it is illustrated in Fig.17. We have assumed that the scattering of the two dipoles, the quark-(antiquark part of the gluon) and the (quark part of the gluon)-antiquark dipoles, factorize when scattering off the hadron. This was clear in the model Kovchegov considered where the hadron was a large nucleus. This factorization is less obvious in the general case and the Kovchegov equation may be a sort of mean field approximation to a more complete equation. Also the final term on the right-hand side of (58), corresponding to the last two graphs in Fig.17, give the virtual contributions necessary to normalize the wavefunction[22]. The necessity of this last term can be seen by considering the weak interaction limit where $S \to 1$. Then the final term on the righthand side of (58) is necessary to get $\frac{dS}{dY} = 0$ when $S = 1$.

There are two interesting limits to (58). First suppose that S is near 1 and write $S = 1 - iT$. One easily finds, keeping only linear terms in T,

$$\frac{dT(\underline{x}_1-\underline{x}_2,Y)}{dy} = \frac{\alpha N_c}{\pi^2}\int d^2z \frac{(\underline{x}_1-\underline{x}_2)^2}{(\underline{x}_1-\underline{z})^2(\underline{x}_2-\underline{z})^2}[T(\underline{x}_1-\underline{z},Y) - \frac{1}{2}T(\underline{x}_1-\underline{x}_2,Y)] \qquad (12.59)$$

which is the dipole form of the BFKL equation. In this case the factorized form of the scattering is justified by the large N_c limit and the weak coupling approximation.

The other interesting limit is where S is small in which case one need only keep the second term on the righthand side of (58) giving

$$\frac{dS(\underline{x}_1-\underline{x}_2,Y)}{dY} = -\frac{\alpha N_c}{2\pi^2}\int \frac{d^2z(\underline{x}_1-\underline{x}_2)^2}{(\underline{x}_1-\underline{z})^2(\underline{x}_2-\underline{z})^2} S(\underline{x}_1-\underline{x}_2 Y). \qquad (12.60)$$

Of course (60) as written cannot be valid. The assumption that S be small can be true only when the dipole size is large compared to $1/Q_s$. Thus we should restrict the integration in (60) to the region $(\underline{x}_1 - \underline{x}_2)^2 >> 1/Q_S^2$ as well as to the region $(\underline{x}_1 - \underline{z})^2, (\underline{x}_2 - \underline{z})^2 >> 1/Q_S^2$ so that the nonlinear term in (58) not cancel the linear term. In the logarithmic regions of integration one can rewrite the integral (60) as

$$\frac{dS(\underline{x}_1 - \underline{x}_2, Y)}{dY} = -2\frac{\alpha N_c}{2\pi^2}\int_{1/Q_S^2}^{(\underline{x}_1-\underline{x}_2)^2} \frac{\pi d(\underline{x}_1 - \underline{z})^2}{(\underline{x}_1 - \underline{z})^2} S(\underline{x}_1 - \underline{x}_2, Y) \quad (12.61)$$

giving

$$\frac{dS(\underline{x}, Y)}{dY} = -\frac{\alpha N_c}{\pi} \ell n(Q_S^2 \underline{x}^2) S(\underline{x}, Y) \quad (12.62)$$

whose solution is

$$S(\underline{x}, Y) = e^{-\frac{\alpha N_c}{\pi}\int_{Y_0}^{Y} dy\ \ell n[Q_S^2(y)\underline{x}^2]} S(\underline{x}, Y_0). \quad (12.63)$$

If Q_S^2 is exponentially behaved

$$Q_S^2(y) = e^{c\frac{\alpha N_c}{\pi}(y - Y_0)} Q_S^2(Y_0)$$

then

$$S(\underline{x}, Y) = e^{-\frac{c}{2}(\frac{\alpha N_c}{\pi})^2 (Y - Y_0)^2} S(\underline{x}, Y_0) \quad (12.64)$$

where Y_0 should be chosen to satisfy

$$Q_S^2(Y_0)\underline{x}^2 = 1. \quad (12.65)$$

Eq.(64) is the result found in Ref.23 and argued heuristically already some time ago[24], although without an evaluation of the coefficient of the Y^2 term in the exponent.
The Kovchegov equation is an interesting equation for studying scattering when one is near unitarity limits. In the next section we shall use a procedure very similar to what we have done here to derive an equation whose content is presumably equivalent to the Balitsky equation[5], a somewhat more general form than the Kovchegov equation. The advantage of (58) is its simplicity and its likely qualitative correctness in QCD.

13. A Simple Derivation of the JIMWLK equation[14,25–27]

Over the past seven years or so there has been an ambitious program dedicated to finding appropriate equations for dealing with high density wavefunctions in QCD. This program has been quite successful and a renormalization group equation in the form of a functional Fokker-Planck equation for the wavefunction of a high-energy hadron has been given by the authors of Refs.14 and 25-27. (JIMWLK). The most complete derivation is given in Ref.27 where the equation is written in terms of a covariant gauge potential, α, coming from light-cone gauge quanta in a high-energy hadron. Here we give an alternative simple derivation[28]. We can imitate this mixture of gauges used in Ref.27 by taking Coulomb gauge[29] which has a gluon propagator

$$D_{\alpha\beta}(k) = -\frac{i}{k^2}[g_{\alpha\beta} - \frac{N\cdot k(N_\alpha k_\beta + N_\beta k_\alpha) - k_\alpha k_\beta}{(\vec{k})^2}] \tag{13.66}$$

where $N \cdot v = v_0$ for any vector v. Suppose the propagator has $k_+^2 >> \underline{k}^2, k_-^2$ and connects two highly right moving lines. Then the Coulomb gauge propagator is equivalent to the $A_+ = 0$ light-cone gauge propagator

$$D_{\alpha\beta}(k) = \frac{-i}{k^2}[g_{\alpha\beta} - \frac{\eta_\alpha k_\beta + \eta_\beta k_\alpha}{\eta \cdot k}] \tag{13.67}$$

Problem 12(E): Show that D_{--}, and D_{-i} as given in (66) and (67) agree when $k_+^2 >> \underline{k}^2, k_-^2$. For a right moving system these are the important components of the gluon propagator.

Similarly for a left moving system Coulomb gauge is equivalent to $A_- = 0$ gauge while for gluon lines which connect left moving systems to right moving systems the dominant component in (66) is $D_{+-} = 1/(\vec{k})^2$ which looks like covariant gauge when $k_+^2, k_-^2 << \underline{k}^2$.

We are going to consider the scattering of a set of left moving quanta, quarks and gluons, on some high-energy, right moving hadron. These quarks and gluons may be parts of a hadronic wavefunction which are frozen in the passage over the right moving hadron or they may come from a current as in our discussion of deep inelastic scattering given in Sec.8. For simplicity we shall limit our discussion to left moving quark and antiquark lines, but this is simply to avoid too cumbersome notation. Then a left moving quark interacting with the right moving hadron can be represented by

$$V^\dagger(\underline{x}) = P\, exp\{ig \int_{-\infty}^{\infty} dx_- A_+(\underline{x}, x_-) \tag{13.68}$$

where we have taken $x_+ = 0$ and fixed the left moving quark to have transverse coordinate $\underline{x}$. Except for the change of right moving quark to left moving quark

(68) is the same as the operator in the matrix element in (26) where we showed how quarks could be identified with Wilson lines in the fundamental representation. By taking gauge invariant combinations of V's and $V^\dagger$'s we can form observables which depend on A_+ and which correspond to the scattering of quite general left moving systems on the right moving hadron. We denote a general such observable by $O(A_+)$.
Although we have put the integration in (68) exactly on the light-cone we in fact are going to assume that the left moving observable has rapidity y obeying $\alpha y << 1$ so that transverse gluons are unlikely to be emitted by the left movers allowing us to identify the left moving system at $x_- = \pm\infty$. If the relative rapidity of the scattering is Y then we imagine that $\alpha(Y - y) \approx \alpha Y >> 1$ so that the right moving hadron has, in general, a wavefunction including many gluons. If the right moving hadron has momentum p then the scattering amplitude is

$$< O >_Y = < p|O|p > = \int D[\alpha(\underline{x}, x_-)]O(\alpha)W_Y[\alpha] \tag{13.69}$$

where the weight function W_Y is given by

$$W_Y[\alpha] = \int D[A_\mu]\delta(A_+ - \alpha)\delta(F(A))\Delta_F[A]e^{iS[A]} \tag{13.70}$$

where D indicates a functional integral, F is a gauge fixing, and Δ_F is the corresponding Fadeev-Popov determinant times an operator which projects out the state $|p >$ initially and finally. We suppose W_Y is normalized to

$$\int D[\alpha]W_Y[\alpha] = 1. \tag{13.71}$$

Now consider the $Y-$dependence of $< O >_Y$. From (69) one can clearly write

$$\frac{d}{dY} < O >_Y = \int D[\alpha]O(\alpha)\frac{d}{dY}W_Y[\alpha]. \tag{13.72}$$

However, one can equally well imagine calculating $\frac{d}{dY} < O >_Y$ by evaluating the change of the left moving system, due to an additional gluon emission, as one increases the rapidity of the left moving system by an amount dY. The change in the left moving state is given by the gluon emissions and absorptions shown in Figs.18 and 19 where spectator quark and antiquark lines are not shown. In Fig.18 we have assumed that a gluon connects a quark and an antiquark line. This is for definiteness. We equally well could have assumed a connection to two quark lines or to two antiquark lines. The vertical line in the figures represents the ."time," $x_- = 0$, at which the left moving system passes the right moving hadron. This view of the $Y-$dependence of $< O >_Y$ in terms of a change of rapidity of the (rather simple) left movers is in the spirit of work previously done by Balitsky[5], Kovchegov[21], and Weigert[26].

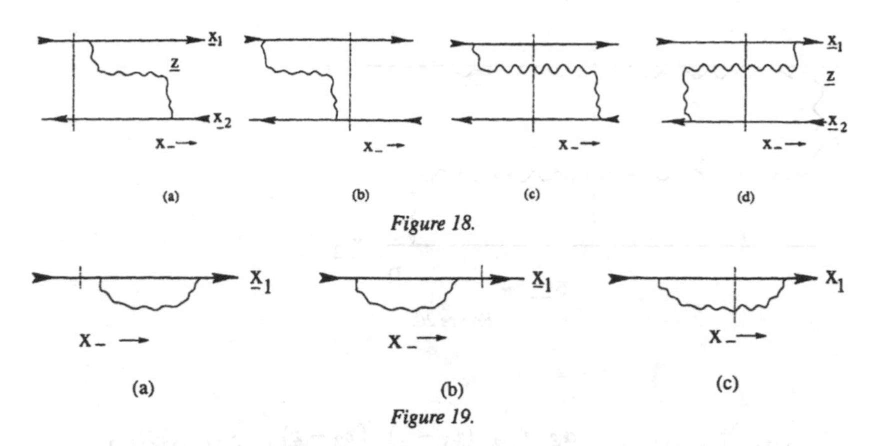

Figure 18.

Figure 19.

We shall examine in some detail the graphs of Fig.18 before stating the complete result including the graphs of Fig.19. Begin with the graph shown in Fig.18c. We do the calculation in $A_- = 0$ light-cone gauge, which for left movers is equivalent to our Coulomb gauge choice. This graph is exactly the same as has been calculated in Sec.11 except for the fact that the quark and antiquark lines are not necessarily in a color singlet. The result is the same as $\frac{1}{2}$ the first term on the righthand side of (56) except for the color factors which we put in separately. The result is

$$V^\dagger(\underline{x}_1)\otimes V(\underline{x}_2) \to -\frac{\alpha_S}{\pi^2}\int d^2z \frac{(\underline{x}_1-\underline{z})\cdot(\underline{x}_2-\underline{z})}{(\underline{x}_1-\underline{z})^2(\underline{x}_2-\underline{z})^2}\tilde{V}_{cd}(\underline{z})V^\dagger(\underline{x}_1)T^c\otimes V(\underline{x}_2)T^d \tag{13.73}$$

in going from O to $\frac{dO}{dY}$. The additional factors, as compared to the corresponding term in (56), are T^c which comes to the right of $V^\dagger(\underline{x}_1)$ because the emission off the $\underline{x}_1-$ line is at early values of x_-, the T^d which comes to the right of $V(\underline{x}_2)$ because the absorption on the $\underline{x}_2-$line is at late values of x_-, and the factor $\tilde{V}_{cd}(\underline{z})$ giving the interaction of the gluon with the hadron as a Wilson line in the adjoint representation.
Now write

$$V^\dagger(\underline{x}_1)T^c = V^\dagger(\underline{x}_1)T^cV(\underline{x}_1)V^\dagger(\underline{x}_1) = \tilde{V}_{ca}(\underline{x}_1)T^aV^\dagger(\underline{x}_1). \tag{13.74}$$

Now T^a comes to the left of $V^\dagger(\underline{x}_1)$ as if the emission of the gluon were at late values of x_-. Indeed, we may view the graph in Fig.18c as being given by the "mnemonic" graph shown in Fig.20 where the adjoint line integral starts at large positive values of x_- and proceeds to large negative values of x_- at a transverse coordinate $\underline{x}_1$ then back again to large positive values of x_- at a transverse coordinate $\underline{z}$. The result of the graph of Fig.18c then is

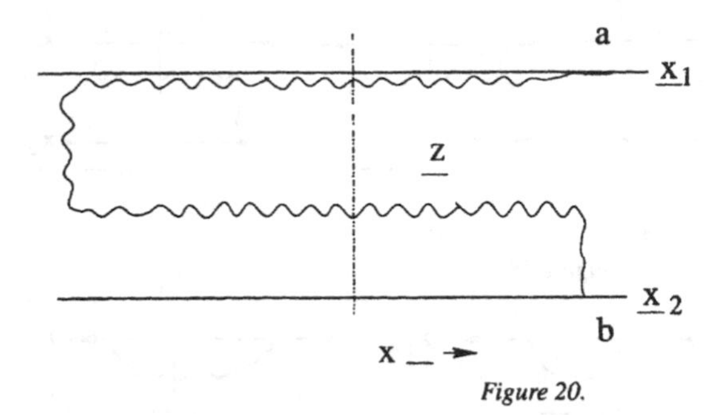

Figure 20.

$$V^\dagger(\underline{x}_1) \otimes V(\underline{x}_2) \to \frac{\alpha_S}{\pi^2}\int d^2z \frac{(\underline{x}_1 - \underline{z})\cdot(\underline{x}_2 - \underline{z})}{(\underline{x}_1 - \underline{z})^2(\underline{x}_2 - \underline{z})^2}\{-\tilde{V}^\dagger(x_1)\tilde{V}(\underline{z})\}_{ab}$$

$$T^aV^\dagger(\underline{x}_1) \otimes V(\underline{x}_2)T^b \tag{13.75}$$

Now it is straightforward to add in the other terms of Fig.18 and those of Fig.19 to get

$$V^\dagger(\underline{x}_1)\otimes V(\underline{x}_2) \to \alpha_S\{\frac{1}{2}\int d^2xd^2y\eta^{ab}_{\underline{xy}}\frac{\delta^2}{\delta\alpha^a(\underline{x},x_-)\delta\alpha^b(\underline{y},y_-)} + \int d^2x\nu^a_{\underline{x}}\frac{\delta}{\delta\alpha^a(\underline{x},x_-)}\}\cdot$$

$$\cdot V^\dagger(\underline{x}_1) \otimes V(\underline{x}_2) \tag{13.76}$$

where, if one takes large positive values of x_- and y_-, the functional derivatives in (76) simply insert color matrices in the appropriate places, as, for example, in (75). η and ν are given by

$$g^2\eta^{ab}_{\underline{xy}} = 4\int \frac{d^2z}{4\pi^2}\frac{(\underline{x}-\underline{z})\cdot(\underline{y}-\underline{z})}{(\underline{x}-\underline{z})^2(\underline{y}-\underline{z})^2}\cdot$$

$$\cdot\{1 + \tilde{V}^\dagger(\underline{x})\tilde{V}(\underline{y}) - \tilde{V}^\dagger(\underline{x})\tilde{V}(\underline{z}) - \tilde{V}^\dagger(\underline{z})\tilde{V}(\underline{y})\}_{ab} \tag{13.77}$$

and

$$g\nu^a_{\underline{x}} = 2i\int \frac{d^2z}{(\underline{x}-\underline{z})^2} tr[T^a\tilde{V}^\dagger(\underline{x}_1)\tilde{V}(\underline{z})]. \tag{13.78}$$

For a general scattering one simply replaces $V^\dagger(\underline{x}_1) \otimes V(\underline{x}_2)$ in (76) by $O(\alpha)$. After multiplying by $W_Y[\alpha]$ and integrating over $D[\alpha]$ one gets[28]

$$\int D[\alpha]O(\alpha)\frac{dW_Y[\alpha]}{dY} = \int D[\alpha]O(\alpha)\alpha_S\{\frac{1}{2}d^2xd^2y\frac{\delta^2}{\delta\alpha^a(\underline{x},x_-)\delta\alpha^b(\underline{y},y_-)}$$

$$\cdot [W_Y \eta^{ab}_{\underline{xy}}] - \int d^2x \frac{\delta}{\delta \alpha^a(\underline{x}, x_-)} [W_Y \nu^a_{\underline{x}}]\} \tag{13.79}$$

where an integration by parts in α has been done on the righthand side of (79). To the extent that the $O(\alpha)$ form a complete set of observables, and it is not clear how close this is to being true, one can equate the integrands of (79) and arrive at the JIMWLK equation

$$\frac{dW_Y[\alpha]}{dY} = \alpha_S\{\frac{1}{2}\int d^2x d^2y \frac{\delta^2}{\delta\alpha^a(\underline{x}, x_-)\delta\alpha^b(\underline{y}, y_-)} [W_Y \eta^{ab}_{\underline{xy}}]$$

$$- \int d^2x \frac{\delta}{\delta \alpha^a(\underline{x}, x_-)} [W_Y \nu^a_{\underline{x}}]\}. \tag{13.80}$$

The exact values of x_- and y_- appear to have some arbitrariness as discussed in some detail in Ref.28. Eq.(80) is an elegant equation of a functional Fokker-Planck type the nature of whose solutions is now under investigation.

References

1. Yu.V. Kovchegov, Phys.Rev.**D54** (1996) 5463; **D55** (1997) 5445.
2. Yu. V. Kovchegov and A.H. Mueller,*Nucl.* Phys.**B529** (1998) 451.
3. W. Buchmüller, M.F. McDermott and A. Hebecker, *Nucl.Phys.* **B487** (1997) 283; **B500** (1997) 621 (E)
4. W. Buchmüller, T. Gehrman and A. Hebecker, *Nucl.Phys.* **B538** (1999) 477.
5. I. Balitsky, Nucl. Phys.**B463** (1996) 99.
6. A.H. Mueller, *Nucl.Phys.* **B558** (1999) 285.
7. J.D. Bjorken in Proceedings of the International Symposium on Electron and Photon Interactions at High Energies, pages 281-297, Cornell (1971).
8. A.H. Mueller, Nucl. Phys.**B335** (1990) 115.
9. L.L. Frankfurt and M. Strikman, Phys.Rep.**160** (1998) 235.
10. B. Blättel, G. Baym, L.L. Frankfurt and M. Strikman, Phys. Rev.Lett. **70** (1993) 896.
11. B.G. Zakharov, JETP Lett.**63** (1996) 952.
12. L.McLerran and R. Venugopalan,Phys.Rev. **D49**(1994) 2233;**D49**(1994) 3352; **D50**(1994) 2225.
13. J. Jalilian-Marian, A. Kovner, L. McLerran and H. Weigert, Phys. Rev. **D55** (1997) 5414.
14. E. Iancu and L. McLerran,Phys.Lett.**B510** (2001) 145.
15. K. Golec-Biernat and M. Wüsthoff, Phys. Rev.**D59** (1999) 014017; Phys.Rev.**D60** (1999) 114023.
16. S. Munier, A.M. Stasto and A.H. Mueller,Nucl.Phys. **B603** (2001) 427.
17. U. Amaldi and K. R. Schubert,Nucl.Phys.**B166** (1980) 301.
18. H.G. Dosch, T. Gousset, G. Kulzinger and H.J. Pirner, Phys. Rev.**D55** (1997) 2602.
19. J. Nemchik, N.N. Nikolaev, E. Predazzi and B.G. Zakharov, Z.Phys.**C75** (1997) 71.
20. L. Frankfurt, N. Koepf and M. Strikman, Phys.Rev.**D54** (1996) 3194.
21. Yu. Kovchegov, Phys.Rev.**D60** (1999) 034008; Phys.Rev.**D61** (2000) 074018.
22. A.H.Mueller, Nucl. Phys. **B415** (1994) 373.
23. E. Levin and K. Tuchin, Nucl.Phys.**B573** (2000) 83; hep-ph/0101275.

24. A.H. Mueller and G.P. Salam, Nucl.Phys.**B475** (1996) 293.
25. J. Jalilian-Marian, A. Kovner, A. Leonidov and H. Weigert, Nucl. Phys.**B504** (1997) 415;Phys. Rev. **D59** (1999) 014014.
26. H. Weigert, hep-ph/0004044.
27. E. Iancu, A. Leonidov and L. McLerran, Nucl.Phys. **A692** (2001) 583; Phys.Lett.**B510** (2001) 133; E. Ferreiro, E. Iancu, A. Leonidov and L. McLerran, hep-ph/0109115.
28. A.H. Mueller, Phys. Lett. **B523** (2001) 243.
29. T. Jaroszewicz, Acta Physica Polonica **B11** (1980) 965.

THE COLOUR GLASS CONDENSATE: AN INTRODUCTION

EDMOND IANCU (eiancu@cea.fr)
Service de Physique Théorique, CEA/DSM/SPhT
CEA Saclay, 91191 Gif-sur-Yvette cedex, France

ANDREI LEONIDOV (leonidov@td.lpi.ac.ru)
P. N. Lebedev Physical Institute, Moscow, Russia

LARRY MCLERRAN (mclerran@quark.phy.bnl.gov)
Nuclear Theory Group, Brookhaven National Laboratory, Upton, NY 11793, USA

Abstract. In these lectures, we develop the theory of the Colour Glass Condensate. This is the matter made of gluons in the high density environment characteristic of deep inelastic scattering or hadron-hadron collisions at very high energy. The lectures are self contained and comprehensive. They start with a phenomenological introduction, develop the theory of classical gluon fields appropriate for the Colour Glass, and end with a derivation and discussion of the renormalization group equations which determine this effective theory. In particular, these equations predict gluon saturation at sufficiently high energy, with a maximal density of order $1/\alpha_s$.

1. General Considerations

1.1. INTRODUCTION

The goal of these lectures is to convince you that the average properties of hadronic interactions at very high energies are controlled by a new form of matter, a dense condensate of gluons. This is called the Colour Glass Condensate since

- Colour: The gluons are coloured.
- Glass: The associated fields evolve very slowly relative to natural time scales, and are disordered. This is like a glass which is disordered and is a liquid on long time scales but seems to be a solid on short time scales.
- Condensate: There is a very high density of massless gluons. These gluons can be packed until their phase space density is so high that interactions prevent more gluon occupation. With increasing energy, this forces the gluons to occupy higher momenta, so that the coupling becomes weak. The gluon density saturates at a value of order $1/\alpha_s \gg 1$, corresponding to a multiparticle state which is a Bose condensate.

J.-P. Blaizot and E. Iancu (eds.), QCD Perspectives on Hot and Dense Matter, 73–145.

In these lectures, we will try to explain why the above is very plausible.
Before doing this, however, it is useful to review some of the typical features of hadronic interactions, and some unanswered theoretical questions which are associate with these phenomena. This will motivate much of the later discussion.

1.2. TOTAL CROSS SECTIONS AT ASYMPTOTIC ENERGY

Computing total cross sections as $E \to \infty$ is one of the great unsolved problems of QCD. Unlike for processes which are computed in perturbation theory, it is not required that any energy transfer become large as the total collision energy $E \to \infty$. Computing a total cross section for hadronic scattering therefore appears to be intrinsically non-perturbative. In the 60's and early 70's, Regge theory was extensively developed in an attempt to understand the total cross section. The results of this analysis were to our mind inconclusive, and certainly can not be claimed to be a first principles understanding from QCD.
Typically, it is assumed that the total cross section (say, for pp collisions) grows as $\ln^2 E2$ as $E \to \infty$. This is the so called Froissart bound, which corresponds to the maximal growth allowed by the unitarity of the scattering matrix. Is this correct? Is the coefficient of $\ln^2 E$ universal for all hadronic precesses? Why is the unitarity limit saturated? Can we understand the total cross section from first principles in QCD? Is it understandable in weakly coupled QCD, or is it an intrinsically non-perturbative phenomenon?

1.3. HOW ARE PARTICLES PRODUCED IN HIGH ENERGY COLLISIONS ?

In order to discuss particle production, it is useful to introduce some kinematical variables adapted for high energy collisions: the light cone coordinates. Let z be the longitudinal axis of the collision. For an arbitrary 4-vector $v^\mu = (v^0, v^1, v^2, v^3)$ ($v^3 = v_z$, etc.), we define its light-cone (LC) coordinates as

$$v^+ \equiv \frac{1}{\sqrt{2}}(v^0 + v^3), \qquad v^- \equiv \frac{1}{\sqrt{2}}(v^0 - v^3), \qquad v_\perp \equiv (v^1, v^2). \tag{1.1}$$

In particular, we shall refer to $x^+ = (t+z)/\sqrt{2}$ as the LC "time", and to $x^- = (t-z)/\sqrt{2}$ as the LC "longitudinal coordinate". The invariant dot product reads:

$$p \cdot x = p^- x^+ + p^+ x^- - p_\perp \cdot x_\perp, \tag{1.2}$$

which suggests that p^- — the momentum variable conjugate to the "time" x^+ — should be interpreted as the LC energy, and p^+ as the (LC) longitudinal momentum. In particular, for particles on the mass-shell: $p^\pm = (E \pm p_z)/\sqrt{2}$, with $E = (m^2 + \mathbf{p}^2)^{1/2}$, and therefore:

$$p^+ p^- = \frac{1}{2}(E^2 - p_z^2) = \frac{1}{2}(p_\perp^2 + m^2) = \frac{1}{2} m_\perp^2 . \tag{1.3}$$

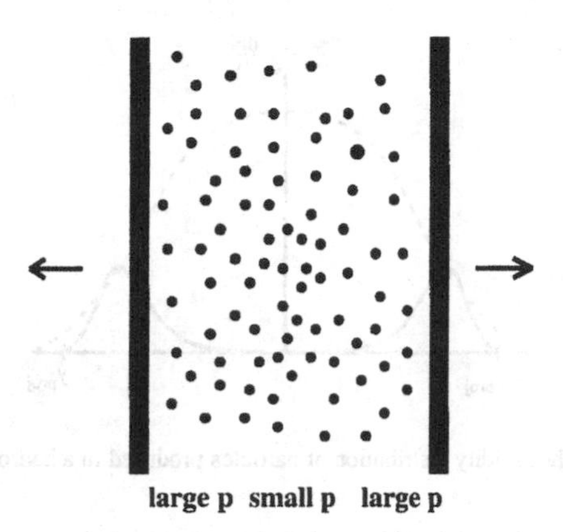

Figure 1. A hadron-hadron collision. The produced particles are shown as circles.

This equation defines the transverse mass $m_\perp$. We shall also need the *rapidity* :

$$\mathrm{y} \equiv \frac{1}{2}\ln\frac{p^+}{p^-} = \frac{1}{2}\ln\frac{2p^{+2}}{m_\perp^2}. \tag{1.4}$$

These definitions are useful, among other reasons, because of their simple properties under longitudinal Lorentz boosts: $p^+ \to \kappa p^+, p^- \to (1/\kappa)p^-$, where κ is a constant. Under boosts, the rapidity is just shifted by a constant: $\mathrm{y} \to \mathrm{y} + \kappa$. Consider now the collision of two identical hadrons in the center of mass frame, as shown in Fig. 1. In this figure, we have assumed that the colliding hadrons have a transverse extent which is large compared to the size of the produced particles. This is true for nuclei, or if the typical transverse momenta of the produced particles is large compared to Λ_{QCD}, since the corresponding size will be much smaller than a Fermi. We have also assumed that the colliding particles have an energy which is large enough so that they pass through one another and produce mesons in their wake. This is known to happen experimentally: the particles which carry the quantum numbers of the colliding particles typically lose only some finite fraction of their momenta in the collision. Because of their large energy, the incoming hadrons propagate nearly at the speed of light, and therefore are Lorentz contracted in the longitudinal direction, as suggested by the figure.
In LC coordinates, the right moving particle ("the projectile") has a 4-momentum $p_1^\mu = (p_1^+, p_1^-, 0_\perp)$ with $p_1^+ \simeq \sqrt{2}p_z$ and $p_1^- = M^2/2p_1^+$ (since $p_z \gg M$, with $M =$ the projectile mass). Similarly, for the left moving hadron ("the target"), we have $p_2^+ = p_1^-$ and $p_2^- = p_1^+$. The invariant energy squared is $s = (p_1 + p_2)^2 = 2p_1 \cdot p_2 \simeq 2p_1^+ p_2^- \simeq 4p_z^2$, and coincides, at it should, with the total energy squared $(E_1 + E_2)^2$ in the center of mass frame.

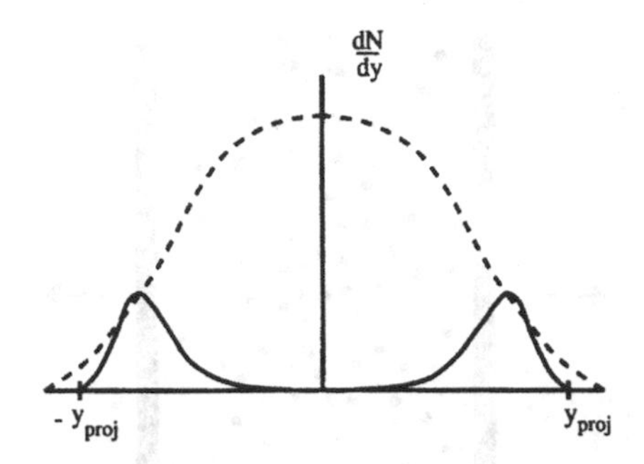

Figure 2. The rapidity distribution of particles produced in a hadronic collision.

Consider a pion produced in this collision and which is moving in the positive z direction. For such a pion, we define the longitudinal momentum fraction, or Feynman's x, as :

$$\mathrm{x} \equiv \frac{p_\pi^+}{p_1^+} \qquad \text{(right mover)}, \tag{1.5}$$

which implies $m_\perp/\sqrt{2}p_1^+ \leq \mathrm{x} \leq 1$. The rapidity of the pion is then

$$\mathrm{y} = \frac{1}{2}\ln\frac{p_\pi^+}{p_\pi^-} = \frac{1}{2}\ln\frac{2p_\pi^{+2}}{m_\perp^2} = \mathrm{y}_{proj} - \ln\frac{1}{\mathrm{x}} + \ln\frac{M}{m_\perp}, \tag{1.6}$$

($\mathrm{y}_{proj} = \ln(\sqrt{2}p_1^+/M) \simeq \ln(\sqrt{s}/M)$), and lies in the range $0 \leq \mathrm{y} \leq \mathrm{y}_{proj} + \ln(M/m_\perp)$. For a left moving pion ($p_\pi^z < 0$), we use similar definitions where p^+ and p^- are exchanged. This gives a symmetric range for y, as in Fig. 2.
In this figure we plot a typical distribution of the particles produced in the hadronic collision. We denote by dN/dy the number of produced particles per unit rapidity. The leading particles are shown by the solid line and are clustered around the projectile and target rapidities. For example, in a heavy ion collision, this is where the nucleons would be. The dashed line is the distribution of produced mesons. Several theoretical issues arise in multiparticle production:
Can we compute dN/dy ? Or even dN/dy at $\mathrm{y} = 0$ ("central rapidity") ? How does the average transverse momentum of produced particles $\langle p_\perp \rangle$ behave with energy? What is the ratio of produced strange/nonstrange mesons, and corresponding ratios of charm, top, bottom etc at $\mathrm{y} = 0$ as the center of mass energy approaches infinity? Does multiparticle production as $s \to \infty$ at $\mathrm{y} = 0$ become simple, understandable and computable?
Note that $\mathrm{y} = 0$ corresponds to particles with $p_z = 0$ or $p^+ = m_\perp/\sqrt{2}$, for which $\mathrm{x} = m_\perp/(\sqrt{2}p_1^+) = m_\perp/\sqrt{s}$ is small, $\mathrm{x} \ll 1$, in the high-energy limit of interest. Thus, presumably, the multiparticle production at central rapidity reflects properties of the small-x degrees of freedom in the colliding hadron wavefunctions.

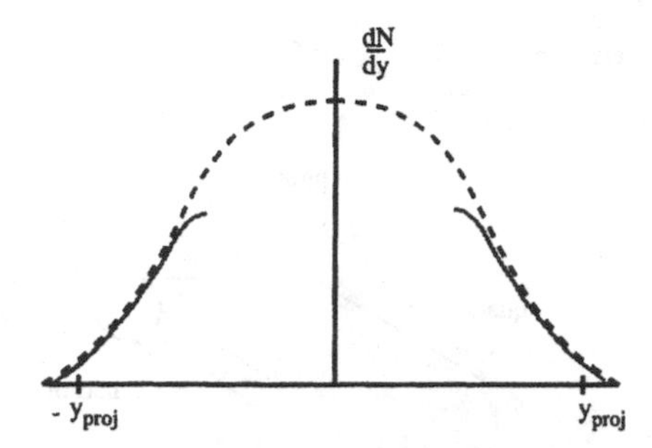

Figure 3. Feynman scaling of rapidity distributions. The two different lines correspond to rapidity distributions at different energies.

There is a remarkable feature of rapidity distributions of produced hadrons, which we shall refer to as Feynman scaling. If we plot rapidity distributions of produced hadrons at different energies, then as function of $y - y_{proj}$, the rapidity distributions are to a good approximation independent of energy. This is illustrated in Fig. 3, where the rapidity distribution measured at one energy is shown with a solid line and the rapidity distribution at a different, higher, energy is shown with a dotted line. (In this plot, the rapidity distribution at the lower energy has been shifted by an amount so that particles of positive rapidity begin their distribution at the same y_{proj} as the high energy particles, and correspondingly for the negative rapidity particles. This of course leads to a gap in the center for the low energy particles due to this mapping.)
This means that as we go to higher and higher energies, the new physics is associated with the additional degrees of freedom at small rapidities in the center of mass frame (small-x degrees of freedom). The large x degrees of freedom do not change much. This suggests that there may be some sort of renormalization group description in rapidity where the degrees of freedom at larger x are held fixed as we go to smaller values of x. We shall see that in fact these large x degrees of freedom act as sources for the small x degrees of freedom, and the renormalization group is generated by integrating out degrees of freedom at relatively large x to generate these sources.

1.4. DEEP INELASTIC SCATTERING

In Fig. 4, deep inelastic scattering is shown. Here an electron emits a virtual photon which scatters from a quark in a hadron. The momentum and energy transfer of the electron is measured, but the results of the hadron break up are not. In these lectures, we do not have sufficient time to develop the theory of deep inelastic scattering (see, e.g., [1] for more details). For the present purposes, it is enough to

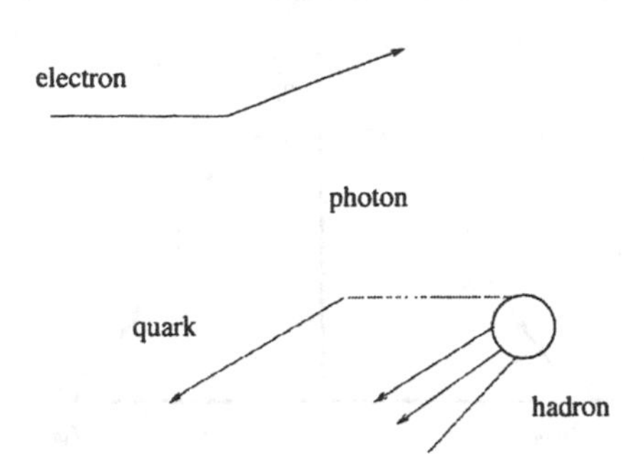

Figure 4. Deep inelastic scattering of an electron on a hadron.

say that, at large momentum transfer $Q^2 \gg \Lambda^2_{QCD}$, this experiment can be used to measure the distributions of quarks in the hadron.

To describe the quark distributions, it is convenient to work in a reference frame where the hadron has a large light-cone longitudinal momentum $P^+ \gg M$ ("infinite momentum frame"). In this frame, one can describe the hadron as a collection of constituents ("partons"), which are nearly on-shell excitations carrying some fraction x of the total longitudinal momentum P^+. Thus, the longitudinal momentum of a parton is $p^+ = xP^+$, with $0 \leq x < 1$.

For the struck quark in Fig. 4, this x variable ("Feynman's x") is equal to the Bjorken variable x_{Bj}, which is defined in a frame independent way as $x_{Bj} = Q^2/2P \cdot q$, and is directly measured in the experiment. In this definition, $Q^2 = -q^\mu q_\mu$, with q^μ the (space-like) 4-momentum of the exchanged photon. The condition that $x = x_{Bj}$ is what maximizes the spatial overlap between the struck quark and the virtual photon, thus making the interaction favourable.

The Bjorken variable scales like $x_{Bj} \sim Q^2/s$, with s = the invariant energy squared. Thus, in deep inelastic scattering at high energy (large s at fixed Q^2) one measures quark distributions dN_{quark}/dx at small x ($x \ll 1$).

It is useful to think about these distributions as a function of rapidity. We define the rapidity in deep inelastic scattering as

$$y = y_{hadron} - \ln(1/x), \tag{1.7}$$

and the invariant rapidity distribution as

$$\frac{dN}{dy} = x\frac{dN}{dx}. \tag{1.8}$$

In Fig. 5, a typical dN/dy distribution for constituent gluons of a hadron is shown. This plot is similar to the rapidity distribution of produced particles in hadron-hadron collisions (see Fig. 2). The main difference is that, now, we have only half

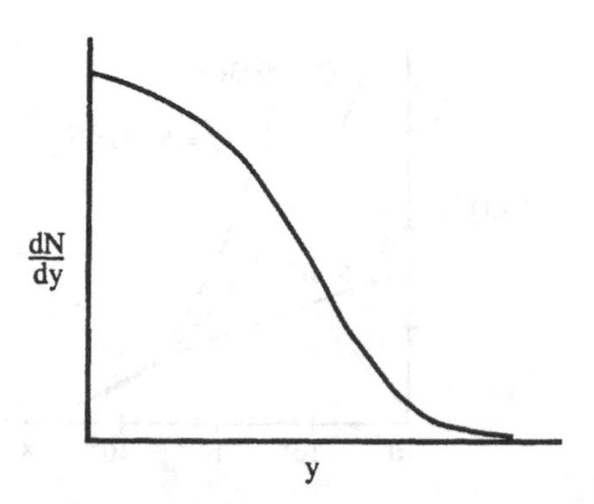

Figure 5. The rapidity distribution of gluons inside of a hadron.

of the plot, corresponding to the right moving hadron in a collision in the center of mass frame.

One may in fact argue that there is indeed a relationship between the structure functions as measured in deep inelastic scattering and the rapidity distributions for particle production. We expect, for instance, the gluon distribution function to be proportional to the pion rapidity distribution. This is what comes out in many models of particle production. It is further plausible, since the degrees of freedom of the gluons should not be lost, but rather converted into the degrees of freedom of the produced hadrons.

The small x problem is that in experiments at HERA, the rapidity distributions for quarks and gluons grow rapidly as the rapidity difference

$$\tau \equiv \ln(1/x) = y_{hadron} - y \tag{1.9}$$

between the quark and the hadron increases [2]. This growth appears to be more rapid than τ or τ^2, and various theoretical models based on the original considerations by Lipatov and colleagues [3] suggest it may grow as an exponential in τ [3, 4]. The more established DGLAP evolution equation [5] predicts a less rapide growth, like an exponential in $\sqrt{\tau}$, but this is still exceeding the Froissart unitarity bound, which requires rapidity distributions to grow at most as τ^2 (since $\tau \sim \ln s$).

In Fig. 6, the ZEUS data for the gluon distribution are plotted for $Q^2 = 5$ GeV^2, 20 GeV^2 and 200 GeV^2 [2]. The gluon distribution is the number of gluons per unit rapidity in the hadron wavefunction, $xG(x, Q^2) = dN_{gluons}/dy$. Experimentally, it is extracted from the data for the quark structure functions, by exploiting the dependence of the latter upon the resolution of the probe, that is, upon the transferred momentum Q^2. Note the rise of $xG(x, Q^2)$ at small x: this is the small x problem. If one had plotted the total multiplicity of produced particles

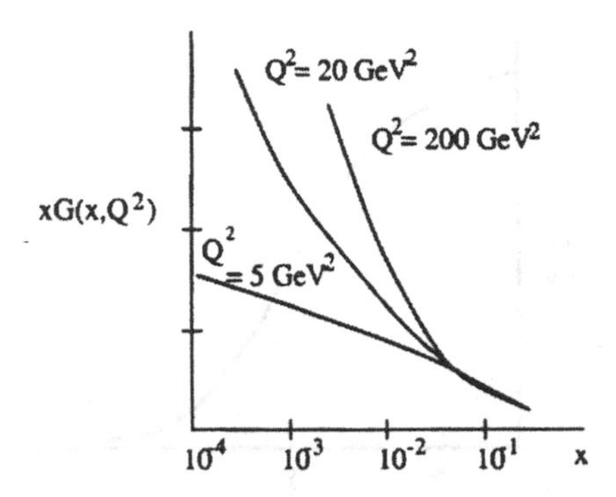

Figure 6. The Zeus data for the gluon structure functions.

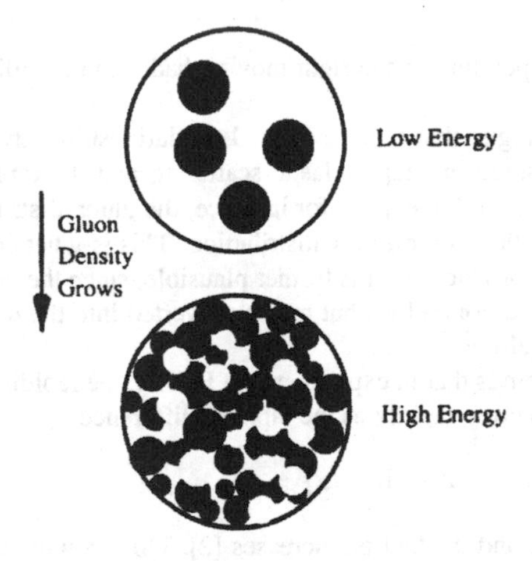

Figure 7. Saturation of gluons in a hadron. A view of a hadron head on as x decreases.

in pp and $\bar{p}p$ collisions on the same plot, one would have found rough agreement in the shape of the curves.

Why is the small x rise in the gluon distribution a problem? Consider Fig. 7, where we view the hadron head on. The constituents are the valence quarks, gluons and sea quarks shown as coloured circles. As we add more and more constituents, the hadron becomes more and more crowded. If we were to try to measure these constituents with say an elementary photon probe, as we do in deep inelastic scattering, we might expect that the hadron would become so crowded that we could not ignore the shadowing effects of constituents as we make the measurement. (Shadowing means that some of the partons are obscured by virtue of having another parton in front of them. This would result in a decrease of the scattering cross section relative to what is expected from incoherent independent scattering.)

We shall later argue that the distribution functions at fixed Q^2 *saturate*, which means that they cease growing so rapidly at high energy [6, 7, 8, 9, 10]. (See also Refs. [11, 12, 13, 14] for recent reviews and more references.) This saturation will be seen to occur at transverse momenta below some intrinsic scale, the "saturation scale", which is estimated as:

$$Q_s^2 = \alpha_s N_c \frac{1}{\pi R^2} \frac{dN}{dy}, \tag{1.10}$$

where dN/dy is the gluon distribution. Only gluons matter since, at small x, the gluon density grows faster then the quark density, and is the driving force towards saturation. This is why in the forthcoming considerations we shall ignore the (sea) quarks, but focus on the gluons alone. Furthermore, πR^2 — with R the hadron radius — is the area of the hadron in the transverse plane. (This is well defined as long as the wavelengths of the external probes are small compared to R.) Finally, $\alpha_s N_c$ is the colour charge squared of a single gluon. Thus, the "saturation scale" (1.10) has the meaning of the average colour charge squared of the gluons in the hadron wavefunction per unit transverse area.

Since the gluon distribution increases rapidly with the energy, as shown by the HERA data, so does the saturation scale. We shall use the rapidity difference $\tau = \ln(1/x) \sim \ln s$, eq. (1.9), to characterize this increase, and write $Q_s^2 \equiv Q_s^2(\tau)$. For sufficiently large τ (i.e., high enough energy, or small enough x),

$$Q_s^2(\tau) \gg \Lambda_{QCD}^2, \tag{1.11}$$

and $\alpha_s(Q_s^2) \ll 1$. Then we are dealing with *weakly coupled* QCD, so we should be able to perform a first principle calculation of, e.g.,

- the gluon distribution function;
- the quark and heavy quark distribution functions;
- the intrinsic $p_\perp$ distributions of quarks and gluons.

But weak coupling does not necessarily mean that the physics is perturbative. There are many examples of nonperturbative phenomena at weak coupling. An example is instantons in electroweak theory, which lead to the violation of baryon number. Another example is the atomic physics of highly charged nuclei, where the electron propagates in the background of a strong nuclear Coulomb field. Also, at very high temperature, QCD becomes a weakly coupled quark-gluon plasma, but it exhibits nonperturbative phenomena on large distances $r \gg 1/T$ (with T the temperature), due to the collective behaviour of many quanta [15].

Returning to our small-x gluons, we notice that, at low transverse momenta $Q^2 \leq Q_s^2(\tau)$, they make a high density system, in which the interaction probability

$$\frac{\alpha_s(Q^2)}{Q^2}\,\frac{1}{\pi R^2}\,\frac{dN}{dy} \sim 1 \quad \text{when} \quad Q^2 \sim Q_s^2(\tau) \tag{1.12}$$

is of order one [6, 7, 16]. That is, although the coupling is small, $\alpha_s(Q^2) \ll 1$, the effects of the interactions are amplified by the large gluon density (we shall see that $dN/dy \sim 1/\alpha_s$ at saturation), and ordinary perturbation theory breaks down. To cope with this, a resummation of the high density effects is necessary. Our strategy to do so — to be described at length in these lectures — will be to construct an *effective theory* in which the small-x gluons are described as the classical colour fields radiated by "colour sources" at higher rapidity. Physically, these sources are the "fast" partons, i.e., the hadron constituents with larger longitudinal momenta $p^+ \gg xP^+$. The properties of the colour sources will be obtained via a renormalization group analysis, in which the "fast" partons are integrated out in steps of rapidity and in the background of the classical field generated at the previous steps.

The advantage of this strategy is that the non-linear effects are dealt with in a classical context, which makes exact calculations possible. Specifically, (a) the classical field problem will be solved exactly, and (b) at each step in the renormalization group analysis, the non-linear effects associated with the classical fields will be treated exactly. On the other hand, the mutual interactions of the fast partons will be treated in perturbation theory, in a "leading-logarithmic" approximation which resums the most important quantum corrections at high energy (namely, those which are enhanced by the large logarithm $\ln(1/x)$).

As we shall see, the resulting effective theory describes the saturated gluons as a *Colour Glass Condensate*. The classical field approximation is appropriate for these saturated gluons, because of the large occupation number $N_k \sim 1/\alpha_s \gg 1$ of their true quantum state. In this limit, the Heisenberg commutators between particle creation and annihilation operators become negligible:

$$[a_k, a_k^\dagger] = 1 \ll a_k^\dagger a_k = N_k\,, \tag{1.13}$$

which corresponds indeed to a classical regime. The classical field language is also well adapted to describe the *coherence* of these small-x gluons, which overlap with each other because of their large longitudinal wavelengths.

The phenomenon of saturation provides also a natural solution to the unitarity problem alluded to before. We shall see that, with increasing energy, the new partons are produced preponderently at momenta $p_\perp \gtrsim Q_s$. Thus, these new partons have a typical transverse size $\sim 1/p_\perp \lesssim 1/Q_s$. Smaller is x (i.e., larger is τ), larger is $Q_s(\tau)$, and therefore smaller are the newly produced partons. An external probe of transverse resolution $\Delta x_\perp \sim 1/Q$ will not see partons smaller than this resolution size. For τ large enough, $Q^2 < Q_s^2(\tau)$, so that the partons produced when further increasing the energy will not contribute to the cross section at fixed Q^2. Thus, although the gluon distribution keeps increasing with τ, there is nevertheless no contradiction with unitarity.

1.5. GEOMETRICAL SCALING

Another striking feature of the experimental data at HERA is *geometrical scaling* at Bjorken x < 0.01 [17]. In general, one expects the structure functions extracted from deep inelastic scattering to depend upon two dimensionless kinematical variables, x and Q^2/Λ^2, where Λ^2 is some arbitrary momentum scale of reference, which is fixed. The striking feature alluded to before is the observation that the x dependence measured at HERA at x < 0.01 and for a broad region of Q^2 (between 0.045 and 450 Gev^2) can be entirely accounted for by a corresponding dependence of the reference scale $\Lambda^2 \to 1/R^2(\mathrm{x})$ alone. That is, rather than being functions of two independent variables x and Q^2/Λ^2, the measured structure functions at x < 0.01 depend effectively only upon the scaling variable

$$\mathcal{T} \equiv Q^2 R^2(\mathrm{x}) \tag{1.14}$$

where $R^2(\mathrm{x}) \sim \mathrm{x}^\lambda$ and $\lambda \sim 0.3 - 0.4$ in order to fit the data. This is illustrated in Fig. 8 [17]. Such a scaling behaviour is consistent with the saturation scenario [18, 10, 19], as we shall discuss towards the end of these lectures. Note however that the experimentally observed scaling extends to relatively large values of x and Q^2, above all the estimates for the saturation scale. Thus, this feature seems to be more general than the phenomenon of saturation.

1.6. UNIVERSALITY

There are two separate formulations of universality which are important in understanding small x physics.
a) The first is a weak universality [8, 10]. This is the statement that at sufficiently high energy, physics should depend upon the specific properties of the hadron at hand (like its size or atomic number A) only via the saturation scale $Q_s(\tau, A)$. Thus, at high energy, there should be some equivalence between nuclei and protons: When their Q_s^2 values are the same, their properties must be the same. An empirical parameterization of the gluon structure function in eq. (1.10) is

$$\frac{1}{\pi R^2}\frac{dN}{dy} \sim \frac{A^{1/3}}{x^\delta} \tag{1.15}$$

where $\delta \sim 0.2 - 0.3$ [2]. This suggests the following correspondences:

- RHIC with nuclei $\sim$ HERA with protons;

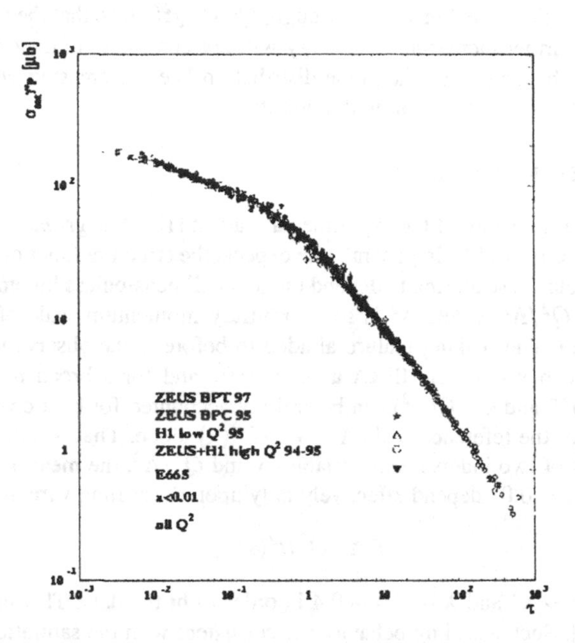

Figure 8. Experimental data on the cross section for virtual photon-proton deep inelastic scattering from the region x < 0.01 plotted verses the scaling variable $\mathcal{T} = Q^2 R^2(x)$.

– LHC with nuclei ∼ HERA with nuclei.

Estimates of the saturation scale for nuclei at RHIC energies give $\sim 1 - 2$ Gev, and at LHC $Q_s \sim 2 - 3$ Gev.

b) The second is a strong universality which is meant in a statistical mechanical sense. This is the statement that the effective action which describes small x distribution function is critical and at a fixed point of some renormalization group. This means that the behavior of correlation functions is given by universal critical exponents, which depend only on general properties of the theory such as its symmetries and dimensionality.

1.7. SOME APPLICATIONS

We conclude these introductory considerations with a (non-exhaustive) enumeration of recent applications of the concept of saturation and the Colour Glass Condensate (CGC) to phenomenology.

Consider deep inelastic scattering first. It has been shown in Refs. [18] that the HERA data for (both inclusive and diffractive) structure functions can be well accounted for by a phenomenological model which incorporates saturation. The same model has motivated the search for geometrical scaling in the data, as explained in Sect. 1.5.
Coming to ultrarelativistic heavy ion collisions, as experimentally realized at RHIC and, in perspective, at LHC, we note that the CGC should be the appropriate description of the initial conditions. Indeed, most of the multiparticle production at central rapidities is from the small-x ($x \lesssim 10^{-3}$) partons in the nuclear wavefunctions, which are in a high-density, semi-classical, regime. The early stages of a nuclear collision, up to times $\sim 1/Q_s$, can thus be described as the melting of the Colour Glass Condensates in the two nuclei. In Refs. [20], this melting has been systematically studied, and the multiparticle production computed, via numerical simulations of the classical effective theory [8, 21]. After they form, the particles scatter with each other, and their subsequent evolution can be described by transport theory [22].
The first experimental data at RHIC [23] have been analyzed from the perspective of the CGC in Refs. [24, 25, 26]. Specifically, the multiparticle production has been studied with respect to its dependence upon centrality ("number of participants") [24], rapidity [25] and transverse momentum distribution [26].
The charm production from the CGC in peripheral heavy-ion collisions has been investigated in [27].
Electron-nucleus (eA) deeply inelastic scattering has been recently summarized in [28]. Some implications of the Colour Glass Condensate for the central region of $p + A$ collisions have been explored in Refs. [29, 30].
Instantons in the saturation environment have been considered in Ref. [31].

2. The classical effective theory

With this section, we start the study of an effective theory for the small x component of the hadron wavefunction [8, 10, 32, 33, 34, 35, 36, 37] (see also the previous review papers [12, 38]). Motivated by the physical arguments exposed before, in particular, by the separation of scales between *fast* partons and *soft* (i.e., small-x) gluons, in the infinite momentum frame, this effective theory admits a rigourous derivation from QCD, to be described in Sect. 3. Here, we shall rather rely on simple kinematical considerations to motivate its general structure.

2.1. A STOCHASTIC YANG-MILLS THEORY

In brief, the effective theory is a classical Yang-Mills theory with a random colour source which has only a "plus" component [1] :

$$(D_\nu F^{\nu\mu})_a(x) = \delta^{\mu +}\rho_a(x). \tag{2.1}$$

The classical gauge fields A_a^μ represent the *soft* gluons in the hadron wavefunction, i.e., the gluons with small longitudinal momenta ($k^+ = xP^+$ with $x \ll 1$). For these gluons, the classical approximation should be appropriate since they are in a multiparticle state with large occupation numbers.
The *fast* partons, with momenta $p^+ \gg k^+$, are not dynamical fields anylonger, but they have been rather replaced by the colour current $J_a^\mu = \delta^{\mu +}\rho_a$ which acts as a source for the soft gluon fields. This is quite intuitive: the soft gluons in the hadron wavefunction are radiated by typically fast partons, via the parton cascades shown in Fig. 9. It is in fact well known that, for the tree-level radiative process shown in Fig. 9.a, classical and quantum calculations give identical results in the limit where the emitted gluon is soft [1]. What is less obvious, but will be demonstrated by the analysis in Sect. 3, is that quantum corrections like those displayed in Fig. 9.b do not invalidate this classical description, but simply renormalize the properties of the classical source, in particular, its correlations.

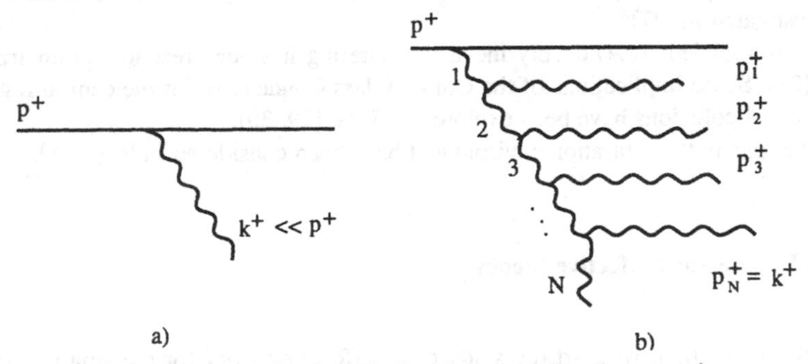

Figure 9. a) Soft gluon emission by a fast parton; b) a gluon cascade.

The gross properties of this source follow from kinematics. The fast partons move along the z axis at nearly the speed of light. They can emit, or absorb, soft gluons, but in a first approximation they preserve straightline trajectories along the light-cone ($z = t$). In terms of LC coordinates, they propagate in the positive x^+ direction, while sitting at $x^- = 0$. Their colour current is proportional to their velocity, which implies $J_a^\mu = \delta^{\mu +}\rho_a$, with a charge density $\rho_a(x)$ which is

[1] Written as it stands, eq. (2.1) is correct only for field configurations having $A^- = 0$; in general, when $A^- \neq 0$, the source ρ in its r.h.s. gets rotated by Wilson lines built from A^- [37].

localized near $x^- = 0$. More precisely, as quantum fields, the fast partons are truly delocalized over a longitudinal distance $\Delta x^- \sim 1/p^+$, as required by the uncertainty principle. But since $1/p^+ \ll 1/k^+$, they still look as sharply localized when "seen" by the soft gluons, which have long wavelengths and therefore a poor longitudinal resolution.

The separation of scales in longitudinal momenta implies a corresponding separation in time: Softer partons have larger energies, and therefore shorter lifetimes. Consider indeed the radiative process in Fig. 9.a, where $k^+ \ll p^+$. This is a virtual excitation whose lifetime (in units of LC time x^+) can be estimated from the uncertainty principle as

$$\Delta x^+ = \frac{1}{\varepsilon_{p-k} + \varepsilon_k - \varepsilon_p} \simeq \frac{1}{\varepsilon_k} \ll \frac{1}{\varepsilon_p}. \tag{2.2}$$

This is small as compared to the typical time scale $1/\varepsilon_p$ for the dynamics of the fast partons. [In eq. (2.2), $\varepsilon_p \equiv p_\perp^2/2p^+$ is the LC energy of the on-shell gluon with momentum $\vec{p} = (p^+, \mathbf{p}_\perp)$, and we have used the fact that, for $k^+ \ll p^+$ and comparable transverse momenta $k_\perp$ and $p_\perp$, $\varepsilon_k \gg \varepsilon_p, \varepsilon_{p-k}$.] Thus, the "fast" degrees of freedom are effectively frozen over the short lifetime of the soft gluon, and can be described by a *time-independent* (i.e., independent of x^+) colour source $\rho_a(x^-, x_\perp)$.

Still, this colour source is eventually changing over the larger time scale $1/\varepsilon_p$. Thus, if another soft gluon is emitted after a time interval $\gtrsim 1/\varepsilon_p$, it will "see" a different configuration of ρ, without quantum interference between the different configurations. This can be any of the configurations allowed by the dynamics of the fast partons. We are thus led to treat $\rho_a(x^-, x_\perp)$ as a *classical random* variable (here, a *field* variable), with some probability density, or *weight function*, $W_{k^+}[\rho]$, which is a functional of ρ.

As suggested by its notation, the weight function depends upon the soft scale k^+ at which we measure correlations. Indeed, as we shall see in Sect. 3, $W_{k^+}[\rho]$ is obtained by integrating out degrees of freedom with longitudinal momenta larger than k^+. It turns out that it is more convenient to use the *rapidity*[2]

$$\tau \equiv \ln(P^+/k^+) = \ln(1/\mathrm{x}) \tag{2.3}$$

to indicate this dependence, and thus write $W_\tau[\rho] \equiv W_{k^+}[\rho]$.

To deal with field variables and functionals of them, it is convenient to consider a discretized (or lattice) version of the 3-dimensional configuration space, with lattice points $(x^-, x_\perp)$. (We use the same notations for discrete and continuous coordinates, to avoid a proliferation of symbols.) A configuration of the colour

[2] Strictly speaking, this is the rapidity *difference* between the small-x gluon and the hadron, as defined previously in eq. (1.9). But this difference is the relevant quantity for what follows, so from now on it will be simply referred to as "the rapidity", for brevity.

source is specified by giving its values $\rho^a(x^-, x_\perp)$ at the N lattice points. The functional $W_\tau[\rho]$ is a (real) function of these N values. To have a meaningful probabilistic interpretation, this function must be positive semi-definite ($W_\tau[\rho] \geq 0$ for any ρ), and normalized to unity:

$$\int D[\rho]\, W_\tau[\rho] = 1\,, \tag{2.4}$$

with the following functional measure:

$$D[\rho] \equiv \prod_a \prod_{x^-} \prod_{x_\perp} d\rho^a(x^-, x_\perp)\,. \tag{2.5}$$

Gluon correlation functions at the soft scale $k^+ = xP^+ = P^+ e^{-\tau}$ are obtained by first solving the classical equations of motion (2.1) and then averaging the solution over ρ with the weight function $W_\tau[\rho]$ (below $\vec{x} \equiv (x^-, x_\perp)$) :

$$\langle A_a^i(x^+, \vec{x}) A_b^j(x^+, \vec{y}) \cdots \rangle_\tau = \int D[\rho]\, W_\tau[\rho]\, \mathcal{A}_a^i(\vec{x}) \mathcal{A}_b^j(\vec{y}) \cdots\,, \tag{2.6}$$

where $\mathcal{A}_a^i \equiv \mathcal{A}_a^i[\rho]$ is the solution to the classical Yang-Mills equations with static source ρ_a, and is itself independent of time (cf. Sect. 2.3 below). Note that only equal-time correlators can be computed in this way; but these are precisely the correlators that are measured by a small-x external probe, which is absorbed almost instantaneously by the hadron (cf. eq. (2.2)).
The formula (2.6) is readily extended to any operator which can be related to ρ. To guarantee that only the physical, gauge-invariant, operators acquire a non-vanishing expectation value, we shall require $W_\tau[\rho]$ to be gauge-invariant. In practical calculations, one generally has to fix a gauge, so the gauge symmetry of $W_\tau[\rho]$ may not be always manifest.
To summarize, the effective theory is defined by eqs. (2.1) and (2.6) together with the (so far, unspecified) weight function $W_\tau[\rho]$. In what follows, we shall devote much effort to derive this theory from QCD, and construct the weight function $W_\tau[\rho]$ in the process (in Sects. 3–5). But before doing that, let us gain more experience with the classical theory by solving the equations of motion (2.1) (in Sect. 2.3), and then using the result to compute the gluon distribution of a large nucleus (in Sect. 2.4). In performing these calculations, we shall need a more precise definition of the gluon distribution function and, more generally, of the relevant physical observables, so we start by discussing that.

2.2. SOME USEFUL OBSERVABLES

In subsequent applications of the effective theory, we shall mainly focus on two observables which, because of their physical content and of the specific structure of the effective theory, are particularly suggestive for studies of non-linear phenomena like saturation. These observables, that we introduce now, are the gluon distribution function and the cross-section for the scattering of a "colour dipole" off the hadron.

2.2.1. *The gluon distribution function*

We denote by $G(\mathrm{x}, Q^2)dx$ the number of gluons in the hadron wavefunction having longitudinal momenta between $\mathrm{x}P^+$ and $(\mathrm{x}+dx)P^+$, and a transverse size $\Delta x_\perp \sim 1/Q$. In other terms, the *gluon distribution* $\mathrm{x}G(\mathrm{x}, Q^2)$ is the number of gluons with transverse momenta $k_\perp \lesssim Q$ per unit rapidity :

$$\begin{aligned} \mathrm{x}G(\mathrm{x}, Q^2) &= \int^{Q^2} d^2k_\perp\, k^+ \frac{dN}{dk^+ d^2k_\perp}\bigg|_{k^+=xP^+} \\ &= \int d^3k\, \Theta(Q^2 - k_\perp^2)\, \mathrm{x}\delta(\mathrm{x} - k^+/P^+)\, \frac{dN}{d^3k}\,, \end{aligned} \tag{2.7}$$

where $\vec{k} \equiv (k^+, \mathbf{k}_\perp)$ and

$$\frac{dN}{d^3k} = \frac{dN}{dk^+ d^2k_\perp}\,, \tag{2.8}$$

is the Fock space gluon density, i.e., the number of gluons per unit of volume in momentum space. The difficulty is, however, that this number depends upon the gauge, so in general it is not a physical observables. Still, as we shall shortly argue, this quantity can be given a gauge-invariant meaning when computed in the light-cone (LC) gauge

$$A_a^+ = 0\,. \tag{2.9}$$

(We define the light-cone components of A_a^μ in the standard way, as $A_a^\pm = (A_a^0 \pm A_a^3)/\sqrt{2}$.) In this gauge, the equations of motion[3]

$$D_\mu F^{\mu\nu} = 0, \tag{2.10}$$

imply for the + component

$$D_i F^{i+} + D^+ F^{-+} = 0, \tag{2.11}$$

which allows one to compute A^- in terms of A^i as

$$A^- = \frac{1}{\partial^{+2}} D^i \partial^+ A^i\,. \tag{2.12}$$

This equation says that we can express the longitudinal field in terms of the transverse degrees of freedom which are specified by the transverse fields entirely and

[3] For the purposes of LC quantization we use the equations of motion without sources; that is, we consider real QCD, and not the effective theory (2.1).

explicitly. These degrees of freedom correspond to the two polarization states of the gluons. The quantization of these degrees of freedom proceeds by writing [39]:

$$A^i_c(x^+,\vec{x}) = \int_{k^+>0} \frac{d^3k}{(2\pi)^3 2k^+} \left(e^{i\vec{k}\cdot\vec{x}} a^i_c(x^+,\vec{k}) + e^{-i\vec{k}\cdot\vec{x}} a^{i\dagger}_c(x^+,\vec{k}) \right) \quad (2.13)$$

($\vec{x}\cdot\vec{k} = x^-k^+ - \mathbf{x}_\perp\cdot\mathbf{k}_\perp$) with the creation and annihilation operators satisfying the following commutation relation at equal LC time x^+ :

$$[a^i_b(x^+,\vec{k}), a^{j\dagger}_c(x^+,\vec{q})] = \delta^{ij}\delta_{bc}\, 2k^+(2\pi)^3\delta^{(3)}(k-q). \quad (2.14)$$

In terms of these Fock space operators, the gluon density is computed as:

$$\frac{dN}{d^3k} = \langle a^{i\dagger}_c(x^+,\vec{k})\, a^i_c(x^+,\vec{k})\rangle = \frac{2k^+}{(2\pi)^3} \langle A^i_c(x^+,\vec{k}) A^i_c(x^+,-\vec{k})\rangle\,, \quad (2.15)$$

where the average is over the hadron wavefunction. By homogeneity in time, this equal-time average is independent of the coordinate x^+, which will be therefore omitted in what follows. By inserting this into eq. (2.7) and using the fact that, in the LC-gauge, $F^{i+}_a(k) = ik^+A^i_a(k)$, one obtains (with $k^+ = \mathrm{x}P^+$ from now on):

$$\mathrm{x}G(\mathrm{x},Q^2) = \frac{1}{\pi}\int \frac{d^2k_\perp}{(2\pi)^2}\, \Theta(Q^2-k_\perp^2) \big\langle F^{i+}_a(\vec{k}) F^{i+}_a(-\vec{k})\big\rangle. \quad (2.16)$$

As anticipated, this does not look gauge invariant. In coordinate space:

$$F^{i+}_a(\vec{k}) F^{i+}_a(-\vec{k}) = \int d^3x \int d^3y\, e^{i(\vec{x}-\vec{y})\cdot\vec{k}}\, F^{i+}_a(\vec{x}) F^{i+}_a(\vec{y}) \quad (2.17)$$

involves the electric fields[4] at different spatial points $\vec{x}$ and $\vec{y}$. A manifestly gauge invariant operator can be constructed by appropriately inserting Wilson lines. Specifically, in some arbitrary gauge, we define

$$\mathcal{O}_\gamma(\vec{x},\vec{y}) \equiv \mathrm{Tr}\, \left\{ F^{i+}(\vec{x})\, U_\gamma(\vec{x},\vec{y})\, F^{i+}(\vec{y})\, U_\gamma(\vec{y},\vec{x}) \right\}, \quad (2.18)$$

where (with $\vec{A}_a \equiv (A^+_a, \mathbf{A}^a_\perp)$, $\vec{A} \equiv \vec{A}_a T^a$)

$$U_\gamma(\vec{x},\vec{y}) = \mathrm{P}\exp\left\{ ig\int_\gamma d\vec{z}\cdot\vec{A}(\vec{z}) \right\}, \quad (2.19)$$

and γ is an arbitrary oriented path going from $\vec{y}$ to $\vec{x}$. The (omitted) temporal coordinates x^+ are the same for all fields. For any path γ, the operator in eq. (2.18) is gauge-invariant, since the chain of operators there makes a closed loop.

[4] The component $F^{i+}_a = -\partial^+ A^i_a$ is usually referred to as the (LC) "electric field" by analogy with the standard electric field $E^i_a = F^{i0}_a = -\partial^0 A^i_a$ (in the temporal gauge $A^0_a = 0$).

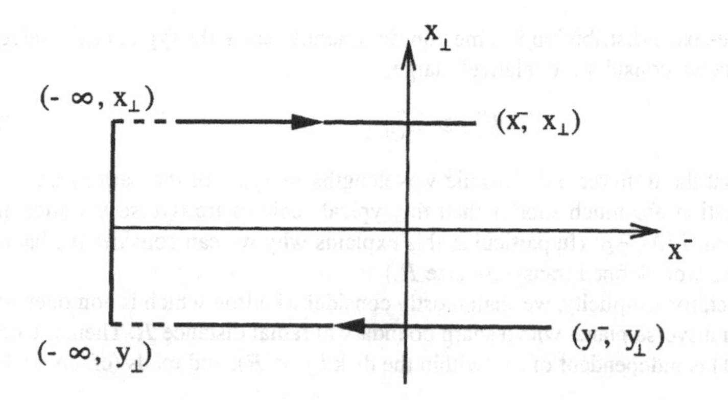

Figure 10. The path γ used for the evaluation of the gauge-invariant operator (2.18).

We now show that, by appropriately chosing the path, the gauge, and the boundary conditions, the gauge-invariant operator (2.18) can be made to coincide with the simple 2-point function (2.17). Specifically, consider the path shown in Fig. 10, with the the following three elements: two "horizontal" pieces going along the x^- axis from $(y^-, y_\perp)$ to $(-\infty, y_\perp)$, and, respectively, from $(-\infty, x_\perp)$ to $(x^-, x_\perp)$, and a "vertical" piece from $(-\infty, y_\perp)$ to $(-\infty, x_\perp)$. Along the horizontal pieces, $d\vec{z} \cdot \vec{A} = dz^- A^+$, so these pieces do not matter in the LC gauge. Along the vertical piece, $d\vec{z} \cdot \vec{A} = d\mathbf{z}_\perp \cdot \mathbf{A}_\perp(-\infty, z_\perp)$, and the path γ between $y_\perp$ and $x_\perp$ is still arbitrary. But the contribution of any such a path to the Wilson line vanishes once we impose the following, "retarded", boundary condition:

$$A_a^i(x) \to 0 \quad \text{as} \quad x^- \to -\infty. \tag{2.20}$$

(Note that the "retardation" property refers here to x^-, and not to time.)
To summarize, for the particular class of paths mentioned above, in the LC gauge $A^+ = 0$, and with the boundary condition (2.20), $U_\gamma(\vec{x}, \vec{y}) \to 1$, and the manifestly gauge-invariant operator in eq. (2.18) reduces to the simpler operator (2.17) which defines the number of gluons in this gauge. Converserly, the latter quantity has a gauge-invariant meaning, as the expression of a gauge-invariant operator in a specific gauge.
We shall need later also the gluon distribution function in the transverse phase-space (in short, the "gluon density"), i.e., the number of gluons per unit rapidity per unit transverse momentum per unit transverse area:

$$\mathcal{N}_\tau(k_\perp, b_\perp) \equiv \frac{d^5 N}{d\tau d^2 k_\perp d^2 b_\perp} = \frac{d^2\, xG(x, k_\perp^2)}{d^2 k_\perp d^2 b_\perp}, \tag{2.21}$$

where $\tau = \ln(1/x) = \ln(P^+/k^+)$ and $b_\perp$ is the impact parameter in the transverse plane (i.e., the central coordinate $b_\perp = (x_\perp + y_\perp)/2$ in eq. (2.17)). This phase-space distribution is a meaningful quantity since the typical transverse momenta we consider are relatively large,

$$k_\perp^2 \gg \Lambda_{QCD}^2 \sim 1/R^2, \tag{2.22}$$

so that the transverse de Broglie wavelengths $\sim 1/k_\perp$ of the partons under consideration are much shorter than the typical scale of transverse variation in the hadron, $1/\Lambda_{QCD}$. (In particular, this explains why we can consider the hadron to have a well defined transverse size R.)

In fact, for simplicity, we shall mostly consider a hadron which is homogeneous in the transverse plane, with a sharp boundary at radial distance R. Then, the density (2.21) is independent of $b_\perp$ (within the disk $b_\perp < R$), and reads (cf. eq. (2.16)) :

$$\mathcal{N}_\tau(k_\perp) = \frac{1}{\pi R^2}\frac{d^3N}{d\tau d^2k_\perp} = \frac{1}{4\pi^4 R^2}\langle F_a^{i+}(\vec{k})F_a^{i+}(-\vec{k})\rangle. \tag{2.23}$$

2.2.2. *The dipole-hadron cross-section*

Consider high energy deep inelastic scattering (DIS) in a special frame — the "dipole frame" — in which the virtual photon γ^* is moving very fast, say, in the negative z direction, but most of the total energy is still carried by the hadron, which moves nearly at the speed of light in the positive z direction. Thus, the rapidity gap between the hadron and the virtual photon is

$$\tau = y_{hadron} - y_{\gamma^*}, \qquad \text{with} \qquad |y_{\gamma^*}| \ll y_{hadron}. \tag{2.24}$$

(As in Sect. 1.4, $\tau = \ln(1/x) \approx \ln(s/Q^2)$, where Q^2 is the virtuality of γ^* and s is the invariant energy squared. Note also that $y_{\gamma^*} < 0$, since γ^* is a left mover.)

The dipole frame is special in two respects [14] (and references therein):

i) The DIS looks like a two step process, in which γ^* fluctuates first into a quark–antiquark pair, which then scatters off the hadron. The $q\bar{q}$ pair is in a colour singlet state, so it forms a *colour dipole*.

ii) The essential of the quantum evolution is put in the hadron wavefunction, which carries most of the energy. The dipole wavefunction, on the other hand, is simple and given by lowest order perturbation theory. More precisely, if $\alpha_s|y_{\gamma^*}| \ll 1$, then the dipole is just a quark–antiquark pair, without additional gluons.

Thus, in this frame, all the non-trivial dynamics is in the dipole-hadron scattering. Because of the high energy of the $q\bar{q}$ pair, this scattering can be treated in the eikonal approximation [40, 41, 42, 44] : the quark (and the antiquark) follows a straight line trajectory with $z = -t$ (or $x^+ = 0$), and the effect of its interactions with the colour field of the hadronic target is contained in the Wilson line:

$$V^\dagger(x_\perp) = \mathrm{P}\exp\left(ig\int_{-\infty}^{\infty} dx^- A_a^+(x^-, x_\perp)t^a\right), \tag{2.25}$$

where $x_\perp$ is the transverse coordinate of the quark, t^a's are the generators of the colour group in the fundamental representation, and the symbol P denotes the ordering of the colour matrices $A^+(\vec{x}) = A_a^+(\vec{x})t^a$ in the exponent from right to left in increasing order of their x^- arguments. Note that A^+ is the projection of A^μ along the trajectory of the fermion. For an antiquark with transverse coordinate $y_\perp$ the corresponding gauge factor is $V(y_\perp)$. Clearly, we adopt here a gauge where $A_a^+ \neq 0$ (e.g., the covariant gauge to be discussed at length in Sect. 2.3).
It can then be shown that the S-matrix element for the dipole-hadron scattering is obtained by averaging the total gauge factor $\mathrm{tr}(V^\dagger(x_\perp)V(y_\perp))$ (the colour trace occurs since we consider a colourless $q\bar{q}$ state) over all the colour field configurations in the hadron wavefunction:

$$S_\tau(x_\perp, y_\perp) \equiv \frac{1}{N_c}\left\langle \mathrm{tr}(V^\dagger(x_\perp)V(y_\perp))\right\rangle_\tau. \tag{2.26}$$

The dipole frame is like the hadron infinite momentum frame in that $y_{hadron} \approx \tau$, cf. eq. (2.24), so the average in eq. (2.26) can be computed within the effective theory of Sect. 2.1, that is, like in eq. (2.6).
The dipole–hadron cross section for a dipole of size $r_\perp = x_\perp - y_\perp$ is obtained by integrating $2(1 - S_\tau(x_\perp, y_\perp))$ over all the impact parameters $b_\perp = (x_\perp + y_\perp)/2$:

$$\sigma_{dipole}(\tau, r_\perp) = 2\int d^2b_\perp \frac{1}{N_c}\left\langle \mathrm{tr}\Big(1 - V^\dagger(x_\perp)V(y_\perp)\Big)\right\rangle_\tau. \tag{2.27}$$

Finally, the γ^*–hadron cross-section is obtained by convoluting the dipole cross-section (2.27) with the probability that the incoming photon splits into a $q\bar{q}$ pair:

$$\sigma_{\gamma^* h}(\tau, Q^2) = \int_0^1 dz \int d^2r_\perp\, |\Psi(z, r_\perp; Q^2)|^2\, \sigma_{dipole}(\tau, r_\perp). \tag{2.28}$$

Here, $\Psi(z, r_\perp; Q^2)$ is the light-cone wavefunction for a photon splitting into a $q\bar{q}$ pair with transverse size $r_\perp$ and a fraction z of the photon's longitudinal momentum carried by the quark [40, 41].

2.3. THE CLASSICAL COLOUR FIELD

From the point of view of the effective theory, the high density regime at small x is characterized by strong classical colour fields, whose non-linear dynamics must be treated exactly. Indeed, we shall soon discover that, at saturation, $xG(x, Q^2) \sim 1/\alpha_s$, which via eqs. (2.16) and (2.6) implies classical fields with amplitudes $\mathcal{A}^i \sim 1/g$. Such strong fields cannot be expanded out from the covariant derivative $D^i = \partial^i - igA^i$. Thus, we need the exact solution to the classical equations of motion (2.1), that we shall now construct.
We note first that, for a large class of gauges, it is consistent to look for solutions having the following properties:

$$F_a^{ij} = 0, \qquad A_a^- = 0, \qquad A_a^+, A_a^i \ : \ \text{static}, \tag{2.29}$$

where "static" means independent of x^+. (In fact, once such a static solution is found in a given gauge, then the properties (2.29) will be preserved by any time-independent gauge transformation.) This follows from the specific structure of the colour source which has just a "+" component, and is static. For instance, the component $\mu = i$ of eq. (2.1) reads:

$$0 = D_\nu F^{\nu i} = D_j F^{ji} + D_+ F^{+i} + D_- F^{-i}. \tag{2.30}$$

But $D_+ = D^- = \partial^- - igA^-$ vanishes by eq. (2.29), and so does F^{-i}. Thus eq. (2.30) reduces to $D_j F^{ji} = 0$, which implies $F^{ij} = 0$, as indicated in eq. (2.29). This further implies that the transverse fields A^i form a two-dimensional pure gauge. That is, there exists a gauge rotation $U(x^-, x_\perp) \in \mathrm{SU}(N)$ such that (in matrix notations appropriate for the adjoint representation: $A^i = A^i_a T^a$, etc) :

$$A^i(x^-, x_\perp) = \frac{i}{g} U(x^-, x_\perp)\, \partial^i U^\dagger(x^-, x_\perp). \tag{2.31}$$

Thus, the requirements (2.29) leave just two independent field degrees of freedom, $A^+(\vec{x})$ and $U(\vec{x})$, which are further reduced to one (either A^+ or U) by imposing a gauge-fixing condition.
We consider first the covariant gauge (COV-gauge) $\partial_\mu A^\mu = 0$. By eqs. (2.29) and (2.31), this implies $\partial_i A^i = 0$, or $U = 0$. Thus, in this gauge:

$$\tilde{\mathcal{A}}^\mu_a(x) = \delta^{\mu +} \alpha_a(x^-, x_\perp), \tag{2.32}$$

with $\alpha_a(\vec{x})$ linearly related to the colour source $\tilde{\rho}_a$ in the COV-gauge :

$$-\nabla^2_\perp \alpha_a(\vec{x}) = \tilde{\rho}_a(\vec{x}). \tag{2.33}$$

Note that we use curly letters to denote solutions to the classical field equations (as we did already in eq. (2.6)). Besides, we generally use a tilde to indicate quantities in the COV-gauge, although we keep the simple notation $\alpha_a(\vec{x})$ for the classical field in this gauge, since this quantity will be frequently used.
Eq. (2.33) has the solution :

$$\begin{aligned} \alpha_a(x^-, x_\perp) &= \int d^2 y_\perp \langle x_\perp | \frac{1}{-\nabla^2_\perp} | y_\perp \rangle\, \tilde{\rho}_a(x^-, y_\perp) \\ &= \int \frac{d^2 y_\perp}{4\pi} \ln \frac{1}{(x_\perp - y_\perp)^2 \mu^2}\, \tilde{\rho}_a(x^-, y_\perp), \end{aligned} \tag{2.34}$$

where the infrared cutoff μ is necessary to invert the Laplacean operator in two dimensions, but it will eventually disappear from (or get replaced by the confinement scale Λ_{QCD} in) our subsequent formulae.
The only non-trivial field strength is the electric field:

$$\tilde{\mathcal{F}}^{+i}_a = -\partial^i \alpha_a. \tag{2.35}$$

In terms of the usual electric ($\mathbf{E}$) and magnetic ($\mathbf{B}$) fields, this solution is characterized by purely transverse fields, $\mathbf{E}_\perp = (E^1, E^2)$ and $\mathbf{B}_\perp = (B^1, B^2)$, which are orthogonal to each other: $\mathbf{E}_\perp \cdot \mathbf{B}_\perp = 0$ (since $B^1 = -E^2$ and $B^2 = E^1$).
To compute the gluon distribution (2.16), one needs the classical solution in the LC-gauge $A^+ = 0$. This is of the form $\mathcal{A}^\mu_a = \delta^{\mu i}\mathcal{A}^i_a$ with $\mathcal{A}^i_a(x^-, x_\perp)$ a "pure gauge", cf. eq. (2.31). The gauge rotation $U(\vec{x})$ can be obtained by inserting the Ansatz (2.31) in eq. (2.1) with $\mu = +$ to deduce an equation for U. Alternatively, and simpler, the LC-gauge solution can be obtained by a gauge rotation of the solution (2.32) in the COV-gauge:

$$\mathcal{A}^\mu = U\Big(\tilde{\mathcal{A}}^\mu + \frac{i}{g}\partial^\mu\Big)U^\dagger, \tag{2.36}$$

where the gauge rotation $U(\vec{x})$ is chosen such that $\mathcal{A}^+ = 0$, i.e.,

$$\alpha = \frac{i}{g} U^\dagger \left(\partial^+ U\right). \tag{2.37}$$

Eq. (2.37) is easily inverted to give

$$U^\dagger(x^-, x_\perp) = \mathrm{P}\exp\left\{ ig \int_{-\infty}^{x^-} dz^- \,\alpha(z^-, x_\perp)\right\}. \tag{2.38}$$

From eq. (2.36), $\mathcal{A}^i$ is obtained indeed in the form (2.31), with U given in eq. (2.38). The lower limit $x_0^- \to -\infty$ in the integral over x^- in eq. (2.38) has been chosen such as to impose the "retarded" boundary condition (2.20). Furthermore:

$$\mathcal{F}^{+i}(\vec{x}) \equiv \partial^+ \mathcal{A}^i(\vec{x}) = U(\vec{x})\tilde{\mathcal{F}}^{+i}(\vec{x})U^\dagger(\vec{x}). \tag{2.39}$$

Together, eqs. (2.31), (2.34) and (2.38) provide an explicit expression for the LC-gauge solution $\mathcal{A}^i$ in terms of the colour source $\tilde{\rho}$ in the COV-gauge. The corresponding expression in terms of the colour source in the LC-gauge ρ cannot be easily obtained: Eq. (2.33) implies indeed

$$-\nabla^2_\perp \alpha = U^\dagger \rho U, \tag{2.40}$$

which implicitly determines α (and thus U) in terms of ρ, but which we don't know how to solve explicitly. But this is not a difficulty, as we argue now:
Recall indeed that the classical source is just a "dummy" variable which is integrated out in computing correlations according to eq. (2.6). Both the measure and the weight function in eq. (2.6) are gauge invariant. Thus, one can compute correlation functions in the LC-gauge by performing a change of variables $\rho \to \tilde{\rho}$, and thus replacing the a priori unknown functionals $\mathcal{A}^i[\rho]$ by the functionals $\mathcal{A}^i[\tilde{\rho}]$, which are known explicitly. In other terms, one can replace eq. (2.6) by

$$\langle A^\mu(x^+, \vec{x})A^\nu(x^+, \vec{y})\cdots\rangle_\tau = \int \mathcal{D}[\tilde{\rho}]\; W_\tau[\tilde{\rho}]\; \mathcal{A}^\mu_x[\tilde{\rho}]\, \mathcal{A}^\nu_y[\tilde{\rho}]\cdots, \tag{2.41}$$

where $\mathcal{A}^\mu[\tilde{\rho}]$ is the classical solution in some *generic* gauge (e.g., the LC-gauge), but expressed as a functional of the colour source $\tilde{\rho}$ in the *COV*-gauge.
Moreover, the gauge-invariant observables can be expressed directly in terms of the gauge fields in the COV-gauge, although the corresponding expressions may look more complicated than in the LC-gauge. For instance, the operator which enters the gluon distribution can be written as (cf. eq. (2.39))

$$\mathrm{Tr}\left\{\mathcal{F}^{+i}(\vec{x})\mathcal{F}^{+i}(\vec{y})\right\} = \mathrm{Tr}\left\{U(\vec{x})\tilde{\mathcal{F}}^{+i}(\vec{x})U^\dagger(\vec{x})\, U(\vec{y})\tilde{\mathcal{F}}^{+i}(\vec{y})\, U^\dagger(\vec{y})\right\}, \quad (2.42)$$

where the classical fields are in the LC-gauge in the l.h.s. and in the COV-gauge in the r.h.s, and U and $U^\dagger$ are given by eq. (2.38). Both writings express the gauge-invariant operator (2.18) (with the path γ shown in Fig. 10) in the indicated gauges. (Indeed, $U_\gamma(\vec{x},\vec{y}) = U^\dagger(\vec{x})U(\vec{y})$ for the COV-gauge field $\tilde{A}^\mu = \delta^{\mu+}\alpha$.)
Note that, while in the LC-gauge the non-linear effects are encoded in the electric fields $\mathcal{F}^{+i}$, in the COV-gauge they are rather encoded in the Wilson lines U and $U^\dagger$ (the corresponding field $\tilde{\mathcal{F}}_a^{+i} = -\partial^i\alpha_a$ being linear in $\tilde{\rho}_a$).
Up to this point, the longitudinal structure of the source has been arbitrary: the solutions written above hold for any function $\rho^a(x^-)$. For what follows, however, it is useful to recall, from Sect. 2.1, that ρ has is localized near $x^- = 0$. More precisely, the quantum analysis in Sect. 3.4 will demonstrate that the classical source at the longitudinal scale k^+ has support at positive x^-, with $0 \leq x^- \leq 1/k^+$. From eqs. (2.33)–(2.34), it is clear that this is also the longitudinal support of the "Coulomb field" $\alpha(\vec{x})$. Thus, integrals over x^- as that in eq. (2.38) receive contributions only from x^- in this limited range. The resulting longitudinal structure for the classical solution is illustrated in Fig. 11, and can be approximated as follows:

$$\mathcal{A}^i(x^-,x_\perp) \;\approx\; \theta(x^-)\frac{i}{g}V(\partial^i V^\dagger) \;\equiv\; \theta(x^-)\mathcal{A}^i_\infty(x_\perp), \quad (2.43)$$

$$\mathcal{F}^{+i}(\vec{x}) \;\equiv\; \partial^+\mathcal{A}^i \;\approx\; \delta(x^-)\,\mathcal{A}^i_\infty(x_\perp). \quad (2.44)$$

It is here understood that the $\delta - -$ and $\theta - -$functions of x^- are smeared over a distance $\Delta x^- \sim 1/k^+$. In the equations above, V and $V^\dagger$ are the asymptotic values of the respective gauge rotations as $x^- \to \infty$:

$$V^\dagger(x_\perp) \equiv \mathrm{P}\exp\left\{ig\int_{-\infty}^{\infty} dz^-\,\alpha(z^-,x_\perp)\right\}. \quad (2.45)$$

In practice, $U(x^-,x_\perp) = V(x_\perp)$ for any $x^- \gg 1/k^+$. Note that (2.45) is the same Wilson line as in the discussion of the eikonal approximation in Sect. 2.2.2 (compare to eq. (2.25) there). In the present context, the eikonal approximation is implicit in the special geometry of the colour source in eq. (2.1), which is created by fast moving particles.

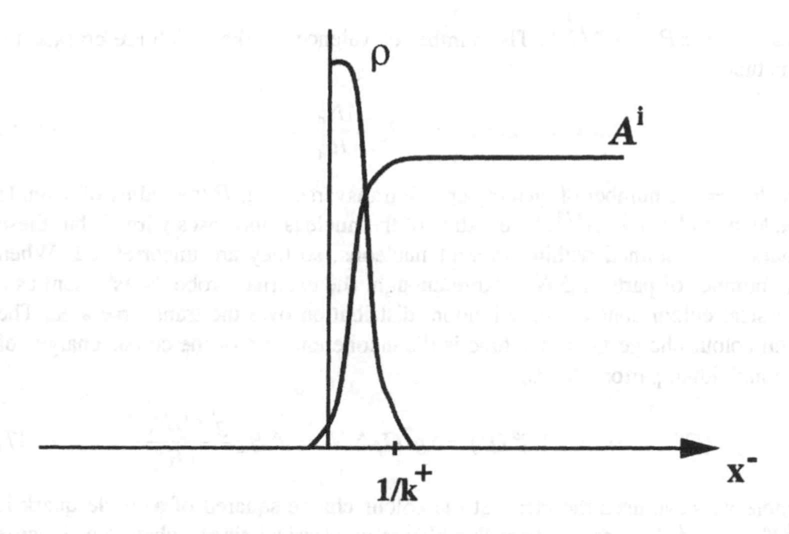

Figure 11. The longitudinal structure of the colour source ρ and of the classical field solution $\mathcal{A}^i$ for the effective theory at the scale k^+. As functions of x^-, α and $\mathcal{F}^{+i}$ are as localized as ρ.

2.4. THE GLUON DISTRIBUTION OF THE VALENCE QUARKS

To compute observables in the effective theory, one still needs an expression for the weight function $W_\tau[\rho]$. Before discussing the general construction of $W_\tau[\rho]$ in Sect. 3, let us present a simple model for it, due to McLerran-Venugopalan (MV) [8], which takes into account the colour charge of the valence quarks alone. That is, it ignores the quantum evolution of the colour sources with τ. This model is expected to work better for a large nucleus, with atomic number $A \gg 1$; indeed, this has many valence quarks ($A \times N_c$), and therefore as many colour sources, which can create a strong colour field already at moderate values of x, where the quantum evolution can be still neglected. In this model, τ is fixed, but one can study the strong field effects (in particular, gluon saturation) in the limit where A is large. Besides, the MV model provides a reasonable initial condition for the quantum evolution towards small x, to be described later.
The main assumption of the MV model is that the $A \times N_c$ valence quarks can be treated as *independent* colour sources. This relies on confinement. Note first that confinement plays no role for the dynamics in the transverse plane: Indeed, we probe the nucleus with large transerse momenta $Q^2 \gg \Lambda_{QCD}^2$, that is, over distance scales much shorter than those where confinement sets in. On the other hand, even at moderate values of x, we are still probing an integrated version of the hadron in the longitudinal direction, i.e., we measure all the "partons" (here, valence quarks) in a tube of transverse area $\Delta S_\perp \sim 1/Q^2$ and longitudinal extent

$\Delta x^- \sim 1/xP^+ > 1/P^+$. The number of valence quarks which are crossed by this tube,

$$\Delta N \approx n\,\Delta S_\perp = \Delta S_\perp \frac{AN_c}{\pi R_A^2} \sim A^{1/3}, \tag{2.46}$$

(with $n =$ the number of quarks per unit transverse area, R the radius of a single nucleon, and $R_A = A^{1/3}R$ the radius of the nucleus) increases with A, but these quarks are confined within different nucleons, so they are uncorrelated. When the number of partons ΔN is large enough, the external probe "sees" them as a classical colour source with a random distribution over the transverse area. The total colour charge $\mathcal{Q}^a$ in the tube is the incoherent sum of the colour charges of the individual partons. Thus,

$$\langle \mathcal{Q}^a \rangle = 0, \qquad \langle \mathcal{Q}^a \mathcal{Q}^a \rangle = g^2 C_f \Delta N = \Delta S_\perp \frac{g^2 C_f N_c A}{\pi R_A^2}, \tag{2.47}$$

where we have used the fact that the colour charge squared of a single quark is $g^2 t^a t^a = g^2 C_f$. One can treat this charge as classical since, when ΔN is large enough, we can ignore commutators of charges:

$$|\,[\mathcal{Q}^a, \mathcal{Q}^b]\,| = |\, i f^{abc} \mathcal{Q}^c \,| \ll \mathcal{Q}^2 . \tag{2.48}$$

In order to take the continuum limit (i.e., the limit where the transverse area $\Delta S_\perp$ of the tube is small[5]), it is convenient to introduce the colour charge densities $\rho^a(x^-, x_\perp)$ (with the same meaning as in Sect. 2.1) and

$$\rho^a(x_\perp) \equiv \int dx^- \rho^a(x^-, x_\perp) \tag{2.49}$$

(the colour charge per unit area in the transverse plane). Then,

$$\mathcal{Q}^a = \int_{\Delta S_\perp} d^2x_\perp\, \rho^a(x_\perp) = \int_{\Delta S_\perp} d^2x_\perp \int dx^-\, \rho^a(x^-, x_\perp), \tag{2.50}$$

and eqs. (2.47) imply (recall that $C_f = (N_c^2 - 1)/2N_c$) :

$$\begin{aligned} \langle \rho_a(x_\perp)\rho_b(y_\perp)\rangle_A &= \delta_{ab}\delta^{(2)}(x_\perp - y_\perp)\,\mu_A, \qquad \mu_A \equiv \frac{g^2 A}{2\pi R_A^2}, \\ \langle \rho_a(x^-, x_\perp)\rho_b(y^-, y_\perp)\rangle_A &= \delta_{ab}\delta^{(2)}(x_\perp - y_\perp)\delta(x^- - y^-)\,\lambda_A(x^-), \\ \int dx^-\, \lambda_A(x^-) &= \mu_A . \end{aligned} \tag{2.51}$$

[5] This amounts to increasing Q^2, so, strictly speaking, at this step one should also include the DGLAP quantum evolution (i.e., the fact that, with increasing transverse resolution, the original "quark" is resolved into a set of smaller constituents). The quantum analysis to be discussed later will include that in the "double-log approximation"; see the discussion in Sect. 5.3.

Here, $\mu_A \sim A^{1/3}$ is the average colour charge squared of the valence quarks per unit transverse area and per colour, and $\lambda_A(x^-)$ is the corresponding density per unit volume. The latter has some dependence upon x^-, whose precise form is, however, not important since the final formulae will involve only the integrated density μ_A. There is no explicit dependence upon $x_\perp$ in μ_A or $\lambda_A(x^-)$ since we assume transverse homogeneity within the nuclear disk of radius R_A. Finally, the correlations are local in x^- since, as argued before, colour sources at different values of x^- belong to different nucleons, so they are uncorrelated. All the higher-point, connected, correlation functions of $\rho_a(\vec{x})$ are assumed to vanish. The non-zero correlators (2.51) are generated by the following weight function [8] :

$$W_A[\rho] = \mathcal{N}\exp\left\{-\frac{1}{2}\int d^3x \frac{\rho_a(\vec{x})\rho_a(\vec{x})}{\lambda_A(x^-)}\right\}, \tag{2.52}$$

which is a Gaussian in ρ_a, with a local kernel. This is gauge-invariant, so the variable ρ_a in this expression can be the colour source in any gauge. The integral over $x_\perp$ in eq. (2.52) is effectively cutoff at R_A. By using this weight function, we shall now compute the observables introduced in Sect. 2.2.
Consider first the gluon distribution in the low density regime, i.e., when the atomic number A is not too high, so that the corresponding classical field is weak and can be computed in the linear approximation. By expanding the general solution (2.31) to linear order in ρ, or, equivalently, by directly solving the linearized version of eq. (2.1), one easily obtains:

$$\mathcal{A}_a^i(k) \simeq -\frac{k^i}{k^+ + i\varepsilon}\frac{\rho_a(k^+, k_\perp)}{k_\perp^2}, \qquad \mathcal{F}_a^{+i}(k) \simeq i\frac{k^i}{k_\perp^2}\rho_a(\vec{k}), \tag{2.53}$$

which together with eq. (2.51) implies:

$$\langle \mathcal{F}_a^{i+}(\vec{k})\mathcal{F}_a^{i+}(-\vec{k})\rangle_A \simeq \frac{1}{k_\perp^2}\langle \rho_a(\vec{k})\rho_a(-\vec{k})\rangle_A = \pi R_A^2(N_c^2-1)\frac{\mu_A}{k_\perp^2}. \tag{2.54}$$

By inserting this approximation in eqs. (2.23) and (2.16), one obtains the following estimates for the gluon density and distribution function:

$$\begin{aligned} \mathcal{N}_A(k_\perp) &\simeq \frac{N_c^2-1}{4\pi^3}\frac{\mu_A}{k_\perp^2}, \\ xG(x, Q^2) &\simeq \frac{(N_c^2-1)R_A^2}{4\pi}\mu_A\int_{\Lambda_{QCD}^2}^{Q^2}\frac{dk_\perp^2}{k_\perp^2} = \frac{\alpha_s AN_cC_f}{\pi}\ln\frac{Q^2}{\Lambda_{QCD}^2}, \end{aligned} \tag{2.55}$$

(with $\alpha_s = g^2/4\pi$). The integral over $k_\perp$ in the second line has a logarithmic infrared divergence which has been cut by hand at the scale Λ_{QCD} since we know that, because of confinement, there cannot be gluon modes with transverse wavelengths larger than $1/\Lambda_{QCD}$ (see also Ref. [35]).

We recognize in eq. (2.55) the standard bremsstrahlung spectrum of soft "photons" radiated by fast moving charges [1]. In deriving this result, we have however neglected the non-Abelian nature of the radiated fields, i.e., the fact that they represent gluons, and not photons. This will be corrected in the next subsection.

2.5. GLUON SATURATION IN A LARGE NUCLEUS

According to eq. (2.55), the gluon density in the transverse phase-space is proportional to $A^{1/3}$, and becomes arbitrarily large when A increases. This is however an artifact of our previous approximations which have neglected the interactions among the radiated gluons, i.e., the non-linear effects in the classical field equations. To see this, one needs to recompute the gluon distribution by using the exact, non-linear solution for the classical field, as obtained in Sect. 2.3. This involves the following LC-gauge field-field correlator:

$$\langle \mathcal{F}_a^{+i}(\vec{x})\mathcal{F}_a^{+i}(\vec{y})\rangle_A = \left\langle \left(U_{ab}^{\dagger}\partial^i\alpha^b\right)_{\vec{x}} \left(U_{ac}^{\dagger}\partial^i\alpha^c\right)_{\vec{y}}\right\rangle_A, \tag{2.56}$$

which, in view of the non-linear calculation, has been rewritten in terms of the classical field in the COV-gauge (cf. eq. (2.42)), where $\tilde{\mathcal{F}}_a^{+i} = -\partial^i\alpha_a$. To evaluate (2.56), one expands the Wilson lines in powers of α and then contracts the α fields in all the possible ways with the following propagator:

$$\begin{aligned}\langle \alpha_a(\vec{x})\alpha_b(\vec{y})\rangle_A &= \delta_{ab}\delta(x^- - y^-)\,\gamma_A(x^-, x_\perp - y_\perp),\\ \gamma_A(x^-, k_\perp) &\equiv \frac{1}{k_\perp^4}\,\lambda_A(x^-).\end{aligned}\tag{2.57}$$

We have used here $\tilde{\rho}^a(x^-, k_\perp) = k_\perp^2\alpha^a(x^-, k_\perp)$, cf. eq. (2.34), together with eq. (2.51) which holds in any gauge and, in particular, in the COV-gauge. The propagator (2.57) is very singular as $k_\perp \to 0$, but this turns out to be (almost) harmless for the considerations to follow.

The fact that the fields α are uncorrelated in x^- greatly simplifies the calculation of the correlator (2.56). Indeed, this implies that the two COV-gauge electric fields $\partial^i\alpha_b(\vec{x})$ and $\partial^i\alpha_c(\vec{y})$ can be contracted only together, and not with the other fields α generated when expanding the Wilson lines. That is:

$$\begin{aligned}\left\langle \left(U_{ab}^{\dagger}\partial^i\alpha^b\right)_{\vec{x}} \left(U_{ac}^{\dagger}\partial^i\alpha^c\right)_{\vec{y}}\right\rangle &= \left\langle \partial^i\alpha^b(\vec{x})\partial^i\alpha^c(\vec{y})\right\rangle \left\langle U_{ab}^{\dagger}(\vec{x})U_{ca}(\vec{y})\right\rangle\\ &= \delta(x^- - y^-)\langle \mathrm{Tr}\, U^{\dagger}(\vec{x})U(\vec{y})\rangle(-\nabla_\perp^2\gamma_A(x^-, x_\perp - y_\perp)),\end{aligned}\tag{2.58}$$

where we have used $U_{ac}^{\dagger} = U_{ca}$ in the adjoint representation. Eq. (2.58) can be proven as follows: i) By rotational symmetry, $\partial^i\alpha(\vec{x})$ cannot be contracted with a field $\alpha(z^-, x_\perp)$ resulting from the expansion of $U^{\dagger}(\vec{x})$; indeed:

$$\langle \alpha(z^-, x_\perp)\partial^i\alpha(x^-, x_\perp)\rangle \propto \partial^i\gamma_A(x^-, r_\perp)\Big|_{r_\perp=0} \propto \int \frac{d^2k_\perp}{(2\pi)^2}\frac{-ik^i}{k_\perp^2} = 0\,.$$

ii) Contractions of the type

$$\langle\alpha(z^-,y_\perp)\partial^i\alpha(x^-,x_\perp)\rangle\,\langle\alpha(u^-,x_\perp)\partial^i\alpha(y^-,y_\perp)\rangle \propto \delta(x^- - z^-)\delta(u^- - y^-)$$

are not allowed by the ordering of the Wilson lines in x^-: $\alpha(z^-,y_\perp)$ has been generated by expanding $U^\dagger(\vec{y})$, which requires $z^- < y^-$ (and similarly $u^- < x^-$). Then, the first contraction in (2.59) implies $x^- = z^- < y^-$, while the second one leads to the contradictory requirement $y^- = u^- < x^-$.
The allowed contractions in eq. (2.58) involve:

$$S_A(x^-,x_\perp - y_\perp) \equiv \frac{1}{N_c^2-1}\,\langle \mathrm{Tr}\, U^\dagger(x^-,x_\perp)U(x^-,y_\perp)\rangle_A, \qquad (2.59)$$

which is like the S-matrix element (2.26) for the dipole-hadron scattering, but now for a colour dipole in the adjoint representation (i.e., a dipole made of two gluons). This can be computed by expanding the Wilson lines, performing contractions with the help of eq. (2.57), and recognizing the result as the expansion of an ordinary exponential. One thus finds (see also Sect. 5.1 for a more rapid derivation):

$$\begin{aligned} S_A(x^-,r_\perp) &= \exp\Big\{-g^2N_c[\xi_A(x^-,0_\perp)-\xi_A(x^-,r_\perp)]\Big\},\\ \xi_A(x^-,r_\perp) &\equiv \int_{-\infty}^{x^-} dz^-\,\gamma_A(z^-,r_\perp), \end{aligned} \qquad (2.60)$$

where the exponent can be easily understood: It arises as

$$\Big\langle gT^a(\alpha_a(\vec{x})-\alpha_a(\vec{y}))\,gT^b(\alpha_b(\vec{x})-\alpha_b(\vec{y}))\Big\rangle \qquad (2.61)$$

where $igT^a(\alpha_a(\vec{x})-\alpha_a(\vec{y}))$ is the amplitude for the dipole scattering off the "Coulomb" field α_a, to lowest order in this field (i.e., the amplitude for a single scattering). Then, (2.61) is the amplitude times the complex conjugate amplitude, that is, the *cross section* for such a single scattering. This appears as an exponent in eq. (2.60) since this equation resums multiple scatterings to all orders, and, in the eikonal approximation, the all-order result is simply the exponential of the lowest order result. Since, moreover, α_a is the field created by the colour sources in the hadron (here, the valence quarks), we deduce that eq. (2.60) describes the multiple scattering of the colour dipole off these colour sources.
If the field α_a is slowly varying over the transverse size $r_\perp = x_\perp - y_\perp$ of the dipole ("small dipole"), one can expand

$$gT^a(\alpha_a(\vec{x})-\alpha_a(\vec{y})) \approx -gT^a(x^i-y^i)\partial^i\alpha_a(\vec{x}) = gT^a(x^i-y^i)\tilde{\mathcal{F}}_a^{+i}(\vec{x}), \qquad (2.62)$$

and then eq. (2.61) involves the correlator of two (COV-gauge) electric fields. This is indeed the case, at it can be seen by an analysis of the exponent in eq. (2.60) :

$$\xi_A(x^-,0_\perp)-\xi_A(x^-,r_\perp) = \mu_A(x^-)\int\frac{d^2k_\perp}{(2\pi)^2}\frac{1}{k_\perp^4}\left[1-e^{ik_\perp\cdot r_\perp}\right],$$

$$\mu_A(x^-) \equiv \int_{-\infty}^{x^-} dz^- \lambda_A(z^-). \tag{2.63}$$

The above integral over $k_\perp$ is dominated by soft momenta, and has even a logarithmic divergence which reflects the lack of confinement in our model (see also [35]). Note, however, that the dominant, quadratic, infrared divergence $\sim \int (d^2k_\perp/k_\perp^4)$, which would characterize the scattering of a *coloured* particle (a single gluon) off the hadronic field[6], has cancelled between the two components of the *colourless* dipole. The remaining, logarithmic, divergence can be cut off by hand, by introducing an infrared cutoff Λ_{QCD}. Then one can expand:

$$\int \frac{d^2k_\perp}{(2\pi)^2}\frac{1-e^{ik_\perp\cdot r_\perp}}{k_\perp^4} \simeq \int^{1/r_\perp^2} \frac{d^2k_\perp}{(2\pi)^2}\frac{1}{k_\perp^4}\frac{(k_\perp\cdot r_\perp)^2}{2} \simeq \frac{r_\perp^2}{16\pi}\ln\frac{1}{r_\perp^2\Lambda_{QCD}^2}. \tag{2.64}$$

(This is valid to leading logarithmic accuracy, since the terms neglected in this way are not enhanced by a large transverse logarithm.) We thus obtain:

$$S_A(x^-, r_\perp) \simeq \exp\left\{-\frac{\alpha_s N_c}{4} r_\perp^2\, \mu_A(x^-)\ln\frac{1}{r_\perp^2\Lambda_{QCD}^2}\right\}, \tag{2.65}$$

which together with eq. (2.58) can be used to finally evaluate the gluon density (2.23). This requires a double Fourier transform (to k^+ and $k_\perp$), as shown in eq. (2.17). The presence of the δ-function in eq. (2.58) makes the Fourier transform to k^+ trivial, and one gets:

$$\mathcal{N}_A(k_\perp) = \frac{N_c^2-1}{4\pi^3}\int d^2r_\perp e^{-ik_\perp\cdot r_\perp}\int dx^- S_A(x^-, r_\perp)(-\nabla_\perp^2\gamma_A(x^-, r_\perp)), \tag{2.66}$$

where (cf. eqs. (2.57) and (2.63)) :

$$-\nabla_\perp^2\gamma_A(x^-, r_\perp) = \lambda_A(x^-)\int\frac{d^2p_\perp}{(2\pi)^2}\frac{e^{ip_\perp\cdot x_\perp}}{p_\perp^2} = \frac{1}{4\pi}\ln\frac{1}{r_\perp^2\Lambda_{QCD}^2}\frac{\partial\mu_A(x^-)}{\partial x^-}. \tag{2.67}$$

The non-linear effects in eq. (2.66) are encoded in the quantity $S_A(x^-, r_\perp)$, which finds its origin in the gauge rotations in the r.h.s. of eq. (2.56). In fact, by replacing $S_A(x^-, r_\perp) \to 1$ in eq. (2.66), one would recover the linear approximation of eq. (2.55). To perform the integral over x^- in eq. (2.66), we note that the quantity (2.67) is essentially the derivative w.r.t. x^- of the exponent in $S_A(x^-, r_\perp)$, eq. (2.65). Therefore:

$$\mathcal{N}_A(k_\perp) = \frac{N_c^2-1}{4\pi^4}\int d^2r_\perp e^{-ik_\perp\cdot r_\perp}\frac{1-\exp\left\{-\frac{1}{4}r_\perp^2 Q_A^2\ln\frac{1}{r_\perp^2\Lambda_{QCD}^2}\right\}}{\alpha_s N_c r_\perp^2}, \tag{2.68}$$

[6] Such a divergence would occur in $\langle U^\dagger(x)\rangle_A$, which describes the scattering of a single gluon.

where the integral is restricted to $r_\perp < 1/\Lambda_{QCD}$, and

$$Q_A^2 \equiv \alpha_s N_c \mu_A = \alpha_s N_c \int dx^- \lambda_A(x^-) \sim A^{1/3}. \tag{2.69}$$

Eq. (2.68) is the complet result for the gluon density of a large nucleus in the MV model [33, 34]. To study its dependence upon $k_\perp$, one must still perform the Fourier transform, but the result can be easily anticipated:
i) At high momenta $k_\perp \gg Q_A$, the integral is dominated by small distances $r_\perp \ll 1/Q_A$, and can be evaluated by expanding out the exponential. To lowest non-trivial order (which corresponds to the linear approximation), one obtains the bremsstrahlung spectrum of eq. (2.55):

$$\mathcal{N}_A(k_\perp) \propto \frac{1}{\alpha_s N_c} \frac{Q_A^2}{k_\perp^2} = \frac{\mu_A}{k_\perp^2} \qquad \text{for} \quad k_\perp \gg Q_A. \tag{2.70}$$

ii) At small momenta, $k_\perp \ll Q_A$, the dominant contribution comes from large distances $r_\perp \gg 1/Q_A$, where one can simply neglect the exponential in the numerator and recognize $1/r_\perp^2$ as the Fourier transform[7] of $\ln k_\perp^2$:

$$\mathcal{N}_A(k_\perp) \approx \frac{N_c^2 - 1}{4\pi^3} \frac{1}{\alpha_s N_c} \ln \frac{Q_A^2}{k_\perp^2} \qquad \text{for} \quad k_\perp \ll Q_A. \tag{2.71}$$

There are two fundamental differences between eqs. (2.70) and (2.71), which refer both to a *saturation* of the increase of the gluon density: either with $1/k_\perp^2$ (at fixed atomic number A), or with A (at fixed transverse momentum $k_\perp$). In both cases, this saturation is only *marginal* : in the low–$k_\perp$ regime, eq. (2.71), the gluon density keeps increasing with $1/k_\perp^2$, and also with A, but this increase is only *logarithmic*, in contrast to the strong, power-like, increase $\propto (A^{1/3}/k_\perp^2)$ in the high–$k_\perp$ regime, eq. (2.70).
Moreover, the gluon density at low $k_\perp$ is of order $1/\alpha_s$, which is the maximum density allowed by the repulsive interactions between the strong colour fields $\bar{A}^i = \sqrt{\langle A^i A^i \rangle} \sim 1/g$. When increasing the atomic number A, the new gluons are produced preponderently at large transverse momenta $\gtrsim Q_A$, where this repulsion is less important. This is illustrated in Fig. 12.
To be more precise, the true scale which separates between the two regimes (2.70) and (2.71) is not Q_A, but rather the *saturation momentum* $Q_s(A)$ which is the reciprocal of the distance $1/r_\perp$ where the exponent in eq. (2.68) becomes of order one. Thus, this is defined as the solution to the following equation:

$$Q_s^2(A) = \frac{1}{4}\, \alpha_s N_c \mu_A \ln \frac{Q_s^2(A)}{\Lambda_{QCD}^2}. \tag{2.72}$$

[7] The saturation scale provides the ultraviolet cutoff for the logarithm in eq. (2.71) since the short distances $r_\perp \ll 1/Q_A$ are cut off by the exponential in eq. (2.68).

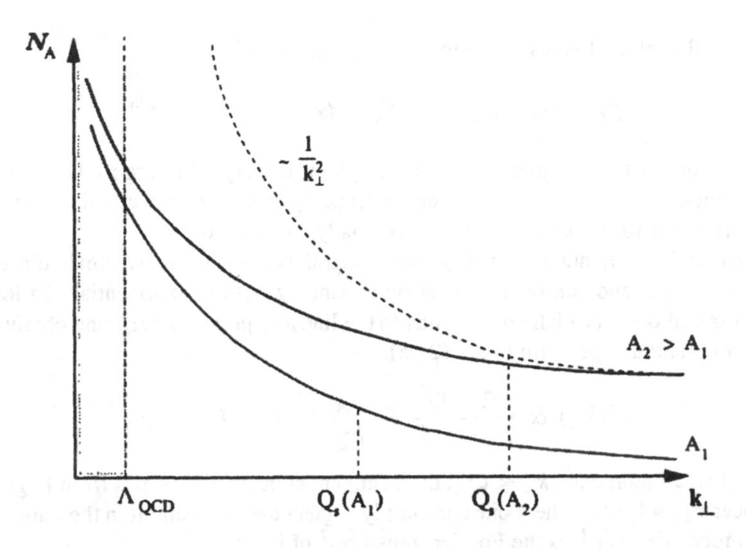

Figure 12. The gluon phase-space density $\mathcal{N}_A(k_\perp)$ of a large nucleus (as described by the MV model) plotted as a function of $k_\perp$ for two values of A. Notice the change from a $1/k_\perp^2$ behaviour at large momenta $k_\perp > Q_s$ to a logarithmic behaviour at small momenta $k_\perp < Q_s$.

To clarify its physical interpretation, note that, at short-distances $r_\perp \ll 1/Q_A$,

$$\mu_A \ln \frac{1}{r_\perp^2 \Lambda_{QCD}^2} \propto \frac{xG(x, 1/r_\perp^2)}{(N_c^2-1)\pi R_A^2} \tag{2.73}$$

is the number of gluons (of each colour) having tranverse size $r_\perp$ per unit of transverse area (cf. eq. (2.55)). Since each such a gluon carries a colour charge squared $(gT^a)(gT^a) = g^2 N_c$, we deduce that

$$\alpha_s N_c \mu_A \ln \frac{1}{r_\perp^2 \Lambda_{QCD}^2} \tag{2.74}$$

is the average colour charge squared of the gluons having tranverse size $r_\perp$ per unit area and per colour. Then, eq. (2.72) is the condition that the total colour charge squared within the area occupied by each gluon is of order one. This is the original criterion of saturation by Gribov, Levin and Ryskin [6], for which the MV model offers an explicit realization.

To conclude this discussion of the MV model, note that, in the previous computation, we have also obtained the S-matrix element $S_A(r_\perp)$ for the dipole-hadron scattering (cf. Sect. 2.2.2). This is given by eq. (2.65) with $\mu_A(x^-) \to \mu_A$ and $N_c = T^a T^a$ replaced in general by the colour Casimir $t^a t^a$ for the representation of interest (e.g., $C_f = (N_c^2-1)/2N_c$ for the fundamental representation). As discussed after eq. (2.61), this describes the multiple scattering of the colour dipole

on the colour field in the hadron (here, the field of the valence quarks). According to eq. (2.65), one can distinguish, here too, between a short-distance and a large-distance regime, which moreover are separated by the same "saturation scale" as for the gluon distribution:
i) A small-size dipole $r_\perp \ll 1/Q_s$ is only weakly interacting with the hadron:

$$1 - S_A(r_\perp) \approx \frac{1}{4} r_\perp^2 Q_A^2 \ln \frac{1}{r_\perp^2 \Lambda_{QCD}^2} \qquad \text{for} \quad r_\perp \ll 1/Q_s(A), \tag{2.75}$$

a phenomenon usually referred to as "colour transparency".
ii) A relatively large dipole, with $r_\perp \gg 1/Q_s$, is strongly absorbed:

$$S_A(r_\perp) \approx 0 \qquad \text{for} \quad r_\perp \gg 1/Q_s(A), \tag{2.76}$$

a situation commonly referred to as the "black disk", or "unitarity", limit.
The remarkable fact that the critical dipole size is set by the saturation scale Q_s can be understood as follows: A small dipole — small as compared to the typical variation scale of the external Coulomb field — couples to the associated electric field $\tilde{\mathcal{F}}^{+i}$ (cf. eq. (2.62)), so its cross-section for one scattering, eq. (2.61), is proportional to the number of gluons $\langle \tilde{\mathcal{F}}^{+i} \tilde{\mathcal{F}}^{+i} \rangle$ within the transverse area $r_\perp^2$ explored by the dipole. This is manifest on eq. (2.65), whose exponent is precisely the colour charge squared of the gluons within that area (cf. the remark after eq. (2.74)). At saturation, this charge becomes of order one, and the dipole is strongly interacting. The important lesson is that the unitarity limit (2.76) for the scattering of a small dipole on a high energy hadron is equivalent to gluon saturation in the hadron wavefunction [40, 9, 10, 45, 14].

3. Quantum evolution and the Colour Glass Condensate

In this section, we show that the classical Yang-Mills theory described in Sect. 2 can be actually derived from QCD as an effective theory at small x. This requires integrating out quantum fluctuations in layers of p^+, which can be done with the help of a renormalization group equation (RGE) for the weight function $W_\tau[\rho]$. We shall not present all the calculations leading to this RGE; this would require heavy technical developments going far beyond the purpose of these lectures. (See Ref. [37] for more details.) Rather, we shall emphasize the general strategy of this construction and the physical picture behind it (that of the colour glass), together with those elements of the calculation which are important to understand the structure of the final equation.

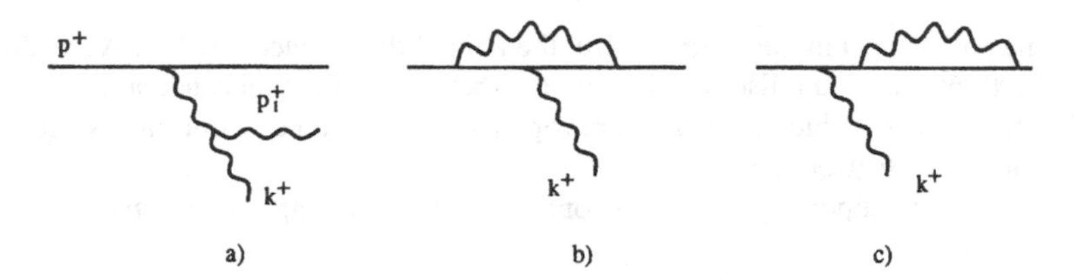

Figure 13. Lowest-order quantum corrections to the emission of a soft gluon by a fast parton: a) a real-gluon emission; b) a vertex correction; c) a self-energy correction.

3.1. THE BFKL CASCADE

In Sect. 2.1, we have argued that the radiation of a soft gluon by a fast parton via the tree-level graph shown in Fig. 9.a can be described as a classical process with a colour source whose structure is largerly fixed by the kinematics. Our main goal in this section will be to show that this picture is not spoilt by quantum corrections. We start by showing that the dominant quantum corrections, those which will be resummed in what follows, preserve indeed the separation of scales which lies at the basis of the effective theory developed in Sect. 2.

Consider first the lowest-order radiative correction to the tree-level graph in Fig. 9.a, namely, the emission of one additional (quantum) gluon, as shown in Fig. 13.a. At the same level of accuracy, one should include also the vertex and self-energy corrections illustrated in Fig. 13.b, c. This will be done in the complete calculation presented in Sect. 3.4. But in order to get a simple order-of-magnitude estimate for the quantum corrections — which is our purpose in this subsection — it is enough to consider the radiative process in Fig. 13.a.

The probability for the emission of a quantum gluon with longitudinal momentum p_1^+ in the range $p^+ > p_1^+ > k^+$ is

$$\Delta P \propto \frac{\alpha_s N_c}{\pi} \int_{k^+}^{p^+} \frac{dp_1^+}{p_1^+} = \frac{\alpha_s N_c}{\pi} \ln \frac{p^+}{k^+} \sim \alpha_s \ln \frac{1}{x} . \tag{3.1}$$

This becomes large when the available interval of rapidity $\Delta\tau = \ln(1/x)$ is large. This is the typical kind of quantum correction that we would like to resum here. A calculation which includes effects of order $(\alpha_s \ln(1/x))^n$ to all orders in n is said to be valid to "leading logarithmic accuracy" (LLA).

The typical contributions to the logarithmic integration in eq. (3.1) come from modes with momenta p_1^+ *deeply* inside the strip: $p^+ \gg p_1^+ \gg k^+$. Thus, in Fig. 13.a, the soft final gluon with momentum k^+ is emitted typically from a relatively fast gluon, with momentum $p_1^+ \gg k^+$. This latter gluon can therefore be seen as a component of the *effective* colour source at the soft scale k^+. In other terms, one can visualise the combined effect of the tree-level process, Fig. 9.a, and the first-order radiative correction, Fig. 13.a, as the generation of a modified colour source at the scale k^+, which receives contributions *only* from the modes with longitudinal momenta much larger than k^+. This is illustrated in Fig. 14.

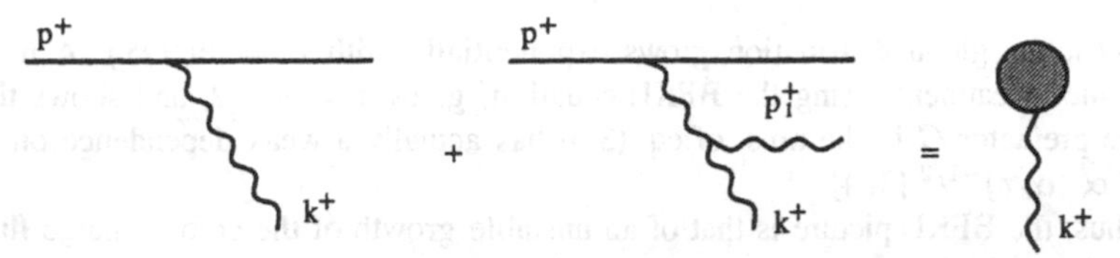

Figure 14. Effective colour source after including the lowest-order radiative correction.

Clearly, when x is small enough, $\ln(1/x) \sim 1/\alpha_s$, the "correction" (3.1) becomes of $\mathcal{O}(1)$, and it is highly probable that more gluons will be emitted along the way. This gives birth to the gluon cascade depicted in Fig. 9.b, whose dominant contribution, for a fixed number of "rungs" N, is of order $(\alpha_s \ln(1/x))^N$, and comes from the kinematical domain where the longitudinal momenta are strongly ordered:

$$p^+ \equiv p_0^+ \gg p_1^+ \gg p_2^+ \gg \cdots \gg p_N^+ \equiv k^+. \tag{3.2}$$

(Other momentum orderings give contributions which are suppressed by, at least, one factor of $1/\ln(1/x)$, and thus can be neglected to LLA.) With this ordering, this is the famous BFKL cascade, that we would like to include in our effective source. This should be possible since the hierarchy of scales in eq. (3.2) is indeed consistent with the kinematical assumptions in Sect. 2.
Note first that, the strong ordering (3.2) in longitudinal momenta implies a corresponding ordering in the lifetimes of the emitted gluons (cf. eq. (2.2)):

$$\Delta x_0^+ \gg \Delta x_1^+ \gg \Delta x_2^+ \gg \cdots \gg \Delta x_N^+. \tag{3.3}$$

Because of this, any newly emitted gluon lives too shortly to notice the dynamics of the gluons above it. This is true in particular for the last emitted gluon, with momentum k^+, which "sees" the N previous gluons in the cascade as a frozen colour charge distribution, with an average colour charge $\mathcal{Q} \equiv \sqrt{\langle \mathcal{Q}_a \mathcal{Q}_a \rangle} \sim N$. Thus, this $(N+1)$th gluon is emitted *coherently* off the colour charge fluctuations of the N previous ones, with a differential probability (compare to eq. (3.1)) :

$$dP_N \propto \frac{\alpha_s N_c}{\pi} N(\tau)\, d\tau. \tag{3.4}$$

When increasing the rapidity by one more step, $\tau \to \tau + d\tau$, the number of radiated gluons changes according to

$$N(\tau + d\tau) = (1 + N(\tau))dP_N + N(\tau)(1 - dP_N), \tag{3.5}$$

which together with eq. (3.4) implies (with $\bar{\alpha}_s \equiv \alpha_s N_c/\pi$)

$$xG(x, Q^2) \equiv \frac{dN}{d\tau} \sim C\alpha_s e^{\kappa\bar{\alpha}_s\tau}. \tag{3.6}$$

Thus, the gluon distribution grows exponentially with $\tau = \ln(1/x)$. A more refined treatment, using the BFKL equation, gives $\kappa = 4\ln 2$, and shows that the prefactor C in the r.h.s. of eq. (3.6) has actually a weak dependence on τ: $C \propto (\alpha_s\tau)^{-1/2}$ [3, 4].

Thus, the BFKL picture is that of an unstable growth of the colour charge fluctuations as x becomes smaller and smaller. However, this evolution assumes the radiated gluons to behave as free particles, so it ceases to be valid at very low x, where the gluon density becomes so large that their mutual interactions cannot be neglected anylonger. This happens, typically, when the interaction probability for the radiated gluons becomes of order one, cf. eq. (1.12), which is also the criterion for the saturation effects to be important (compare in this respect eq. (1.12) and eqs. (2.72)–(2.73)). Thus one cannot study saturation consistently without including non-linear effects in the quantum evolution. It is our main objective in what follows to explain how to do that.

3.2. THE QUANTUM EFFECTIVE THEORY

To the accuracy of interest, quantum corrections can be incorporated in the effective theory by renormalizing the source ρ_a and its correlation functions (i.e., the weight function $W_\tau[\rho]$). The argument proceeds by induction: We assume the effective theory to exist at some scale Λ^+ and show that it can be extended at the lower scale $b\Lambda^+ \ll \Lambda^+$. Specifically:

I) We assume that a *quantum* effective theory exists at some original scale Λ^+ with $\Lambda^+ \ll P^+$. That is, we assume that the *fast* quantum modes with momenta $p^+ \gg \Lambda^+$ can be replaced, as far as their effects on the correlation functions at the scale Λ^+ are concerned, by a classical random source ρ_a with weight function $W_{\Lambda^+}[\rho]$. (We shall eventually convert Λ^+ into the rapidity τ by using $\tau = \ln(P^+/\Lambda^+)$.) On the other hand, the *soft* gluons, with momenta $p^+ < \Lambda^+$, are still explicitely present in the theory, as quantum gauge fields. Thus, this effective theory includes both the classical field $\mathcal{A}^i[\rho]$ generated by ρ, and the soft quantum gluons.
Within this theory, the correlation functions of the soft ($k^+ \le \Lambda^+$) fields are obtained as (e.g., for the 2-point function)

$$\langle \mathrm{T} A^\mu(x) A^\nu(y) \rangle = \int \mathcal{D}\rho\, W_{\Lambda^+}[\rho] \left\{ \frac{\int^{\Lambda^+} \mathcal{D}A\, \delta(A^+)\, A^\mu(x) A^\nu(y)\, \mathrm{e}^{iS[A,\rho]}}{\int^{\Lambda^+} \mathcal{D}A\, \delta(A^+)\, \mathrm{e}^{iS[A,\rho]}} \right\}, \tag{3.7}$$

where T stays for time ordering (i.e. ordering in x^+). This is written in the LC-gauge $A_a^+ = 0$, and involves two functional integrals:

a) a quantum path integral over the soft gluon fields A^μ at fixed ρ:

$$\langle \mathrm{T} A^\mu(x) A^\nu(y) \rangle_\rho = \frac{\int^{\Lambda^+} \mathcal{D}A\, \delta(A^+)\, A^\mu(x) A^\nu(y)\, \mathrm{e}^{iS[A,\rho]}}{\int^{\Lambda^+} \mathcal{D}A\, \delta(A^+)\, \mathrm{e}^{iS[A,\rho]}}, \tag{3.8}$$

b) a classical average over ρ, like in eq. (2.6) :

$$\langle \mathrm{T} A^\mu(x) A^\nu(y) \rangle = \int \mathcal{D}\rho W_{\Lambda^+}[\rho] \langle \mathrm{T} A^\mu(x) A^\nu(y) \rangle_\rho \,. \tag{3.9}$$

The upper script "Λ^+" on the quantum path integral is to recall the restriction to soft ($|p^+| < \Lambda^+$) longitudinal momenta[8]. The action $S[A, \rho]$ is chosen such as to generate the classical field equations (2.1) in the saddle point approximation $\delta S/\delta A^\mu = 0$. This requirement, together with gauge symmetry and the eikonal approximation, single out the following action [36] :

$$S[A,\rho] = -\int d^4x \, \frac{1}{4} \, F^a_{\mu\nu} F^{\mu\nu}_a + \frac{i}{gN_c} \int d^3\vec{x} \, \mathrm{Tr} \left\{ \rho(\vec{x}) \, W[A^-](\vec{x}) \right\}, \tag{3.10}$$

where $W[A^-]$ is a Wilson line in the temporal direction:

$$W[A^-](\vec{x}) = \mathrm{T} \exp \left\{ ig \int dx^+ A^-(x) \right\}. \tag{3.11}$$

With this action, the condition $\delta S/\delta A^\mu = 0$ implies indeed eq. (2.1) for field configurations having $A^-_a = 0$. Thus, the classical solution $\mathcal{A}^\mu_a = \delta^{\mu i} \mathcal{A}^i_a[\rho]$ found in Sect. 2.3 is the tree-level field in the present quantum theory.
As long as we are interested in correlation functions at the scale Λ^+, or slightly below it, we can satisfy ourselves with this classical (or saddle point) approximation. That is, to the accuracy to which holds the effective theory in eq. (3.7), the gluon correlations at the scale Λ^+ can be computed from the classical field solution, as in eq. (2.6). But quantum corrections become important when we consider correlations at a much softer scale $k^+ \ll \Lambda^+$, such that $\alpha_s \ln(\Lambda^+/k^+) \sim 1$.

II) Within the quantum effective theory, we integrate out the *semi-fast* quantum fluctuations, i.e., the fields with longitudinal momenta inside the strip:

$$b\Lambda^+ \ll |p^+| \ll \Lambda^+, \text{ with } b \ll 1 \text{ and } \alpha_s \ln(1/b) < 1. \tag{3.12}$$

This generates quantum corrections to the correlation functions at the softer scale $b\Lambda^+$, which can be computed by decomposing the total gluon field as follows:

$$A^\mu_c = \mathcal{A}^\mu_c[\rho] + a^\mu_c + \delta A^\mu_c. \tag{3.13}$$

[8] The separation between fast and soft degrees of freedom according to their longitudinal momenta has a gauge-invariant meaning (within the LC-gauge) since the residual gauge transformations, being independent of x^-, cannot change the p^+ momenta.

Here, $\mathcal{A}_c^\mu$ is the tree-level field, a_c^μ are the semi-fast fluctuations to be integrated out, and δA_c^μ are the *soft* modes with momenta $|p^+| \le b\Lambda^+$ whose correlations receive quantum corrections from the semi-fast gluons.
These *induced* correlations must be computed to leading order in $\alpha_s \ln(1/b)$ (LLA), but to *all* orders in the classical fields $\mathcal{A}^i[\rho]$ (since we expect $\mathcal{A}^i \sim 1/g$ at saturation). This amounts to an one-loop calculation, but with the exact background field propagator $\langle a^\mu(x) a^\nu(y)\rangle_\rho$ of the semi-fast gluons.
For instance, the quantum corrections to the 2-point function read schematically:

$$\begin{aligned} \langle(\mathcal{A}^i[\rho] + \delta A^i)(\mathcal{A}^j[\rho] + \delta A^j)\rangle_\rho \, - \, \mathcal{A}^i[\rho]\mathcal{A}^j[\rho] = \\ = \mathcal{A}^i[\rho]\langle\delta A^j\rangle_\rho + \langle\delta A^i\rangle_\rho \mathcal{A}^j[\rho] + \langle\delta A^i \delta A^j\rangle_\rho \end{aligned} \tag{3.14}$$

where the brackets $\langle\cdots\rangle_\rho$ stand for the quantum average over the semi-fast fields in the background of ρ; this average is defined as in eq. (3.8), but with the functional integral now restricted to the fields a_c^μ. The purpose of the quantum calculation is to provide explicit expressions for the 1-point function $\langle\delta A^i\rangle_\rho$ and the 2-point function $\langle\delta A^i \delta A^j\rangle_\rho$ as functionals of ρ (to the indicated accuracy). Once these expressions are known, the 2-point function $\langle A^i(x) A^j(y)\rangle$ at the scale $b\Lambda^+$ can be finally computed as:

$$\langle A^i A^j\rangle = \Big\langle \langle(\mathcal{A}^i[\rho] + \delta A^i)(\mathcal{A}^j[\rho] + \delta A^j)\rangle_\rho \Big\rangle_{W_\Lambda}, \tag{3.15}$$

where the external brackets $\langle\cdots\rangle_{W_\Lambda}$ denote the classical average over ρ with weight function $W_\Lambda[\rho]$, as in eq. (3.9).

III) We finally show that the induced correlations can be absorbed into a functional change $W_{\Lambda^+}[\rho] \to W_{b\Lambda^+}[\rho]$ in the weight function for ρ. That is, the result (3.15) of the classical+quantum calculation in the effective theory at the scale Λ^+ can be reproduced by a purely classical calculation, but with a modified weight function $W_{b\Lambda^+}[\rho]$, corresponding to a new effective theory:

$$\Big\langle \langle(\mathcal{A}^i[\rho] + \delta A^i)(\mathcal{A}^j[\rho] + \delta A^j)\rangle_\rho \Big\rangle_{W_\Lambda} = \langle \mathcal{A}^i[\rho]\mathcal{A}^i[\rho]\rangle_{W_{b\Lambda}}, \tag{3.16}$$

where the average in the r.h.s. is defined as in eq. (2.6), or (3.9), but with weight function $W_{b\Lambda^+}[\rho]$. This demonstrates the existence of the effective theory at the softer scale $b\Lambda^+$.
Since $\Delta W \equiv W_{b\Lambda^+} - W_{\Lambda^+} \propto \alpha_s \ln(1/b)$, the evolution of the weight function is best written in terms of rapidity: $W_{\tau+\Delta\tau} - W_\tau = -\Delta\tau H W_\tau$, where $\tau = \ln(P^+/\Lambda^+)$, $\Delta\tau = \ln(1/b)$, and $H \equiv H[\rho, \frac{\delta}{\delta\rho}]$ is a functional differential operator acting on W_τ (generally, a non-linear functional of ρ). In the limit $\Delta\tau \to 0$, this gives a *renormalization group equation* (RGE) describing the flow of the weight function with τ [33, 36] :

$$\frac{\partial W_\tau[\rho]}{\partial\tau} = -H\Big[\rho, \frac{\delta}{\delta\rho}\Big] W_\tau[\rho]\,. \tag{3.17}$$

By integrating this equation with initial conditions at $\tau \ll 1$ (i.e., at $\Lambda^+ \sim P^+$), one can obtain the weight function at the rapidity τ of interest. The initial conditions are not really perturbative, but one can rely on some non-perturbative model, like the MV model discussed in Sects. 2.4–2.5.

A key ingredient in this approach, which makes the difference w.r.t. the BFKL equation, are the non-linear effects encoded in the background field calculation. Recall that ρ, and therefore the classical fields $\mathcal{A}^i[\rho]$, are random variables whose correlators (2.6) reproduce the gluon density and, more generally, the n-point correlation functions of the gluon fields at the scale Λ^+. Thus by computing quantum corrections in the presence of these background fields, and then averaging over the latter, one is effectively studying quantum evolution in a medium with high gluon density. After each step in this evolution, the properties of the medium (i.e., the correlators of ρ) are updated, by including the latest quantum corrections. In terms of Feynman graphs of the ordinary perturbation theory, this corresponds to a complicated resummation of diagrams describing the interactions between the gluons radiated in different parton cascades and at different rapidities. A typical such a diagram is shown in Fig. 15. At low density, where the non-linear effects can be neglected, eq. (3.17) correctly reproduces the BFKL equation [36], as it should.

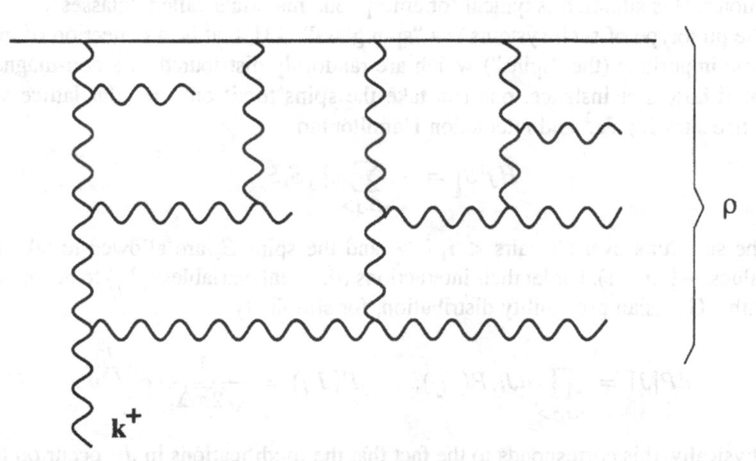

Figure 15. A typical Feynman diagram that is implicitely resummed in the quantum evolution of the effective theory.

3.3. THE COLOUR GLASS CONDENSATE

Note the special form of the average in eq. (3.7). This is *not* the same as :

$$\frac{\int \mathcal{D}\rho\, W_\Lambda[\rho] \int^\Lambda \mathcal{D}A\, A^\mu(x)A^\nu(y)\ \mathrm{e}^{iS[A,\rho]}}{\int \mathcal{D}\rho\, W_\Lambda[\rho] \int^\Lambda \mathcal{D}A\ \mathrm{e}^{iS[A,\rho]}}. \tag{3.18}$$

In eq. (3.18), both the colour source ρ_a and the gauge fields A_a^μ are dynamical variables that are summed over on the same footing. They are free to take on values which extremize the total "effective action" :

$$S_{eff}[A,\rho] = S[A,\rho] - i\ln W_\Lambda[\rho]. \tag{3.19}$$

By contrast, in eq. (3.7), the average over A^μ is taken at fixed ρ : the gauge fields can vary in response to ρ, but ρ cannot vary in response to the gauge fields. That is, ρ is not a dynamical variable, but rather an "external" source. Giving a colour charge distribution $\rho_a(\vec{x})$ specifies a medium in which propagate the quantum gluons. But this medium is, by itself, random, so after performing the quantum analysis at fixed ρ, one must also perform an average over ρ. The reason for treating ρ and A^μ differently lies is the separation of scales in the problem: the changes in ρ happens on time scales much larger than the lifetime of the soft gluons. This situation is typical for amorphous materials called "glasses".
The prototype of such systems is a "spin glass" [43], that is, a collection of magnetic impurities (the "spins") which are randomly distributed in a non-magnetic metal host. For instance, one can take the spins to sit on a regular lattice with lattice sites $i, j, \dots$, and interaction Hamiltonian

$$H_J[S] = -\sum_{<i,j>} J_{ij}S_iS_j, \tag{3.20}$$

(the sum runs over all pairs $< i,j >$, and the spins S_i are allowed to take two values, $+1$ or -1), but let their interactions (the "link variables" J_{ij}) to be random, with a Gaussian probability distribution, for simplicity:

$$dP[J] = \prod_{<i,j>} dJ_{ij}P(J_{ij}), \qquad P(J_{ij}) = \frac{1}{\sqrt{2\pi\Delta_{ij}}}\,\mathrm{e}^{-\frac{J_{ij}^2}{2\Delta_{ij}}}. \tag{3.21}$$

Physically, this corresponds to the fact that the modifications in J_{ij} occur on time scales much larger than the time scales characterizing the dynamics of the spins (e.g., their thermalization when the system is brought in contact with a thermal bath). In practice, the J_{ij}'s are frozen into their fixed values by rapid cooling when the sample is prepared. This kind or rapid cooling is called "quenching", and one says that the J_{ij}'s are "quenched variables", as opposed to the "dynamical variables", the spins S_i. This procedure selects random values for the J_{ij}'s, with the probability distribution (3.21).

Thus, the spins thermalize for a given set of "quenched variables", and for each such a set one can compute the thermal partition function and the free energy:

$$Z[J] = \sum_{\{S\}} \mathrm{e}^{-\beta H_J[S]}, \qquad F[J] = -T \ln Z[J]. \tag{3.22}$$

But the J_{ij}'s are themselves random, so the experimentally relevant quantity is the following average

$$F = \langle F[J] \rangle \equiv \int dP[J]\, F[J] = -T \int dP[J]\, \ln Z[J]. \tag{3.23}$$

Note that it is $\ln Z[J]$, not $Z[J]$ itself, which should be averaged ("quenched average"). Similarly, (connected) correlation functions are generated by the free energy in the presence of a site-dependent external magnetic field:

$$\langle S_i S_j \rangle - \langle S_i \rangle \langle S_j \rangle = T^2 \int dP[J]\, \frac{\partial^2 \ln Z[J,h]}{\partial h_i \partial h_j}, \tag{3.24}$$

with $\ln Z[J,h]$ defined as in eq. (3.22), but with $H_J[S] \to H_J[S] - \sum_i h_i S_i$. Eqs. (3.23)–(3.24) are the analogs of eq. (3.7) for the problem at hand: the colour source ρ_a is our "quenched variable", and the quantum average over the fields A^μ at fixed ρ, eq. (3.8), corresponds to the thermal average at fixed J_{ij}'s, eq. (3.22). As in eq. (3.23), it is $\ln Z$, and not Z, which is effectively averaged in eq. (3.7) (the average of Z would rather corresponds to eq. (3.18)). In fact, the *connected* correlation functions of the soft gluons in the effective theory are obtained from the following generating functional:

$$F[j_a^\mu] = \int \mathcal{D}\rho\, W_\Lambda[\rho]\, \ln \left(\int^\Lambda \mathcal{D}A\, \delta(A^+)\, \mathrm{e}^{iS[A,\rho] - i\int j\cdot A} \right), \tag{3.25}$$

which is the analog of eq. (3.23) with $\ln Z[J] \to \ln Z[J,h]$. (The external current j_a^μ in (3.25) is just a device to generate Green's functions via differentiations, and should not be confused with the physical source ρ_a.)

We are thus naturally led to interprete the small-x component of the hadron wavefunction as a *glass*, with the colour charge density playing the role of the spin for spin glasses. Thus, this is a *colour* glass. Unlike what happens for spin glasses, which may have a non-zero value for the average magnetization $\langle S_i \rangle$ (at least locally, i.e., at a given site), the *average* colour charge must be zero,

$$\langle \rho_a(\vec{x}) \rangle = 0 \quad \text{at any } \vec{x}, \tag{3.26}$$

by gauge symmetry. In practice, this is insured by the fact that we sum over all the possible configurations of $\rho_a(\vec{x})$ with a gauge-invariant weight function. Let us however examine a particular configuration $\rho_a(\vec{x})$ from this ensemble. We now argue that, at sufficiently small x (or large atomic number A), this configuration describes typically a *Bose condensate*.

This applies to the *saturated* modes, i.e., the modes with transverse momenta $\Lambda_{QCD} \ll k_\perp \ll Q_s(\tau)$ and longitudinal momenta $k^+ = xP^+ \ll P^+$. As argued in Sect. 2.5, these modes are characterized by a high gluon number density in the transverse phase-space, $\mathcal{N}_\tau(k_\perp) \sim 1/\alpha_s$. (This prediction of the classical MV model remains valid after including the quantum evolution, as we shall see in Sect. 5.4 below.) Microscopically, these modes correspond to bosonic states with large occupation numbers $\sim 1/\alpha_s$. Each such a state is a Bose condensate.

More precisely, the general definition of a Bose condensate is that of a quantum state in which the Fock space annihilation operator $a_c^i(\vec{k})$ (cf. eq. (2.13)), or, equivalently, the field operator $A_c^i(x)$, takes on a non-zero expectation value. This situation may be characterized as the spontaneous generation of a classical field. Of course, this cannot happen for gluons in the vacuum, as it would violate gauge symmetry. And, in an absolute sense, this cannot happen in a hadron neither, since the average colour charge vanishes there too (cf. eq. (3.26)), and therefore so does the associated classical field: $\langle \mathcal{A}_c^i[\rho]\rangle = 0$. But in the hadron there *are* colour sources, and, as argued before, they can be even treated as a classical charge distribution which is frozen during the short lifetime of the small-x gluons. Thus, over such a short time scale (short as compared to the typical time scale for changes in the colour distribution), one effectively has a non-trivial classical field $\mathcal{A}_c^i[\rho]$. At saturation, this field is typically strong (cf. eqs. (2.68) and (2.44)) :

$$\left\langle \mathcal{A}_\infty^{ia}(x_\perp)\, \mathcal{A}_\infty^{ia}(y_\perp)\right\rangle \approx \frac{N_c^2-1}{\pi N_c}\frac{1-\mathrm{e}^{-r_\perp^2 Q_s^2}}{\alpha_s r_\perp^2}\,, \tag{3.27}$$

$$\bar{\mathcal{A}}^i \sim \sqrt{\langle \mathcal{A}_\infty^i \mathcal{A}_\infty^i\rangle} \sim \frac{1}{\sqrt{\alpha_s r_\perp^2}} \qquad \text{for } r_\perp \gg 1/Q_s\,, \tag{3.28}$$

and its typical amplitude (3.28) at large $r_\perp$ is even independent of the actual strength $\bar{\rho} \sim \sqrt{\langle \rho_a \rho_a\rangle} \sim Q_s$ of the colour source. This can be thus characterized as a Bose condensate.

We thus see that it is the same fundamental separation in time scales which allows us to speak about both the *colour glass* and the *Bose condensate*, although these two concepts seem at a first sight contradictory: the notion of a "glass" makes explicit reference to the *average* over ρ, while the "condensate" rather refers to a specific realization of ρ, before averaging.

3.4. THE RENORMALIZATION GROUP EQUATION

As explained in Sect. 3.2, the quantum evolution of the effective theory is obtained by matching correlations computed in two ways: (a) via a classical+quantum calculation in the effective theory at the scale Λ^+, and (b) via a purely classical

calculation within the effective theory at the scale $b\Lambda^+$. The quantum corrections that are included in this way are those generated by the coupling between the "semi-fast" gluons with p^+ momenta in the strip (3.12) and the "soft" gluons δA_c^μ with momenta $|p^+| \le b\Lambda^+$. To the accuracy of interest, it is sufficient to consider the eikonal coupling $\delta A_a^- \delta\hat{\rho}_a$ to the plus component $\delta\hat{J}_a^+ \equiv \delta\hat{\rho}_a$ of the colour current of the semi-fast gluons. Indeed, these gluons are relatively fast moving in the x^+ direction, so $\delta\hat{\rho}_a$ is the large component of their current.
The results of the matching can be summarized as follows:
i) To $\mathcal{O}(\alpha_s \ln(1/b))$, the induced correlations of the transverse fields A_a^i (see eq. (3.14) for an example) can be all related to the following 1-point and 2-point functions of $\delta\hat{\rho}$ (with $\Delta\tau = \ln(1/b)$) :

$$\sigma_a(x_\perp) \equiv \frac{1}{\Delta\tau}\int dx^- \langle \delta\hat{\rho}_a(x)\rangle_\rho , \tag{3.29}$$

$$\chi_{ab}(x_\perp, y_\perp) \equiv \frac{1}{\Delta\tau}\int dx^- \int dy^- \langle \delta\hat{\rho}_a(x^+,\vec{x})\, \delta\hat{\rho}_b(x^+,\vec{y})\rangle_\rho , \tag{3.30}$$

where, as in eq. (3.14), $\langle\cdots\rangle_\rho$ denotes the average over semi-fast quantum fluctuations in the background of the tree-level source ρ.
Thus the quantum evolution consists in adding new correlations σ and χ to ρ.
ii) These new correlations can be included in the weight function $W_\tau[\rho]$ by allowing this to evolve with τ according to the following RGE [36, 37] :

$$\frac{\partial W_\tau[\rho]}{\partial\tau} = \frac{1}{2}\frac{\delta^2}{\delta\rho_\tau^a(x_\perp)\delta\rho_\tau^b(y_\perp)}[W_\tau \chi_{xy}^{ab}] - \frac{\delta}{\delta\rho_\tau^a(x_\perp)}[W_\tau \sigma_x^a]. \tag{3.31}$$

We use here compact notations where $\sigma_x^a \equiv \sigma_a(x_\perp)$, $\chi_{xy}^{ab} \equiv \chi_{ab}(x_\perp, y_\perp)$, and repeated colour indices (and coordinates) are understood to be summed (integrated) over. The notation $\rho_\tau^a(x_\perp)$ will be explained later (see eq. (3.46)).
A complete proof of the statements above would require the lengthy analysis of Refs. [37]. But assuming them to be true, it is easy to understand the general structure of the RGE (3.31). Indeed, according to eqs. (3.29)–(3.30), the induced correlations that we need to take into account are (with the notations of eq. (3.15)):

$$\Big\langle \langle(\rho_a + \delta\hat{\rho}_a)_{x_\perp}(\rho_b + \delta\hat{\rho}_b)_{y_\perp}\rangle_\rho \Big\rangle_{W_\tau} - \Big\langle \rho_a(x_\perp)\rho_b(y_\perp)\Big\rangle_{W_\tau} \tag{3.32}$$

$$= \int D[\rho]\, W_\tau[\rho]\, \Delta\tau \,\{\sigma_a(x_\perp)\rho_b(y_\perp) + \rho_a(x_\perp)\sigma_b(y_\perp) + \chi_{ab}(x_\perp, y_\perp)\} =$$

$$= \int D[\rho]\, W_\tau[\rho]\, \Delta\tau \Big\{\sigma_z^c \frac{\delta}{\delta\rho^c(z_\perp)} + \frac{1}{2}\chi_{zu}^{cd}\frac{\delta^2}{\delta\rho^c(z_\perp)\delta\rho^d(u_\perp)}\Big\}\rho_a(x_\perp)\rho_b(y_\perp),$$

where the colour indices c, d (the transverse coordinates $z_\perp, u_\perp$) in the last line are to be summed (integrated) over. After a few integrations by parts w.r.t. ρ, the last expression can be recast into the form:

$$\int D[\rho]\; \rho_a(x_\perp)\rho_b(y_\perp)\, \Delta W_\tau[\rho], \tag{3.33}$$

with $\Delta W_\tau[\rho]$ given by the finite-difference version of eq. (3.31).
In eqs. (3.32)–(3.33), we have considered only correlators of *two*–dimensional (or "integrated") charge densities, like

$$\rho_a(x_\perp) \equiv \int dx^- \, \rho_a(x^-, x_\perp), \tag{3.34}$$

and similarly $\delta\hat{\rho}_a(x_\perp)$. This is in agreement with eqs. (3.29)–(3.30), which show that only such *integrated* (over x^-) quantum corrections are relevant to the order of interest, and is moreover physically intuitive: The soft gluons ($k^+ \lesssim b\Lambda^+$) to which applies the effective theory are unable to discriminate the internal longitudinal structure of their sources, which are localized in x^- over relatively short distances $\ll 1/b\Lambda^+$, because of their large p^+ momenta. Although essentially correct, this argument is a little too simplistic as shown by the fact that some of the quantities encountered before *are* in fact sensitive to the longitudinal structure of ρ (i.e., they are not simply functionals of the integrated charge density (3.34)). A generic example is the background field $\mathcal{A}^i[\rho]$, or any other quantity built with the Wilson lines (2.38) or (2.45). Such quantities are sensitive to the x^- dependence of ρ because of the path-ordering of the Wilson lines in x^-. The ordering is important since colour matrices $\rho(x^-) = \rho_a(x^-)T^a$ at different values of x^- do not commute with each other. This suggests that the correct way to think of an "integrated" version of the hadron (over x^-) is in terms of Wilson lines — which take into account the colour precession in the colour field of the hadron, with the proper ordering of colour matrices —, and not of 2-dimensional charge densities like (3.34). This will be confirmed by the subsequent analysis of the quantum corrections.

3.4.1. *The quantum colour source*

For the purposes of the quantum calculation, it is useful to expand the action $S[A, \rho] \equiv S[\mathcal{A} + a + \delta A, \rho]$ to quadratic order in the small fluctuations a^μ, and retain only their eikonal coupling to the component δA_a^- of the soft fields:

$$S[\mathcal{A} + \delta A + a, \rho] \approx S[\mathcal{A} + \delta A, \rho] + \frac{1}{2} a^\mu(x) G^{-1}_{\mu\nu}(x, y)[\rho] a^\nu(y) - \delta A_a^- \delta\hat{\rho}_a, \tag{3.35}$$

with

$$G^{-1\,ab}_{\mu\nu}(x, y)[\rho] \equiv \left.\frac{\delta^2 S[A, \rho]}{\delta A_a^\mu(x) \delta A_b^\nu(y)}\right|_{\mathcal{A}}, \tag{3.36}$$

$$\delta\hat{\rho}_a(x) \equiv - \left.\frac{\delta^2 S}{\delta A_a^-(x) \delta A_b^\nu(y)}\right|_{\mathcal{A}} a_b^\nu(y) - \frac{1}{2} \left.\frac{\delta^3 S}{\delta A_a^-(x) \delta A_b^\nu(y) \delta A_c^\lambda(z)}\right|_{\mathcal{A}} a_b^\nu(y) a_c^\lambda(z), \tag{3.37}$$

where it is understood that only the soft modes with $k^+ \lesssim b\Lambda^+$ are kept in the products of fields.

The expansion (3.35) corresponds to a one-loop approximation for the soft correlation functions like $\langle\delta A^i\rangle_\rho$ and $\langle\delta A^i \delta A^j\rangle_\rho$ (cf. eq. (3.14)), but where the propagator $iG^{\mu\nu}(x,y) = \langle \mathrm{T} a^\mu(x) a^\nu(y)\rangle_\rho$ of the semi-fast gluons running along the loop is computed in the background of the tree-level field $\mathcal{A}^i[\rho]$, by inverting the differential operator in eq. (3.36).

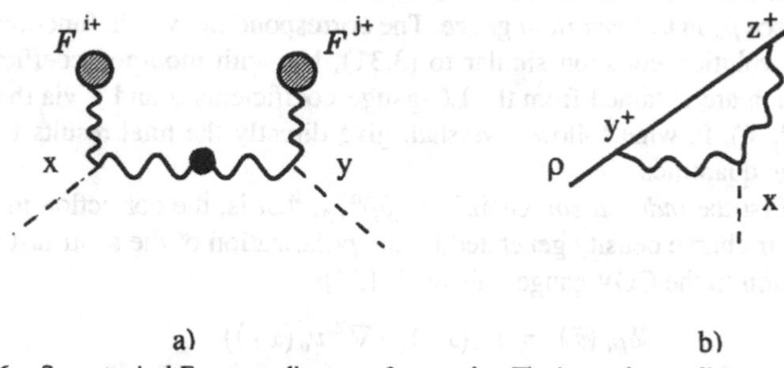

Figure 16. Some typical Feynman diagrams for χ and σ. The internal wavy lines are propagators of the semi-fast gluons; the external dotted lines carry soft momenta, and couple to the fields δA^-. (a) A contribution to χ. The external blobs denote insertions of the electric field $\mathcal{F}^{+i}$; the internal line with a blob denotes the background field propagator. (b) A contribution to σ to linear order in ρ. The continuous line represents the source ρ.

To gain some more intuition, we use as an example the contributions to $\delta\hat{\rho}_a$ coming from the Yang-Mills piece of the action, $S_{YM} = \int d^4x(-F^2_{\mu\nu}/4)$:

$$\delta\hat{\rho}_a(x)|_{YM} = 2gf^{abc}\mathcal{F}^{+i}_b(\vec{x})a^i_c(x) + gf^{abc}(\partial^+ a^i_b(x))a^i_c(x). \tag{3.38}$$

The first term in the r.h.s., which is linear in a^i, is the only one to contribute to χ, eq. (3.30), to leading order in α_s. It generates the tree-like diagram in Fig. 16.a, where the internal line with a blob represents the background field propagator $G^{ij}(x,y)$ of the semi-fast gluons. Physically, Fig. 16.a describes the emission of an on-shell (or "real") semi-fast gluon by the classical source.

Since $\langle a^i\rangle = 0$, it is only the second, quadratic term in the r.h.s. of eq. (3.38) which contributes to σ, eq. (3.29). In Fig. 16.b we show such a contribution of lowest order in ρ. (This involves also vertices from the Wilson line piece of the action, eqs. (3.10)–(3.11).) Obviously, this represents a vertex correction to the tree-level emission in Fig. 9.a.

The structures illustrated by Figs. 16.a and b are generic: χ is the "real" correction, whose iteration generates the gluon cascades; σ is the "virtual" correction, which provides one-loop corrections to the emission vertices in these cascades. Both χ and σ include terms non-linear in ρ which describe interactions among gluons at different rapidities in different cascades. In general, real and virtual corrections

are related by gauge symmetry, and this is also the case for χ and σ, as we shall discuss later.
The diagrams contributing to σ and χ in the general case, together with their explicit evaluation, can be found in Ref. [37]. Here, we shall present only the final results of this calculation.

3.4.2. *The induced colour source and field*

For the reasons explained in Sect. 2.3, it is more convenient to work with the colour source $\tilde{\rho}_a$ in the *covariant* gauge. The corresponding weight function $W_\tau[\tilde{\rho}]$ obeys an evolution equation similar to (3.31), but with modified coefficients $\tilde{\sigma}$ and $\tilde{\chi}$, which are obtained from the LC-gauge coefficients σ and χ via the gauge rotation (2.36). In what follows, we shall give directly the final results for these COV-gauge quantities.
Consider first the *induced source* $\delta\rho^a = \langle\delta\hat{\rho}^a\rangle_\rho$, that is, the correction to the average colour charge density generated by the polarization of the semi-fast gluons. After rotation to the COV-gauge, this reads [37] :

$$\delta\tilde{\rho}_a(\vec{x}) \,=\, F_\Lambda(x^-)(-\nabla_\perp^2\nu_a(x_\perp)), \tag{3.39}$$

where the "form factor"

$$F_\Lambda(x^-) \equiv \theta(x^-)\,\frac{e^{-ib\Lambda x^-} - e^{-i\Lambda x^-}}{x^-}\,, \tag{3.40}$$

specifies the longitudinal profile of $\delta\tilde{\rho}_a$, while ($V_x^\dagger \equiv V^\dagger(x_\perp)$, cf. eq. (2.45))

$$\nu^a(x_\perp) \,=\, \frac{ig}{2\pi}\int\frac{d^2z_\perp}{(2\pi)^2}\,\frac{1}{(x_\perp - z_\perp)^2}\,\mathrm{Tr}\Big(T^a V_x^\dagger V_z\Big) \tag{3.41}$$

contains the dependence upon the background field α_a (via the Wilson lines V and $V^\dagger$), together with the transverse and colour structure of $\delta\tilde{\rho}_a$. By comparing eqs. (3.39) and (2.33), we deduce that $F(x^-)\nu^a(x_\perp)$ is the *induced field* in the COV-gauge, i.e., the quantum correction to the tree-level field α_a. Since:

$$\int dx^-\,F_\Lambda(x^-) \,=\, \ln\frac{1}{b} \,=\, \Delta\tau, \tag{3.42}$$

eqs. (3.29) and (3.39) immediately imply :

$$\tilde{\sigma}_a(x_\perp) \,=\, -\nabla_\perp^2\nu_a(x_\perp). \tag{3.43}$$

This is the coefficient of the virtual term in the RGE for $W_\tau[\tilde{\rho}]$.
But the longitudinal structure of $\delta\tilde{\rho}_a$ is also interesting. Eq. (3.39) shows that the induced source and field have typically support at[9]

$$1/\Lambda^+ \,\lesssim\, x^- \,\lesssim\, 1/(b\Lambda^+)\,. \tag{3.44}$$

[9] Indeed, $F_\Lambda(x^-) \approx 0$ both for small $x^- \ll 1/\Lambda^+$ (since in this case the two exponentials mutually cancel), and for large $x^- \gg 1/b\Lambda^+$ (where the two exponentials are individually small).

Recall that $\delta\tilde{\rho}_a$ has been generated by integrating out quantum fluctuations in the strip $b\Lambda^+ \ll |p^+| \ll \Lambda^+$. Thus, when integrating out quantum gluons in layers of p^+, one builds the classical source ρ (or field α) in layers of x^-, with a one-to-one correspondence between the x^- coordinate of a given layer and the p^+ momenta of the modes that have been integrated out to generate that layer. By induction, we deduce that $\rho_a(\vec{x})$ ($\equiv$ the colour source generated by the quantum evolution down to Λ^+) has support at $0 \le x^- \lesssim 1/\Lambda^+$, as anticipated in Sect. 2. This allows us to consider only positive values for x^- in what follows.
To exploit this tight correspondence between p^+ and x^-, it is convenient to use the *space-time rapidity* y,

$$\mathrm{y} \equiv \ln(x^-/x_0^-), \quad x_0^- \equiv 1/P^+, \quad -\infty < \mathrm{y} < \infty, \tag{3.45}$$

to indicate the longitudinal coordinate of a field. We shall set, e.g.,

$$\begin{aligned} \rho_{\mathrm{y}}^a(x_\perp) \,&\equiv\, x^- \rho^a(x^-, x_\perp) \qquad \text{for} \quad x^- = x_{\mathrm{y}}^- \equiv x_0^- e^{\mathrm{y}}, \\ \int d\mathrm{y}\, \rho_{\mathrm{y}}^a(x_\perp) \,&=\, \int dx^-\, \rho^a(x^-, x_\perp), \end{aligned} \tag{3.46}$$

and similarly for the other fields ($\tilde{\rho}$, α, etc.). The previous discussion on the longitudinal structure can then be summarized as follows:
The source $\rho_{\mathrm{y}}^a(x_\perp)$ generated by the quantum evolution from $\tau' = 0$ up to τ has support at y in the interval $0 \le \mathrm{y} \le \tau$. When new quantum modes, with rapidities τ' in the interval $\tau < \tau' < \tau + \Delta\tau$, are integrated out, the preexisting colour source at $\mathrm{y} \le \tau$ is not changed, but some new contribution is added to it, in the rapidity bin $\tau < \mathrm{y} < \tau + \Delta\tau$. Because of that, $\Delta W \equiv W_{\tau+\Delta\tau} - W_\tau$ involves only the change in ρ_{y} within that last bin. In the continuum limit $\Delta\tau \to 0$, this generates the functional derivatives of W_τ with respect to ρ_{y}^a at $\mathrm{y} = \tau$, as shown in eq. (3.31). This clarifies the longitudinal structure of the RGE.
Consider also the transverse and colour structure of the induced field (3.41). This can be understood by reference to Fig. 16.b. The transverse kernel in eq. (3.41) has been generated as:

$$\frac{1}{(2\pi)^2}\frac{1}{(x_\perp - z_\perp)^2} = \frac{1}{2\pi}\frac{x^i - z^i}{(x_\perp - z_\perp)^2}\,\frac{1}{2\pi}\frac{x^i - z^i}{(x_\perp - z_\perp)^2}, \tag{3.47}$$

where (compare with eqs. (2.34) and (2.53))

$$\frac{1}{2\pi}\frac{x^i - z^i}{(x_\perp - z_\perp)^2} = \partial_x^i \langle x_\perp | \frac{1}{-\nabla_\perp^2} | z_\perp \rangle = \int \frac{d^2p_\perp}{(2\pi)^2}\,\frac{-ip^i}{p_\perp^2}\, e^{ip_\perp\cdot(x_\perp - z_\perp)} \tag{3.48}$$

is the propagator of the semi-fast gluon emitted by the source ρ (recall that $\mathcal{F}^{+j} \approx (ip^j/p_\perp^2)\rho$ to linear order). The two Wilson lines in eq. (3.41) account for the scattering of this semi-fast gluon off the background field at $z_\perp$ (this brings in a factor $V^\dagger(z_\perp)$ in the eikonal approximation), and for its gauge rotation by the classical field $\mathcal{A}^i(\vec{x})$ at $x^- > 1/\Lambda^+$ (cf. eq. (3.44)), which is a pure gauge (cf. eq. (2.43)).

3.4.3. *The RGE in the α–representation*

Eq. (3.43) suggests that it may be technically simpler and physically more transparent to work directly with the classical field α_a and the quantum corrections to it (like ν_a), rather than with the colour source $\tilde{\rho}_a$ and the corresponding corrections (like $\tilde{\sigma}_a$). This point of view is also supported by the fact that the LC-gauge field and the related observables are primarily related to α_a (cf. Sects. 2.2 and 2.3), and reexpressing them in terms of $\tilde{\rho}$ — with the help of eq. (2.34) — would introduce a dependence upon the unphysical infrared cutoff μ.

For these reasons, we prefer to work in the *α–representation*, in which observables are expressed in terms of α, and the average is performed with the weight function $W_\tau[\alpha] \equiv W_\tau[\tilde{\rho} = -\nabla_\perp^2 \alpha]$. This satisfies the following RGE, which is obtained after a change of variables in eq. (3.31) :

$$\frac{\partial W_\tau[\alpha]}{\partial \tau} = \frac{1}{2}\frac{\delta^2}{\delta \alpha_\tau^a(x_\perp)\delta\alpha_\tau^b(y_\perp)}[W_\tau \eta_{xy}^{ab}] - \frac{\delta}{\delta\alpha_\tau^a(x_\perp)}[W_\tau \nu_x^a], \tag{3.49}$$

where $\nu_x^a \equiv \nu^a(x_\perp)$, cf. eq. (3.41), and $\eta_{xy}^{ab} \equiv \eta^{ab}(x_\perp, y_\perp)$, with

$$\eta^{ab}(x_\perp, y_\perp) \equiv \int d^2 z_\perp d^2 u_\perp \langle x_\perp|\frac{1}{-\nabla_\perp^2}|z_\perp\rangle\, \tilde{\chi}^{ab}(z_\perp, u_\perp)\, \langle u_\perp|\frac{1}{-\nabla_\perp^2}|y_\perp\rangle. \tag{3.50}$$

It is thus sufficient to give the result for the "real correction" $\langle \delta\hat{\rho}_a \delta\hat{\rho}_b\rangle_\rho$ directly in the α-representation (cf. eqs. (3.30) and (3.50)). This reads [37]:

$$\eta^{ab}(x_\perp, y_\perp) = \frac{1}{\pi}\int \frac{d^2 z_\perp}{(2\pi)^2}\frac{(x^i - z^i)(y^i - z^i)}{(x_\perp - z_\perp)^2 (y_\perp - z_\perp)^2} \times \left\{1 + V_x^\dagger V_y - V_x^\dagger V_z - V_z^\dagger V_y\right\}^{ab}. \tag{3.51}$$

The transverse and colour structure of η have the same pattern as discussed after eq. (3.53) in connection with ν.

The r.h.s. of eq. (3.49) involves functional derivatives w.r.t. the colour field $\alpha_\tau^a(x_\perp)$ at the end point $\mathrm{y} = \tau$. When applied to the coefficients η and ν, this requires the corresponding derivatives of the Wilson lines V and $V^\dagger$, that we compute now. Note first that, since $\alpha_\mathrm{y} = 0$ for $\mathrm{y} > \tau$, we can rewrite

$$V^\dagger(x_\perp) \equiv \mathrm{P}\, \mathrm{e}^{ig\int_{-\infty}^{\infty} d\mathrm{y}\, \alpha_\mathrm{y}(x_\perp)} = \mathrm{P}\, \mathrm{e}^{ig\int_{-\infty}^{\tau} d\mathrm{y}\, \alpha_\mathrm{y}(x_\perp)}. \tag{3.52}$$

Therefore (with $\delta_{xy} \equiv \delta^{(2)}(x_\perp - y_\perp)$):

$$\frac{\delta V^\dagger(x_\perp)}{\delta\alpha_\tau^a(y_\perp)} = ig\delta_{xy} T^a V^\dagger(x_\perp), \qquad \frac{\delta V(x_\perp)}{\delta\alpha_\tau^a(y_\perp)} = -ig\delta_{xy} V(x_\perp) T^a. \tag{3.53}$$

A simple interpretation of the four terms in eq. (3.51) follows from the dual picture of the dipole-hadron scattering, in which the quantum evolution is put in the dipole wavefunction, and, more generally, in the Wilson line operators through which a generic external projectile scatters off the hadronic target [44, 45, 46, 47, 48, 49, 50, 51]. (See also the lectures notes by Al Mueller in this volume [14].) Recent analyses of the high energy scattering from this dual perspective have led to a set of coupled evolution equations for the correlation functions of Wilson lines, originally derived by Balitsky [44] (see also [45, 50]), and subsequently reformulated by Weigert [47] in a compact way, as a functional evolution equation for the generating functional of these correlation functions. It turns out that Weigert's equation is equivalent to the RGE (3.49) [37, 52], which demonstrates the equivalence between the two descriptions — the target picture and the projectile picture — of the nonlinear evolution in QCD at small x. We shall say more on Balitsky's equations in Sect. 4.3.

4. A functional Fokker-Planck equation

We now dispose of a powerful tool — the functional RGE (3.49) — to construct the effective theory by integrating out quantum fluctuations in perturbation theory. Eq. (3.49) has a rich and elegant mathematical structure, to be described in Sects. 4.1 and 4.2. Then, in Sects. 4.3 and 5, we shall indicate two strategies to make use of this equation:

i) One can use it to derive ordinary (i.e., non-functional) evolution equations for the correlation functions of interest. When specialized to correlation functions of the Wilson lines, this strategy leads to a system of equations originally derived by Balitsky [44]. In the weak field limit, the equations linearize, and one recovers the BFKL equation [3]. This will be discussed in Sect. 4.3.

A difficulty with this approach is that it generally leads to *coupled* equations (the 2-point function is coupled to the 4-point one, etc.), so that one has to follow simultaneously the evolution of infinitely many correlators. Still, some progress has been done, by using functional techniques [53], and, especially, by recognizing that, in the large N_c limit, a closed equation can be written for the 2-point function: this is the Kovchegov equation [45].

ii) One can try and solve directly the functional RGE, with appropriate initial conditions. An exact but formal solution can be written in the form of a path integral [52]. This is well suited for lattice simulations in 2+1 dimensions. But approximate analytic solutions, which allow for a more direct physical insight, have been found as well [10, 57]. These solutions will be described in Sect. 5.

4.1. GENERAL PROPERTIES AND CONSEQUENCES OF THE RGE

We start with a summary of the most important properties of the RGE (3.49).
i) The coefficients η and ν are real quantities. Moreover, η is symmetric: $\eta_{ab}(x_\perp, y_\perp) = \eta_{ba}(y_\perp, x_\perp)$, and positive semi-definite.
ii) The RGE preserves the correct normalization of the weight function:

$$\int \mathcal{D}\alpha \; W_\tau[\alpha] = 1 \qquad \text{at any } \tau. \tag{4.1}$$

Indeed, the r.h.s. of eq. (3.49) is a total derivative with respect to α. Thus, if eq. (4.1) is satisfied by the initial condition at τ_0, it remains true at any $\tau > \tau_0$. Properties (i) and (ii) guarantee that the solution $W_\tau[\alpha]$ to the RGE has a meaningful probabilistic interpretation (cf. the discussion prior to eq. (2.4)).
iii) The momentum rapidity τ and the space-time rapidity y are identified by the quantum evolution. That is, the field $\alpha_{\rm y}$ in the rapidity bin $({\rm y}, {\rm y} + d{\rm y})$ is generated by the quantum evolution from $\tau = {\rm y}$ up to $\tau = {\rm y} + d{\rm y}$. This follows from the discussion in Sect. 3.4.2, and implies that the two rapidities can be treated as only one variable, the "evolution time".
With this interpretation, the function $\{\alpha^a_{\rm y}(x_\perp) \,|\, -\infty < {\rm y} < \infty\}$ — which physically represents the longitudinal profile of the 3-dimensional field $\alpha^a(x^-, x_\perp)$ in units of rapidity (cf. eq. (3.46)) — is viewed as a *trajectory* in the functional space spanned by the 2-dimensional fields $\alpha^a(x_\perp)$. Quantum evolution then appears as the progression of the "point" $\alpha^a(x_\perp)$ along this trajectory. Thus, eq. (3.49) describes effectively a field theory in 2+1 dimensions (the transverse coordinates and the "evolution time"), which is however *non-local* in both $x_\perp$ and y (since the coefficients (3.41) and (3.51) of the RGE involve $\alpha_{\rm y}$ at all the "times" ${\rm y} \le \tau$, via the Wilson lines (3.52)).
iv) The initial condition. Let the quantum evolution proceed from some original "time" τ_0 up to the actual "time" τ. The "trajectory" $\{\alpha^a_{\rm y}(x_\perp) \,|\, -\infty < {\rm y} < \infty\}$ can be decomposed into three pieces: a) The field $\alpha_{\rm y}$ at ${\rm y} \le \tau_0$ belongs to the initial conditions. b) The field $\alpha_{\rm y}$ at $\tau_0 < {\rm y} \le \tau$ is generated by the quantum evolution. c) There is no field at all at larger y: $\alpha_{\rm y} = 0$ for any ${\rm y} > \tau$. Thus:

$$W_\tau[\alpha] \,=\, \delta[\alpha^>]\,\mathcal{W}_\tau[\alpha^<], \tag{4.2}$$

where $\alpha^<_{\rm y}$ ($\alpha^>_{\rm y}$) is the function $\alpha_{\rm y}$ for ${\rm y} < \tau$ (respectively, ${\rm y} > \tau$) :

$$\alpha_{\rm y}(x_\perp) \,\equiv\, \theta(\tau - {\rm y})\alpha^<_{\rm y}(x_\perp) + \theta({\rm y} - \tau)\alpha^>_{\rm y}(x_\perp), \tag{4.3}$$

and the δ–functional $\delta[\alpha^>]$ should be understood with a discretization of the configuration space, as in eq. (2.5):

$$\delta[\alpha^>] \,\equiv\, \prod_{{\rm y}>\tau}\prod_a\prod_{x_\perp} \delta(\alpha^a_{\rm y}(x_\perp)). \tag{4.4}$$

Moreover, it can be shown [10, 52] that $\mathcal{W}_\tau[\alpha^<]$ has the factorized structure:

$$\mathcal{W}_\tau[\alpha^<] = \mathcal{W}_{\tau,\tau_0}[\alpha|V_0]\,\mathcal{W}_{\tau_0}[\alpha]\,, \tag{4.5}$$

where $\mathcal{W}_{\tau_0}[\alpha]$ is the initial weight function at τ_0, and

$$V_0^\dagger(x_\perp) \equiv \mathrm{P}\,\mathrm{e}^{ig\int_{-\infty}^{\tau_0} dy\,\alpha_y(x_\perp)} \tag{4.6}$$

is the Wilson line built with the initial field. In eq. (4.5), it is understood that, in $\mathcal{W}_{\tau_0}$, the field argument α_y has support at $y \le \tau_0$, while in $\mathcal{W}_{\tau,\tau_0}$ it has support at $\tau_0 < y \le \tau$. The "propagator" $\mathcal{W}_{\tau,\tau_0}$ from τ_0 to τ depends also upon the initial field at $y \le \tau_0$, but only in an integrated way, via the Wilson lines V_0 and $V_0^\dagger$. From eq. (4.5) we deduce that $\mathcal{W}_{\tau,\tau_0}[\alpha|V_0] \to 1$ when $\tau \to \tau_0$.
The initial weight function $\mathcal{W}_{\tau_0}$ cannot be obtained within the present formalism, but rather requires some model for the hadron wavefunction at rapidity τ_0. It is convenient to choose a moderate value for $\tau_0 = \ln(1/x_0)$, e.g., $x_0 \simeq 10^{-2}$. This x_0 is small enough for the LLA to apply, but still large enough for the non-linear effects to remain negligible. Then one can use initial conditions which are consistent with the standard, linear, evolution equations (cf. Sect. 5.3 below). Once a convenient value for x_0 has been chosen, one can always redefine $\tau \equiv \ln(x_0/x)$ so that the initial condition is formulated at $\tau_0 = 0$. With this choice, the field α_y at positive rapidities $y > 0$ is generated by the quantum evolution, while the field at negative rapidities $y < 0$ must be specified by the initial condition.
v) The Hamiltonian structure of the RGE. Eq. (3.49) can be rewritten as:

$$\frac{\partial W_\tau[\alpha]}{\partial\tau} = \frac{\delta}{\delta\alpha_\tau^a(x_\perp)}\left\{\frac{1}{2}\eta_{xy}^{ab}\,\frac{\delta W_\tau}{\delta\alpha_\tau^b(y_\perp)} + \left(\frac{1}{2}\,\frac{\delta\eta_{xy}^{ab}}{\delta\alpha_\tau^b(y_\perp)} - \nu_x^a\right)W_\tau\right\}, \tag{4.7}$$

A crucial property, with many consequences, is that the second term within the braces is actually zero. Indeed, the following relation holds between the coefficients of the RGE [47, 37]:

$$\frac{1}{2}\int d^2y_\perp\,\frac{\delta\eta^{ab}(x_\perp,y_\perp)}{\delta\alpha_\tau^b(y_\perp)} = \nu^a(x_\perp)\,. \tag{4.8}$$

It is easy to prove this relation by using eq. (3.53) to act with $\delta/\delta\alpha_\tau^b(y_\perp)$ on $\eta_{ab}(x_\perp,y_\perp)$, eq. (3.51). This yields, e.g.,

$$\begin{aligned}\frac{\delta}{\delta\alpha_\tau^b(y_\perp)}\,(V_x^\dagger V_y)^{ab} &= \frac{\delta V_x^{\dagger\,ac}}{\delta\alpha_\tau^b(y_\perp)}\,V_y^{cb} + V_x^{\dagger\,ac}\,\frac{\delta V_y^{cb}}{\delta\alpha_\tau^b(y_\perp)} \\ &= ig\delta_{xy}\big(T^bV_y^\dagger\big)_{ac}V_y^{cb} - ig\delta^{(2)}(0_\perp)V_x^{\dagger\,ac}\big(V_yT^b\big)_{cb} = 0,\end{aligned} \tag{4.9}$$

where both terms in the second line vanish because of the antisymmetry of the colour group generators in the adjoint representation (e.g., $(T^b)_{ab} = 0$). The only nonvanishing contribution is

$$-\frac{\delta}{\delta\alpha_\tau^b(y_\perp)}(V_x^\dagger V_z)^{ab} = -ig\delta_{xy}\Big(T^b V_x^\dagger V_z\Big)_{ab} = ig\delta_{xy}\mathrm{Tr}\Big(T^a V_x^\dagger V_z\Big), \quad (4.10)$$

which reproduces indeed eq. (3.41) after integration over $y_\perp$, since:

$$\mathcal{K}(x_\perp, y_\perp, z_\perp) \equiv \frac{(x^i - z^i)(y^i - z^i)}{(x_\perp - z_\perp)^2 (y_\perp - z_\perp)^2} \to \frac{1}{(x_\perp - z_\perp)^2} \quad \text{for } y_\perp \to x_\perp. \quad (4.11)$$

With eqs. (4.7) and (4.8), the RGE can be brought into a Hamiltonian form:

$$\frac{\partial W_\tau[\alpha]}{\partial\tau} = -HW_\tau[\alpha], \quad (4.12)$$

with the following Hamiltonian:

$$H \equiv \frac{1}{2}\int d^2x_\perp \int d^2y_\perp \frac{i\delta}{\delta\alpha_\tau^a(x_\perp)}\eta_{xy}^{ab}\frac{i\delta}{\delta\alpha_\tau^b(y_\perp)} = \int\frac{d^2z_\perp}{2\pi}J_a^i(z_\perp)J_a^i(z_\perp),$$
$$J_a^i(z_\perp) \equiv \int\frac{d^2x_\perp}{2\pi}\frac{z^i - x^i}{(z_\perp - x_\perp)^2}(1 - V_z^\dagger V_x)_{ab}\frac{i\delta}{\delta\alpha_\tau^b(x_\perp)}, \quad (4.13)$$

which is Hermitian (since η_{xy}^{ab} is real and symmetric) and positive semi-definite (since the "current" $J_a^i(z_\perp)$ is itself Hermitian).

vi) The infrared and ultraviolet behaviours of the RGE. These are determined by the kernel $\eta^{ab}(x_\perp, y_\perp)$ in the Hamiltonian. In the infrared limit, where $z_\perp$ is much larger than both $x_\perp$ and $y_\perp$ (see eq. (3.51)), $\mathcal{K}(x_\perp, y_\perp, z_\perp) \approx 1/z_\perp^2$, and the ensuing integral $(d^2z_\perp/z_\perp^2)$ has a logarithmic infrared divergence[10]. Thus, there is potentially an IR problem in the RGE. This is not necessarily a real difficulty, since IR problems are expected to be absent only for the *gauge-invariant* observables. We shall see indeed, on specific examples, that the IR divergences cancel when the RGE is used to derive evolution equations for gauge-invariant quantities. This cancellation relies in a crucial way on the property (4.8).

Coming now to the ultraviolet, or short-distance, behaviour, it is easy to see on eq. (3.51) that no UV problem is to be anticipated. For instance, the would-be linear pole of $\mathcal{K}(x_\perp, y_\perp, z_\perp)$ at $|z_\perp - x_\perp| \to 0$ is actually cancelled by the factor $1 - V_z^\dagger V_x$ which vanishes in the same limit.

4.2. QUANTUM EVOLUTION AS BROWNIAN MOTION

To clarify the probabilistic interpretation of the RGE (3.49), we start by recalling the simplest example of a stochastic process, namely the Brownian motion of a

[10] This infrared behaviour is not modified by the $z_\perp$ dependence of the Wilson lines since, e.g., $\langle V_x^\dagger V_z\rangle \to 0$ as $|z_\perp - x_\perp| \to \infty$; cf. Sect. 5.2 below.

small particle in a viscous liquid and in the presence of some external force, like gravitation [59]. The particle is so small that it can feel the collisions with the molecules in the liquid; after each such a collision, the velocity of the particle changes randomly. And the liquid is so viscous that, after each collision, the particle enters immediately a constant velocity regime in which the friction force $\propto v^i$ (with v^i the velocity of the particle) is equilibrated by the random force due to collisions together with the external force $F^i(x)$. In these conditions, the particle executes a random walk whose description is necessary statistical. The relevant quantity is the probability density $P(x,t)$ to find the particle at point x at time t. This is normalized as:

$$\int d^3x\, P(x,t) = 1, \tag{4.14}$$

and obeys an evolution equation of the diffusion type, known as the Fokker-Planck equation [59] :

$$\frac{\partial P(x,t)}{\partial t} = D\frac{\partial^2}{\partial x^i \partial x^i} P(x,t) - \frac{\partial}{\partial x^i}\big(F^i(x)P(x,t)\big). \tag{4.15}$$

Here, D is the diffusion coefficient, which is a measure of the strength of the random force; for simplicity, we assume this to be a constant, i.e., independent of x or t. The solution to eq. (4.15) corresponding to some arbitrary initial condition $P(x,t_0)$ can be written as

$$P(x,t) = \int d^3x_0\, P(x,t|x_0,t_0)\, P(x_0,t_0), \tag{4.16}$$

where $P(x,t|x_0,t_0)$ is the solution to (4.15) with the initial condition:

$$P(x,t_0|x_0,t_0) = \delta^{(3)}(x-x_0). \tag{4.17}$$

Physically, this is the probability density to find the particle at point x at time t knowing that it was at x_0 at time t_0.
If $F^i = 0$, this solution is immediately obtained by going to momentum space: The Fourier transform $\tilde{P}(k,t)$ of $P(x,t_0|x_0,0) \equiv P(x-x_0,t)$ obeys to:

$$\frac{\partial \tilde{P}(k,t)}{\partial t} = -Dk^2\,\tilde{P}(k,t), \qquad \tilde{P}(k,t=0) = 1, \tag{4.18}$$

with the obvious solution $\tilde{P}(k,t) = \mathrm{e}^{-Dk^2 t}$, or, finally,

$$P(x-x_0,t) = \frac{1}{(4\pi Dt)^{3/2}}\,\mathrm{e}^{-\frac{(x-x_0)^2}{4Dt}}. \tag{4.19}$$

This shows a purely diffusive behaviour: the probability to find the particle within a fixed volume centered at some point x goes smoothly to zero as $t \to \infty$ for any

x (runaway solution). The correlations of x reflect this behaviour too; for instance:

$$\overline{r^2}(t) \equiv \langle (x - x_0)^2 \rangle (t) \equiv \int d^3x \, (x - x_0)^2 \, P(x - x_0, t) \; = \; 6Dt, \quad (4.20)$$

showing that, on the average, the particle gets further and further away from the original point x_0, but along a non-differentiable trajectory: $\bar{r}(t) \propto \sqrt{t}$, so the average velocity $\bar{v} = \bar{r}(\Delta t)/\Delta t$ has no well-defined limit when $\Delta t \to 0$.
This situation may change, however, if the motion of the particle is biased by an external force. Assume this force to be derived from a potential: $F^i = -\partial V/\partial x^i$. Then one can check that the time-independent distribution $P_0(x) \sim \exp[-\beta V(x)]$ is a stationary solution to eq. (4.15) provided $\beta D = 1$. Of course, this solution is acceptable as a probability density only if it is normalizable, which puts some constraints on the form of the potential. But assuming this to be the case, then $P_0(x) \sim e^{-\beta V}$ represents an equilibrium distribution which is (asymptotically) reached by the system at large times [59]. Once this is done, all the correlations become independent of time (unlike (4.20)). This solution is a "fixed point" in the functional space of all (acceptable) distributions.
Returning to our RGE (3.49), it should be clear by now that this is a functional Fokker-Planck equation which describes a random walk in the functional space of the colour fields $\alpha^a(x_\perp)$. In this equation, η plays the role of the "diffusion coefficient", while ν is like a "force term", although this identification is somehow ambiguous since η is itself a functional of α, so its derivatives can generate other contributions to the force term, as shown in eq. (4.7). (In the analogous problem of the Brownian motion, this would correspond to a diffusion coefficient which depends on x and has a tensorial structure: $D \to D_{ij}(x)$. This situation occurs, e.g., in the description of a random walk on a curved manifold [59].) In fact, it is more correct to identify the combination $\frac{1}{2}(\delta\eta/\delta\alpha_\tau) - \nu$ as the effective "force term", since the remaining second-order differential operator in eq. (4.7) — which describes diffusion — is then Hermitian and positive semi-definite.
A fixed point of the quantum evolution would be a solution $W[\alpha]$ to eq. (3.49) which is normalizable and independent of "time" τ. If such a solution existed, then the high energy limit of QCD scattering would be trivial (at least, within the present approximations): At sufficiently high energies, all the cross sections would become independent of energy (recall that $\tau \sim \ln s$). The relation (4.8) between the coefficients in the RGE guarantees, however, that such a "fixed point" does not exist: The effective force in eq. (4.7) vanishes, and the corresponding evolution Hamiltonian (4.13) is just a kinetic operator, which describes pure diffusion. We thus expect gluon correlations to keep growing with $\tau \sim \ln s$ even at asymptotically large energies. In Sect. 5, we shall find approximate solutions to eq. (4.12) which show indeed such a behaviour [10].

4.3. THE BALITSKY-KOVCHEGOV EQUATION

If $\langle O[\alpha] \rangle_\tau$ is any observable which can be computed as an average over α:

$$\langle O[\alpha] \rangle_\tau = \int \mathcal{D}[\alpha] \, O[\alpha] \, W_\tau[\alpha], \tag{4.21}$$

(cf. eq. (2.41)), then its evolution with τ is governed by the following equation:

$$\begin{aligned}\frac{\partial}{\partial\tau}\langle O[\alpha] \rangle_\tau &= \int \mathcal{D}\alpha \, O[\alpha] \, \frac{\partial W_\tau[\alpha]}{\partial\tau} \\ &= \left\langle \frac{1}{2} \frac{\delta}{\delta\alpha_\tau^a(x_\perp)} \eta_{xy}^{ab} \frac{\delta}{\delta\alpha_\tau^b(y_\perp)} O[\alpha] \right\rangle_\tau ,\end{aligned} \tag{4.22}$$

where, in writing the second line, we have used eq. (4.12) for $\partial W_\tau/\partial\tau$ and then integrated twice by parts within the functional integral over α.

Let us apply this to the 2-point function (2.26) of the Wilson lines in the fundamental representation. We recall that, physically, this is the S-matrix element for dipole-hadron scattering (cf. Sect. 2.2.2). A straightforward calculation yields (see [37] for details):

$$\begin{aligned}\frac{\partial}{\partial\tau}\langle \mathrm{tr}(V_x^\dagger V_y)\rangle_\tau &= -\frac{\alpha_s}{2\pi^2}\int d^2z_\perp \frac{(x_\perp - y_\perp)^2}{(x_\perp - z_\perp)^2(y_\perp - z_\perp)^2} \\ &\times \left\langle N_c \mathrm{tr}(V_x^\dagger V_y) - \mathrm{tr}(V_x^\dagger V_z)\mathrm{tr}(V_z^\dagger V_y)\right\rangle_\tau .\end{aligned} \tag{4.23}$$

This is the equation originally obtained by Balitsky [44], within a quite different formalism : by an analysis of the quantum evolution of the dipole itself.

Not that the above equation is not closed: It relates the 2-point function to the 4-point function $\langle \mathrm{tr}(V_x^\dagger V_z)\mathrm{tr}(V_z^\dagger V_y)\rangle$. One can similarly derive an evolution equation for the latter [44], but this will in turn couple the 4-point function to a 6-point function, and so on. That is, eq. (4.23) is just the first in an infinite hierarchy of coupled equations [44].

A closed equation can still be obtained in the large N_c limit, in which the 4-point function in eq. (4.23) factorizes:

$$\left\langle \mathrm{tr}(V_x^\dagger V_z)\,\mathrm{tr}(V_z^\dagger V_y)\right\rangle_\tau \longrightarrow \left\langle \mathrm{tr}(V_x^\dagger V_z)\right\rangle_\tau \left\langle \mathrm{tr}(V_z^\dagger V_y)\right\rangle_\tau \quad \text{for } N_c \to \infty.$$

Then eq. (4.23) reduces to a closed equation for $S_\tau(x_\perp, y_\perp) = \langle \mathrm{tr}(V_x^\dagger V_y)\rangle_\tau / N_c$:

$$\begin{aligned}\frac{\partial}{\partial\tau} S_\tau(x_\perp, y_\perp) &= -\frac{\alpha_s N_c}{2\pi^2}\int d^2z_\perp \frac{(x_\perp - y_\perp)^2}{(x_\perp - z_\perp)^2(y_\perp - z_\perp)^2} \\ &\times \{S_\tau(x_\perp, y_\perp) - S_\tau(x_\perp, z_\perp)S_\tau(z_\perp, y_\perp)\} .\end{aligned} \tag{4.24}$$

The same equation has been independently obtained by Kovchegov [45] within Mueller's dipole model [46, 14]. (See also Ref. [50] for another derivation.)

An important observation refers to the transverse kernel in eqs. (4.23) or (4.24): This is not the same as the original kernel $\mathcal{K}(x_\perp, y_\perp, z_\perp)$, eq. (4.11), of the RGE. Rather, this has been generated as

$$\mathcal{K}(x_\perp, x_\perp, z_\perp) + \mathcal{K}(y_\perp, y_\perp, z_\perp) - 2\mathcal{K}(x_\perp, y_\perp, z_\perp) = \frac{(x_\perp - y_\perp)^2}{(x_\perp - z_\perp)^2 (y_\perp - z_\perp)^2},$$

and has the remarkable feature to show a better infrared behaviour than eq. (4.11): When $z_\perp \gg x_\perp$, $y_\perp$, the kernel above decreases like $(x_\perp - y_\perp)^2/z_\perp^4$, so its integral over $z_\perp$ is actually finite.
For a small dipole, $r_\perp \equiv |x_\perp - y_\perp| \ll 1/Q_s$, $S_\tau(x_\perp, y_\perp)$ is close to one (cf. eq. (2.75)), and one can linearize eq. (4.24) with respect to $\mathcal{N}_\tau(r_\perp) \equiv 1 - S_\tau(r_\perp)$ (the dipole scattering amplitudes). The ensuing linear equation for $\mathcal{N}_{xy} \equiv \mathcal{N}_\tau(x_\perp, y_\perp)$, namely:

$$\frac{\partial}{\partial \tau} \mathcal{N}_{xy} = \frac{\alpha_s N_c}{2\pi^2} \int d^2 z_\perp \frac{(x_\perp - y_\perp)^2}{(x_\perp - z_\perp)^2 (y_\perp - z_\perp)^2} \left\{ \mathcal{N}_{xz} + \mathcal{N}_{zy} - \mathcal{N}_{xy} \right\}, \quad (4.25)$$

can be recognized as the (coordinate space form of the) BFKL equation [3, 46]. There is currently a large interest in the solutions to eq. (4.24), and significant progress has been achieved by combining analytic and numerical methods [19, 45, 50, 54, 55, 58]. The conclusions reached in this way are equivalent to those obtained from direct investigations of the RGE (3.49) [10, 57] that we shall review in what follows.

5. Approximate solutions to the Renormalization Group Equation

We shall now construct approximate solutions to the RGE (4.12) and study their physical implications [10, 57].

5.1. THE MEAN FIELD APPROXIMATION

As compared to the standard diffusion equation (4.15), the main complication with the RGE (4.12) comes from the fact that its kernel η is itself dependent on α. In this respect, eq. (4.12) is similar to the following diffusion equation:

$$\frac{\partial P(x,t)}{\partial t} = \frac{\partial}{\partial x^i} D_{ij}(x) \frac{\partial}{\partial x^j} P(x,t), \quad (5.1)$$

in which the diffusivities $D_{ij}(x)$ are allowed to depend upon the position x of the particle. This dependence makes eq. (5.1) difficult to solve in general (i.e., for some arbitrary tensor field $D_{ij}(x)$). But since x is a random variable, with probability density $P(x,t)$, a reasonable approximation is obtained by replacing $D_{ij}(x)$ in eq. (5.1) by its expectation value:

$$D_{ij}(x) \longrightarrow \langle D_{ij}(x) \rangle(t) \equiv \int d^3x \, \bar{P}(x,t) \, D_{ij}(x) \equiv \delta_{ij} \bar{D}(t), \quad (5.2)$$

which is independent of x, but a function of time. We denote with a bar quantities evaluated in this "mean field approximation" (MFA). In particular, $\bar{P}(x,t)$ is itself related to $\bar{D}(t)$, as the solution to the following approximate equation:

$$\frac{\partial \bar{P}(x,t)}{\partial t} = \bar{D}(t)\frac{\partial^2}{\partial x^i \partial x^i}\bar{P}(x,t). \tag{5.3}$$

Thus, eq. (5.2) is actually a *self-consistent* equation for $\bar{D}(t)$. Being homogeneous in x, eq. (5.3) is easily solved by Fourier transform, as in eqs. (4.18)–(4.19). For the initial condition $\bar{P}(x,t=0) = \delta^{(3)}(x)$, one thus obtains:

$$\bar{P}(x,t) = \frac{1}{(4\pi\xi(t))^{3/2}}\,\mathrm{e}^{-\frac{x^2}{4\xi(t)}}, \qquad \xi(t) \equiv \int_0^t dt'\,\bar{D}(t'). \tag{5.4}$$

By inserting this solution in eq. (5.2), one can compute the average there (as a functional of $\bar{D}(t)$), and then solve the self-consistent equation for $\bar{D}(t)$, thus completely specifying the approximate solution (5.4).
This is the strategy that we shall use to obtain approximate solutions to the functional diffusion equation (4.12). The corresponding MFA reads:

$$\frac{\partial \bar{W}_\tau[\alpha]}{\partial \tau} = \frac{1}{2}\int_{x_\perp, y_\perp} \gamma_\tau(x_\perp, y_\perp)\frac{\delta^2 \bar{W}_\tau[\alpha]}{\delta\alpha_\tau^a(x_\perp)\delta\alpha_\tau^a(y_\perp)}, \tag{5.5}$$

with

$$\delta^{ab}\gamma_\tau(x_\perp, y_\perp) \equiv \langle \eta^{ab}(x_\perp, y_\perp)\rangle_\tau \equiv \int D[\alpha]\,\eta^{ab}(x_\perp, y_\perp)\,\bar{W}_\tau[\alpha], \tag{5.6}$$

where the trivial colour structure in the l.h.s. follows from gauge symmetry. By the same argument, $\langle \nu^a(x_\perp)\rangle_\tau = 0$, which is indeed consistent with the MFA (5.6) for η and the condition (4.8).
Eq. (5.5) is homogeneous in the functional variable $\alpha_\tau^a(x_\perp)$ (since its kernel γ_τ is independent of α), so it can be solved by functional Fourier analysis. This is the straightforward extension of the corresponding analysis for ordinary functions, and can be more rigourously introduced by using a discretized version of the 3-dimensional configuration space $(\mathrm{y}, x_\perp)$, as in eqs. (2.5) or (4.4). We write, e.g.,

$$\begin{aligned}\delta[\alpha] &= \int D[\pi]\,\mathrm{e}^{-i\int d\mathrm{y}\int d^2x_\perp\,\pi_\mathrm{y}^a(x_\perp)\alpha_\mathrm{y}^a(x_\perp)}, \\ \bar{W}_\tau[\alpha] &= \int D[\pi]\,\mathrm{e}^{-i\int d\mathrm{y}\int d^2x_\perp\,\pi_\mathrm{y}^a(x_\perp)\alpha_\mathrm{y}^a(x_\perp)}\,\tilde{W}_\tau[\pi].\end{aligned} \tag{5.7}$$

By inserting this representation for $\bar{W}_\tau[\alpha]$ in eq. (5.5), and using

$$\frac{\delta}{\delta\alpha_\tau^a(x_\perp)}\int d\mathrm{y}\int d^2z_\perp\,\pi_\mathrm{y}^c(z_\perp)\alpha_\mathrm{y}^c(z_\perp) = \pi_\tau^a(x_\perp), \tag{5.8}$$

one obtains the following equation for $\tilde{W}_\tau[\pi]$ (compare to eq. (4.18)):

$$\frac{\partial \tilde{W}_\tau[\pi]}{\partial \tau} = -\frac{1}{2}\int_{x_\perp, y_\perp} \gamma_\tau(x_\perp, y_\perp)\, \pi^a_\tau(x_\perp)\pi^a_\tau(y_\perp)\tilde{W}_\tau[\pi], \tag{5.9}$$

with the immediate solution (transverse coordinates are omitted, for simplicity):

$$\tilde{W}_\tau[\pi] = \mathrm{e}^{-\frac{1}{2}\int_0^\tau dy \gamma_y \pi^a_y \pi^a_y}\, \tilde{W}_0[\pi]. \tag{5.10}$$

The argument π_y of the initial weight function $\tilde{W}_0[\pi]$ has support only at $y < 0$. After insertion in eq. (5.7), this yields:

$$\bar{W}_\tau[\alpha] = \int D[\pi]\, \mathrm{e}^{-i\int_{-\infty}^{\infty} dy\, \pi^a_y \alpha^a_y}\, \mathrm{e}^{-\frac{1}{2}\int_0^\tau dy\, \gamma_y \pi^a_y \pi^a_y} = \delta[\alpha^>]\, \bar{\mathcal{W}}_\tau[\alpha^<], \tag{5.11}$$

with $\delta[\alpha^>]$ defined in eq. (4.4) (this has been generated by the functional integral over π_y with $y > \tau$) and

$$\bar{\mathcal{W}}_\tau[\alpha^<] = \mathcal{N}_\tau \exp\left\{-\frac{1}{2}\int_0^\tau dy \int_{x_\perp, y_\perp} \frac{\alpha^a_y(x_\perp)\alpha^a_y(y_\perp)}{\gamma_y(x_\perp, y_\perp)}\right\} \mathcal{W}_0[\alpha]. \tag{5.12}$$

In this equation, $\mathcal{W}_0[\alpha]$ is the original weight function at $\tau = 0$, and is a functional of the field α_y with $y \le 0$. ($\mathcal{N}_\tau$ is an irrelevant normalization factor.)
The solution (5.11)–(5.12) has the general structure anticipated in eqs. (4.2)–(4.5). If the initial conditions are described by the MV model, or any other MFA, then $\mathcal{W}_0[\alpha]$ is a Gaussian too (see, e.g., eq. (2.52)), and eq. (5.12) can be rewritten as:

$$\bar{\mathcal{W}}_\tau[\alpha^<] = \mathcal{N}_\tau \exp\left\{-\frac{1}{2}\int_{-\infty}^\tau dy \int_{x_\perp, y_\perp} \frac{\alpha^a_y(x_\perp)\alpha^a_y(y_\perp)}{\gamma_y(x_\perp, y_\perp)}\right\}. \tag{5.13}$$

For $y \le 0$, the width γ_y is specified by the initial conditions, while at positive rapidities $0 < y \le \tau$, it is determined by the quantum evolution, as we shall see. The fact that the weight function (5.13) is a Gaussian does not necessarily mean that the present approximation describes a system of independent colour sources (like the MV model). It just means that, in the MFA, all the correlations are encoded in the width of the Gaussian, or, equivalently, in the 2-point function

$$\langle \alpha^a_y(x_\perp)\, \alpha^b_{y'}(y_\perp)\rangle_\tau = \delta^{ab}\delta(y - y')\theta(\tau - y)\,\gamma_y(x_\perp, y_\perp). \tag{5.14}$$

But this 2-point function contains also information on the higher-point correlations, although just in an averaged way, because it is determined by the following, *non-linear*, self-consistency equation:

$$\gamma_\tau(x_\perp, y_\perp) = \frac{1}{\pi}\int \frac{d^2 z_\perp}{(2\pi)^2}\, \mathcal{K}(x_\perp, y_\perp, z_\perp) \tag{5.15}$$
$$\times \Big(1 + S_\tau(x_\perp, y_\perp) - S_\tau(x_\perp, z_\perp) - S_\tau(z_\perp, y_\perp)\Big),$$

which follows from eqs. (5.6) and (3.51) together with the fact that, for a Gaussian weight function[11],

$$\langle (V_x^\dagger V_y)^{ab} \rangle_\tau = \frac{\delta^{ab}}{N_c^2-1} \Big\langle \mathrm{Tr}\,(V^\dagger(x_\perp)V(y_\perp)) \Big\rangle_\tau \equiv \delta^{ab} S_\tau(x_\perp, y_\perp), \quad (5.16)$$

with S_τ a (non-linear) functional of γ_y, to be constructed shortly.
The correlation function (5.14) is local in y : colour sources located at different space-time rapidities appear to be statistically independent. This is, of course, just an artifact of the MFA. The complete RGE generates correlations in rapidity, via the Wilson lines in its coefficients. But the only trace of these correlations in the MFA is the fact that the self-consistency equation (5.15) is non-local in y.
To perform the average in eq. (5.16), we first derive an evolution equation for S_τ, by using the corresponding equation (5.5) for $\bar{W}_\tau$:

$$\begin{aligned} \frac{\partial}{\partial\tau} S_\tau(x_\perp, y_\perp) &= \int D[\alpha]\, \frac{\partial \bar{W}_\tau[\alpha]}{\partial\tau}\, V_x^\dagger V_y \qquad (5.17) \\ &= \int D[\alpha]\, \bar{W}_\tau[\alpha] \int_{u_\perp, v_\perp} \frac{1}{2}\, \gamma_\tau(u_\perp, v_\perp)\, \frac{\delta^2}{\delta\alpha_\tau^a(u_\perp)\delta\alpha_\tau^a(v_\perp)}\, V_x^\dagger V_y, \\ &= -\frac{g^2 N_c}{2}\Big[\gamma_\tau(x_\perp, x_\perp) + \gamma_\tau(y_\perp, y_\perp) - 2\gamma_\tau(x_\perp, y_\perp)\Big] S_\tau(x_\perp, y_\perp). \end{aligned}$$

(The functional derivatives of the Wilson lines have been evaluated as

$$\frac{\delta^2}{\delta\alpha_\tau^a(u_\perp)\delta\alpha_\tau^a(v_\perp)} \mathrm{Tr}(V_x^\dagger V_y) = -g^2 N_c\, \mathrm{Tr}(V_x^\dagger V_y)(\delta_{xv} - \delta_{yv})(\delta_{xu} - \delta_{yu}), \quad (5.18)$$

where we have used eq. (3.53) and $T^a T^a = N_c$.) Eq. (5.17) can be trivially integrated. To simplify the calculations, we assume homogeneity in the transverse plane within the hadron disk of radius R; then $\gamma_y(x_\perp, y_\perp) = \gamma_y(x_\perp - y_\perp)$ and

$$S_\tau(r_\perp) = \mathrm{e}^{-g^2 N_c \int_0^\tau dy[\gamma_y(0_\perp) - \gamma_y(r_\perp)]} S_0(r_\perp) = \mathrm{e}^{-g^2 N_c[\xi_\tau(0_\perp) - \xi_\tau(r_\perp)]}, \quad (5.19)$$

where $r_\perp = x_\perp - y_\perp$,

$$\xi_\tau(r_\perp) \equiv \xi_0(r_\perp) + \int_0^\tau dy\, \gamma_y(r_\perp), \quad (5.20)$$

and in writing the second equality in (5.19) we have assumed that the initial condition $S_0(r_\perp)$ can be written in the form $S_0(r_\perp) = \mathrm{e}^{-g^2 N_c[\xi_0(0_\perp) - \xi_0(r_\perp)]}$. This is indeed the case for the weight function in eq. (5.13) — in particular, for the MV model, cf. eq. (2.60) —, which yields :

$$\xi_0(r_\perp) = \int_{-\infty}^{0} dy\, \gamma_y(r_\perp). \quad (5.21)$$

[11] Note that, as compared to eq. (2.26), S_τ is now written in the adjoint representation.

By combining eqs. (5.15), (5.19) and (5.20), one can finally rewrite the self-consistency equation as an evolution equation for $\xi_\tau(r_\perp)$:

$$\begin{aligned} \frac{\partial \xi_\tau(x_\perp - y_\perp)}{\partial \tau} &= \frac{1}{\pi}\int \frac{d^2 z_\perp}{(2\pi)^2}\,\mathcal{K}(x_\perp, y_\perp, z_\perp) \\ &\times \Big(1 + S_\tau(x_\perp - y_\perp) - S_\tau(x_\perp - z_\perp) - S_\tau(z_\perp - y_\perp)\Big), \end{aligned} \tag{5.22}$$

with $S_\tau(r_\perp) = \mathrm{e}^{-g^2 N_c[\xi_\tau(0_\perp)-\xi_\tau(r_\perp)]}$. As anticipated, this equation is highly non linear in ξ_τ. It is furthermore non-local in the transverse coordinates, but local in the "evolution time" τ. (The original non-locality of eq. (5.15) in y has been now absorbed in the relation (5.20) between ξ_τ and γ_y.)
In the next sections, we shall develop further approximations, which rely on the kinematics and allow us to make progress with eq. (5.22).

5.2. SATURATION SCALE AND KINEMATICAL APPROXIMATIONS

Both the non-local and the non-linear structure of the evolution equation (5.22) depend crucially upon the behaviour of $S_\tau(r_\perp)$ with the transverse separation $r_\perp$. From its definition (5.16), it is clear that $S_\tau(r_\perp) \to 1$ as $r_\perp \to 0$ for any τ. Moreover, since a large dipole is strongly absorbed by a hadronic target, we expect that $S_\tau(r_\perp) \ll 1$ for sufficiently large $r_\perp$, where what we mean by "sufficiently large" will generally depend on τ. For instance, we have seen in Sect. 2.5, within the MV model, that $S_\tau(r_\perp) \ll 1$ for $r_\perp \gg 1/Q_s$, with Q_s the *saturation scale* for gluons in the hadron wavefunction (cf. eq. (2.76)). In that classical model, Q_s was independent of energy, but in general we expect it to increase with τ, because of the quantum evolution (cf. the discussion in Sect. 1.4 and Sect. 5.3 below). At a formal level, this intimate connection between the strong absorbtion limit for a colour dipole and gluon saturation is based on the fact that, in both problems, the non-linear effects are encoded in Wilson lines.
So, let us introduce the correlation length $1/Q_s(\tau)$ of $S_\tau(r_\perp)$:

$$S_\tau(r_\perp) \approx \begin{cases} 1, & \text{for} \quad r_\perp \ll 1/Q_s(\tau) \\ 0, & \text{for} \quad r_\perp \gg 1/Q_s(\tau) \end{cases} \tag{5.23}$$

which, as its notation suggests, will play also the role of the saturation scale. This behaviour of $S_\tau(r_\perp)$, with an unique separation scale between a short-range regime and a long-range one, is confirmed by numerical studies of the Kovchegov equation, which also show a rapid increase of Q_s with τ [19, 50, 54, 58].
Eq. (5.23), together with the expression (5.19) for $S_\tau(r_\perp)$ in the MFA, imply the following condition:

$$g^2 N_c[\xi_\tau(0_\perp) - \xi_\tau(r_\perp)] \sim 1 \quad \text{for} \quad r_\perp \sim 1/Q_s(\tau), \tag{5.24}$$

that we shall use later to obtain an estimate for $Q_s(\tau)$.

An external probe with transverse momentum $k_\perp$ will measure correlations in the hadron over a typical transverse size $r_\perp \sim 1/k_\perp$. Thus, short distances $r_\perp \ll 1/Q_s(\tau)$ correspond to high $k_\perp$ momenta, $k_\perp \gg Q_s(\tau)$, while large separations $r_\perp \gg 1/Q_s(\tau)$ correspond to low transverse momenta $k_\perp \ll Q_s(\tau)$. In what follows, we shall not aim at a precise description of the physics around the saturation scale, but rather focus on the two limiting regimes — high–$k_\perp$ and low–$k_\perp$ — and perform appropriate simplifications on the evolution equation (5.22).

a) High–$k_\perp$. It is convenient to rewrite eq. (5.19) in momentum space as:

$$S_\tau(r_\perp) = \exp\left\{-g^2 N_c \int \frac{d^2p_\perp}{(2\pi)^2}\, \xi_\tau(p_\perp)\left[1 - e^{ip_\perp \cdot r_\perp}\right]\right\}. \tag{5.25}$$

For $r_\perp \ll 1/Q_s(\tau)$, the integral over $p_\perp$ is dominated by momenta within the range $Q_s(\tau) \ll p_\perp \ll 1/r_\perp$. This holds to leading *transverse*–log accuracy: In this range, $\xi_\tau(p_\perp) \sim 1/p_\perp^4$ (up to logs), so the integral over $p_\perp$ produces the large logarithm $\ln(1/r_\perp^2 Q_s^2(\tau))$. To the same logarithmic accuracy, one can expand the exponential in (5.25) in powers of $p_\perp \cdot r_\perp$, like in eq. (2.64), and thus obtain:

$$S_\tau(r_\perp) \simeq \exp\left\{-\frac{g^2 N_c}{4} r_\perp^2 \int^{1/r_\perp^2} \frac{d^2p_\perp}{(2\pi)^2}\, p_\perp^2\, \xi_\tau(p_\perp)\right\}. \tag{5.26}$$

When extrapolated to $r_\perp \sim 1/Q_s(\tau)$, this expression gives us the following estimate for the correlation length $1/Q_s(\tau)$ (cf. eq. (5.24)) :

$$Q_s^2(\tau) \simeq \frac{\alpha_s N_c}{4} \int^{Q_s^2(\tau)} dp_\perp^2\, p_\perp^2\, \xi_\tau(p_\perp). \tag{5.27}$$

For $r_\perp \ll 1/Q_s(\tau)$, where eq. (5.26) is strictly valid, the exponential there can be expanded to lowest order:

$$S_\tau(r_\perp) \simeq 1 - \frac{g^2 N_c}{4} r_\perp^2 \left(-\nabla_\perp^2 \xi_\tau(r_\perp)\right)_{r_\perp = 0}. \tag{5.28}$$

(The ultraviolet cutoff $1/r_\perp$ is implicit in the momentum representation of $\xi_\tau(0)$.) By inserting this into (5.22), we obtain a *linear* evolution equation for $\xi_\tau(r_\perp)$:

$$\begin{aligned} \frac{\partial \xi_\tau(x_\perp - y_\perp)}{\partial \tau} &= \alpha_s N_c \int \frac{d^2z_\perp}{(2\pi)^2} \frac{(x^i - z^i)(y^i - z^i)}{(x_\perp - z_\perp)^2 (y_\perp - z_\perp)^2} \\ &\times \left((x_\perp - y_\perp)^2 - (x_\perp - z_\perp)^2 - (z_\perp - y_\perp)^2\right) \nabla_\perp^2 \xi_\tau(0)\,. \end{aligned} \tag{5.29}$$

Thus, the short-distance approximation is automatically a linear, or weak-field, approximation. This is to be expected since, at high $k_\perp$, the gluon density is low.

To perform the integral over $z_\perp$ in eq. (5.29), it is useful to recall eq. (3.48) and then notice that, within the integrand of (5.29), one can effectively replace:

$$\frac{1}{(2\pi)^2}\frac{(x^i-z^i)(y^i-z^i)}{(x_\perp-z_\perp)^2(y_\perp-z_\perp)^2} \longrightarrow \frac{1}{2}\nabla_z^2\left(\langle x_\perp|\frac{1}{-\nabla_\perp^2}|z_\perp\rangle\,\langle y_\perp|\frac{1}{-\nabla_\perp^2}|z_\perp\rangle\right).$$

(The additional terms in the r.h.s. are δ-functions at $z_\perp = x_\perp$ or $z_\perp = y_\perp$, which vanish when multiplied by the remaining factor in (5.29).) By using this, together with a couple of integrations by parts w.r.t. $z_\perp$, and a Fourier transform to momentum space, we finally obtain the following evolution equation:

$$\frac{\partial\mu_\tau(k_\perp)}{\partial\tau} = \frac{\alpha_s N_c}{\pi}\int^{k_\perp^2}\frac{dp_\perp^2}{p_\perp^2}\,\mu_\tau(p_\perp), \tag{5.30}$$

for the quantity:

$$\mu_\tau(k_\perp) \equiv k_\perp^4\,\xi_\tau(k_\perp), \tag{5.31}$$

which, physically, is the 2-point function of the colour charge density in the transverse plane $\rho^a(x_\perp)$:

$$\langle\rho^a(x_\perp)\rho^b(y_\perp)\rangle_\tau = \delta^{ab}\mu_\tau(x_\perp - y_\perp), \qquad \rho^a(x_\perp) = \int dy\,\rho^a_y(x_\perp). \tag{5.32}$$

The initial condition for eq. (5.30) can be taken from the MV model: $\mu_\tau(k_\perp) = \mu_A$ for $\tau = 0$, cf. eq. (2.51). This initial condition is independent of $k_\perp$ and, together with eq. (5.30), it implies that $\mu_\tau(k_\perp)$ remains a rather slowly varying function of $k_\perp$ in this high momentum regime. This will be manifest on the solutions to eq. (5.30) that we shall write in the next subsection.

b) Low–$k_\perp$. For large distances $r_\perp \gg 1/Q_s(\tau)$, $S_\tau(r_\perp) \ll 1$, and the 2-point functions of the Wilson lines can be simply neglected in the self-consistency equations (5.15) or (5.22) [10, 57]. Eq. (5.22) then simplifies to (see also eq. (3.48))

$$\begin{aligned}\frac{\partial\xi_\tau(x_\perp - y_\perp)}{\partial\tau} &\approx \frac{1}{\pi}\int d^2z_\perp\,\partial_z^i\langle x_\perp|\frac{1}{-\nabla_\perp^2}|z_\perp\rangle\,\partial_z^i\langle y_\perp|\frac{1}{-\nabla_\perp^2}|z_\perp\rangle\\ &= \frac{1}{\pi}\langle x_\perp|\frac{1}{-\nabla_\perp^2}|y_\perp\rangle,\end{aligned} \tag{5.33}$$

or in momentum space (cf. eq. (5.20)):

$$\gamma_\tau(k_\perp) \equiv \frac{\partial\xi_\tau(k_\perp)}{\partial\tau} = \frac{1}{\pi}\frac{1}{k_\perp^2}. \tag{5.34}$$

This is not an equation anylonger, but rather an explicit, and rather simple, expression for the propagator $\gamma_\tau(k_\perp)$ of the fields α: this is just the 2-dimensional Coulomb propagator.

Remarkably, the QCD coupling constant g has dropped out from eqs. (5.33) and (5.34). (This should be contrasted with the corresponding equation at high $k_\perp$, eq. (5.29), whose r.h.s. is explicitly proportional to $\alpha_s = g^2/4\pi$.) The same property holds then for the corresponding mean-field Hamiltonian (cf. eq. (5.5)) :

$$\bar{H}_{\text{low}-k_\perp} \approx -\frac{1}{2\pi}\int \frac{d^2k_\perp}{(2\pi)^2}\,\frac{1}{k_\perp^2}\,\frac{\delta^2}{\delta\alpha_\tau^a(k_\perp)\delta\alpha_\tau^a(-k_\perp)}, \tag{5.35}$$

which is quite remarkable since at low $k_\perp$ we are effectively in a strong coupling regime (in the sense that the COV-gauge fields are strong: $\alpha^a \sim 1/g$; see Sect. 5.4). If g nevertheless drops out in this limit, it is because of the special way it enters the evolution Hamiltonian: via the exponent of the Wilson lines. That is, the relevant degrees of freedom in the non-linear regime are not the (strong) colour fields by themselves, but rather the Wilson lines built with these fields. The Wilson lines are rapidly oscillating over distances $r_\perp \gg 1/Q_s(\tau)$ (since their exponent is of order one, and the typical scale for variations is $1/Q_s(\tau)$), and thus average to zero ("random phase approximation").

For what follows, it is useful to summarize the previous kinematical approximations into the following, factorized, form for the weight function (5.13), which is most conveniently written as a weight function for[12] $\rho_y^a(k_\perp) = k_\perp^2 \alpha_y^a(k_\perp)$:

$$\bar{\mathcal{W}}_\tau[\rho] \approx \mathcal{W}_\tau^{\text{high}}[\rho]\mathcal{W}_\tau^{\text{low}}[\rho], \tag{5.36}$$

$$\mathcal{W}_\tau^{\text{low}}[\rho] \equiv \mathcal{N}_\tau \exp\left\{-\frac{\pi}{2}\int_{-\infty}^{\tau} dy \int^{Q_s(y)} \frac{d^2k_\perp}{(2\pi)^2}\,\frac{\rho_y^a(k_\perp)\rho_y^a(-k_\perp)}{k_\perp^2}\right\}, \tag{5.37}$$

$$\mathcal{W}_\tau^{\text{high}}[\rho] \equiv \mathcal{N}_\tau \exp\left\{-\frac{1}{2}\int_{-\infty}^{\tau} dy \int_{Q_s(y)} \frac{d^2k_\perp}{(2\pi)^2}\,\frac{\rho_y^a(k_\perp)\rho_y^a(-k_\perp)}{\lambda_y(k_\perp)}\right\}. \tag{5.38}$$

In writing this equation, we have separated, for each rapidity y, the low-momentum ($k_\perp < Q_s(y)$) modes of ρ from the high-momentum ($k_\perp > Q_s(y)$) ones, we have used the approximation (5.34) for the width of the Gaussian at low momenta, and we have written $\lambda_y(k_\perp) \equiv \partial\mu_y(k_\perp)/\partial y$, with $\mu_\tau(k_\perp)$ determined by eq. (5.30), at high momenta. Note that the modes with $k_\perp \sim Q_s(y)$ are not correctly described by the present approximations, but we shall assume that they give only small contributions to the quantities to be computed below.

[12] This is the colour charge density in the COV-gauge, but we omit the tilde symbol on ρ, to simplify writing.

5.3. HIGH–$K_\perp$: RECOVERING THE PERTURBATIVE EVOLUTION

We now consider the implications of eqs. (5.30) and (5.38) for the physics at high transverse momenta $k_\perp \gg Q_s(\tau)$. To this aim, we compute the gluon density (2.23) in this low density regime, where one can use the linear approximation $\mathcal{F}^{+j}(\vec{k}) \simeq (ik^j/k_\perp^2)\rho(\vec{k})$. The calculation is similar to that already performed in eqs. (2.53)–(2.55). Specifically, by using (cf. eq. (5.38)) :

$$\langle \rho^a_{\rm y}(x_\perp)\, \rho^b_{\rm y'}(y_\perp)\rangle_\tau = \delta^{ab}\delta({\rm y}-{\rm y'})\lambda_{\rm y}(x_\perp - y_\perp), \quad \lambda_{\rm y}(r_\perp) = \frac{\partial \mu_{\rm y}(r_\perp)}{\partial {\rm y}}, \tag{5.39}$$

one eventually obtains:

$$\mathcal{N}_\tau(k_\perp) \simeq \frac{N_c^2-1}{4\pi^3}\frac{\mu_\tau(k_\perp)}{k_\perp^2}, \tag{5.40}$$

$${\rm x}G({\rm x},Q^2) \simeq \frac{N_c^2-1}{4\pi} R^2 \int_{Q_s^2(\tau)}^{Q^2} \frac{dk_\perp^2}{k_\perp^2}\,\mu_\tau(k_\perp)\,. \tag{5.41}$$

Note the lower limit $Q_s(\tau)$ in the integral giving ${\rm x}G({\rm x},Q^2)$: for $Q^2 \gg Q_s^2(\tau)$, and to leading transverse–log accuracy, it is sufficient to consider the contribution of the high–$k_\perp$ modes of ρ to the gluon distribution. We shall check later that the corresponding contribution of the modes with $k_\perp \ll Q_s(\tau)$ is infrared finite, although subleading as compared to eq. (5.41) [10, 57]. This cures the infrared problem that we have faced in the classical calculation of Sects. 2.4–2.5.

Physically, $\mu_\tau(k_\perp)$ plays the same role as μ_A in the MV model: It measures the density of the colour sources in the transverse plane, and, in the linear regime at high–$k_\perp$, it is also proportional to the unintegrated gluon distribution: $\mu_\tau(Q^2) \propto \partial\, {\rm x}G({\rm x},Q^2)/\partial \ln Q^2$. But unlike μ_A, which is constant for a given atomic number A, $\mu_\tau(k_\perp)$ has non-trivial dependences upon both τ and $k_\perp$, as determined by its quantum evolution according to eq. (5.30). The dependence on τ describes the increase in the density of the colour sources via soft gluon radiation. The dependence on $k_\perp$ corresponds in coordinate space to correlations in the transverse plane, which occur via the exchange of quantum gluons (see Fig. 16.a).

Eq. (5.30) can be recognized as the standard, linear evolution equation in the double-logarithmic approximation (DLA) [5], i.e., in the limit in which BFKL and DGLAP coincide with each other. The emergence of DLA is natural, given the approximations that we performed in deriving eq. (5.30): we have kept only terms of leading-log accuracy in both $\tau = \ln(1/{\rm x})$ (in the construction of the effective theory), and $\ln(k_\perp^2/Q_s^2(\tau))$ (in the short-range expansion at high $k_\perp$). Eqs. (5.30) and (5.41) imply the more standard form of the DLA equation [5] :

$$\frac{\partial^2}{\partial\tau\,\partial\ln Q^2}\,{\rm x}G({\rm x},Q^2) = \frac{\alpha_s N_c}{\pi}\,{\rm x}G({\rm x},Q^2). \tag{5.42}$$

At large τ and/or Q^2, the solution to this equation increases like (with $\bar{\alpha}_s \equiv \alpha_s N_c/\pi$ and Q_0^2 some scale of reference) [5]

$$xG(x, Q^2) \propto \exp\left\{2\sqrt{\bar{\alpha}_s\,\tau\,\ln(Q^2/Q_0^2)}\right\}, \tag{5.43}$$

where we have assumed α_s to be independent of Q^2. If instead one takes the running of the coupling into account, by writing $\alpha_s(Q^2) = b_0/\ln(Q^2/\Lambda_{QCD}^2)$, then the dependence of the solution upon Q^2 gets softer [5] :

$$xG(x, Q^2) \propto \exp\left\{2\sqrt{b_0\,\tau\ln\left(\ln(Q^2/\Lambda_{QCD}^2)\right)}\right\}. \tag{5.44}$$

In any case, eqs. (5.43) and (5.44) show that, at high transverse momenta $Q^2 \gg Q_s^2(\tau)$, the gluon distribution $xG(x, Q^2)$ grows rapidly with τ. This is the standard picture of parton evolution, which, if extrapolated to arbitrarily high energies, would predict violations of the unitarity bound[13]. But from the previous analysis, we know that the approximations leading to eq. (5.42) will break down at sufficiently large energies, where the non-linear effects in the quantum evolution cannot be neglected anylonger. Alternatively, for fixed rapidity τ, the linear approximation breaks down at low transverse momenta $k_\perp \ll Q_s(\tau)$, with $Q_s(\tau)$ the saturation scale. An estimate for this scale has been given in eq. (5.27), which, together with eqs. (5.31) and (5.41), implies:

$$Q_s^2(\tau) \simeq \frac{\alpha_s N_c}{4}\int^{Q_s^2(\tau)}\frac{dp_\perp^2}{p_\perp^2}\,\mu_\tau(p_\perp) = \frac{\pi\alpha_s N_c}{N_c^2-1}\frac{1}{R^2}\,xG(x, Q_s^2(\tau)). \tag{5.45}$$

By further combining this result with eq. (5.43) or (5.44), one can deduce the τ-dependence of the saturation scale in the DLA. One thus obtains:

$$Q_s^2(\tau) = Q_0^2\,\mathrm{e}^{4\bar{\alpha}_s\tau}, \qquad \text{(fixed coupling)}, \tag{5.46}$$

and, respectively,

$$Q_s^2(\tau) = \Lambda_{QCD}^2\,\mathrm{e}^{\sqrt{2b_0\tau\ln\tau}}, \qquad \text{(running coupling)}. \tag{5.47}$$

Eq. (5.46) (or (5.47)) defines a curve in the $\tau - k_\perp$ plane, which divides this plane in two (see Fig. 17) : Points on its right are effectively in the high momentum regime; they correspond to a dilute gas of weakly correlated colour sources whose density is rapidly increasing with τ. Points on the left of the saturation curve correspond to the low momentum regime, to be discussed in the next subsection.

[13] Note that, although slower than for the BFKL solution (3.6), the growth with τ of the DLA solution (5.43) or (5.44) is still faster than that of any power of $\tau \sim \ln s$.

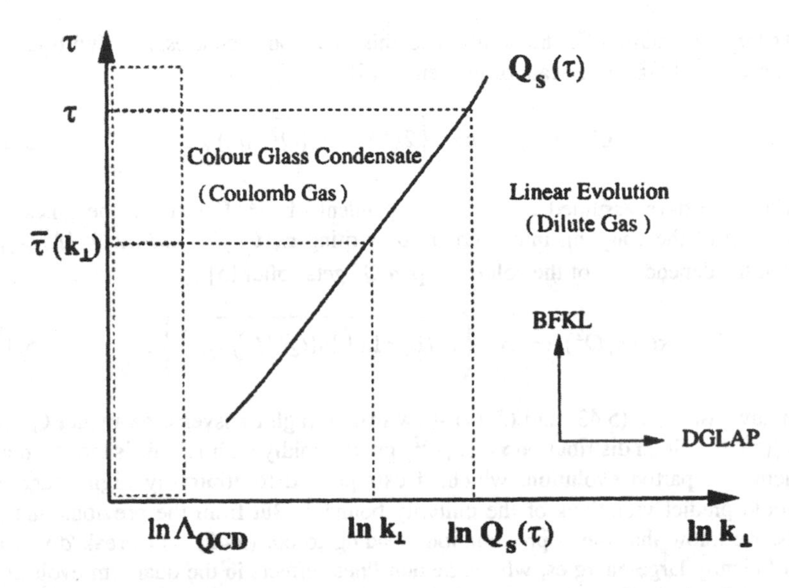

Figure 17. A "phase-diagram" of the various regions for evolution in the $\tau - k_\perp$ plane.

5.4. LOW-$K_\perp$: COULOMB GAS AND GLUON SATURATION

We finally turn to the most interesting physical regime, that of the non-linear physics at small transverse momenta $k_\perp \ll Q_s(\tau)$ (with $k_\perp \gg \Lambda_{QCD}$, though), whose understanding was a main motivation for all the previous developments. Within the effective theory, the low-momentum modes of the colour source are described by the weight function $\mathcal{W}_\tau^{\rm low}$, eq. (5.37), which is equivalently rewritten as (cf. Fig. 17):

$$\mathcal{W}_\tau^{\rm low}[\rho] \,=\, \mathcal{N}_\tau \exp\left\{-\,\frac{\pi}{2}\int^{Q_s^2(\tau)}\frac{d^2k_\perp}{(2\pi)^2}\int_{\bar\tau(k_\perp)}^{\tau} dy\,\frac{\rho_{\rm y}^a(k_\perp)\rho_{\rm y}^a(-k_\perp)}{k_\perp^2}\right\}, \tag{5.48}$$

with $\bar\tau(k_\perp)$ = the rapidity at which the saturation momentum is equal to $k_\perp$:

$$Q_s^2(\bar\tau(k_\perp)) \,=\, k_\perp^2\,. \tag{5.49}$$

There are several noteworthy features about eq. (5.48) :

i) This describes a *Coulomb gas*, i.e., a system of colour charges interacting via long-range Coulomb forces. The colour source $\rho_{\rm y}^a(x_\perp)$ at $x_\perp$ feels the Coulomb field $\alpha_{\rm y}^a(x_\perp)$ created at $x_\perp$ by all the other sources:

$$\int_{k_\perp}\frac{\rho_{\rm y}^a(k_\perp)\rho_{\rm y}^a(-k_\perp)}{k_\perp^2} = \int_{x_\perp,y_\perp}\rho_{\rm y}^a(x_\perp)\langle x_\perp|\,\frac{1}{-\nabla_\perp^2}\,|y_\perp\rangle\rho_{\rm y}^a(y_\perp) = \int_{x_\perp}\rho_{\rm y}^a(x_\perp)\alpha_{\rm y}^a(x_\perp).$$

The fact that the charge-charge correlator appears to vanish when $k_\perp \to 0$ is in agreement with gauge symmetry: The colour source $\rho^a_{\rm y}(k_\perp)$ at low $k_\perp$ is an *induced* source, whose global strength must vanish:

$$\langle Q^2 \rangle \equiv \int_{x_\perp, y_\perp} \int_{\rm y, y'} \langle \rho^a_{\rm y}(x_\perp) \rho^a_{\rm y'}(y_\perp) \rangle \propto \langle \rho^a_{\rm y}(k_\perp) \rho^a_{\rm y}(-k_\perp) \rangle \Big|_{k_\perp = 0} = 0. \quad (5.50)$$

ii) The colour charge correlations are *local in rapidity* : the Coulomb forces couple only sources located in the same layer of y (or x^-). At low–$k_\perp$, this property is not just an artifact of the MFA, but rather has a deep physical meaning: In the quantum evolution, the colour sources at different rapidities get correlated with each other because of the presence of Wilson lines in the evolution Hamiltonian (4.13). But these correlations are washed out on a large scale $r_\perp \gg 1/Q_s(\tau)$, on which the Wilson lines average to zero. In particular, this explains why the width $\propto k_\perp^2$ of the Gaussian (5.48) is independent of the initial conditions at $\tau \simeq 0$. (By contrast, at high momenta, the width $\lambda_{\rm y}(k_\perp) = \partial \mu_{\rm y}(k_\perp)/\partial {\rm y}$ in eq. (5.38) is sensitive to the initial conditions, since determined by solving eq. (5.30).)
iii) According to eq. (5.48), the low-momentum modes of ρ are *uniformly* distributed in rapidity, within the interval $\bar{\tau}(k_\perp) < {\rm y} < \tau$. It follows that the integrated quantity:

$$\mu_\tau(k_\perp) = \int_{\bar{\tau}(k_\perp)}^{\tau} d{\rm y}\, \frac{k_\perp^2}{\pi} = \big(\tau - \bar{\tau}(k_\perp)\big) \frac{k_\perp^2}{\pi}, \quad (5.51)$$

which measures the density of sources (with given $k_\perp$) in the transverse plane, grows only *linearly* with τ, that is, logarithmically with the energy. This is to be contrasted with the strong, quasi-exponential, increase of $\mu_\tau(k_\perp)$ in the high-momentum regime (cf. eqs. (5.43) and (5.44)). We conclude that, at low momenta $k_\perp \ll Q_s(\tau)$, the colour sources *saturate*, because of the strong non-linear effects in the quantum evolution.
iv) The saturated sources form the outermost layers of the hadron in the longitudinal direction: for given $k_\perp$, they are located at $x^- \geq x_0^- {\rm e}^{\bar{\tau}(k_\perp)}$. In particular,

$$\tau - \bar{\tau}(k_\perp) \simeq \frac{1}{4\bar{\alpha}_s} \ln \frac{Q_s^2(\tau)}{k_\perp^2}, \quad (5.52)$$

is the longitudinal extent of the saturated part of the hadron, in units of rapidity (for modes with tranverse momentum $k_\perp$). In writing (5.52), we have used the DLA estimate (5.46) for the τ-dependence of the saturation scale.
v) Note the factor $1/\alpha_s$ in the r.h.s. of (5.52); this implies that, at saturation, the *integrated* charge density $\rho^a(x_\perp)$ has typically large amplitudes: $\bar{\rho} \sim \sqrt{\langle \rho\rho \rangle} \sim 1/g$. The same is therefore true for the COV-gauge field $\alpha^a(x_\perp) : \bar{\alpha} \sim 1/g$.

Since the colour sources at low–$k_\perp$ are saturated, there should be no surprise that the gluons emitted by these sources are saturated as well, and this independently of their mutual interactions (i.e., of the non-linear effects in the classical Yang-Mills equations). Indeed, a quasi-Abelian calculation of the gluon distribution, based on the linearized solution $\mathcal{F}^{+j}(k) \approx (ik^j/k_\perp^2)\rho$, yields the following gluon density (cf. eqs. (5.40) and eq. (5.51)) :

$$\mathcal{N}_\tau(k_\perp) \simeq \frac{N_c^2-1}{4\pi^4 c}\Big(\tau - \bar{\tau}(k_\perp)\Big) \simeq \frac{N_c^2-1}{16\pi^4 c}\frac{1}{\bar{\alpha}_s}\ln\frac{Q_s^2(\tau)}{k_\perp^2}, \tag{5.53}$$

which already exhibits saturation ! In fact, as argued in Refs. [10], the only effect of the non-linearities in the classical Yang-Mills equations in this low–$k_\perp$ regime is to modify the overall normalization of the linear-order result. In anticipation of this, we have inserted in eq. (5.53) a corrective factor c, which cannot be accurately determined in the present approximations (since sensitive to the physics around Q_s), but should be smaller than one (although not much smaller).
Note the striking similarity between eq. (5.53) and the corresponding prediction (2.71) of the classical MV model. Despite of the differences in the physical mechanism leading to saturation — non-linear quantum evolution for eq. (5.53), as opposed to non-linear classical dynamics for eq. (2.71) —, the final results look very much the same. So, the earlier discussion of eq. (2.71) can be immediately adapted to eq. (5.53), after replacing $A \to s$: Eq. (5.53) shows *marginal saturation* (in the sense of a logarithmic increase only) with both s and $1/k_\perp^2$, with a typical amplitude of order $1/\alpha_s$. This is illustrated in Fig. 18, which should be compared to Fig. 12. (The high–$k_\perp$ behaviour in Fig. 18 is taken from eq. (5.40).)

Aside from saturation, eq. (5.53) has also other important consequences, which all reflect the proportionality to the rapidity window[14] $\tau - \bar{\tau}(k_\perp)$, eq. (5.52):
a) Scaling. The gluon density at saturation depends upon the energy s and the transverse momentum $k_\perp$ only via the scaling variable

$$\mathcal{T} \equiv Q_s^2(\tau)/k_\perp^2\,. \tag{5.54}$$

A similar scaling is observed in the solutions to the Kovchegov equation [19, 56, 58]. As mentioned in the Introduction, such a scaling has been actually observed in DIS at HERA [17].
b) Universality. Eq. (5.53) is only weakly sensitive — via its logarithmic dependence upon the saturation scale — to the initial conditions for quantum evolution, and therefore to the specific properties of the hadron under consideration (e.g., its size and atomic number). Thus, eq. (5.53) not only provides arguments in the favour of hadron universality at high energy, but also predicts what should be the pattern of its violation.

[14] These properties are therefore generic: They hold for any quantity which receives his dominant contributions from the saturated gluons.

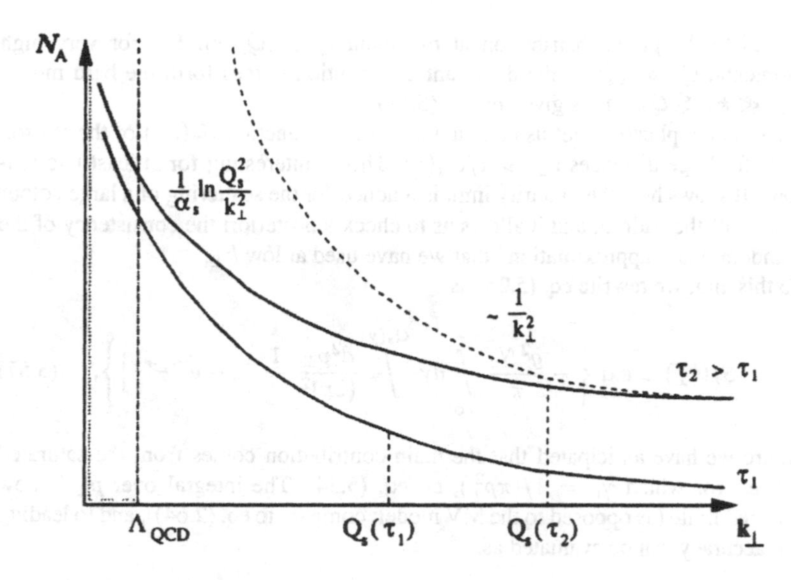

Figure 18. The gluon phase-density $\mathcal{N}_\tau(k_\perp)$ in the effective theory plotted as a function of $k_\perp$ for two values of $\tau = \ln(1/x)$.

The gluon distribution $xG(x, Q^2)$ at $Q^2 \ll Q_s^2(\tau)$ is immediately obtained by integration in eq. (5.53):

$$\begin{aligned} xG(x, Q^2) &\simeq \frac{N_c^2 - 1}{16\pi^3 c} \frac{1}{\bar{\alpha}_s} \pi R^2 \int_0^{Q^2} dk_\perp^2 \ln \frac{Q_s^2(\tau)}{k_\perp^2} \\ &= \frac{N_c^2 - 1}{16\pi^2 c} \frac{1}{\bar{\alpha}_s} R^2 Q^2 \left[\ln(Q_s^2(\tau)/Q^2) + 1 \right]. \end{aligned} \qquad (5.55)$$

Note that, since $\mu_\tau(k_\perp) \sim k_\perp^2$ in the saturation regime (cf. eq. (5.51)), the above integral is almost insensitive to the soft modes $k_\perp \lesssim \Lambda_{QCD}$. This has allowed us to extend the integration down to $k_\perp = 0$ without loss of accuracy. As anticipated, the phenomenon of saturation reduces the sensitivity of physical quantities to the infrared gauge fields, thus making the weak coupling expansion reliable. (In Ref. [58] a similar conclusion is drawn on the basis of Kovchegov equation.)
If extrapolated up to $Q \sim Q_s$, eq. (5.55) yields

$$xG(x, Q_s^2(\tau)) \simeq \frac{N_c^2 - 1}{16\pi^2 c} \frac{1}{\bar{\alpha}_s} R^2 Q_s^2(\tau), \qquad (5.56)$$

in rough agreement with the corresponding extrapolation from the high momentum regime, eq. (5.45). Eq. (5.56) gives also the contribution of the saturated

modes to the gluon distribution at momenta $Q > Q_s(\tau)$. But for very high momenta, $Q \gg Q_s(\tau)$, the dominant contribution comes form the hard modes ($Q_s \ll k_\perp \lesssim Q$), and is given by eq. (5.41).
As a final application, let us compute the 2-point function $S_\tau(r_\perp)$ of the Wilson lines for large distances $r_\perp \gg 1/Q_s(\tau)$. This is interesting for at least two reasons: It shows how the unitarity limit is reached for the scattering of a large colour dipole off the hadron, and it allows us to check a posteriori the consistency of the "random phase approximation" that we have used at low $k_\perp$.
To this aim, we rewrite eq. (5.25) as

$$S_\tau(r_\perp) \simeq \exp\Bigg\{ -\frac{g^2 N_c}{\pi} \int_{-\infty}^{\tau} dy \int^{Q_s(y)} \frac{d^2p_\perp}{(2\pi)^2} \frac{1}{p_\perp^2} \Big[1 - e^{ip_\perp \cdot r_\perp}\Big]\Bigg\}, \qquad (5.57)$$

where we have anticipated that the main contribution comes from the saturated modes, for which $\gamma_\tau = 1/(\pi p_\perp^2)$, cf. eq. (5.34). The integral over $p_\perp$ is now infrared finite (as opposed to the MV model: compare to eq. (2.64)), and to leading log accuracy can be evaluated as:

$$\int^{Q_s(y)} \frac{d^2p_\perp}{(2\pi)^2} \frac{1}{p_\perp^2} \Big[1 - e^{ip_\perp \cdot r_\perp}\Big] \simeq \theta(y - \bar{\tau}(r_\perp)) \frac{1}{4\pi} \ln\Big(Q_s^2(y) r_\perp^2\Big). \qquad (5.58)$$

The result can be understood as follows: as long as $1/r_\perp \gg Q_s(y)$, or $y < \bar{\tau}(r_\perp)$, $e^{ip_\perp \cdot r_\perp} \approx 1$, and the integral vanishes. Bur for $y > \bar{\tau}(r_\perp)$, or $1/r_\perp \ll Q_s(y)$, the integrals corresponding to the two terms in the brackets are cut off at different ultraviolet scales: $Q_s(y)$ for the first term, and $1/r_\perp$ for the second one. Their difference gives the log in the r.h.s. By also using $\ln(Q_s^2(y)r_\perp^2) = 4\bar{\alpha}_s(y - \bar{\tau}(r_\perp))$, cf. eq. (5.52), and performing the integral over y, we finally deduce:

$$S_\tau(r_\perp) \simeq \exp\Big\{-2\bar{\alpha}_s^2(\tau - \bar{\tau}(r_\perp))^2\Big\} = \exp\Big\{ -\frac{1}{8}\Big[\ln(Q_s^2(\tau)r_\perp^2)\Big]^2\Big\}, \qquad (5.59)$$

which coincides with the result obtained from the Kovchegov equation [19, 14]. Eq. (5.59) shows that the correlator of the Wilson lines is rapidly decreasing when $Q_s^2(\tau)r_\perp^2 \gg 1$, so that the RPA is indeed justified, at least as a mean field approximation.
More details and further applications of the mean field approximation will be presented in Ref. [57], where the results obtained in this way will be also compared to the corresponding predictions of the Kovchegov equation. It would be also interesting (especially in view of applications to phenomenology) to take into account the transverse inhomogeneity of the hadron (i.e., the dependence upon the impact parameter in the transverse plane). This can be done already in the framework of the MFA, but, more generally, it would be important to understand

the limitations of the latter, and to be able to solve the complete RGE. This might be done, for instance, via numerical simulations on a lattice.

6. Acknowledgments

We would like to thank our colleagues Alejandro Ayala-Mercado, Jean-Paul Blaizot, Elena Ferreiro, Miklos Gyulassy, Kazunori Itakura, Dima Kharzeev, Yuri Kovchegov, Alex Kovner, Jamal Jalilian-Marian, Genya Levin, Al Mueller, Robi Peschanski, Raju Venugopalan and Heribert Weigert with whom many of the ideas presented in these lectures were developed, or discussed. We particularly thank Ulrich Heinz, Kazunori Itakura and Michaela Oswald for a careful and critical reading of the manuscript.
This manuscript has been authorized under Contract No. DE-AC02-98H10886 with the U. S. Department of Energy.

References

1. G. Sterman, *An Introduction to Quantum Field Theory*, Cambridge University Press, Cambridge, 1993; M.E. Peskin and D.V. Schroeder, *An Introduction to Quantum Field Theory*, Addison-Wesley, New York, 1995; R.K. Ellis, W.J. Stirling and B.R. Webber, *QCD and Collider Physics*, Cambridge University Press, Cambridge, 1996.
2. J. Breitweg et. al., *Eur. Phys. J.* **67**, 609 (1999) and references therein.
3. L.N. Lipatov, *Sov. J. Nucl. Phys.* **23** (1976), 338;
E.A. Kuraev, L.N. Lipatov and V.S. Fadin, *Sov. Phys. JETP* **45** (1977), 199;
Ya.Ya. Balitsky and L.N. Lipatov, *Sov. J. Nucl. Phys.* **28** (1978), 822.
4. J.R. Forshaw and D.A. Ross, *Quantum Chromodynamics and the Pomeron*, Cambridge University Press, Cambridge, 1997.
5. V.N. Gribov and L.N. Lipatov, *Sov. Journ. Nucl. Phys.* **15** (1972), 438; G. Altarelli and G. Parisi, *Nucl. Phys.* **B126** (1977), 298; Yu. L. Dokshitzer, *Sov. Phys. JETP* **46** (1977), 641. See also Yu. L. Dokshitzer, V. A. Khoze, A. H. Mueller, and S.I. Troyan, *"Basics of Perturbative QCD"* (Editions Frontières, Gif-sur-Yvette, 1991).
6. L. V. Gribov, E. M. Levin, and M. G. Ryskin, *Phys. Rept.* **100** (1983) 1.
7. A. H. Mueller and Jian-wei Qiu, *Nucl. Phys.* **B268** (1986) 427.
8. L.McLerran and R. Venugopalan, *Phys. Rev.* **D49** (1994) 2233; *ibid.* **49** (1994) 3352; *ibid.* **50** (1994) 2225.
9. A. H. Mueller, *Nucl. Phys.* **B558** (1999), 285.
10. E. Iancu and L. McLerran, *Phys. Lett.* **B510** (2001) 145.
11. A. H. Mueller, *Small-x Physics, High Parton Densities and Parton Saturation in QCD*, hep-ph/9911289. Published in *Lisbon 1999, QCD: Perturbative or nonperturbative?*, pp. 180-209.
12. L. McLerran, *The color glass condensate and small x physics: 4 lectures*, Lectures given at 40th Internationale Universitatswochen fuer Theoretische Physik: Dense Matter (IUKT 40), Schladming, Austria, 3-10 Mar 2001, hep-ph/0104285.
13. E. Levin, *Saturation at low x*, hep-ph/0105205.
14. A. H. Mueller, *Parton Saturation-An Overview*, hep-ph/0111244 (this volume).
15. J.-P. Blaizot and E. Iancu, *The quark-gluon plasma: Collective dynamics and hard thermal loops*, *Physics Reports* **359** (2002) 355, hep-ph/0101103.

16. J.-P. Blaizot and A. H. Mueller, *Nucl. Phys.* **B289** (1987) 847.
17. A. M. Stasto, K. Golec-Biernat and J. Kwieczinski, *Phys. Rev. Lett.*, **86**, 596 (2001).
18. K. Golec-Biernat and M. Wüsthoff, *Phys. Rev.* **D59** (1999), 014017; *ibid.* **D60** (1999) 114023; *Eur. Phys. J.* **C20** (2001) 313.
19. E. Levin and K. Tuchin, *Nucl. Phys.* **B573** (2000) 833; *Nucl.Phys.* **A691** (2001) 779; *Nucl.Phys.* **A693** (2001) 787.
20. A. Krasnitz, R. Venugopalan, *Phys. Rev. Lett.* **84** (2000) 4309; *ibid.* **86** (2001) 1717; A. Krasnitz, Y. Nara, and R. Venugopalan, *Phys. Rev. Lett.* **87** (2001) 192302.
21. A. Kovner, L. McLerran and H. Weigert, *Phys. Rev.* **D52** (1995) 3809; *ibid.* 6231.
22. R. Baier, A.H. Mueller, D. Schiff and D.T. Son, *Phys. Lett.* **B502** (2001) 51.
23. See, e.g., the proceedings of 15th International Conference on Ultrarelativistic Nucleus-Nucleus Collisions (QM2001), Stony Brook, New York, 15-20 Jan 2001. Published in *Nucl. Phys.* **A698** (2002).
24. D. Kharzeev, M. Nardi, *Phys. Lett.* **B507** (2001) 121.
25. D. E. Kharzeev, E. Levin, *Phys. Lett.* **B523** (2001) 79.
26. L. McLerran, J. Schaffner-Bielich, *Phys. Lett.* **B514** (2001) 29; J. Schaffner-Bielich, D. Kharzeev, L. McLerran and R. Venugopalan, nucl-th/0108048.
27. F. Gelis and A. Peshier, *Nucl. Phys.* **A697** (2002) 879; hep-ph/0111227.
28. R. Venugopalan, *Deeply inelastic scattering off nuclei at RHIC*, hep-ph/0102087.
29. A. Dumitru, L. McLerran, *How protons shatter colored glass*, hep-ph/0105268.
30. Yu. V. Kovchegov, *Nucl.Phys.* **A692** (2001) 557; *Phys.Rev.* **D64** (2001) 114016.
31. D. E. Kharzeev, Yu. V. Kovchegov, and E. Levin, *Nucl. Phys.* **A699** (2002) 745.
32. Yu.V. Kovchegov, *Phys. Rev.* **D54** (1996), 5463; *Phys. Rev.* **D55** (1997), 5445.
33. J. Jalilian-Marian, A. Kovner, L. McLerran and H. Weigert, *Phys. Rev.* **D55** (1997), 5414.
34. Yu.V. Kovchegov and A.H. Mueller, *Nucl. Phys.* **B529** (1998), 451.
35. C. S. Lam and G. Mahlon, *Phys. Rev.* **D62** (2000) 114023; *ibid.* **D64** (2001) 016004.
36. J. Jalilian-Marian, A. Kovner, A. Leonidov and H. Weigert, *Nucl. Phys.* **B504** (1997), 415; *Phys. Rev.* **D59** (1999), 014014.
37. E. Iancu, A. Leonidov and L. McLerran, *Nucl. Phys.* **A692** (2001), 583; *Phys. Lett.* **B510** (2001) 133; E. Ferreiro, E. Iancu, A. Leonidov and L. McLerran, hep-ph/0109115.
38. R. Venugopalan, *Classical methods in DIS and nuclear scattering at small x*, Lectures at the XXXIX Cracow School of Theoretical Physics, Zakopane, Poland, May29th-June8th, 1999. Published in *Acta Phys. Polon.* **B30** (1999) 3731.
39. J.B. Kogut and D.E. Soper, *Phys.Rev.* **D1** (1970) 2901.
40. A. H. Mueller, *Nucl. Phys.* **B335** (1990) 115.
41. N.N. Nikolaev and B.G. Zakharov, *Z. Phys.* **C49** (1991) 607, *ibid.* **C53** (1992) 331.
42. W. Buchmuller, M.F. McDermott and A. Hebecker, *Nucl. Phys.* **B487** (1997) 283, Erratum-ibid. **B500** (1997) 621; W. Buchmuller, T. Gehrmann, and A. Hebecker *Nucl. Phys.* **B537** (1999) 477.
43. K. H. Fischer and J. A. Hertz, *Spin glasses*, Cambridge University Press, Cambridge, 1991.
44. I. Balitsky, *Nucl. Phys.* **B463** (1996), 99; *High-energy QCD and Wilson lines*, hep-ph/0101042.
45. Yu. V. Kovchegov, *Phys. Rev.* **D60** (1999), 034008; *ibid.* **D61** (2000), 074018.
46. A. H. Mueller, *Nucl. Phys.* **B415** (1994), 373; *ibid.* **B437** (1995), 107.
47. H. Weigert, *Unitarity at small Bjorken x*, hep-ph/0004044.
48. A. H. Mueller, *Phys. Lett.* **B523** (2001) 243.
49. A. Kovner, J. G. Milhano and H. Weigert, *Phys. Rev.* **D62** (2000), 114005.
50. M. Braun, *Eur. Phys. J.* **C16** (2000) 337; N. Armesto and M. Braun, *ibid.* **C20** (2001) 517.
51. I. Balitsky and A.V. Belitsky, *Nonlinear evolution in high-density QCD*, hep-ph/0110158.
52. J.-P. Blaizot, E. Iancu and H. Weigert, in preparation.

53. I. Balitsky, *Phys. Lett.* **B518** (2001) 235.
54. M. Lublinsky, E. Gotsman, E. Levin, and U. Maor, *Nucl.Phys.* **A696** (2001) 851; E. Levin and M. Lublinsky, *Nucl.Phys.* **A696** (2001) 833.
55. E. Levin and M. Lublinsky, *Phys. Lett.* **B521** (2001) 233; *Eur. Phys. J.* **C22** (2002) 647.
56. M. Lublinsky, *Eur. Phys. J.* **C21** (2001) 513.
57. E. Iancu, K. Itakura, and L. McLerran, in preparation.
58. K. Golec-Biernat, L. Motyka, and A. M. Stasto, *Diffusion into infra-red and unitarization of the BFKL pomeron*, hep-ph/0110325.
59. G. Parisi, *Statistical Field Theory*, Addison-Wesley, New York, 1988;
J. Zinn-Justin, *Quantum field theory and critical phenomena*, (International series of monographs on physics, 77), Clarendon, Oxford, 1989.

RECENT PROGRESS IN COMPUTING INITIAL CONDITIONS FOR HIGH ENERGY HEAVY ION COLLISIONS

R. VENUGOPALAN (raju@bnl.gov)
RIKEN-BNL Research Center & Brookhaven National Laboratory, Upton, NY-11973, USA

Abstract.
Classical methods are used to compute multi-particle production in the initial instants of a high energy heavy ion collision. Non-perturbative expressions are derived relating the distributions of produced partons to those of wee partons in the wavefunctions of the colliding nuclei. The time evolution of components of the stress–energy tensor and the impact parameter dependence of elliptic flow is studied. We discuss the space-time picture that emerges and interpret the RHIC data within this framework.

1. Introduction

At the Relativistic Heavy Ion Collider (RHIC), beams of Gold ions are currently being collided at center of mass energies of $\sqrt{s_{NN}} = 200$ GeV/nucleon. The hope is to create briefly an equilibrated state of quarks and gluons called the quark gluon plasma and to study its statistical properties [1], in particular its change of phase to hadronic matter.
It was understood very early on that the likelihood of creating this novel state of matter depended crucially on the initial conditions for the collision [2, 3, 4, 5]. There are several time scales in the problem and propituous values of these are determined by the initial conditions.
It was also understood very early on that the initial conditions for high energy collions are determined by the "wee partons" (partons that carry a very small fraction x of the nuclear momentum) in the wavefunctions of the colliding nuclei [2]. This is because, in the language of quantum mechanics, small x refers to Fock components of the nuclear wavefunction that contain a large number of partons (mostly gluons) [6]. In a nuclear collision, these virtual excitations of the vacuum go on-shell and are therefore responsible for multi-particle production. Thus an understanding of small x physics is essential to any formulation of a theory of heavy ion collisions.

J.-P. Blaizot and E. Iancu (eds.), QCD Perspectives on Hot and Dense Matter, 147–158.

The problem of initial conditions is a difficult one because the behavior of the wee partons is mysterious and defies our naive intuition. For instance, wee partons are long wavelength excitations of the vacuum but they last for very short times [1]. The large coherence length of the excitations is also why a probabilistic picture of multi-particle production (as implemented, for instance, in parton cascade models) must fail at high energies.

A traditional view is that the physics of wee partons is intrinsically non-perturbative: for example, multi-particle production is believed to be determined by non-perturbative excitations, called Pomerons, with vacuum quantum numbers [8, 9]. It was believed that Pomerons could be constructed in perturbation theory (the BFKL Pomeron [7]) but the status of that approach is at present unclear [10]. An alternative, increasingly popular, viewpoint is that small x physics is weak coupling physics. This approach is motivated by the idea of saturation [11], namely, that at small x the density of partons could be sufficiently large that recombination and screening effects are significant enough to halt the growth of parton distributions [2]. The large parton density provides a semi-hard scale-the saturation scale Λ_s- that controls the running of the QCD coupling constant- thereby making weak coupling methods feasible. Another consequence of this approach is that small x physics is classical because the occupation number of partons is $\sim 1/\alpha_S(\Lambda_s) >> 1$ [13].

Both of these ideas, weak coupling due to high parton densities and the applicability of classical methods can be cast in the framework of an effective field theory (EFT) [13] which treats partons at large x as static sources of color charges for the wee partons at small x. For a large nucleus, from the central limit theorem, these sources of color charge are Gaussian weights, $P[\rho] = \exp\left(-\frac{1}{\Lambda_s^2}\int d^2x_t\, \mathrm{Tr}(\rho^2)\right)$ where the color charge charge density, interestingly, is the saturation scale we mentioned previously. The classical theory for a single nucleus is solvable analytically and the distributions of partons computed [14].

What is *ad hoc* in the classical picture is the separation between static sources and dynamical fields. Remarkably, a Wilsonian renormalization group procedure has been developed which quantifies this separation of scales in x in a systematic way [15]. The structure of the classical field is preserved under evolution; it is the weight function $P[\rho]$ in the effective action that obeys a renormalization group equation. The saturation scale (whose validity extends beyond the Gaussian model) now acquires energy dependence-it is a function $\Lambda_s(x)$ of x. The analogy of the EFT to spin glasses, and the high occupation number of fields with the momenta peaked at Λ_s suggests that matter in this state is a Color Glass Condensate (CGC)[13, 16, 15].

[1] This is why, from the uncertainity principle, one needs very high energies to probe these excitations.

[2] For a discussion of an alternative "final state" saturation scenario, see Refs. [12]

Our focus in this talk is on applying the classical EFT to nuclear collisions. In the following section, we will outline the classical formalism for nuclear collisions. In section 3, we will apply this formalism to compute energy and number distributions of the gluons produced in a heavy ion collision. Next, we will discuss elliptic and radial flow. In the final section we will briefly discuss an interpretation of the RHIC data and shall conclude with brief outline of open problems and potential solutions in the classical approach.

2. Classical formalism for nuclear collisions

The classical EFT was first applied to the study of collisions of large nuclei by Kovner, McLerran and Weigert [17]. The model, as applied to nuclear collisions, may be summarized as follows. The colliding nuclei are idealized to travel along the light cone The high-x and the low-x modes in the nuclei are treated separately. The former corresponds to valence quarks and hard sea partons and are considered recoilless sources of color charge. Each of the large Lorentz-contracted nuclei (for simplicity, we will consider only collisions of identical nuclei) now has a Gaussian distribution of their color charge density $\rho_{1,2}$ in the transverse plane. The variance Λ_s of the color charge distribution is the only dimensionful parameter of the model, apart from the linear size L of the nucleus. For central impact parameters, Λ_s can be estimated in terms of single-nucleon structure functions [18]. It is assumed, in addition, that the nucleus is infinitely thin in the longitudinal direction. Under this simplifying assumption, the resulting gauge fields are explicitly boost-invariant.

The small x fields are then described by the Yang-Mills equation $D_\mu F_{\mu\nu} = J_\nu$ with the random sources on the two light cones: $J_\nu = \sum_{1,2} \delta_{\nu,\pm}\delta(x_\mp)\rho_{1,2}(r_t)$. The two signs correspond to two possible directions of motion along the beam axis z. As shown by Kovner, McLerran and Weigert (KMW) [17], low-x fields in the central region of the collision obey sourceless Yang-Mills equations (this region is in the forward light cone of both nuclei) with the initial conditions in the $A_\tau = 0$ gauge given by $A^i = A^i_1 + A^i_2$ and $A^\pm = \pm\frac{ig}{2}x^\pm[A^i_1, A^i_2]$. Here the pure gauge fields $A^i_{1,2}$ are solutions of the Yang-Mills equations for each of the two nuclei in the absence of the other nucleus.

In order to obtain the resulting gluon field configuration at late proper times, one needs to solve the YM-equations with the above mentioned initial conditions. Since the latter depends on the random color source, averages over realizations of the source must be performed. KMW showed that in perturbation theory the gluon number distribution by transverse momentum (per unit rapidity) suffers from an infrared divergence and argued that the distribution must have the form $n_{k_\perp} \propto \frac{1}{\alpha_s}\left(\frac{\Lambda_s}{k_\perp}\right)^4 \ln\left(\frac{k_\perp}{\Lambda_s}\right)$ for $k_\perp \gg \Lambda_s$. The log term clearly indicates that the perturbative description breaks down for $k_\perp \sim \Lambda_s$.

A reliable way to go beyond perturbation theory is to re-formulate the EFT on a lattice by discretizing the transverse plane. The resulting lattice theory can then be solved numerically to all orders in the color charge densities ρ_1 and ρ_2. The lattice Hamiltonian is formulated in $A^\tau = 0$ gauge. The real time gluodynamics of gauge fields can then be studied by solving Hamilton's equations on the lattice. We shall not dwell here on the details of the lattice formulation, which is described in detail in Ref. [19, 20]. Keeping in mind that Λ_s and the linear size L of the nucleus [3] are the only physically interesting dimensional parameters of the model [16], we can write any dimensional quantity q as $\Lambda_s^d f_q(\Lambda_s L)$, where d is the dimension of q. All the non-trivial physical information is contained in the dimensionless function $f_q(\Lambda_s L)$. We can estimate the values of the product $\Lambda_s L$ which correspond to key collider experiments. Assuming Au-Au collisions, we take $L = 11.6$ fm (for a square nucleus!) and estimate the saturation scale Λ_s to be 2 GeV for RHIC and 4 GeV for LHC [18]. Also, we have approximately $g = 2$ for energies of interest. The rough estimate is then $\Lambda_s L \approx$ 120-150 for RHIC and $\Lambda_s L \approx$ 240-300 for LHC. Since the gluon distribution in nuclei is not known to great precision, there is a considerable systematic uncertainty in these estimates.

3. Energy and Number distributions of produced gluons

The classical formalism has been applied to study classical gluon production arising from the "melting" of the Color Glass Condensate. The energy and number distributions have been computed numerically, and the dependence of these quantities on Λ_s has been determined [20, 21]. The initial simulations were performed for an SU(2) gauge theory. Recently, we have extended the work of Refs. [20, 21] to an SU(3) gauge theory [22]. Our results can now in principle be compared to available and forthcoming data from RHIC.
For the transverse energy of gluons, we obtain the relation

$$\frac{1}{\pi R^2}\frac{dE_T}{d\eta}|_{\eta=0} = \frac{1}{g^2} f_E(\Lambda_s R)\Lambda_s^3, \tag{3.1}$$

The function f_E is determined non-perturbatively as follows. In Figure 1 (a), we plot the Hamiltonian density, for a particular fixed value of $\Lambda_s R = 83.7$ (on a 512×512 lattice) in dimensionless units as a function of the proper time in dimensionless units. We note that in the SU(3) case, as in SU(2), $\varepsilon\tau$ converges very rapidly to a constant value. The form of $\varepsilon\tau$ is well parametrized by the functional form $\varepsilon\tau = \alpha + \beta\exp(-\gamma\tau)$. Here $dE_T/d\eta/\pi R^2 = \alpha$ has the proper interpretation of being the energy density of produced gluons, while $\tau_D = 1/\gamma/\Lambda_s$ is the "formation time" of the produced glue.
In Figure 1 (b), the convergence of α to the continuum limit is shown as a function of the lattice spacing in dimensionless units for two values of $\Lambda_s R$. In Ref. [20],

[3] L is the length scale for a cylindrical nucleus; $L^2 = \pi R^2$ where R is the radius of the nucleus.

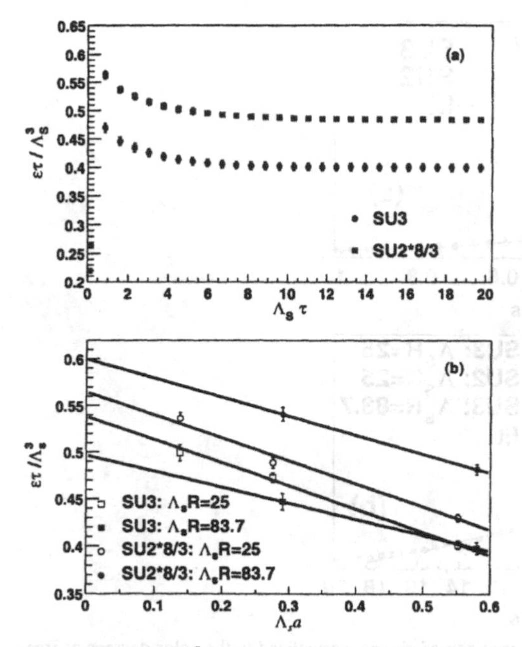

Figure 1. (a) $\varepsilon\tau/\Lambda_s^3$ as a function of $\tau\Lambda_s$ for $\Lambda_s R = 83.7$. (b) $\varepsilon\tau/\Lambda_s^3$ as a function of $\Lambda_s a$ for $\Lambda_s R = 83.7$ (squares) and 25(circles), where a is the lattice spacing. Lines are fits of the form $a - bx$.

this convergence to the continuum limit was studied extensively for very large lattices (up to 1024×1024 sites) and shown to be linear. The trend is the same for the SU(3) results. Thus, despite being further from the continuum limit for SU(3) (due to the significant increase in computer time), a linear extrapolation is justified. We can therefore extract the continuum value for α. We find $f_E(25) = 0.537$ and $f_E(83.7) = 0.497$. The RHIC value likely lies in this range of $\Lambda_s R$. The formation time $\tau_D = 1/\gamma/\Lambda_s$ is essentially the same for SU(2)-for $\Lambda_s R = 83.7$, $\gamma = 0.362 \pm 0.023$. As discussed in Ref. [20], it is ~ 0.3 fm for RHIC and ~ 0.13 fm for LHC (taking $\Lambda_s = 2$ GeV and 4 GeV respectively).
We now combine our expression in Eq. (3.1) with our non-perturbative expression for the formation time to obtain a non-perturbative formula for the initial energy density, $\varepsilon = \frac{0.17}{g^2} \Lambda_s^4$ This formula gives a rough estimate of the initial energy density, at a formation time of $\tau_D = 1/\bar{\gamma}/\Lambda_s R$ where we have taken the average value of the slowly varying function γ to be $\bar{\gamma} = 0.34$.
To determine the gluon number per unit rapidity, we first compute the gluon transverse momentum distributions. The procedure followed is identical to that described in Ref. [21] -we compute the number distribution in Coulomb gauge,

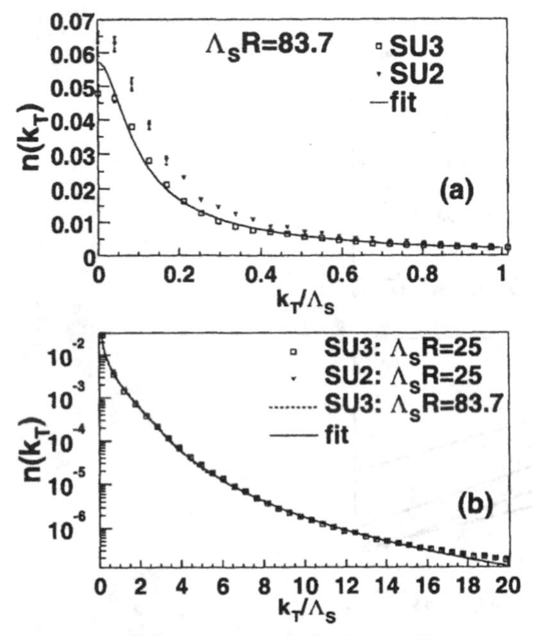

Figure 2. Transverse momentum distribution of gluons, normalized to the color degrees of freedom, $n(k_T) = \tilde{f}_n/(N_c^2-1)$ (see Eq. (3.2)) as a function of $\Lambda_S R$ for SU(3) (squares) and SU(2) (diamonds). Solid lines correspond to the fit in Eq.(3.3).

$\nabla_\perp \cdot A_\perp = 0$. In Fig. 2(a), we plot the normalized gluon transverse momentum distributions versus k_T/Λ_s with the value $\Lambda_s R = 83.7$, together with SU(2) result. Clearly, we see that the normalized result for SU(3) is suppressed relative to the SU(2) result in the low momentum region. In Fig. 2(b), we plot the same quantity over a wider range in k_T/Λ_s for two values of $\Lambda_s R$. At large transverse momentum, we see that the distributions scale exactly as $N_c^2 - 1$, the number of color degrees of freedom. This is as expected since at large transverse momentum, the modes are nearly those of non–interacting harmonic oscillators. At smaller momenta, the suppression is due to non-linearities, whose effects, we have confirmed, are greater for larger values of the effective coupling $\Lambda_s R$.
The SU(3) gluon momentum distribution can be fitted by the following function,

$$\frac{1}{\pi R^2}\frac{dN}{d\eta d^2k_T} = \frac{1}{g^2}\tilde{f}_n(k_T/\Lambda_s)\,, \tag{3.2}$$

where $\tilde{f}_n(k_T/\Lambda_s)$ is

$$\tilde{f}_n = \begin{cases} a_1 \left[\exp\left(\sqrt{k_T^2 + m^2}/T_{\text{eff}}\right) - 1\right]^{-1} & (k_T/\Lambda_s \le 3) \\ a_2 \, \Lambda_s^4 \log(4\pi k_T/\Lambda_s) k_T^{-4} & (k_T/\Lambda_s > 3) \end{cases} \tag{3.3}$$

with $a_1 = 0.0295$, $m = 0.067\Lambda_s$, $T_{\text{eff}} = 0.93\Lambda_s$, and $a_2 = 0.0343$. At low momenta, the functional form is approximately that of a Bose-Einstein distribution in two dimensions even though the underlying dynamics is that of classical fields. The functional form at high momentum is motivated by the lowest order perturbative calculations [18, 17, 23].

Integrating our results over all momenta, we obtain for the gluon number per unit rapidity, the non-perturbative result,

$$\frac{1}{\pi R^2} \frac{dN}{d\eta}|_{\eta=0} = \frac{1}{g^2} f_N(\Lambda_s R) \Lambda_s^2 \,. \tag{3.4}$$

We find that $f_N(83.7) = 0.3$. The results for a wide range of $\Lambda_s R$ vary on the order of 10% in the case of SU(2).

The results described in this section were obtained for the idealized (and simple!) case of cylindrical nuclei. In this case, the saturation scale Λ_s is a constant across the nucleus. For nuclei with realistic nuclear density profiles, $\Lambda_s \equiv \Lambda_s(x_t)$, where $\Lambda_s(x_t)$ varies smoothly, from its maximum value at the center of nuclei to zero at the edges. The average value of $\Lambda_s(x_t)$ is therefore smaller than the values used in our estimates. The effects of the spatial dependence of Λ_s is taken into account in the phenomenological analyses of Kharzeev and collaborators-see the discussion in the last section. These effects have also been taken into account in our numerical simulations-in particular, as discussed in the following section, to treat elliptic flow in heavy ion collisions.

4. Elliptic Flow

The azimuthal anisotropy in the transverse momentum distribution has been proposed as a sensitive probe of the hot and dense matter produced in ultra-relativistic heavy ion collisions [24]. A measure of the azimuthal anisotropy is the second Fourier coefficient of the azimuthal distribution, the elliptic flow parameter v_2. Its definition [25] is

$$v_2 = \langle\cos(2\phi)\rangle = \left\langle \frac{p_x^2 - p_y^2}{p_x^2 + p_y^2} \right\rangle = \frac{\int_{-\pi}^{\pi} d\phi \cos(2\phi) \int p_T dp_T \frac{d^3N}{dy p_T dp_T d\phi}}{\int_{-\pi}^{\pi} d\phi \int p_T dp_T \frac{d^3N}{dy p_T dp_T d\phi}}. \tag{4.5}$$

The first measurements of elliptic flow from RHIC, at center of mass energy $\sqrt{s_{NN}}$, have been reported recently [26]. Hydrodynamic model calculations provide good agreement, for large centralities, and for particular initial conditions and equations of state, with the measured centrality dependence of the data. The agreement at smaller centralities is less good, perhaps reflecting the breakdown of a hydrodynamic description in smaller systems. Hydrodynamic models are also in excellent agreement with the p_t dependence of the unintegrated elliptic flow parameter $v_2(p_t)$ up to 1.5 GeV/c at mid-rapidity [27]. However, above 1.5 GeV, the experimental distribution appears to saturate, while the hydrodynamic model distribution continues to rise. It has been argued recently that jet quenching might explain this saturated behaviour of $v_2(p_t)$ [28] albeit its success requires extremely large values of partonic cross-sections. We should also note here that hadronic transport model calculations underestimate the RHIC v_2 data[26, 29].

We will now apply the classical Yang–Mills approach to discuss the elliptic flow generated in a nuclear collision. As previously, we assume boost invariance–the lattice Hamiltonian is the Kogut-Susskind Hamiltonian in 2+1-dimensions coupled to an adjoint scalar field in $A^\tau = 0$ gauge [19]. In our earlier work, periodic boundary conditions were imposed to compute the space–time evolution of the gauge fields after the collision [20, 21, 22]. Since, as discussed previously, elliptic flow is a consequence of an initial spatial anisotropy, periodic boundary conditions are inadequate and open boundary conditions are required. This technical improvement has been implemented in the work described here.

Our numerical procedure is as follows (see Refs. [19, 20, 21, 22] for more details). Gaussian color charge distributions, as described in the previous section, are generated for each of the (identical) nuclei. At $\tau = 0$, the local color charge squared densities per unit area for each nucleus A are sampled in the transverse overlap region according to the following distribution, $\Lambda_s^2(\mathbf{s}) = \Lambda_{s0}^2 T_A(\mathbf{s} \pm \mathbf{b}/2)$, where $T_A(\mathbf{s}) = \int_{-\infty}^{\infty} dz\, \kappa(\mathbf{r})$ is a thickness function with the normalization $\int d^2\mathbf{s} T_A(\mathbf{s}) = 3\pi R^2/4$, $\mathbf{b}$ is the impact parameter, $\kappa(\mathbf{r})$ is the nuclear density profile, and Λ_{s0}^2 is the color charge squared per unit area in the center of each nucleus. In our preliminary studies, we have considered only hard sphere nuclei. For hard spheres, $T_A(\mathbf{s}) = \sqrt{1 - \frac{\mathbf{s}^2}{R^2}}$.

The rest of the numerical procedure is as discussed in our previous work [19, 20, 21, 22]. For each configuration of color charges sampled by the above equation for Λ_s, we solve Hamilton's equations on the lattice for the gauge fields and their conjugate canonical momenta. We compute the space-time evolution of the components of the Stress–Energy tensor, in particular, the two transverse components of the pressure T^{xx} and T^{yy} as well as the energy density T^{00}. In order to calculate v_2 within our model, we apply the cooling method which was proposed in our previous work [21]. There we obtained, for the total number of classically produced gluons, the equation $N = \sqrt{\frac{8}{\pi}} \int_0^\infty \frac{dt}{\sqrt{t}} V(t)$, where $V(t)$ is the potential energy for a system of free harmonic oscillators as a function of the *cooling* time t. It is clear that the gluon number defined in this manner is gauge invariant. For v_2, one can similarly prove that

$$v_2 = \frac{\int_0^\infty \frac{dt}{\sqrt{t}}(T^{xx}(t) - T^{yy}(t))}{\int_0^\infty \frac{dt}{\sqrt{t}} V(t)} . \tag{4.6}$$

As in the case of the gluon number, this expression is gauge invariant.
A subtle problem arises due to the use of open boundary conditions. Even if the color charge is concentrated inside the overlap region the gauge fields are not. The solution of Poisson's equation permits non-zero gauge fields outside the region of interaction. Clearly, since QCD is a confining theory, gauge fields must be confined within the region of interaction.
There is no unambiguous way of solving this problem since we don't understand how to incorporate confinement uniquely. A convenient approach is to introduce a mass. This has the effect of causing an exponential decay of the gauge fields over a distance $1/m \sim 1/\Lambda_{QCD}$. Thus if $m << \Lambda_s$, it should not significantly affect the spectrum and should simultaneously avoid unphysical contributions from outside the region of interaction. Simulations testing this prescription are in progress.
The momentum distribution $v_2(p_t)$ of the elliptic flow can also be compared to the RHIC data. Unlike the integrated distribution, it is not explicitly gauge invariant but must be computed in a physical gauge.

5. The CGC and RHIC data

The classical formalism discussed here is applicable only in the initial instants of a nuclear collision. It is inapplicable once the occupation number $f << 1$. Moreover, the final states observed are hadrons while the CGC predicts only the initial distribution of gluons. Subsequent interactions may lead to a thermalized Quark Gluon Plasma. The possibility that the CGC thermalizes has been discussed extensively [30]. It was argued recently that for asymptotic values of the saturation scale ($\Lambda_s \to \infty$) the CGC matter does indeed thermalize [31]. For realistic values of the saturation scale, the situation is unclear.
We will assume here, minimally, that gluonic matter formed from the CGC interacts very strongly in the transverse plane at early times and then free streams. Since the typical momentum of the gluons ($\sim \Lambda_s$) is large than the hadronization scale ($\sim \Lambda_{QCD}$), the gluons may further fragment independently before hadronizing. Invoking parton-hadron duality at hadronization then enables us to compare our results to the data.
From the numerical simulations described previously we find that if we fit the total hadron multiplicity at $\sqrt{s_{NN}} = 130$ GeV (directly equating initial num. of gluons=final num. of hadrons) we find that we obtain a value of E_t/N that's proportional to Λ_s and numerically nearly six times as large as the observed value. Within the framework of the model, there are two necessary improvements that

will work to reduce the ratio. The first is that the CGC overestimates the contributions from high $p_t > \Lambda_s$-this is more pronounced in E_t since it is a more ultraviolet sensitive quantity. A more careful treatment of this regime will reduce the global ratio of E_t/N. The second is to consider spherical nuclei instead of cylindrical nuclei. This means that $\Lambda_s \equiv \Lambda_s(x_t)$-the average value of the saturation scale is therefore smaller. One still expect that E_t/N to be significantly larger than the measured number. Now E_t/N is not a conserved quantity and will reduce due to both independent fragmentation of the gluon "mini-jets" or hydrodynamic flow or both. Which of these is correct will become clearer once more RHIC data is available.
Kharzeev and Nardi [32] have shown that saturation+parton-hadron duality reproduces the centrality dependence of the RHIC data. Further, Kharzeev and Levin [33] have shown that the rapidity and energy dependence of the RHIC data (going from $\sqrt{s_{NN}} = 130$ GeV to $\sqrt{s_N N} = 200$ GeV) is predicted accurately in the same scenario. Schaffner-Bielich et al. [34] have shown that the RHIC p_t data show an m_t scaling consistent with saturation. (The saturation scale extracted from the m_t scaling of the p_t spectra at different centralities reproduces the centrality dependence of the RHIC data.) Despite these favorable indications, the jury is not in yet. There are several puzzling features of the RHIC data that do not fit neatly into one box or the other. Hopefully, more quantitative numerical studies and the large amount of forthcoming data will help clarify what is at present a rather murky picture.
On a theoretical level, several improvements can be made to the picture presented here. Firstly, it would be interesting to study the effect of rapidity dependence on the gauge fields -is the system stable under rapidity dependent perturbations? Another interesting, if much more difficult, problem is to match the classical field simulations to a kinetic approach when the occupation numbers fall below unity. This would be the appropriate time at which parton cascade type simulations would be relevant [35]. Finally, how does one extend the renormalization group treatment of the single nucleus problem to that of two nuclei. Despite the formidable challenges, we believe that the QCD based classical formalism presented here is a concrete step towards a theory of high energy heavy ion collisions.

References

1. Proceedings of International Symposium on "Statistical Mechanics of Quarks and Hadrons", H. Satz (ed.), Bielefeld, Aug. 24th-31st, 1980, North-Holland Publishers.
2. J. D. Bjorken, Lectures at Int. Summer Instt. in Theoretical Physics, Current Induced Reactions, Hamburg, Germany, Sept. 15th-26th, 1975, SLAC-PUB 1756.
3. R. Anishetty, P. Koehler and L. McLerran, Phys. Rev. **D22**, 2793 (1980).
4. J. D. Bjorken, Phys. Rev. **D27**, 140 (1983).
5. H. Ehtamo, J. Lindfors, and L. McLerran, Z. Phys. **C18**, 341 (1983).
6. J. D. Bjorken, J. Kogut, and D. Soper, Phys. Rev. **D3**, 1382 (1971).
7. E. A. Kuraev, L. N. Lipatov, V. S. Fadin, Sov. Phys. JETP **45** 104 (1977);

I. Balitsky and L. N. Lipatov, Sov. J. Nucl. Phys. **28** 822 (1978).

8. F. Low, Phys. Rev. D12, 163 (1975); S. Nussinov, Phys. Rev. **14**, 246, (1976).
9. A. Donnachie and P. V. Landshoff, Phys. Lett. **B296**, 227 (1992).
10. V. S. Fadin and L. N. Lipatov, Phys. Lett. **B429**, 127 (1998); M. Ciafaloni and G. Camici, Phys. Lett. **B430**, 349, (1998).
11. L.V. Gribov, E. M. Levin and M. G. Ryskin, *Phys. Repts.* **100** (1983) 1; A. H. Mueller and J.-W. Qiu, Nucl. Phys. **B268**(1986) 427; J. P. Blaizot and A. H. Mueller, Nucl. Phys. **B289** (1987) 847.
12. K. J. Eskola, K. Kajantie, P. V. Ruuskanen, and K. Tuominen, *Nucl. Phys.* **B570**, 379 (2000).
13. L. McLerran and R. Venugopalan, *Phys. Rev.* **D49** 2233 (1994); **D49** 3352 (1994); **D50** 2225 (1994).
14. J. Jalilian–Marian, A. Kovner, L. McLerran and H. Weigert, *Phys. Rev.* **D55** 5414 (1997); Y. V. Kovchegov, Phys. Rev. D **54**, 5463 (1996).
15. J. Jalilian-Marian, A. Kovner, A. Leonidov, and H. Weigert, *Nucl. Phys.* **B504** 415 (1997); J. Jalilian-Marian, A. Kovner, and H. Weigert, *Phys. Rev.* **D59** 014015 (1999); L. McLerran and R. Venugopalan, *Phys. Rev.* **D59** 094002 (1999); E. Iancu, A. Leonidov and L. McLerran, Nucl. Phys. **A692**, 583 (2001); Phys. Lett. B **510**, 133 (2001); Phys. Lett. B **510**, 145 (2001); E. Ferreiro, E. Iancu, A. Leonidov and L. McLerran, hep-ph/0109115.
16. R. V. Gavai and R. Venugopalan, Phys. Rev. **D54**, 5795 (1996).
17. A. Kovner, L. McLerran and H. Weigert, *Phys. Rev* **D52** 3809 (1995); **D52** 6231 (1995).
18. M. Gyulassy and L. McLerran, *Phys. Rev.* **C56** (1997) 2219.
19. A. Krasnitz and R. Venugopalan, hep-ph/9706329, hep-ph/9808332; *Nucl. Phys.* **B557** 237 (1999).
20. A. Krasnitz and R. Venugopalan, Phys. Rev. Lett. **84** (2000) 4309.
21. A. Krasnitz and R. Venugopalan, Phys. Rev. Lett. **86** (2001) 1717.
22. A. Krasnitz, Y. Nara and R. Venugopalan, Phys. Rev. Lett. **87**, 192302 (2001).
23. Y. V. Kovchegov and D. H. Rischke, *Phys. Rev.* **C56** (1997) 1084; S. G. Matinyan, B. Müller and D. H. Rischke, *Phys. Rev.* **C56** (1997) 2191; *Phys. Rev.* **C57** (1998) 1927; Xiao-feng Guo, *Phys. Rev.* **D59** 094017 (1999).
24. J.-Y. Ollitraut, Phys. Rev. **46**,229 (1992); Phys. Rev. **D48**,1131 (1993).
25. S. Voloshin and Y. Zhang, Z. Phys. **C70**,665 (1996); A.M. Poskanzer and S. Voloshin, Phys. Rev. **C58**,1671 (1998).
26. STAR Collaboration, K.H. Ackermann et al., Phys. Rev. Lett. **86**,402 (2001). The experimental data have been obtained from http://www.star.bnl.gov/STAR/.
27. P.F.Kolb, P. Huovinen, U. Heinz,and H. Heiselberg, Phys. Lett. **500**, 232 (2001).
28. X. Wang and M. Gyulassy, Phys. Rev. Lett. **68**, 1480 (1992); M. Gyulassy, P. Levai and I. Vitev, Phys. Rev. Lett. **85**, 5535, (2000); U. A. Wiedemann, Nucl. Phys. A **690**, 731 (2001); R. Baier, D. Schiff and B. G. Zakharov, Ann. Rev. Nucl. Part. Sci. **50**, 37 (2000).
29. M. Bleicher and H. Stöcker, hep-ph/0006147.
30. A. H. Mueller, Nucl. Phys. **B572** (2000) 227; A. H. Mueller, Phys. Lett. **B475** 220 (2000); J. Bjoraker and R. Venugopalan, Phys. Rev. **C63** 024609 (2001); A. Dumitru and M. Gyulassy, Phys. Lett. **B494**, 215 (2000).
31. R. Baier, A. H. Mueller, D. Schiff and D. T. Son, Phys. Lett. **B502** 51 (2001).
32. D. Kharzeev and M. Nardi, Phys. Lett. **B507** (2001), 121.

33. D. Kharzeev and E. Levin, nucl-th/0108006.
34. L. McLerran and J. Schaffner-Bielich, Phys. Lett. B **514**, 29 (2001);
J. Schaffner-Bielich, D. Kharzeev, L. McLerran, and R. Venugopalan, nucl-th/0108048.
35. S. A. Bass, nucl-th/0104040;
D. Molnar and M. Gyulassy, Phys. Rev. C **62**, 054907 (2000);
Y. Nara, S. E. Vance and P. Csizmadia, nucl-th/0109018.

REGGE POLES IN QCD AND HEAVY-ION COLLISIONS

A.B. KAIDALOV (kaidalov@vxitep.itep.ru)*
ITEP, B.Cheremushkinskaya 25, 117259 Moscow, Russia

Abstract. The reggeon approach to interactions of hadrons and nuclei at high energies is reviewed. Basic principles of S-matrix formalism are discussed and complex angular momentum method is presented. Basic properties of Regge poles are outlined. Regge poles are considered from both t-channel and s-channel points of view. The main part of the review is devoted to the Regge poles in QCD. The method of the Wilson-loop path integral is used to calculate trajectories of the Regge poles for $q-\bar{q}$ and gluonic states. The problem of the pomeron in QCD is discussed in detail. It is shown how to use the $1/N$ expansion for classification of the reggeon diagrams in QCD. The role of Regge cuts in reggeon theory is discussed and their importance for high-energy phenomenology is emphasized. Models based on the reggeon calculus, $1/N$ expansion in QCD, and string picture of interactions at large distances are reviewed and applied to a broad class of phenomena in strong interactions. It is shown how to apply the reggeon approach to small-x physics in deep inelastic scattering. Special attention is devoted to applications of the reggeon approach to high-energy heavy-ion interactions.

1. Introduction

The Regge poles were introduced in particle physics in the beginning of 1960's [1, 2] and up till the present have been widely used for description of high-energy interactions of hadrons and nuclei. The Regge approach establishes an important connection between the high-energy scattering and the spectrum of particles and resonances. It served as the basis for introduction of dual and string models of hadrons. The derivation of the Regge poles in QCD is a difficult problem closely related to non-perturbative effects in QCD and to the problem of confinement. In this review I shall, first, recall the main properties of the S-matrix approach such as crossing, unitarity and analyticity. They constitute a basis of Regge theory. It will be shown how to obtain high energy theorems (such as Froissart theorem [3] and Pomeranchuk theorem [4] using these general principles. A complex angular momentum method and the Regge pole model will be introduced in Sec. 3. Regge poles will be discussed from both t-channel and s-channel points of view and the connection between Regge poles and multi-particle production will be outlined.

* Partially supported by grants RFBR 00-15-96786, 01-02-17383, NATO PSTCLG-977275, INTAS 00-00366

J.-P. Blaizot and E. Iancu (eds.), QCD Perspectives on Hot and Dense Matter, 159–205.

The problem of the Regge poles in QCD will be considered in Sec. 4. The $1/N$ expansion, where N stands for the number of colors or light flavors will be used. It will be shown how to relate QCD diagrams to the reggeon theory. First attempts to calculate the Regge trajectories in QCD, using a non-perturbative method of Wilson-loop path integral, will be reviewed. Special attention will be devoted to the problem of the pomeron – a Regge pole which determines the asymptotic behavior of high-energy diffractive processes. This problem is closely related to calculation of spectra of glueballs in QCD. An analytic formula for masses of glueballs will be obtained and compared to the results of recent lattice calculations. It will be shown that mixing of gluonic and light $q\bar{-}q$ states is important for the pomeron trajectory in the small t region. The role of short-distance dynamics and results of recent perturbative calculations of the pomeron will be discussed briefly.
In Sec. 5 properties of the Regge cuts will be reviewed and it will be shown that these singularities play an important role at high energies. A classification of different multi-particle configurations in the s-channel and contributions of Regge cuts will be considered and the main properties of the multi-particle production processes will be discussed. Predictions of the models based on reggeon theory, 1/N-expansion in QCD and string models will be given. A description of high-energy hadronic interactions in the framework of these models will be presented.
The section 6 will be devoted to physics of low-x deep inelastic scattering. This interesting kinematical region has been recently studied experimentally at HERA. It will be shown how the reggeon theory and QCD evolution effects allow one to understand the properties of the proton structure function and diffractive dissociation of virtual photons in a broad region of Q^2. The problem of "saturation" of parton densities at very small x will be outlined and the model for its solution will be proposed. The shadowing effects for structure functions of nuclei at small x will be considered and compared to experimental data.
Applications of the reggeon approach for interactions of nuclei at high energies will be reviewed in Sec. 7. The space–time picture of high energy interactions will be discussed. Inclusive spectra of hadrons in nucleus–nucleus collisions will be calculated and compared to the first RHIC data. Implications of these results for the saturation problem will be discussed.

2. S-matrix, amplitudes, cross sections

In S-matrix formalism a transition from an initial state $|i>$ at time $t = -\infty$ to a final state $< f|$ at $t = +\infty$ is characterized by the scattering (or S) matrix. A probability of a given final state f is given by $|S_{fi}|^2$. In absence of interaction $\hat{S} = \hat{I}$, where $\hat{I}$ is the unit matrix. In general

$$S_{fi} = \delta_{fi} + i(2\pi)^4\delta^4(P_f - P_i)T_{fi} \tag{2.1}$$

where T_{fi} is the scattering amplitude.
Cross section for the reaction $1 + 2 \to n$ is given by the formula

$$d\sigma = \frac{|T_{fi}|^2}{4J} d\tau_n \tag{2.2}$$

where $d\tau_n$ is the invariant phase space

$$d\tau_n = \Pi_{a=1}^{n} \frac{d^3 p_a}{(2\pi)^3\, 2E_a} \tag{2.3}$$

and $J = \sqrt{(p_1 p_2)^2 - m_1^2 m_2^2}$ comes from the flux factor. In the c.m. system $J = p \cdot w$ (w is the total energy in c.m.).

2.1. BASIC PROPERTIES OF AMPLITUDES.

a) Lorentz invariance.

Due to the Lorentz invariance of relativistic particle theory scattering amplitudes for spinless particles are functions of invariants which can be constructed from 4-momenta of particles. For example for the reaction $1 + 2 \to 3 + 4$ one can build the following invariants

$$s = (p_1 + p_2)^2 = (p_3 + p_4)^2,$$
$$t = (p_1 - p_3)^2 = (p_2 - p_4)^2,$$
$$u = (p_1 - p_4)^2 = (p_2 - p_3)^2$$

These invariants however are not independent and satisfy to one condition

$$s + t + u = \sum_{i=1}^{4} m_i^2 \equiv h \tag{2.4}$$

where m_i is the mass of the i-th particle ($p_i^2 = m_i^2$). Thus in this case there are 2 independent variables.
In the case of the n-point amplitude there are 3n-10 independent variables.
In the c.m. system $s = w^2$ and t is related to the scattering angle

$$t = m_1^2 + m_3^2 - 2(E_1 E_3 - pp' cos\theta) \tag{2.5}$$

$$p = \frac{\lambda^{1/2}(s, m_1^2, m_2^2)}{2\sqrt{s}}, \; p' = \frac{\lambda^{1/2}(s, m_3^2, m_4^2)}{2\sqrt{s}} \tag{2.6}$$

where $\lambda(s, m_i^2, m_k^2) = [s - (m_i + m_k)^2][s - (m_i - m_k)^2]$
For elastic scattering $t = -2p^2(1 - cos\theta)$.
In terms of the invariant t the differential cross section $d\sigma/dt$ for binary process has the form

$$\frac{d\sigma}{dt} = \frac{\pi}{pp'}\frac{d\sigma}{d\Omega'} = \frac{1}{16\pi}|\frac{1}{2pw}T(s,t)|^2 \tag{2.7}$$

b) Crossing.

In relativistic quantum theory an outgoing particle with the momentum - p (negative energy) corresponds to the incoming anti-particle with the momentum p. Therefore the amplitude $T(s, t, u)$ describes not only the process $1 + 2 \to 3 + 4$ (so called s - channel), but also the processes $1+\bar{3} \to \bar{2}+4$ (t- channel) and $1+\bar{4} \to 3+\bar{2}$ (u - channel). The physical regions of variables s, t, u for these processes are different. For example for the case when all m_i are equal the physical region of the s-channel is $s \geq 4m^2, t \leq, u \leq 0$, for the t-channel $t \geq 4m^2, s \leq 0, u \leq 0$ and for the u-channel $u \geq 4m^2, s \leq 0, t \leq 0$. Analyticity of amplitudes (see below) allows one to connect these different kinematical regions and to describe all three channels by a single analytical function $T(s, t, u)$. These crossing properties of amplitudes in some cases lead to a symmetry of an amplitudes under interchange of variables. For example if $2 = \bar{3}$ (as in $\pi^+ + \pi^- \to \pi^+ + \pi^-$) then s and t-channels are the same and the amplitude $T(s, t)$ should be symmetric under $s \rightleftarrows t$ interchange. For $\bar{4} = 2$ the amplitude is symmetric under $s \rightleftarrows u$.

c) Unitarity

The total probability of a transition of a given state $|i>$ to any final state is equal to unity

$$\sum_f S_{fi}^* S_{fi} = 1 \tag{2.8}$$

and using also the superposition principle we obtain the unitarity condition for the S-matrix

$$S^+ S = \hat{I} \tag{2.9}$$

For the scattering amplitudes this condition leads (t-invariance $T_{if} = T_{fi}$ was also used)

$$ImT_{fi} = \frac{1}{2}\sum_n \int d\tau_n T^*_{nf} T_{ni} \tag{2.10}$$

For forward elastic scattering amplitude (i=f) from eqs. (2.10) and (2) we obtain the relation,

$$ImT(s,o) = 2J\sigma^{(tot)}(s) = 2p\sqrt{s}\sigma^{(tot)} \tag{2.11}$$

which is called the optical theorem.
The unitarity condition has an especially simple form for the amplitudes with a given orbital momentum l (partial wave amplitudes). The partial wave expansion of an elastic amplitude has the form

$$f(s,cos\theta) = \frac{T(s,t)}{8\pi\sqrt{s}} = \frac{1}{p}\sum_{l=0}^{\infty}(2l+1)f_l(s)P_l(cos\theta) \tag{2.12}$$

and

$$f_l(s) = \frac{p}{2}\int_{-1}^{1} f(s,z)P_l(z)dz; \quad z \equiv cos\theta \tag{2.13}$$

For partial waves the unitarity condition (2.10) takes the form

$$Imf_l(s) = |f_l(s)|^2 + G_l^{(in)}(s) \tag{2.14}$$

where $G_l^{(in)}(s) = \sum_n |f_l^{(in)n}|^2 \geq 0$ is the contribution of inelastic processes.
It follows from eq. (2.14) that

$$Imf_l(s) \geq |f_l(s)|^2 \tag{2.15}$$

This condition leads to the inequality

$$|f_l(s)| \leq 1 \tag{2.16}$$

The amplitudes $f_l(s)$ satisfying eqs. (2.14)–(2.16) can be written in the form

$$f_l(s) = \frac{1}{2i}(e^{2i\delta_l(s)} - 1) \tag{2.17}$$

with $Im\delta_l(s) \geq 0$.
For $G_l^{(in)} = 0$ (for example in the region of s below threshold for inelastic processes) $\delta_l(s)$ is real.
In the region, where $Im\delta_l(s) \equiv \eta_l > 0$ the function $G_l^{in}(s)$ has the form

$$G_l^{(in)}(s) = \frac{1}{4}(1 - e^{-4\eta_l(s)}) \tag{2.18}$$

Thus $G_l^{(in)}(s) \leq \frac{1}{4}$.

d) Analyticity

It follows from the unitarity condition (10) that the imaginary part of the scattering amplitude appears at the values of s corresponding to the thresholds $s_{th} = (m_i + m_k)^2$. At these points it behaves as $\tau_2 \sim \sqrt{s - (m_i + m_k)^2}$. For multiparticle production $ImT_n \sim (s - s_n)^{\frac{3n-5}{2}}$ close to threshold of this process. Thus the amplitudes have branch point singularities at the values of s corresponding to thresholds of physical processes in the s-channel. Below the lowest threshold $s = s_2$ amplitudes are real but can have poles, corresponding to possible transitions to stable (relative to strong interactions) particles. For example the amplitude of $\pi^- p$-scattering has the pole at $s = m_N^2$ due to an intermediate neutron state. The pp-scattering amplitude does not have poles in the s-channel, but it has pion poles in the t and u-channels (reactions $\bar{p}p \to \bar{p}p$).
The crossing property and the kinematical relation (2.4) of amplitudes imply that there are also singularities at values of s corresponding to the threshold values in the u-channel at $s = h - t - u_{th}$. Thus the scattering amplitude as a function of s at fixed t has in general the right hand cut (connecting all s-channel branch points.)and left hand cut corresponding to u-channel branch points.
On the real axis between the right hand cut and left hand cut $ImT = 0$, i.e. the amplitude $T(s)$ is real. So in the cases when the right and left hand cuts do not overlap the scattering amplitude satisfy the following condition

$$T(s^*, t) = T^*(s, t) \tag{2.19}$$

If for fixed t and $|s| \to \infty$ the amplitude $T(s,t) \to 0$, then using Cauchy formula we obtain

$$T(s,t) = \sum_i \frac{C_i}{s_i - s} + \sum_j \frac{C'_j}{h - u_j - t - s} + \frac{1}{2\pi i} \oint_C \frac{T(s',t)ds'}{s' - s} \tag{2.20}$$

where the contour of integration C encloses the cuts of the amplitude. This formula can be written in the form

$$T(s,t) = \sum_i \frac{C_i}{s_i - s} + \sum_j \frac{C'_j}{u_j - u} + \frac{1}{\pi} \int_{W_s^2}^{\infty} \frac{A_1(s',t)ds'}{s' - s} +$$

$$+\frac{1}{\pi} \int_{W_u^2}^{\infty} \frac{A_2(u',t)du'}{u' - u} \tag{2.21}$$

where the functions A_1 and A_2 are the discontinuities of the amplitude $T(s,t)$ on the right hand and left hand cuts correspondingly. Equations (2.20), (2.21) are usually called dispersion relations for the amplitude $T(s,t)$ as a function of s for a fixed value of t.
In general case when an amplitude increases for $s \to \infty$ as s^γ it is necessary to write dispersion relation with n "substractions", where n is the smallest integer larger than γ. In this way one obtains

$$T(s,t) = \sum_{k=0}^{n-1} T^{(k)}(s_0)(s - s_0)^k + \frac{1}{\pi} \int_{W_s^2}^{\infty} \frac{(s - s_0)^n}{(s' - s_0)^n} \frac{A_1(s',t)ds'}{s' - s} +$$

$$+\frac{1}{\pi} \int_{W_u^2}^{\infty} \frac{(u - u_0)^n}{(u' - u_0)^n} \frac{A_2(u',t)du'}{u' - u} \tag{2.22}$$

Here we have not written explicitly pole terms. They can be included as δ- type singularities of the functions A_i.
Dispersion relation eq. (2.21) can be also written in the form

$$ReT(s + i\tau, t) = \frac{P}{\pi} \int_{W_s^2}^{\infty} \frac{ImT(s' + i\tau)ds'}{s' - s} + \frac{P}{\pi} \int_{W_u^2}^{\infty} \frac{ImT(u' + i\tau)du'}{u' - u} \tag{2.23}$$

where the symbol P means a principal value of the integral.
Using crossing symmetry of different channels of the processes it is possible to write dispersion relations analogous to eq. (2.21) for fixed values of variables s or u

$$T(s,t) = \frac{1}{\pi} \int_{W_t^2}^{\infty} \frac{A_3(t',s)dt'}{t' - t} + \frac{1}{\pi} \int_{W_u^2}^{\infty} \frac{A_2(u',s)}{u' - u} du' \tag{2.24}$$

The last equation allows one to study analytic properties of $T(s,t) = T(s,z)$ as a function of the scattering angle $z = cos\theta_s$ in the c.m. system of the s-channel. The quantities t and u are related to z by the following relations

$$t = (p_1 - p_3)^2 = t_0 + 2pp'z, \quad u = (p_1 - p_4)^4 = u_0 - 2pp'z \tag{2.25}$$

where $t_0 = -(2\omega_1\omega_3 - m_1^2 - m_3^2)$, $u_0 = -(2\omega_1\omega_4 - m_1^2 - m_4^2)$.
Thus equation (2.24) can be written in terms of the variable z

$$T(s,z) = \frac{1}{\pi}\int\limits_{z_3(s)}^{\infty} \frac{A_3(z',s)}{z'-z}dz' + \frac{1}{\pi}\int\limits_{z_2(s)}^{\infty} \frac{A_2(z',s)dz'}{z'+z} \quad (2.26)$$

where $z_3(s) = \frac{W_t^2 - t_0}{2pp'}$, $z_2(s) = \frac{W_u^2 - U_0}{2pp'}$.

2.2. FROISSART THEOREM

Dispersion relations and unitarity condition allow one to obtain strong bounds on a rate of increase of scattering amplitudes at high energies. Let us derive the bound on total cross section and radius of interaction for $s \to \infty$ obtained by Froissart [3].
Consider the partial wave expansion for the amplitude $T(s,t)$ (2.12).

$$T(s,t) = \sum_{l=0}^{\infty}(2l+1)T_l(s)P_l(z)$$

In a calculation of a partial wave amplitudes $T_l(s) = \frac{1}{2}\int\limits_{-1}^{1} T(s,z)P_l(z)dz$ we will use the dispersion relation for $T(s,z)$ (2.26) and taking into account that
$\frac{1}{2}\int\limits_{-1}^{1} \frac{P_l(z)dz}{z'-z} = Q_l(z')$, and $\frac{1}{2}\int\limits_{-1}^{1} \frac{P_l(z)dz}{z'+z} = (-1)^l Q_l(z')$
where $Q_l(z)$ is the Legandre function of the second type, we obtain

$$T_l(s) = \frac{1}{\pi}\int\limits_{z_3(s)}^{\infty} A_3(z',s)Q_l(z')dz' + (-1)^l\frac{1}{\pi}\int\limits_{z_2(s)}^{\infty} A_2(z',s)Q_l(z')dz' \quad (2.27)$$

If the scattering amplitude increases as s^γ and it is necessary to write dispersion relation with n substractions then the representation (2.27) holds for $l \geq n$.
For large values of l which are important at high energies the functions Q_l decrease exponentially

$$Q_l(z) \simeq \pi[2\pi l(z^2-1)]^{-1/2}exp[-(l+\frac{1}{2})arch(z)]; \; l \to \infty$$

Therefore in the integrals of eq. (2.27) minimal values of z close to the lower integration limit are important

$$T_l(s) \leq s^\gamma exp^{-l arch(z_{min})} \quad (2.28)$$

where $z_{min} = min(z_3, z_2)$, and s^γ corresponds to a behaviour of $A_3(z_3, s)$ or $A_2(z_2, s))$ at $s \to \infty$. Taking into account that at large s $z_{min} = 1 + \frac{c}{s}$, where $c = min(\sqrt{2W_t^2}, \sqrt{2W_u^2})$, and $archz_{min} \simeq \frac{c}{\sqrt{s}}$ we obtain

$$T_l(s) \leq exp^{\frac{-lc}{\sqrt{s}} + \gamma \ln s} \tag{2.29}$$

Thus an effective contribution to the scattering amplitude comes from a limited number of partial wave amplitudes with $l \leq L$, where $L = \frac{N}{c}\sqrt{s} \ln s$, and $N \geq \gamma$. Equation (2.29) means that the region of impact parameters $b = \frac{l+1/2}{p} \approx \frac{2l}{\sqrt{s}}$ where the scattering take place can not increase with energy faster than $\ln s$ and thus an increase of an interaction radius is bounded by $\ln s$ behaviour.
Let us now use for partial wave amplitudes $T_l(s) = \frac{8\pi W}{p} f_l(s)$ an inequality $T_l(s) \leq 8\pi W/p$, which follows from the unitarity bound eq (2.16). We obtain the following bound for the forward scattering amplitude at very high energies

$$|T(s,z)| \leq \sum_{l=0}^{L}(2l+1)|T_l(s)| \leq \sum_{l=0}^{L}(2l+1)\frac{8\pi W}{p} \approx \frac{8\pi W}{p} L^2 = C_1 s \ln^2 s \tag{2.30}$$

where $C_1 = 16\pi N^2/c^2$. This inequality leads to the following bound for the total interaction cross section

$$\sigma^{(tot)}(s) \leq C_1 \ln^2 s \tag{2.31}$$

which is usually called the Froissart theorem.

2.3. POMERANCHUK THEOREM

Crossing and analyticity of scattering amplitudes leads also to interesting relations between asymptotic behaviour of scattering amplitudes for particles and anti-particles. Pomeranchuk [4] has shown that dispersion relations for forward scattering amplitudes with natural extra assumptions:
a) Amplitudes do not oscillate as $s \to \infty$
b) $\frac{|ReT(s,0)|}{ImT(s,0)} \frac{1}{lns} \to 0$ for $s \to \infty$
lead to an important asymptotic relation between total interaction cross sections for particles and anti-particles

$$\sigma_{ab}^{(tot)} = \sigma_{\bar{a}b}^{(tot)}, \quad for \quad s \to \infty \tag{2.32}$$

In reggeon theory described below the Pomeranchuk theorem is automatically satisfied if the leading Regge singularity has vacuum quantum numbers. In this case differences of cross sections for particles and anti-particles decrease as powers of s for $s \to \infty$.

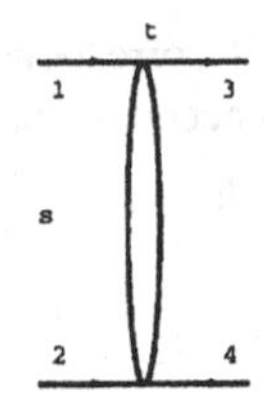

Figure 1. A diagram for a binary reaction.

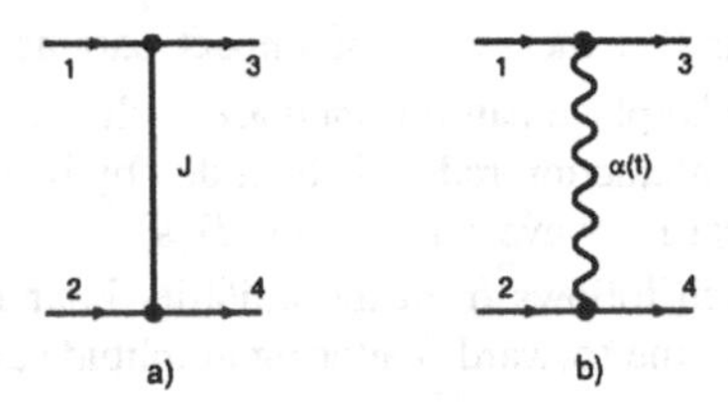

Figure 2. a) Exchange by a particle of spin J in the t-channel. b) Exchange by a Regge pole in the t-channel.

3. Theory of reggeons

3.1. THE REGGEON CONCEPT

The complex angular momentum method was first introduced by Regge in non-relativistic quantum mechanics [5]. In relativistic theory it connects a high energy behaviour of scattering amplitudes with the singularities in the complex angular momentum plane of the partial wave amplitudes in the crossed channel. The simplest singularities are poles (Regge poles). A Regge-pole exchange is a natural generalization of a usual exchange of a particle with spin J to complex values of J. So this method establishes an important connection between high energy scattering and the spectrum of hadrons.

Let us consider reaction $1 + 2 \to 3 + 4$ at high energies $s \gg m^2$ and fixed momentum transfer $t \sim m^2$ (Fig. 1).

An exchange by a particle of spin J in the t-channel (Fig. 2a)) leads to an amplitude of the form

$$T(s,t) = g_1 \cdot g_2 \cdot (s/s_0)^J/(M_J^2 - t) \tag{3.1}$$

where $s_0 = 1\ GeV^2$, g_i are the coupling constants and M_J is the mass of the exchanged particle.

It follows from eq. (3.1) that for particles with spins $J \geq 2$ the amplitude increases faster than s, thus violating Froissart bound. On the other hand we know from experiment that there are many hadrons with spins $J \geq 2$. This problem can be solved by introduction of Regge poles. It should be taken into account that the expression (3.1) for the amplitude is valid, strictly speaking, only close to the pole position $t \approx M_J^2$ and can be strongly modified away from the pole. Regge pole model give an exact form of this modification and absorbs in itself exchanges by states of different spins (Fig 2b)). The corresponding amplitude has the form

$$T(s,t) = f_1(t) \cdot f_2(t) \cdot (s/s_0)^{\alpha(t)} \cdot \eta(\alpha(t)) \quad (3.2)$$

where $\alpha(t)$ is the Regge-trajectory, which is equal to spin J of the corresponding particle at $t = M$. The function $\eta(\alpha(t)) = -(1 + \sigma exp(-i\pi\alpha(t))$ is a signature factor and $\sigma = \pm 1$ is a signature. It appears due to the fact that in relativistic theory it is necessary to consider separately analytic continuation of partial wave amplitudes to complex values of angular momenta J from even $(\sigma = +)$ and odd $(\sigma = -)$ values of J. We will discuss this problem and derivation of eq. (3.2) in detail below. It should be emphasized that single Regge exchange corresponds to an exchange of particles or resonances which are "situated" on the trajectory $\alpha(t)$. For example if $\alpha(t) = J$, where J is an even (odd) integer for $\sigma = +(-)$ for $t = M_J^2$ and M_J^2 is less than the threshold for transition to several hadrons ($4m_\pi^2$ for particles which can decay into two pions), then the Regge amplitude (34) transforms into particle exchange amplitude (3.1) with $g_1 g_2 = f_1(M_J^2) \cdot f_2(M_J^2) 2/\pi\alpha'(M_J^2)$. If M_J is larger than the threshold value than $\alpha(t)$ has a complex value and be written for $t \approx M_J^2$ in the form

$$\alpha(t) = J + \alpha'(M_J^2) \cdot (t - M_J^2) + iIm\alpha(M_J^2) \quad (3.3)$$

In this case for $Im\alpha(M_J^2) \ll 1$ Regge pole amplitude (3.2) corresponds to an exchange in the t–channel by a resonance and has a Breit–Wigner form

$$T(s,t) = -g_1 \cdot g_2 (s/s_0)^J / (t - M_J^2 + iM_J \Gamma_J) \quad (3.4)$$

with a width $\Gamma_J = Im\alpha(M_J^2)/M_J \alpha'(M_J^2)$.

Thus a reggeization of particle exchanges leads to a natural resolution of the above mentioned problem with a violation of the Froissart bound, – Regge trajectories, which correspond to particles with high spins can have $\alpha(t) \leq 1$ in the physical region of high energy scattering $t < 0$ and the corresponding amplitudes will increase with s not faster than s^1, satisfying the Froissart bound. We will see later in this chapter that the experimental information on spectra of hadrons and high energy scattering processes nicely confirms this theoretical expectation. The only exception is the Pomeranchuk pole (or pomeron), which determines high energy behavior of diffractive processes. We will pay a special attention to properties of the pomeron below.

3.2. COMPLEX ANGULAR MOMENTA METHOD

In this section we will give a derivation of the Regge pole amplitude, discussed above and will obtain a general formula, which expresses an asymptotic behavior of scattering amplitudes $T(s,t)$ at high energies $s \gg m^2$ and fixed momentum transfer $t = -q^2$ in terms of singularities in the complex angular momentum plane of the partial wave amplitudes $T_j(t)$ in the crossed (t) channel.
An idea is to consider first the amplitude $T(s,t)$ in the physical region of the t-channel (region III) and to continue it analytically on the variable s to the unphysical region (II) $(s \gg m^2, t > 4m^2)$ and then to continue it, using t variable to the physical region of the s-channel (I) .
This analytical continuation is possible to perform using the Sommerfeld-Watson method of complex angular momenta . For this purpose it is necessary to make an analytic continuation of the partial wave amplitudes $T_l(t)$ to complex values of angular momenta. This can be done, using the same method of dispersion relations, which has been used in the Section 2 to obtain the representation for the partial wave amplitudes in the s-channel $T_l(s)$. In our case the amplitudes $T_l(t)$ can be written in the form

$$T_l(t) = T_l^1(t) + (-1)^l T_l^2(t) \tag{3.5}$$

where

$$T_l^i = \frac{1}{\pi}\int_{z_1(t)}^{\infty} Q_l(z)A_i(z,t)dz, \; i = 1,2 \tag{3.6}$$

We will use eqs. (3.5), (3.6) in order to obtain an analytic continuation of partial wave amplitudes $T_l(t)$ to complex values of $l \equiv j$

$$T(j,t) = \frac{1}{\pi}\int_{z_1(t)}^{\infty} Q_j(z,t)A_1(z,t)dz + \frac{e^{-i\pi j}}{\pi}\int_{z_2(t)}^{\infty} Q_j(z)A_2(z,t)dz \tag{3.7}$$

this equation determines $T(j,t)$ as analytic function of j in the region of j-plane, where integrals over j are convergent. At large z the functions Q_j have the following behavior

$$Q_j(z) \cong \frac{\pi}{C(j)}(2z)^{-(j+1)}, \; C(j) = \pi^{\frac{1}{2}}\Gamma(j+\frac{3}{2})/\Gamma(j+1) \tag{3.8}$$

So, if for a fixed value of t $A_i(z,t)$ behave as z^γ for large z, then integrals converge for $Re j > \gamma$. Thus amplitudes $T(j,t)$ given by eq. (3.8) are analytic function of j in this region and coincide with $T_l(t)$ for integer j. However it is well known, that in order to have a unique analytic continuation of a function from integer values it is also necessary that in the region $Rej > \gamma$ it should satisfy to a condition: $T(j,t) < exp(\pi j)$ as $j \to \infty$ along any direction in this semi-plane. But this condition is not satisfied by the representation (3.7) due to the factor $exp(-i\pi j)$ before the second integral. It increases as $exp(\pi|j|)$ for $Rej = c$ and $Imj \to \infty$. This leads us to necessity to consider separately analytic continuations from even and odd values of angular momenta l. So we introduce two functions

$$T^{\sigma}(j,t) = T^{1}(j,t) + \sigma T^{2}(j,t) \tag{3.9}$$

where $\sigma = \pm$, and

$$T^{i}(j,t) = \frac{1}{\pi}\int_{z_i(t)}^{\infty} Q_j(z) A_i(z,t) dz \tag{3.10}$$

Function $T^{+}(j,t)$ coincides with $T_l(t)$ for even integer values of j and $T^{-}(j,t)$- for odd integers. It is clear that these functions satisfy to all conditions necessary for a unique analytic continuation. The new quantum number σ is usually called signature. It plays an important role in relativistic theory and is closely connected to crossing properties of amplitudes.
The functions $T^{\sigma}(j,t)$ allows us to write the unitarity condition analytically continued to complex values of j. Consider, for example, scattering of two particles with equal masses m. In the region $4m^2 < t < 9m^2$ two-particle intermediate state gives a contribution to the unitarity condition, which for amplitudes $T_l(t)$ has the form

$$ImT_l(t) = \frac{1}{16\pi}\left(\frac{t-4m^2}{t}\right)^{1/2}|T_l(t)|^2 \equiv \rho_2(t)|T_l(t)|^2 \tag{3.11}$$

Taking into account that the function $T_l(t)$ is real on some part of the real axis for $t < 4m^2$ (this can be proved, using the representation (3.5), (3.6)) and thus for $t > 4m^2$ $T_l(t^*) = T_l^*(t)$, unitarity condition (3.11) can be written in the following form

$$T_l(t_+) - T_l(t_-) = 2i\rho_2(t_+)T_l(t_+)T_l(t_-) \tag{3.12}$$

where $t_{\pm} = t \pm i\varepsilon, \varepsilon > 0, \varepsilon \to 0$.
An analytic continuation of this relation to the complex values of j is straightforward

$$T^{\sigma}(j,t_+) - T^{\sigma}(j,t_-) = 2i\rho_2(t_+)T^{\sigma}(j,t_+)T^{\sigma}(j,t_-) \tag{3.13}$$

Now we have everything which is necessary in order to make an analytic continuation of the amplitude $T(s,t)$, written as a sum over partial wave amplitudes,

$$T(s,t) = \sum_{\sigma}\sum_{l=0}^{\infty}(2l+1)\frac{1}{2}(P_l(z_t) + \sigma P_l(-z_t))T_l(t) \tag{3.14}$$

from the physical region of the t-channel to the region II . The cosine of the scattering angle in the t-channel z_t becomes large in the unphysical region II

$$z_t \cong s/2p_{13}(t)p_{24}(t) \gg 1 \tag{3.15}$$

and the sum (3.14) diverges in this region. In order to avoid this difficulty let us write the following representation for the amplitude $T(s,t)$

$$T(s,t) = \sum_{l=0}^{k-1}(2l+1)P_l(z_t)T_l(t) +$$

$$+\sum_{\sigma}\frac{1}{4i}\int_C \frac{2j+1}{sin(\pi j)}(P_j(-z_t)) + \sigma P_j(z_t))T^{\sigma}(j,t)dj \tag{3.16}$$

which is equivalent to the series (3.14) in the physical region. Contour C of the integration in eq. (3.16) encloses a part of the real j axis, starting from $j = k$, where k is the smallest integer larger than γ (γ-was defined above and determines a region of j-plane where an analytic function $T^{\sigma}(j,t)$ exists). Legandre functions $P_j(z)$ are analytic functions of j and the only singularities of the integrand in (3.16) are due to zeros of $sin(\pi j)$ at $j = l$. For $j \to l$ $(1/sin(\pi j)) \cong (-1)^l/\pi(j-l)$ and the integral over contour C which is equal to the sum of residues of this poles reproduces the part of the sum (3.14) from $l = k$ to ∞. This integral can be transformed into the integral over contour $C = L + R$, which consists of the vertical line L and the large semicircle R. In the region III the last contribution vanishes, because the integrand decreases exponentially as $|j| \to \infty$. Thus the integral over the line L in the complex j-plane is equivalent to a part of the sum (3.14) in the region III, where this sum converges. This representation of the amplitude is very convenient for our purpose, because it can be used also in the region II, where $z_t \gg 1$ and both the sum (3.14) and equivalent to it integral over contour C (3.16) diverges. It is not difficult to see that an integral over contour L is absolutely convergent for any complex values of z_t, so it gives an analytic continuation of the amplitude to the region II.
Now let us move the line L to the left in the j-plane enclosing all singular points of the integrand. They include possible singularities of the functions $T^{\sigma}(j,t)$ and simple poles connected to zeros of $sin(\pi j)$ in the points $j = 1 = k-1, k-2, ..., 0$. If in this points $T^{\sigma}(l,t) = T_l(t)$ then this contribution cancels exactly the first sum in eq. (3.16) and, taking into account that $P_j(-z_t) = exp(-i\pi j)P_j(z_t)$, we obtain the following representation of $T(s,t)$

$$T(s,t) = \sum_{\sigma} \int_{\uparrow} \pi\sigma(j+\frac{1}{2}) T^{\sigma}(j,t)\eta_{\sigma}(j) P_j(z_t) \frac{dj}{2\pi i} \tag{3.17}$$

where $\eta(j) = -(1+\sigma exp(-i\pi j))/sin(\pi j)$ is a signature factor and $\uparrow$ indicates the direction of integration a vertical line L' . In the region II, where $p_{13}p_{34} > 0$ and $s \to \infty$ $(s = s + i\varepsilon, \varepsilon > 0)$ we can use an asymptotic formula for functions $P_j(z)$

$$\pi(j+\frac{1}{2}) P_j(z_t) \cong C(j)(2z_t)^J \cong C(j)s^j/(p_{13}p_{24})^j \tag{3.18}$$

and C(j) is the same as in eq. (3.8). So we can write the scattering amplitude as the Mellin-Laplace transform

$$T(s,t) = \int_{\uparrow} \phi(j,t) s^j \frac{dj}{2\pi i} \tag{3.19}$$

of the function $\phi(j,t) = \sum \eta_{\sigma}(j)\phi_{\sigma}(j,t)$, where $\phi_{\sigma}(j,t) = \frac{C(j)\sigma T^{\sigma}(j,t)}{(p_{13}p_{24})^j}$. The amplitudes $\phi_{\sigma}(j,t)$ are convenient, because they do not contain kinematical singularities as $p_{13(24)} \to 0$ for non-integer j and can be easily continued from the region II to the physical region of s-channel (region I).
If the singularities of amplitudes $\phi_{\sigma}(j,t)$ in the j-plane are poles and cuts then we can transform the integral (3.19) into a sum of terms, representing the contributions over contours C_i, which enclose the i-th singularity in the j-plane plus a contribution T_{δ} over the line L'', which is situated in j-plane at $Rej = -\delta$.

$$T(s,t) = \sum_{i} \int_{C_i} \phi_{\sigma}(j,t)\eta_{\sigma}(j) s^j \frac{dj}{2\pi i} + T_{\delta} \tag{3.20}$$

The contribution T_{δ} decreases with s as $s^{-\delta}$ and can be neglected as $s \to \infty$. Representation (3.21) is very important because it allows us to express the asymptotic properties of scattering amplitudes for $s \to \infty$ in terms of the rightmost singularities in the j-plane.

3.3. GENERAL PROPERTIES OF REGGEONS

The most important and physically interesting possibility corresponds to the case when the rightmost singularity in the j-plane is a pole. In general its position depends on t, so this is a moving pole. In this case the function $\phi_{\sigma}(j,t)$ can be written in the form

$$\phi_\sigma(j,t) = \frac{b_a(t)}{j-\alpha_a(t)};\ j \to \alpha_a(t) \tag{3.21}$$

The scattering amplitude $T(s,t)$ according to eq. (3.21) takes the form

$$T(s,t) = \eta_a(t) b_a(t) (s)^{\alpha_a(t)} \tag{3.22}$$

which is equivalent to the Regge-representation (3.2).
Some properties of Regge pole can be inferred from the unitarity condition (3.13). It follows in particular from this equation that amplitudes $T(j,t)$ do not have poles with t-independent positions (fixed poles) and thus have only moving poles. Indeed, if the pole position does not depend on t then the left hand side of eq. (3.13) has a simple pole, while the right hand part- a double pole. This elastic unitarity also shows that the trajectory of the pole $\alpha(t)$ must have a branch point at the threshold $t = 4m^2$, i.e. $\alpha_a(t_+) \neq \alpha_a(t_-)$ for $t > 4m^2$. Below threshold the trajectory is a real function of t. The residue $b(t)$ is a real function of t both below and above threshold.
In general there are several two-body channels and the partial wave amplitude $T_{ik}(j,t)$, which describes a transition in the t-channel from the state i to the state k can be considered as a matrix. The two-particle unitarity condition (3.13) can be written in the matrix form.

$$\hat{T}(j,t_+) - \hat{T}(j,t_-) = 2i\hat{T}(j,t_+)\hat{\rho}(t_+)\hat{T}(j,t_-) \tag{3.23}$$

where $\hat{\rho}(t)$ is the diagonal matrix $\rho_{ii}(t) = p_i/8\pi\sqrt{t}$.
Using this equation it is possible to show that the same Regge pole exists for amplitudes of all coupled channels

$$T_{ik}(j,t) = \frac{r_{ik}(t)}{j-\alpha(t)} \tag{3.24}$$

and the residues r_{ik} have a factorized form

$$r_{ik}(t) = r_i(t) r_k(t) \tag{3.25}$$

A solution of eq. (3.23) can be written in the form

$$\hat{T}(j,t) = |\hat{I} - i\hat{R}(j,t)\hat{\rho}(t)|^{-1}\hat{R}(j,t) \tag{3.26}$$

where R is the matrix, which satisfies to the condition

$$\hat{R}(j,t_+) = \hat{R}(j,t_-) \tag{3.27}$$

A moving pole corresponds to a zero of the determinant of the matrix $\hat{I} - \hat{R}(j,t)i\hat{\rho}(t)$. The unitarity equation (3.23) connects the states with definite values of the conserved quantum numbers such as parity P, isospin I, G-parity G, strangeness S,... So each Regge pole has definite conserved quantum numbers, besides the signature.

The properties of Regge poles formulated in this section mean that the contribution of the Regge pole to the scattering amplitude can be represented by the diagram shown in Fig. 2b). This diagram is a natural generalization of the particle exchange graph of Fig. 2a) and denotes an exchange in the t-channel by a state (reggeon) with arbitrary spin $\alpha(t)$, which has definite signature, parity and other conserved quantum numbers.

3.4. BOSONIC AND FERMIONIC REGGE POLES. THE POMERON

An information on trajectories of Regge poles can be obtained for $t < 0$ from data on two-body reactions at large s and for $t > 0$ from our knowledge of the hadronic spectrum. We have seen that a bosonic trajectory corresponds to particles and resonances for those values of t where it passes integer values ($Re\alpha(t) = n$) even for $\sigma = +$ and odd for $\sigma = -$. While for fermionic trajectories particles correspond to $Re\alpha(t) = \frac{n}{2} = J$ and signature $\sigma = (-1)^{J-\frac{1}{2}}$.

There can be many trajectories with the same quantum numbers indicated above, which differ by a quantum number analogous to the radial quantum number. Such trajectories are usually called "daughter" trajectories and masses of corresponding resonances (with the same value of J) for them are higher than those for the leading trajectory with given quantum numbers.

Trajectories for some well established bosonic Regge poles are shown in Fig.3. Note that all these trajectories have $\alpha_i(0) \leq 0.5$ for $t \leq 0$. One of the most interesting properties of these trajectories is their surprising linearity. This usually interpreted as a manifestation of strong forces between quarks at large distances, which lead to color confinement. The linearity of Regge trajectories indicates to a string picture of the large distance dynamics between quarks and it was a basis of dual models for hadronic interactions. Other properties of mesonic Regge trajectories, which are evident from Fig.3 are exchange and isospin degeneracies, – trajectories with different signatures and I=0 or I=1 (but with the same σP) are degenerate with a good accuracy (at least in the region $t > 0$). This is also in an agreement with dual models or approaches based on 1/N-expansion in QCD. In dual or string models of hadrons the daughter trajectories must be parallel to the leading one and displaced in the j-plane by integers. Experimental information

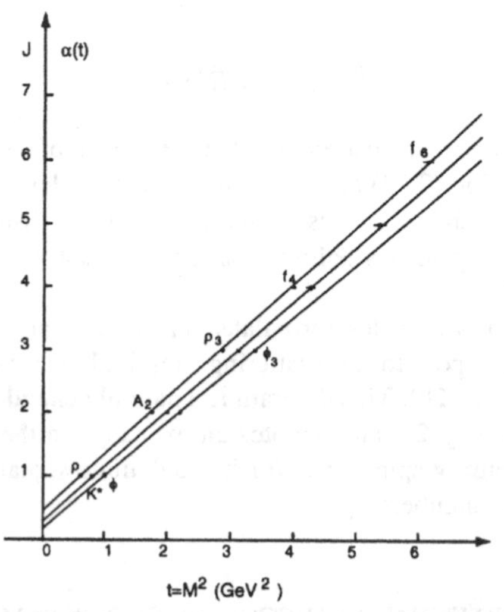

Figure 3. Trajectories for some well-known Regge poles.

on the daughter trajectories is rather limited but it does not contradict to these predictions.

Information on mesonic Regge trajectories in the region of negative t, obtained from an analysis of binary reactions at high energies fits quite well the lines shown in Fig.3 obtained from the spectrum of resonances. The most detailed information exists for ρ and A_2-trajectories, which contribute to the reactions $\pi^- p \to \pi^0 n$ and $\pi^- p \to \eta n$ correspondingly.

Fermionic Regge trajectories are in general analytic functions of $W = \sqrt{t}$ and using the analyticity properties of the corresponding partial-wave amplitudes it is possible to show that there must be pairs of trajectories with different parity, which satisfy to the condition

$$\alpha_+(W) = \alpha_-(-W) \tag{3.28}$$

Experimental data on spectrum of baryons show that baryonic trajectories as well as mesonic ones are nearly straight lines in variable t with the universal slope $\alpha' \approx 1 GeV^{-2}$. This universality of the slopes is natural in the string picture of baryons with a quark and a diquark at the ends. However, according to eq. (3.28) a fermionic trajectory, which is linear in $t = W^2$, does not change as $W \to -W$ and thus should coincide with its parity partner. This in its turn leads to parity doubling of states on this Regge trajectory. Experimentally there are many baryonic states with the same spin and different parity, which are nearly degenerate in mass. However there are no partners for the lowest states (N, Δ) on these trajectories. This pattern of baryonic Regge trajectories is not yet understood theoretically. The relation (3.28) is a consequence of analyticity in relativistic quantum theory, but it is not realized in the existing quark models of baryons.

At the end of this section I shall discuss properties of the pole which has a special status in the Regge approach to particle physics - the Pomeranchuk pole or the pomeron. This pole was introduces into theory in order account for diffractive processes at high energies. In the Regge pole model an amplitude of high energy elastic scattering has the form of eq.(34) and the total interaction cross section, which by the optical theorem is connected to $ImT(s,0)$, can be written as a sum of the Regge poles contributions

$$\sigma^{tot}(s) = \sum_k b_k(0)(s)^{\alpha_k(0)-1} \tag{3.29}$$

The poles, which are shown in Fig.3 have $\alpha_k(0) < 1$ and thus their contributions to $\sigma^{tot}(s)$ decrease as a $s \to \infty$. However experimental data show that at $s \sim 100\ GeV^2$ total cross sections of hadronic interactions have a weak energy dependence and they slowly (logarithmically) increase with energy at higher energies. In the Regge pole model this can be related to the pole, which has an intercept $\alpha_P(0) \approx 1$. This pole is usually called the pomeron or the vacuum pole, because it has the quantum numbers of the vacuum, – positive signature, parity and G (or C) parity and isospin I=0.

It is believed that in QCD this pole is related to gluonic exchanges in the t-channel. So it is usually assumed that gluonium states correspond to this trajectory in the region of positive t. We shall discuss possible relation between QCD and Regge theory in more details in the next chapters.

A value of intercept of the Pomeranchuk pole is of crucial importance for the Regge theory. If $\alpha_P(0) = 1$, as it was assumed initially, then all the total interaction cross sections tend to a constant at very high energies. This theory has however some intrinsic difficulties and must satisfy to many constraints in order to be consistent with unitarity. Besides experimental data indicate that $\sigma_{hN}^{(tot)}$ rise with energy. This logarithmic increase of total cross sections at very high energies is in accord with $ln^2 s$ behavior consistent with the Froissart theorem [3]. Thus at present the super-critical pomeron theory with $\alpha_P(0) > 1$ is widely used. In a model with only Regge poles taken into account an assumption that $\alpha_P(0) > 1$ would lead to a power like increase of total cross sections, thus violating the Froissart bound. However in this case other singularities in the j-plane,– moving branch points should be taken into account and their contributions allow one to restore unitarity and to obtain the high energy behavior of scattering amplitudes which satisfy to the Froissart bound. Properties of these moving cuts are considered below.

Let us note that the Pomeranchuk singularity has a positive signature, so it gives equal contribution to amplitudes for elastic scattering of particle and anti-particle. Thus it automatically satisfies to the Pomeranchuk theorem on asymptotic equal-

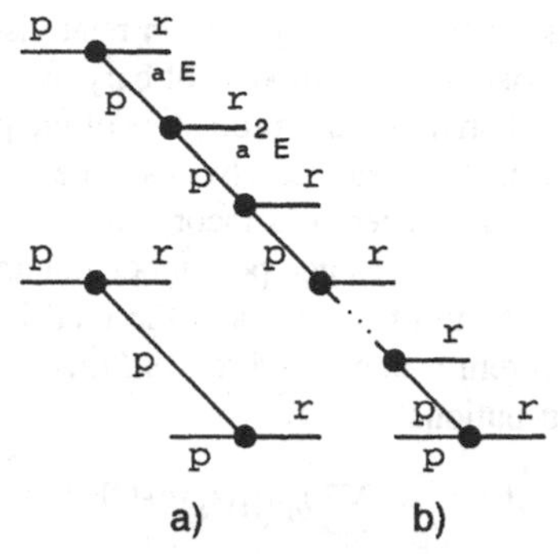

Figure 4. a) Exchange by a pion in the t-channel in the reaction $\pi\pi \rightarrow \rho\rho$. b) Diagram of multi-peripheral production of ρ mesons .

ity of total interaction cross sections for particle and anti-particle. A difference between these amplitudes in the Regge model is connected to poles with negative signature (like ρ and ω). It is usually assumed that the only singularities in the j-plane with negative signature are due to the known Regge poles with $\alpha_k(0) \leq 0.5$. In this case differences of cross sections for aN and $\bar{a}N$-interactions decrease with energy as $1/\sqrt{s}$. This behavior agrees with existing experimental information. In principle it is possible to have a singularity with negative signature at $j \approx 1$. This singularity is usually called "odderon". It appears in perturbative QCD calculations.

3.5. S-CHANNEL PICTURE OF REGGEONS

Regge poles give contributions to imaginary parts of two-body scattering amplitudes. For the Pomeron with $\alpha_P(0) \approx 1$ the amplitude is mostly imaginary. Unitarity relates imaginary parts of two-body amplitudes to sums over intermediate states in the s-channel. So the natural question is: what are the intermediate states connected to reggeons? These are so called multi-peripheral states the properties of which I shall briefly discuss now. Consider as an example the simplest inelastic process of ρ-mesons production in $\pi\pi$ collisions. The diagrams with pion exchange shown in Fig.4 give an important contribution to amplitudes of these processes. They lead to peripheral configurations – all ρ-mesons are produced with rather small momentum transfer $\sim m_\pi$. The diagram of Fig. 4 a) is large only at rather low energies when the average energy of the virtual pion in the lab. system ($\sim \alpha E, \alpha \leq 1$, E-is the lab. energy of the projectile) is low enough to produce a ρ meson with a target pion. At higher energies the cross section of this process decreases with energy as $1/s^2$. At these energies it is necessary to decrease the energy of the virtual pion in several steps as it is shown in Fig.4 b) in order to have a slow virtual pion at the end of the chain. If the energy decreases

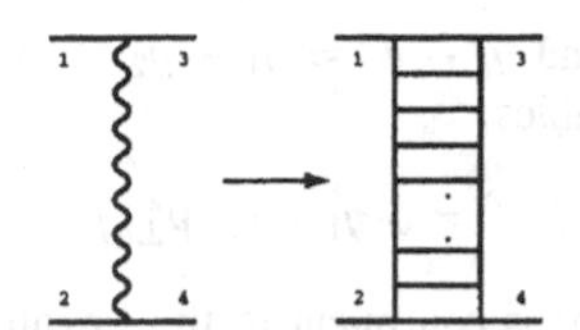

Figure 5. Reggeon as a ladder diagram.

by a factor α at each step then after n steps it will be $\alpha^n E$ and we require that it is $\sim m_\rho$. So on average the number of steps (related to the number of produced resonances) $\langle n \rangle \sim ln\frac{E}{m_\rho}/ln\frac{1}{\alpha}$.
This is an example of the diagrams of the multi-peripheral model [6] of multi-particle production. Summation of these diagrams leads to the Regge-type behavior for an imaginary part of two-body amplitudes. So in this model reggeon corresponds to a sum of ladder-type diagrams shown in Fig.5. Let us consider now the multi-peripheral process in the impact parameter $\vec{b}$ space. Each step has a finite size in $\vec{b} \sim 1/m_\pi$. They add independently, so there is a random walk in the $\vec{b}$-space and the average size of interaction increases with energy as

$$\bar{R}^2 \equiv \langle \vec{b}^2 \rangle \approx \frac{\bar{n}}{m_\pi^2} \sim \frac{ln\frac{E}{m_\rho}}{m_\pi^2} \tag{3.30}$$

The total time of development of the fluctuation in the lab. frame is large, $-\tau \sim E/m^2$.
In the multi-peripheral model the hadronic final states have the following properties.
a) Short range correlations in rapidity. Particles separated by several steps of the multi-peripheral process and having substantially different energies [this means that difference of rapidities $y_1 - y_2$ $(y = ln(E + p_{||})/m_\perp)$ for these particles is large] are uncorrelated. The correlation function exponentially decreases with rapidity difference

$$c(y_1, y_2) = \frac{d\sigma}{\sigma^{in} dy_1 dy_2} - (\frac{d\sigma}{\sigma^{in} dy_1})(\frac{d\sigma}{\sigma^{in} dy_2}) \sim exp[-\lambda(y_1 - y_2)] \tag{3.31}$$

This property of the multi-particle final state leads to many consequences for inclusive cross sections. Consider for example single particle inclusive cross section

$$f^a \equiv E\frac{d^3\sigma^a}{d^3p} = \frac{d^2\sigma(y_1 - y_a, y_a - y_2, p_{\perp a}^2)}{dy_a d^2 p_{\perp a}} \tag{3.32}$$

Short range correlation between particles means in this case that all inclusive densities $n^a \equiv f^a/\sigma^{in}$ are independent of $y_i - y_k$ for $y_i - y_k \gg 1$. For example at high energies when $y_1 - y_2 \simeq ln(s/m^2) \gg 1$ in the fragmentation region of particle 1 $y_1 - y_a \sim 1$ and $y_a - y_2 \approx y_1 - y_2 \gg 1$ the density n^a becomes a function of only two variables

$$n^a = \phi(y_1 - y_a, p_{\perp a}^2) \tag{3.33}$$

b) The property eq. (3.33) is equivalent to the Feynman scaling in the variable $x_F = 2p_{\|a}/\sqrt{s} \approx exp[-(y_1 - y_a)]$

$$n^a = \phi(x_F, p_{\perp a}^2) \tag{3.34}$$

c) In the central rapidity region when both $y_1 - y_a$ and $y_a - y_2$ are large n^a becomes a function of only one variable $(p_{\perp a}^2)$. So in this model rapidity distributions in the central region are flat .
d) The average number of produced particles $\langle n \rangle$ increases logarithmically with energy at large $\ln(s/m^2)$.
e) There is a fast decrease of distributions with $p_\perp$. The form of this function depends on details of the model.
f) Multiplicity distributions of produced particles in these models have Poisson-type behavior.

$$\langle n^2 \rangle - \langle n \rangle^2 \approx c \langle n \rangle \tag{3.35}$$

This is a consequence of short range correlations in rapidity.
The properties of inclusive distributions listed above can be quantified in the Regge model, using Mueller-Kancheli [7] diagrams . There is a relation between a discontinuity of $3 \to 3$ forward scattering amplitude $1\bar{a}2 \to 1\bar{a}2$ and inclusive cross section f^a analogous to the optical theorem which relates a forward elastic scattering amplitude to a total cross section. For large $y_i - y_a$ $(s_{i\bar{a}} \gg m^2)$ it is possible to use the Regge pole theory, which gives predictions on energy (rapidity) dependence of inclusive cross sections. For example consider the central region of rapidities when both $y_1 - y_a \gg 1$ and $y_a - y_2 \gg 1$. In this case diagrams of double Regge limit (Fig.6) can be used and inclusive cross sections can be written as follows

$$f^a = \sum_{i,k} g_{11}^i(0) g_{22}^k(0) g_{ik}^a exp[(\alpha_i(0) - 1)(y_1 - y_a) + (\alpha_k(0) - 1)(y_a - y_2)] \tag{3.36}$$

These results can be easily obtained in the multi-peripheral model.

3.6. DIFFRACTIVE PRODUCTION PROCESSES

Let us consider now diffractive production of particles at high energies. In the Regge pole model these processes are described by the diagrams with the pomeron exchange (Fig.7). It is possible to have excitation of one of the colliding hadrons

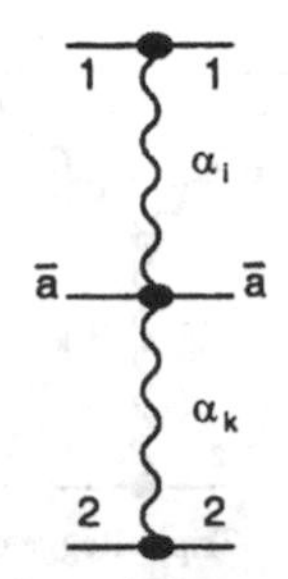

Figure 6. Regge diagram for inclusive cross section in central region.

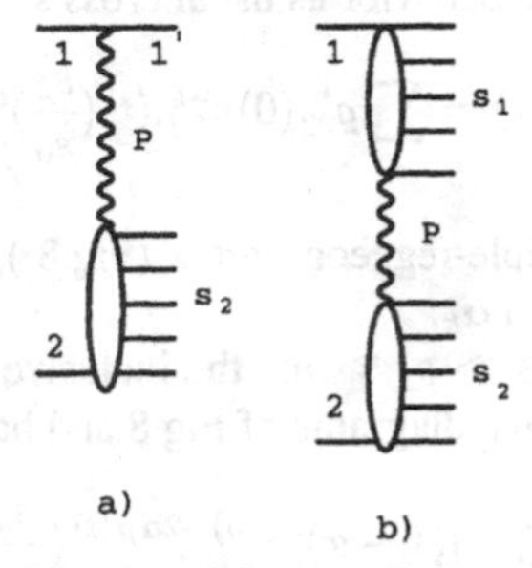

Figure 7. Diagrams for diffractive production of hadrons in the Regge pole model.

(Fig.7 a)), – single diffraction dissociation or excitation of both initial particles, – double diffraction dissociation (Fig.7 b)).

For all diffractive processes there is a large rapidity gap between groups of produced particles. For example for single diffraction dissociation there is a gap between the particle $1'$ and the rest system of hadrons. This rapidity gap $\Delta y \approx ln(1/1-x)$, where x the x_F for hadron $1'$ in Fig.7a). A mass of diffractively excited state at large s can be large. The only condition for diffraction dissociation is $s_i \ll s$. For large masses of excited states $s_2 \approx (1-x)s$ and $\xi' \approx \Delta y \approx ln(s/s_2)$. The cross section for inclusive single diffraction dissociation in the Regge pole model can be written in the following form

$$\frac{d^2\sigma}{d\xi_2 dt} = \frac{(g_{11}(t))^2}{16\pi}|G_p(\xi',t)|^2\sigma_{P2}^{(tot)}(\xi_2,t) \tag{3.37}$$

where $\xi_2 \equiv ln(s_2/s_0)$ and $G_p(\xi',t) = \eta(\alpha_p(t))exp[(\alpha_p(t)-1)\xi']$ is the pomeron Green function. The quantity $\sigma_{P2}^{(tot)}(\xi_2,t)$ can be considered as the pomeron-particle total interaction cross section [8]. Note that this quantity is not directly observable one and it is defined by its relation to the diffraction production cross section eq. (3.37). This definition is useful however because at large s_2 this cross

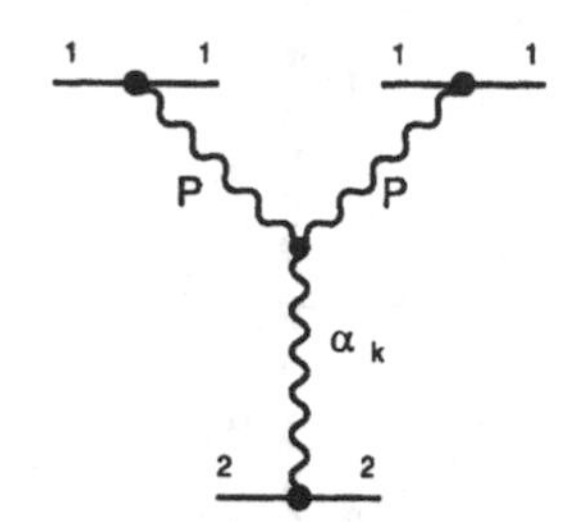

Figure 8. Triple-Regge diagram.

section has the same Regge behavior as usual cross sections

$$\sigma_{P2}^{(tot)}(s_2,t) = \sum_k g_{22}^k(0) r_{PP}^{\alpha_k}(t)(\frac{s_2}{s_0})^{\alpha_k(0)-1} \tag{3.38}$$

where the $r_{PP}^{\alpha_k}(t)$ is the triple-reggeon vertex (Fig.8), which describes coupling of two pomerons to reggeon α_k.
In this kinematical region $s \gg s_2 \gg m^2$ the inclusive diffractive cross section is described by the triple-Regge diagrams of Fig.8 and has the form

$$f^1 = \sum_k G_k(t)(1-x)^{\alpha_k(0)-2\alpha_P(t)}(\frac{s}{s_0})^{\alpha_k(0)-1} \tag{3.39}$$

The pomeron-proton total cross section and triple-Regge vertices r_{PP}^P, r_{PP}^f have been determined from analysis of experimental data on diffractive production of particles in hadronic collisions (see reviews [9]).

4. Mesonic Regge poles in QCD

An astonishing linearity of trajectories for Regge-poles corresponding to the known $q\bar{q}$-states indicates to an essentially nonperturbative, string-like dynamics. A non-perturbative method which can be used in QCD for large distance dynamics is the $1/N$-expansion (or topological expansion) [10, 11].
In this approach the quantities $1/N_c$[10] or $1/N_{lf}$[11] (N_{lf} is the number of light flavors) are considered as small parameters and amplitudes and Green functions are expanded in terms of these quantities. In QCD $N_c = 3$ and $N_{lf} \simeq 3$ and the expansion parameter does not look small enough. However we shall see below that in most cases the expansion parameter is $1/N_c^2 \sim 0.1$.
In the formal limit $N_c \to \infty$ ($N_f/N_c \to 0$) QCD has many interesting properties and has been intensively studied theoretically. There is a hope to obtain an exact solution of the theory in this limit (2-dimensional QCD has been solved in the limit $N_c \to \infty$). However this approximation is rather far from reality, as resonances in

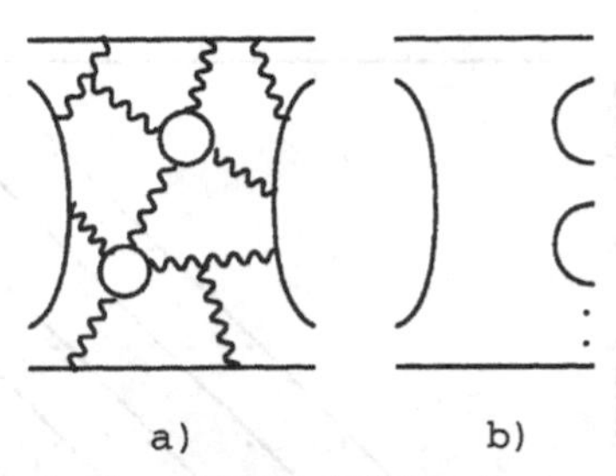

Figure 9. a) Planar diagram for binary reactions. Full lines denote quarks, wavy lines - gluons. b) Same for $12 \to X$. Internal lines of gluons and quarks are not shown.

this limit are infinitely narrow ($\Gamma \sim 1/N_c$). The case when the ratio $N_f/N_c \sim 1$ is fixed and the expansion in $1/N_f$ (or $1/N_c$) is carried out [11] seems more realistic. This approach is called sometimes by the topological expansion, because the given term of this expansion corresponds to an infinite set of Feynman diagrams with definite topology. It should be emphasized that $1/N$-expansion should be applied to Green-functions or amplitudes for white states.

The first term of the expansion corresponds to the planar diagrams of the type shown in Fig.9 for the binary reaction. These diagrams always have as border lines the valence quarks of the colliding hadrons. At high energies they should correspond to exchanges by secondary Regge poles $\alpha_R(\rho, A_2, \omega, ...)$ "made of" light quarks. The s-channel cutting of the planar diagram of Fig.9a) is shown in Fig.9b). Here and in the following we do not show internal lines of gluons and quark loops. This diagram corresponds to a multi-particle production which has the same properties as in the multi-peripheral model. The topological classification of diagrams in QCD leads to many relations between parameters of the reggeon theory, hadronic masses, widths of resonances and total cross sections (for a review see [12]) . All these relations are in a good agreement with experiment.

A contribution of the planar diagrams to the total cross section decreases with energy as $1/s^{(1-\alpha_R(0))} \approx 1/\sqrt{s}$. This decrease is connected to the fact that quarks have spin 1/2 and in the lowest order of perturbation theory an exchange by two quarks in the t-channel leads to the behavior $\sigma \sim 1/s$, which corresponds to the intercept $\alpha_R(0) = 0$. Interaction between quarks should lead to an increase of the intercept to the observed value $\alpha_R(0) \approx 0.5$.

Is it possible to calculate Regge-trajectories from QCD? Even for planar diagrams this is a difficult problem. It was considered in paper [13] using the method of Wilson-loop path integral [14]. It was shown that under a reasonable assumption about large distances dynamics: minimal area law for Wilson loop at large distances, confirmed by numerous lattice data, $\langle W \rangle \sim exp(-\sigma S_{min})$, it is possible to calculate spectrum of $q\bar{q}$-states. In these calculations virtual $q\bar{q}$-pairs and spin

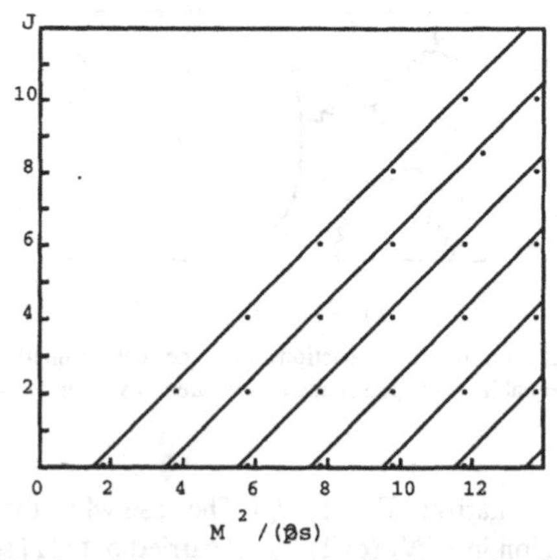

Figure 10. Calculated spectrum of $q - \bar{q}$ states and corresponding Regge-trajectories.

effects were neglected. It was shown that the mass spectrum can be determined from the following effective Hamiltonian H_0

$$H_0 = \frac{p_r^2}{\mu(t)} + \mu(t) + \frac{L(L+1)}{r^2[\mu + 2\int_0^1(\beta - \frac{1}{2})^2 \nu d\beta]} +$$

$$+ \int_0^1 \frac{\sigma^2 d\beta}{2\nu(\beta,t)} r^2 + \frac{1}{2}\int_0^1 \nu(\beta,t) d\beta \tag{4.1}$$

Here $\mu(t)$ and $\nu(\beta, t)$ are positive auxiliary functions which are to be found from the extremum condition.[13] Their extremal values are equal to the effective quark energy $\langle\mu\rangle$ and energy density of the adjoint string $\langle\nu\rangle$.
The resulting spectrum of H_0 for light quarks with a good accuracy is described by a very simple formula

$$\frac{M^2}{2\pi\sigma} = L + 2n_r + c_1 \tag{4.2}$$

where c_1 is a constant ($c_1 \approx 1.55$).
This spectrum is shown in Fig.10 and corresponds to an infinite set of linear Regge trajectories similar to the one of dual and string models.
In order to make realistic calculations of masses of hadrons which can be compared with experiment it is necessary to take into account perturbative interactions at small distances, spin effects and quark loops. For light quarks spin effects are non trivial as the spontaneous violation of the chiral symmetry should be properly taken into account.

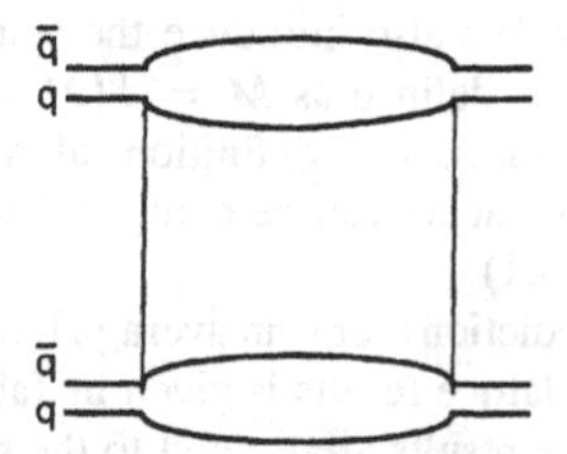

Figure 11. Cylinder-type diagrams for the pomeron.

5. Glueballs and the pomeron in QCD

It was mentioned above that the pomeron in QCD is usually related to gluonic exchanges in the t-channel [15]. This is connected with the fact that gluons naturally lead to the vacuum quantum numbers and the simplest perturbation theory diagram for scattering amplitude with an exchange by 2 gluons leads to a cross section which does not depend on energy (due to spin of gluon $Sg = 1$). Thus in this approximation $\alpha_P(0) = 1$. In perturbation theory an interaction between gluons leads to an increase of the Pomeron intercept.[16]
From the point of view of 1/N-expansion the pomeron is related to the cylinder-type diagrams, shown in Fig.11 with gluonic states (mixed with $q\bar{q}$-pairs) on a surface. In this approach the Pomeron is related to glueballs.
Glueballs are among the most intriguing objects both in experiment and theory. While experimental situation is not yet settled, lattice simulations [17, 18] yield an overall consistent picture of lowest ($< 4\ GeV$) mass spectrum. The mass scale and level ordering of the resulting glueball spectra differ from those of meson spectra, yielding a unique information about the nonperturbative structure of the gluonic vacuum.
The problem of spectra of glueballs and its relation to the pomeron was considered in papers [19, 20], using the method of Wilson-loop path integral discussed above for the case of $q\bar{q}$-Regge poles.
In the approximation when the spin effects and quark loops are neglected the spectrum of two-gluon glueballs is determined by the same Hamiltonian H_0 given by eq. (4.1) with the only difference that the string tension (σ_{fund}) for the $q\bar{q}$ system is changed to σ_{adj}. Thus the mass spectrum for glueballs is given be eq. (4.2) with the change $\sigma \to \sigma_{adj}$.
The value of σ_{adj} in can be found from the string tension σ_{fund} of $q\bar{q}$ system, multiplying it by $\frac{9}{4}$, as it follows from Casimir scaling observed on the lattices. Taking experimental Regge slope for mesons $\alpha' = 0.89\ GeV^{-2}$ one obtains $\sigma_{fund} = 0.18\ GeV^2$ and $\sigma_{adj} \approx 0.40\ GeV^2$.
In order to compare our results with the corresponding lattice calculations [17, 18] it is convenient to consider the quantity $\bar{M}/\sqrt{\sigma_f}$, which is not sensitive to the

choice of string tension σ. We also introduce the spin averaged mass $\bar{M}$ which for $L = 0, n_r = 0$ states is defined as $\bar{M} = \frac{1}{3}(M(0^{++}) + 2M(2^{++}))$, and in a similar way for higher states. This definition takes into account the structure of spin-splitting terms, so that $\bar{M}$ can be compared with the eigenvalues M_0 of spinless Hamiltonian eq. (4.1).

The comparison of our predictions for spin averaged masses of the lowest glueball states with corresponding lattice results is given in Table 1. For the average mass with $L = 2, n_r = 0$ lattice results are limited to the state 3^{++}. An agreement is perfect especially for the mass of the lowest state which is calculated on lattices with highest accuracy.

The spin splittings for glueball masses were calculated in paper [20] assuming that spin effects can be treated as small perturbations. A largest correction is obtained for the lowest state from the spin–spin interaction. The results are compared with lattice calculations in the Table 2 (for $\alpha_s = 0.3$).

Table 1

Spin averaged glueball masses $M_G/\sqrt{\sigma_f}$

Quantum numbers		Ref. [20]	Lattice data	
			Ref. [17]	Ref. [16]
2 gluon states	$l = 0, n_r = 0$	4.68	4.66±0.14	4.55±0.23
	$l = 1, n_r = 0$	6.0	6.36±0.6	6.1 ±0.5
	$l = 0, n_r = 1$	7.0	6.68±0.6	6.45±0.5
	$l = 2, n_r = 0$	7.0	9.0 ±0.7(3^{++})	7.7 ±0.4(3^{++})
	$l = 1, n_r = 1$	8.0		8.14 ±0.4(2^{*-+})
3 gluon state	K=0	7.61		8.19±0.48

Table 2
Comparison of predicted glueball masses with lattice data

J^{PC}	M(GeV)	Lattice data		$M[G]/M[0^{-+}]$		Difference
	[20]	paper [17]	paper [18]	[20]	[17]	
0^{++}	1.58	1.73±0.13	1.74±0.05	0.62	0.67(2)	-7%
0^{++*}	2.71	2.67±0.31	3.14±0.10	1.06	1.03(7)	3%
2^{++}	2.59	2.40±0.15	2.47±0.08	1.01	0.92(1)	9%
2^{++*}	3.73	3.29±0.16	3.21±0.35			
0^{-+}	2.56	2.59±0.17	2.37±0.27			
0^{-+*}	3.77	3.64±0.24		1.47	1.40(2)	5%
2^{-+}	3.03	3.1±0.18	3.37±0.31	1.18	1.20(1)	-1%
2^{-+*}	4.15	3.89±0.23		1.62	1.50(2)	8%
3^{++}	3.58	3.69±0.22	4.3±0.34	1.40	1.42(2)	-2%
1^{--}	3.49	3.85±0.24		1.36	1.49(2)	-8%
2^{--}	3.71	3.93±0.23		1.45	1.52(2)	-1%
3^{--}	4.03	4.13±0.29		1.57	1.59(4)	

Let us consider now the pomeron Regge trajectory in this approach in more details, taking into account both nonperturbative and perturbative contributions to the pomeron dynamics.
The large distance, nonperturbative contribution gives for the leading glueball trajectory ($n_r = 0$)

$$\alpha_P(t) = -c_1 + \alpha'_P t + 2 \tag{5.1}$$

with $\alpha'_P = \frac{1}{2\pi\sigma_a}$.
In eq. (5.1) spins of "constituent" gluons are taken into account, but a small nonperturbative spin–spin interactions were neglected.
For the intercept of this trajectory we obtain $\alpha_P(0) \approx 0.5$, which is substantially below the value found from analysis of high-energy interactions $\alpha_P(0) = 1.08 \div 1.2$. I would like to emphasize that contrary to interactions related to emission of real particles in the s-channel (for example emission of gluons in the perturbative ladder-type diagrams) the confining interaction considered above leads to a decrease of an intercept of a Regge trajectory compared to the Born approximation.
The most important nonperturbative source, which can lead to an increase of the pomeron intercept is the quark–gluon mixing or account of quark loops in the

gluon "medium". In the 1/N -expansion this effect is proportional to N_f/N_c, where N_f is the number of light flavors. In the leading approximation of the $1/N_c$-expansion there are 3 Regge trajectories with vacuum quantum numbers, – $q\bar{q}$–planar trajectories (α_f made of $u\bar{u}$ and $d\bar{d}$ quarks, $\alpha_{f'}$ made of $s\bar{s}$-quarks) and pure gluonic trajectory – α_G. The transitions between quarks and gluons $\sim \frac{1}{N_c}$ will lead to a mixing of these trajectories. Note that for a realistic case of G, f and f'-trajectories (eq.(5.1) and Fig.3) all 3-trajectories before mixing are close to each other in the small t region. In this region mixing between trajectories is essential even for small coupling matrix $g_{ik}(t)$. Lacking calculation of these effects in QCD they were considered in the paper [20] in a semi–phenomenological manner.

Denoting by $\bar{\alpha}_i$ the bare f, f' and G–trajectories and introducing the mixing matrix $g_{ik}(t)(i, k = 1, 2, 3)$ we obtain the following equation for determination of resulting trajectories after mixing [20]:

$$j^3 - j^2 \sum \bar{\alpha}_i + j(\sum_{i \neq k} \bar{\alpha}_i \bar{\alpha}_k - g_{ik}^2) - \bar{\alpha}_1 \bar{\alpha}_2 \bar{\alpha}_3 + \\ + \sum_{i \neq k \neq l} \bar{\alpha}_l g_{ik}^2 - 2 g_{12} g_{13} g_{23} = 0 \quad (5.2)$$

For realistic values of $g_{ik}(t)$ (for details see [20]) the pomeron intercept is shifted to the values $\alpha_P(0) \approx 1$. For $t > 1\ GeV^2$ the pomeron trajectory is very close to the planar f–trajectory, while the second and third vacuum trajectories – to $\alpha_{f'}$ and α_G correspondingly. In the region of large $t > 0$ the effects of mixing are small.

It is interesting that with an account of the quark–gluon mixing the intercept of the pomeron trajectory is close to the value $j = 1$ corresponding to an exchange by 2 noninteracting gluons.

Up to now I have considered mostly nonperturbative, large distance dynamics of the pomeron. A small distance dynamics of the pomeron has been studied in many papers using the QCD perturbation theory (see for example reviews [21]). in the leading log approximation the pomeron corresponds to a sum of the ladder-type diagrams with exchange of reggeized gluons in the t-channel (BFKL [16] pomeron). In this approximation the intercept of the pomeron is equal to

$$\Delta = \alpha_P(0) - 1 = \alpha_s \frac{12}{\pi} ln2 \quad (5.3)$$

It has been found recently [22] that α_s corrections substantially decrease Δ compared to LLA result. The intercept of the pomeron depends on the renormalization scheme and scale for α_s. In the "physical" (BLM) scheme values of Δ are in the region $0.15 \div 0.17$.[23] Unfortunately it is very difficult to calculate higher order corrections in PQCD.

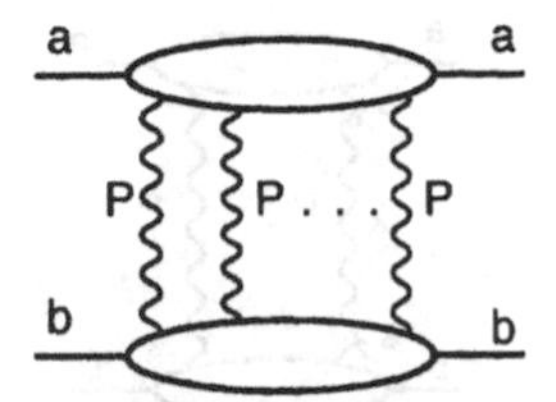

Figure 12. Diagrams with an exchange by several pomerons in the t=channel.

Sometimes the BFKL pomeron is called "hard" pomeron contrary to the "soft" one. However the equation for the pomeron singularity contains both nonperturbative effects discussed above and perturbative dynamics. Thus the resulting "physical" pole is a state due to both "soft" and "hard" interactions.

6. Regge cuts. High-energy hadronic interactions

6.1. MULTI–POMERON EXCHANGES

Regge poles are not the only singularities in the complex angular momentum plane. Exchange by several reggeons in the t-channel shown for the pomeron case in Fig.12 leads to moving branch points (or Regge cuts) in the j-plane. The positions of the branch points for t=0 can be expressed in terms of intercept of the Pomeranchuk pole

$$\alpha_{np}(0) - 1 = n(\alpha_p(0) - 1) = n\Delta \qquad (6.1)$$

Contributions of these singularities to scattering amplitudes $T_n(s,0) \sim s^{1+n\Delta}$ are especially important at high energies for $\Delta > 0$. The whole series of n–pomeron exchanges should be summed. An account of these multi–pomeron exchanges in the t-channel leads to unitarization of scattering amplitudes.
In the framework of 1/N-expansion the n–pomeron exchange amplitudes are due to topologies with $(n-1)$ handles and are of the order $(1/N^2)^n$.
The Gribov reggeon diagrams technique [24] allows one to calculate contributions of Regge cuts to scattering amplitudes. I shall illustrate this using as an example the two–pomeron exchange contribution to the elastic scattering amplitude. The pomeron-particle scattering amplitudes, which enter into the diagram of Fig.13, have usual analyticity properties in variables s_i (poles and cuts) and changing the integration in the diagram from d^4k to $ds_1 ds_2 d^2k_\perp/2s$ one can write the diagram of two–pomeron exchange in the form

$$T_{2P}(s,t) = \frac{i}{2!}\int \frac{d^2k_\perp}{\pi}\eta_P(k_\perp^2)\eta_P((\vec{k}_\perp - \vec{q})^2)$$

$$(\frac{s}{s_0})^{\alpha_P(k_\perp^2)+\alpha_P((\vec{k}_\perp-\vec{q})^2)-2} \times N_a(\vec{k}_\perp,\vec{q})N_b(\vec{k}_\perp,\vec{q}) \qquad (6.2)$$

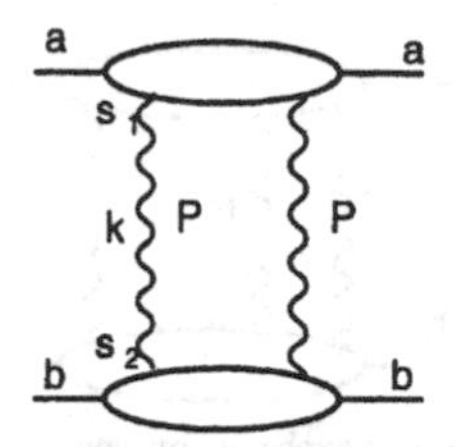

Figure 13. The diagram of two-pomeron exchange for elastic scattering amplitude.

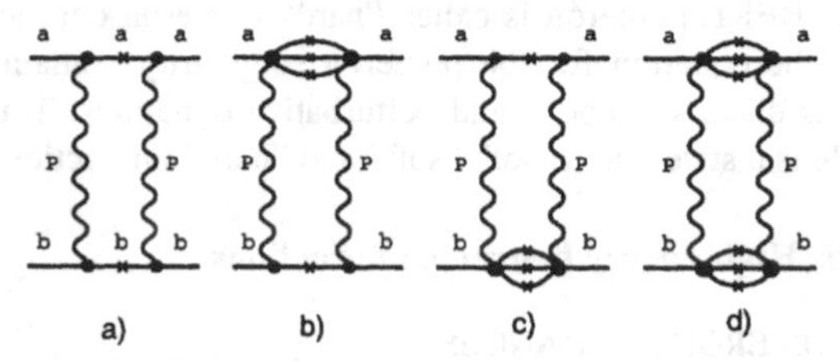

Figure 14. Rescattering diagrams. (Crosses on lines indicate that they are on mass shell).

The amplitudes $T_{aP,bP}$ decrease faster than $1/s_i$ at large s_i and the contours of integrations on s_i can be transformed in such a way that only discontinuities of these amplitudes at the right hand cut enter into N_a, N_b. The pomeron–particle scattering amplitudes satisfy to the unitarity condition, and the discontinuities can be expressed as a sum over contributions of intermediate states on mass shell. Thus the two–pomeron exchange diagram of Fig.13 can be expressed as the sum of diagrams, shown in Fig.14. The diagram of Fig. 14a) corresponds to the P pole contribution to the elastic scattering amplitude, while the diagrams of Fig. 14 b)-d) are P pole contributions to inelastic diffraction. The diagrams with n–pomeron exchange in the t-channel can be treated in the same way. The sum of the elastic rescatterings leads to the eikonal formula for the amplitude of ab-scattering in the impact parameter space

$$f_{ab}(s,b) = \frac{i}{2}(1 - exp[2i\delta_P(s,b)]) \tag{6.3}$$

where $\delta_P(s,b) = \int \frac{d^2q}{2\pi} exp(i\vec{q}\cdot\vec{b})T_P(s,q_\perp^2)$. In general $f(s,b)$ can be written in the form

$$f(s,b) = \frac{1}{2i}\sum_{n=1}^{\infty}\frac{(-v_p)^n}{n!}\gamma_n \tag{6.4}$$

where $v_P(s,b) = -2i\delta_P(s,b)$. In the eikonal approximation all $\gamma_n = 1$. All diagrams in Fig.14 have the same sign. This leads to the constraint $\gamma_2 \equiv C >$

1, which means that $|T_{2P}|$ for the contribution of the PP-cut to the amplitude is always larger than the eikonal value. A simplest generalization of the eikonal model, which takes into account also inelastic diffractive intermediate states, is so called "quasi-eikonal" model, where $\gamma_n = C^{n-1}$ and the amplitude $f(s,b)$ has the form

$$f(s,b) = \frac{i}{2C}(1 - exp(-Cv_P)) \tag{6.5}$$

The function $v_P(s,b) \sim s^{\Delta}$ and it becomes large at very high energies. In this limit the scattering amplitude for elastic scattering $f(s,b) \to i/2$ in the eikonal model (scattering on a black disk) and $f(s,b) \to 1/2C$ in the quasi-eikonal model (scattering on a grey disc). This property of the quasi-eikonal model is closely related to the fact that one of the eigen states for the diffractive matrix $\hat{v}_P$ has in this model a zero eigenvalue. This is a crude approximation, which takes into account a big difference in the interaction cross section of hadrons with different transverse sizes. Configurations of quarks inside hadrons with small transverse size r have total interaction cross sections $\sim r^2$, because hadrons in QCD interact as color dipoles. There is a distribution of quarks and gluons inside colliding hadrons with different values of r so one can expect that there will be a slow approach to the black disk limit for elastic scattering amplitude as $s \to \infty$. In this limit the effective radius of interaction increases as lns. Thus total interaction cross sections for the supercritical pomeron theory have Froissart type behavior $\sigma^{(tot)} \sim ln^2(s)$ as $s \to \infty$.

6.2. AGK-CUTTING RULES

Now we shall discuss the s-channel unitarity for contributions of pomeron cuts. The connection between different s-channel intermediate states and contribution of cuts to the imaginary part of elastic scattering amplitude is given by Abramovsky, Gribov, Kancheli (AGK)-cutting rules [25]. I shall illustrate them using as an example two-Pomeron exchange amplitude (Fig.13). Its contribution to the imaginary part of the forward elastic ab-scattering amplitude is negative and can be denoted as $T_{2P}(s,b) = -A(s)$ and to the $\sigma^{(tot)}$ as $-\sigma_2$.
There are three different classes of diagrams, which can be obtained by the s-channel cuttings of the diagram of Fig.13. They contribute to different classes of physical processes.
a) The cutting between the pomerons leads to diffractive processes (both elastic and inelastic), shown in Fig.15 a). Their contributions to the $ImT_{ab}(s,0)$ and $\sigma^{(tot)}$ according to AGK-rules are $A(s)$ and σ_2 correspondingly.
b) Cutting through one of pomerons (Fig.15 b)) leads to absorptive corrections for the multi-peripheral type inelastic production of particles, corresponding to the s-channel content of the pomeron. AGK rules give for this contribution $-4A(s)$ and $-4\sigma_2$ correspondingly.

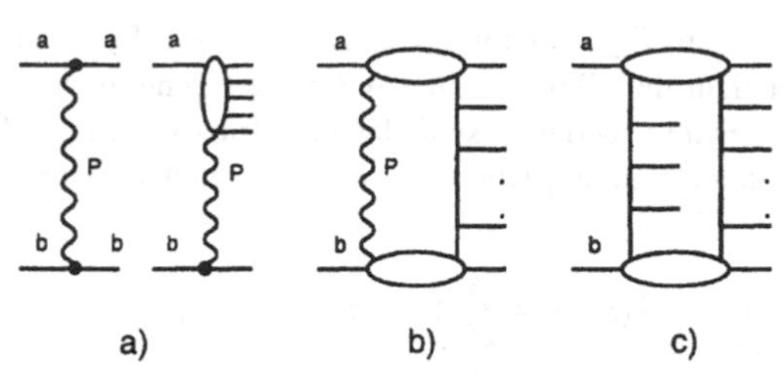

Figure 15. Different contributions to the s-channel cutting of the two-pomeron exchange diagram.

c) Cutting through both of pomerons (Fig.15 c)) leads to a new process, – production of two multi-peripheral chains. According to AGK-rules this contributes $2A(s)$ to the imaginary part of the forward scattering amplitude and $2\sigma_2$ to the total cross section.
The summary contribution of all these process to the total cross section is $(1 - 4 + 2)\sigma_2 = -\sigma_2$.
The AGK rules are formulated [25] for arbitrary diagrams with exchange of n reggeons in the t-channel. They allow to classify all multi-particle configurations and calculate their weights in terms of contributions to forward elastic scattering amplitude (or the corresponding amplitudes in the impact parameter space). For example in the eikonal or quasi-eikonal models all cross sections (diffractive and inelastic with different number of multi-peripheral chains) can be expressed in terms of a single pomeron exchange contribution to an elastic scattering amplitude.
Let us consider some consequences of these rules for particle production at very high energies.
1) For a sum of all n-pomeron exchange diagrams of the eikonal–type (without interaction between pomerons) there is a cancellation of their contributions to the single particle inclusive spectra in the central rapidity region for $n \geq 2$ [25]. So only the pole diagram of Fig.6 contributes and inclusive spectra increase with energy as $f^a \sim (s/s_0)^{\Delta}$. This means in particular that a study of an energy dependence of inclusive spectra in the central rapidity region gives more reliable information on the value of Δ than $\sigma^{(tot)}$, where pomeron cuts strongly modify energy dependence compared to the pole diagram. Inclusive charged particles density at $y = 0$ for $pp(\bar{p}p)$-collisions has a fast increase with energy and can be described in eikonal-type models with the intercept of the pomeron in the range $\Delta = 0.12 - 0.14$. The calculation of the total and elastic scattering cross sections in the same model with these values of Δ also leads to a good description of experiment[26, 27]. It should be noted that the value of Δ becomes larger ($\Delta \approx 0.2$) if the interaction between pomerons is taken into account [28] (see below).

2) Pomeron cuts lead to long range correlations in rapidity. Existence of such correlations (for example long range correlations between number of particles produced in the forward and backward hemispheres in the c.m. system) is now firmly established in particle production at high energies.
3) Existence of poli-peripheral contributions with several multi-peripheral chains leads to broad multiplicity distributions of produced hadrons with the dispersion $D \sim \langle n \rangle$. The s-channel cutting of a single pomeron leads to a Poisson-type distribution with $\langle n \rangle = \langle n \rangle_P$, for cutting of two pomerons (Fig.15 c)) the distribution is of the same type but with $\langle n \rangle \approx 2\langle n \rangle_p$ and so on. The summary contribution for cuttings of all n-pomeron exchange diagrams is broad and has a complicated form. At not too high energies contributions from different n strongly overlap and multiplicity distributions satisfy to a good accuracy KNO-scaling $(\langle n \rangle \sigma_n / \sigma^{(in)} = f(n/\langle n \rangle))$. However as energy increases KNO scaling is violated. The models based on the reggeon diagrams technique and AGK-cutting rules [26, 27] give a good quantitative description of multiplicity distributions .

The theoretical models mentioned above (the Dual Parton Model [27] and the Quark Gluon Strings Model [26] use besides the reggeon theory also 1/N-expansion for interpretation of different reggeon diagrams in QCD. This leads to a very predictive approach to multi-particle production in hadron–hadron, hadron–nucleus and nucleus–nucleus collisions. With a small number of parameters these models allow one to describe total and elastic cross sections, diffraction dissociation, multiplicity distributions, inclusive cross sections for different types of hadrons e.t.c. in a broad region of energies [26, 27]. In the following we shall apply the same approach to DIS in the small x region and to nuclear interactions.

6.3. INTERACTIONS BETWEEN POMERONS

In the eikonal-type models discussed above the diffraction dissociation to the states with not too large masses has been taken into account. The diffractive production of states with large masses is related, as we know from the discussion of diffractive processes, to the diagrams of the type shown in Fig.8 with interaction between pomerons. Neglect in the first approximation by these interactions is justified by a smallness of triple–pomeron and 4–pomeron interaction vertices, found from analysis of diffractive processes [9]. However at very high energies it is necessary to include all these diagrams in order to have a selfconsistent description of high energy hadronic interactions including large mass diffractive production of particles. It was demonstrated in paper [28] that inclusion of these diagrams leads in most of the cases to predictions which are very close to the results of eikonal type models, however the value of the "bare" pomeron intercept increases up to the value $\alpha_P(0) \approx 1.2$.

7. Small-x physics

This section is devoted to a problem of small-x physics in deep inelastic scattering (DIS) . This problem became especially actual due to recent experimental investigation of this region at HERA accelerator.
In DIS it is possible to study different asymptotic limits. For a virtuality of the photon $Q^2 \to \infty$ and $x = Q^2/(W^2 + Q^2) \sim 1$ the usual QCD evolution equations can be applied and Q^2 dependence of the structure functions can be predicted if an initial condition for structure functions at $Q^2 = Q_0^2$ is formulated. On the other hand if Q^2 is fixed and $x \to 0$ (or $\ln(1/x) \to \infty$) the asymptotic Regge limit is relevant. The most interesting question is what is the behavior of DIS in the region where both $\ln(1/x)$ and $\ln(Q^2)$ are large? This is a transition region between perturbative and nonperturbative dynamics in QCD and its study can give an important information on the properties of confinement and its relation to the QCD perturbation theory. The asymptotic Regge limit in DIS can be related to high-energy limit of hadronic interactions and is described in terms of the Pomeranchuk singularity.
Experiments at HERA have found two extremely important properties of small-x physics : a fast increase of parton densities as x decreases [29, 30] and the existence of substantial diffractive production in deep inelastic scattering (DIS).[31, 32]
A fast increase of $\sigma_{\gamma^* p}^{(tot)}$ as $W^2 \equiv s$ increases at large Q^2 observed experimentally [29, 30] raises a question: whether there are two different pomerons – "soft" and "hard"? From the discussion of the pomeron in QCD above it follows that there are no theoretical reasons for such a situation and most probably the rightmost pole in the j–plane is generated by both "soft" and "hard" dynamics. I shall assume that there is one ("physical") pomeron pole with the same $\alpha_P(0)$ as it was determined from high–energy hadronic interactions with an account of many–pomeron cuts. On the other hand the effective intercept, which depends on relative contribution of multi–pomeron diagrams, can be different in different processes.
In paper [33] it was suggested that the increase of the effective intercept of the pomeron, $\alpha_{eff} = 1 + \Delta_{eff}$, as Q^2 increases from zero to several GeV^2 is mostly due to a decrease of shadowing effects with increasing Q^2. A parametrization of the Q^2 dependence of Δ_{eff} such that $\Delta_{eff} \approx 0.1$ for $Q^2 \approx 0$ (as in soft hadronic interactions) and $\Delta_{eff} \approx 0.2$ (bare pomeron intercept) for Q^2 of the order of a few GeV^2, gives a good description of all existing data on $\gamma^* p$ total cross-sections in the region of $Q^2 \leq 5 \div 10$ GeV^2 [33, 34]. At larger Q^2 effects due to QCD evolution become important. Using the above parametrization as the initial condition in the QCD evolution equation, it is possible to describe the data in the whole region of Q^2 studied at HERA [33, 35].
In the reggeon approach discussed above there are good reasons to believe that

the fast increase of the $\sigma_{\gamma^* p}$ with energy in the HERA energy range will change to a milder increase at much higher energies. This is due to multi–pomeron effects, which are related to shadowing in highly dense systems of partons - with eventual "saturation" of densities. This problem has a long history (for reviews see [21, 36]) and has been extensively discussed in recent years [37]. It is closely connected to the problem of the dynamics of very high-energy heavy ion collisions [38].

This problem was investigated recently in our paper [40], where reggeon approach was applied to the processes of diffractive $\gamma^* p$ interaction. It was emphasized in the previous section that in the reggeon calculus [24] the amount of rescatterings is closely related to diffractive production. AGK-cutting rules [25] allow to calculate the cross-section of inelastic diffraction if contributions of multi–pomeron exchanges to the elastic scattering amplitude are known. Thus, it is very important for self-consistency of theoretical models to describe not only total cross sections, but, simultaneously, inelastic diffraction. In particular in the reggeon calculus the variation of Δ_{eff} with Q^2 is related to the corresponding variation of the ratio of diffractive to total cross sections. In the paper [40] an explicit model for the contribution of rescatterings was constructed which leads to the pattern of energy behavior of $\sigma^{(tot)}_{\gamma^* p}(W^2, Q^2)$ for different Q^2 described above. Moreover, it allows to describe simultaneously diffraction production by real and virtual photons. In this model it is possible to study quantitatively a regime of "saturation" of parton densities.

Let us discuss briefly the qualitative picture of diffractive dissociation of a highly virtual photon at high energies. It is convenient to discuss this process in the lab. frame, where the quark-gluon fluctuations of a photon live a long time $\sim 1/x$ (Ioffe time [41]). A virtual photon fluctuates first to $q\bar{q}$ pair. There are two different types of configurations of such pair, depending on transverse distance between quarks (or $k_\perp$).

a) Small size configurations with $k_\perp^2 \sim Q^2$. These small dipoles ($r \sim 1/k_\perp \sim 1/Q$) have a small ($\sim r^2$) total interaction cross section with the proton.

b) Large size configurations with $r \sim 1/\Lambda_{QCD}$ and $k_\perp \sim \Lambda_{QCD} \ll Q$. They have a large total interaction cross section, but contribute with a small phase space at large Q^2, because these configurations are kinematically possible only if the fraction of longitudinal momentum carried by one of the quarks is very small $x_1 \sim k_\perp^2/Q^2 \ll 1$. This configuration corresponds to the "aligned jet", introduced by Bjorken and Kogut.[42]

Both configurations lead to the same behaviour of $\sigma_{\gamma^* p} \sim 1/Q^2$, but they behave differently in the process of the diffraction dissociation of a virtual photon [43, 44]. The cross section of such a process is proportional to a square of modulus of the corresponding diffractive amplitude and for a small size configuration it is small ($\sim 1/Q^4$). For large size configurations a smallness is only due to the phase space and the inclusive cross section for diffractive dissociation of a virtual photon decreases as $1/Q^2$, i.e. in the same way as the total cross section. This

is true only for the total inclusive diffractive cross section, where characteristic masses of produced states are $M^2 \sim Q^2$. For exclusive channels with fixed mass (for example production of vector mesons) situation is different and these cross sections decrease faster than $1/Q^2$ at large Q^2.
Inclusive diffractive production of very large masses ($M^2 \gg Q^2$) can be described in the first approximation by triple–Regge diagrams (Fig.8) [45]. From the point of view of the quark–gluon fluctuation of the fast photon triple–pomeron contribution corresponds to diffractive scattering of very slow (presumably gluonic) parton, which has a small virtuality.
The model [40] uses the picture of diffraction dissociation of a virtual photon outlined above and is a natural generalization of models used for the description of high-energy hadronic interactions. The interaction of the small size component in the wave function of a virtual photon is calculated using QCD perturbation theory. The main parameter of the model – intercept of the pomeron was fixed from a phenomenological study of these interactions discussed above ($\Delta_P = 0.2$) and was found to give a good description of $\gamma^* p$-interactions in a broad range of Q^2 ($0 \leq Q^2 < 10\ GeV^2$). Another important parameter of the theory, the triple–pomeron vertex, obtained from a fit to the data ($r^{(0)}_{PPP}/g^P_{pp}(0) \approx 0.1$) is also in a reasonable agreement with the analysis of soft hadronic interactions [28, 45]. The description of $\sigma^{(tot)}_{\gamma^* p}$ as a function of s for different values of Q^2 (experimental data are from H1 [29], ZEUS [30]) is shown in Fig.16. Diffraction dissociation of a virtual photon is usually presented as a function of Q^2, M^2 (or $\beta = Q^2/(M^2 + Q^2)$) and $x_P = x/\beta = (M^2 + Q^2)/(W^2 + Q^2)$. Description of HERA data on diffractive dissociation [32] in the model is shown in Fig.17. The model reproduces experimental data quite well. It can be used to predict structure functions and partonic distributions at higher energies or smaller x, which will be accessible for experiments at LHC.

7.1. SHADOWING EFFECTS FOR NUCLEAR STRUCTURE FUNCTIONS

A study of the shadowing effects for structure functions of nuclei in the small x region provides a stringent test of the reggeon approach to small-x problem. The shadowing effects are enhanced for nuclei ($\sim A^{1/3}$) and lead to deviations from A^1 behaviour for structure functions of nuclei. Glauber-Gribov [46, 47] approach to interactions of particles with nuclei gives a possibility to calculate rescattering corrections for interaction of a virtual photon with a nucleus in terms of diffractive interaction of a photon with a nucleon, which was discussed above.
A contribution of a double rescattering term to the $\sigma_{\gamma^* A}$ is directly expressed in terms of the differential cross section for a diffraction dissociation of a virtual photon in $\gamma^* N$-interactions

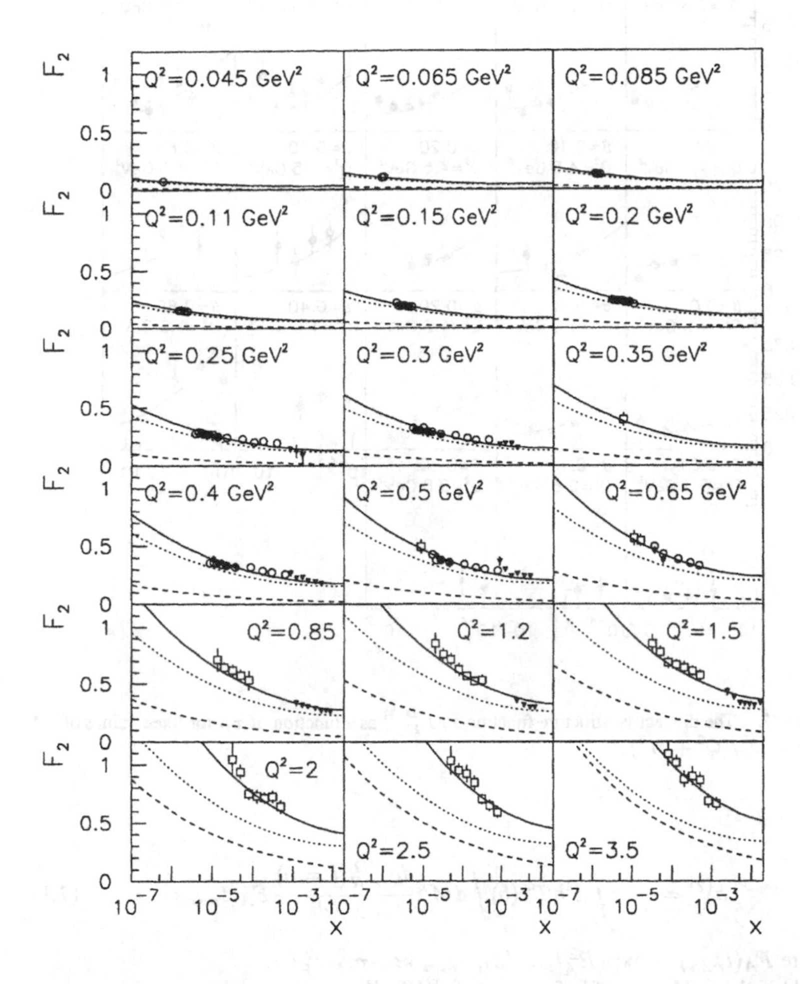

Figure 16. **Structure function F_2 as a function of x for different values of Q^2 compared with experimental data. Dashed lines denote small distance contributions and dotted lines – large distance ones.**

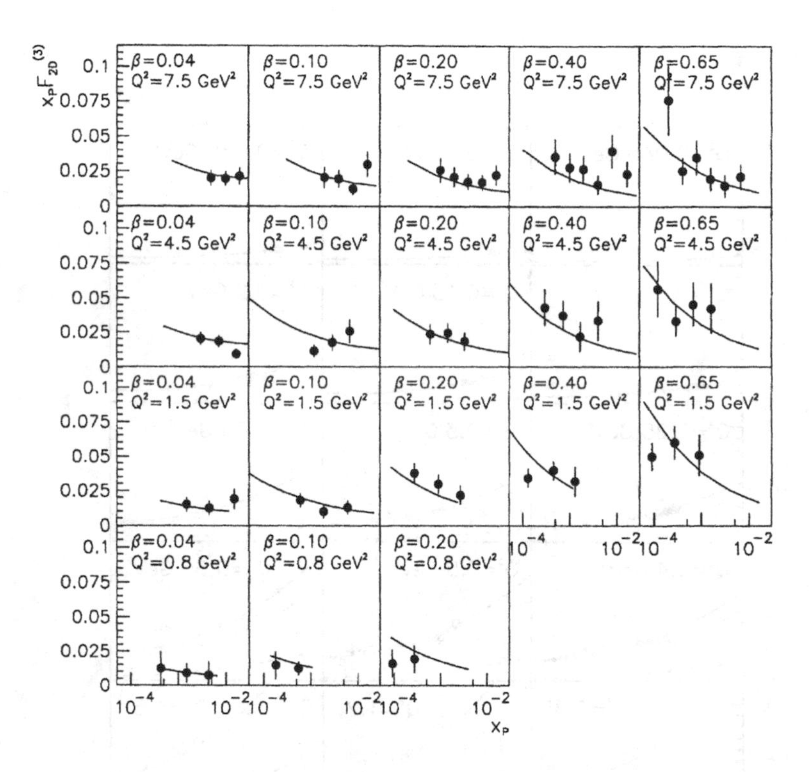

Figure 17. The diffractive structure function $x_P F_2^{D(3)}$ as a function of x_P for fixed values of Q^2 and $\beta = Q^2/(Q^2 + M^2)$.

$$\sigma^{(2)} = -4\pi \int d^2b T_A^2(b) \int dM^2 \frac{d\sigma^{DD}_{\gamma^* N}(t=0)}{dM^2 dt} F_A(t_{min}) \tag{7.1}$$

where $F_A(t_{min}) = \exp(R_A^2 t_{min}/3)$; $t_{min} \approx -m_N^2 x_P^2$;
$T_A(b)$ is the nuclear profile function ($\int d^2b T_A(b) = A$).
Higher order rescatterings are model dependent and in the generalized Schwimmer model [48] we obtain in the region of small x

$$F_{2A}/F_{2N} = \int d^2b \frac{T_A(b)}{1 + F(x, Q^2) T_A(b)} \tag{7.2}$$

with $F(x,Q^2) = 4\pi \int dM^2 \left(d\sigma^{DD}_{\gamma^* N}(t=0)/dM^2 dt\right)$
$\times \left(F_A(t_{min})/\sigma_{\gamma^* N}(x,Q^2)\right)$.
Theoretical predictions [50], based on eq. (7.2) and the model for diffraction dissociation of ref.[45] are in a very good agreement with NMC-data on nuclear structure functions at very small x [49]. We believe that this approach gives reliable predictions for nuclear shadowing effects in the region of smaller x not yet studied experimentally. This region will play an important role in dynamics of heavy ions collisions at super-high (LHC) energies.

8. Heavy ion collisions at high energies

8.1. HIGH-ENERGY NUCLEAR INTERACTIONS

Let us consider first high–energy hadron–nucleus interactions. In the Glauber model it is described by the diagrams, which looks like a successive rescatterings of initial hadron on nucleons of the nucleus. However, as was emphasized by Gribov [47] the space–time picture of the interaction at high–energy $E > m_h \mu R_A$ (μ is a characteristic hadronic scale $\sim 1\ GeV$ and R_A is the radius of the nucleus) is completely different from this simple picture. It corresponds to coherent interactions of a fluctuation of the initial hadron, which is "prepared" long before its interaction with the nucleus. Nevertheless the elastic hA–amplitude can be written as a sum of the diagrams with elastic rescatterings which give the same result as Glauber model, plus all possible diffractive excitations of the initial hadron. At not too high energies $E_L \sim 10^2\ GeV$ these inelastic contributions lead to corrections to the Glauber approximation of $10 - 20\%$ for the total hA cross section. However at very high energies and for inclusive cross sections this modification of the Glauber approximation is very important. The difference between Glauber model and Gribov's theory is essential for understanding shadowing corrections for structure functions of nuclei, as it was shown in the previous section, and for many aspects of multi-particle production on nuclei [51].
An important consequence of the space–time structure of high energy interactions of hadrons with nuclei is the AGK result [25], that for inclusive cross sections all rescatterings cancel and these cross sections are determined by the diagrams shown in Fig. 6 (impulse approximation). Note, however, that this result, valid asymptotically in the central rapidity region, only applies to the diagrams of the Glauber–type, i.e. when masses of intermediate states in rescattering diagrams are limited and do not increase with energy. As a result, the inclusive cross section for the production of a hadron a is expressed, for a given impact parameter b, in terms of inclusive cross section for hN interactions

$$E\frac{d^3\sigma^a_{hA}(b)}{d^3p} = T_A(b) E\frac{d^3\sigma^a_{hN}}{d^3p} \tag{8.1}$$

After integration over b we get

$$E\frac{d^3\sigma^a_{hA}}{d^3p} = A\,E\frac{d^3\sigma^a_{hN}}{d^3p} \tag{8.2}$$

The total and inelastic hA cross sections in the Glauber model can be easily calculated and are given for heavy nuclei by well known expressions. For example

$$\sigma^{in}_{hA} = \int d^2b(1 - \exp(-\sigma^{in}_{hN}T_A(b))) \tag{8.3}$$

The situation for nucleus–nucleus collisions is much more complicated. There are no analytic expressions in the Glauber model for heavy–nuclei elastic scattering amplitudes. The problem stems from a complicated combinatorics and from the existence of dynamical correlations related to "loop diagrams" [52, 53]. Thus, usually optical–type approximation [54, 55] and probabilistic models for multiple rescatterings [56] are used. For inclusive cross sections in AB–collisions the result of the Glauber approximation is very simple to formulate due to the AGK cancellation theorem. It is possible to prove, for an arbitrary number of interactions of nucleons of both nuclei [57], that all rescatterings cancel in the same way as for hA–interactions and only the diagrams of Fig. 6 contribute to the single inclusive spectrum. Thus a natural generalization of eq. (8.1) for inclusive spectra of hadrons produced in the central rapidity region in nucleus–nucleus interactions takes place in the Glauber approximation

$$E\frac{d^3\sigma^a_{AB}(b)}{d^3p} = T_{AB}(b)\,E\frac{d^3\sigma^a_{NN}}{d^3p} \tag{8.4}$$

where $T_{AB}(b) = \int d^2sT_A(\vec{s})T_B(\vec{b}-\vec{s})$. After integration over b eq. (8.4) reads

$$E\frac{d^3\sigma^a_{AB}}{d^3p} = AB\,E\frac{d^3\sigma^a_{NN}}{d^3p} \tag{8.5}$$

The densities of charged particles can be obtained from eqs.(8.4), (8.5) by dividing them by the total inelastic cross section of nucleus–nucleus interaction. For example

$$\frac{dn^{ch}_{AB}(b)}{dy} = \frac{T_{AB}(b)}{\sigma^{in}_{AB}}\frac{d\sigma^{ch}_{NN}}{dy} \tag{8.6}$$

and

$$\frac{dn^{ch}_{AB}}{dy} = \frac{AB}{\sigma^{in}_{AB}}\frac{d\sigma^{ch}_{NN}}{dy} \tag{8.7}$$

In the following we shall use these results to calculate particle densities in the central rapidity region at energies of RHIC and LHC.
Reviews of applications of the Glauber–Gribov approach, $1/N$–expansion in QCD and string model to processes of nucleus–nucleus interactions can be found in refs. [27, 58].

8.2 PARTICLE DENSITIES IN HEAVY-ION COLLISONS AT SUPERHIGH ENERGIES

Now we will address the question of particle densities in heavy–ion collisions. Eq. (8.7) for particle densities integrated over impact parameter (minimum bias events) can be rewritten as

$$\frac{dn^{ch}_{AB}}{dy} = n_{AB}\frac{dn^{ch}_{NN}}{dy} \tag{8.8}$$

where $n_{AB} = \frac{AB\sigma^{in}_{NN}}{\sigma^{in}_{AB}}$. It corresponds to the average number of collisions in the Glauber model. For $A = B >> 1$ n_{AB} behaves as $CA^{4/3}$ with $C \approx \frac{\sigma^{in}_{NN}}{4\pi R_0^2}$ ($R_A = R_0 A^{1/3}$). It is well known that eqs.(8.4), (8.5), (8.8) can be applied to hard processes but in the Glauber approximation they are valid for soft processes as well. We shall see below that for both soft and hard processes these equations have to be modified.
Using eqs.(8.7), (8.8) we obtain for PbPb collisions at LHC at $y = 0$ the following numbers for minimum bias events and central ($b < 3$ fm) collisions, respectively:
$dn^{ch}/dy = 2100$, $dn^{ch}/dy = 8500$.
Thus, the Glauber approximation predicts very large densities of charged hadrons in central heavy–ions collisions at LHC. However, are these predictions realistic ? In order to answer this question we will consider possible limitations of the Glauber approximation and also the corrections to the AGK cancellation theorem which are important at high energies.
There are two types of corrections to eqs.(8.1), (8.4),
a) The effects due to energy–momentum conservation: the energy of the initial hadron is shared by "constituents" and each sub–collision happens at smaller energy. These effects are very important in the fragmentation regions of colliding hadrons (or nuclei) and reduce particle densities. For $y = 0$ this reduction decreases as $(1/s)^{1/4}$. It is very important at SPS energies and has some effect at RHIC ; however, at LHC energies in the central rapidity region this effect is small.
b) Another dynamical effect is important at very high energies when diffractive production of very heavy hadronic states ($M^2 >> m_N^2$) becomes possible. It is related to the triple–pomeron interaction discussed above and corresponds to an interaction between pomerons (strings in the string models of particle production). As the total and inelastic cross sections of hA and AB–interactions at

high energies are close to a black disc limit due to Glauber–type diagrams, these extra interactions have a small influence on total cross sections. However, they are very important for inclusive spectra in the central rapidity region [51], where contributions of Glauber rescatterings cancel due to AGK rules.
Extra shadowing effects related to these interactions modify the A–dependence of the Glauber approximation for inclusive spectra eqs.(8.1),(8.4) in such a way that the behaviour $d\sigma_{AB}/dy \sim AB$ of the Glauber approximation changes to $d\sigma_{AB}/dy \sim A^{\alpha}B^{\alpha}$, where $\alpha < 1$. For very strong interaction between pomerons $\alpha \to 2/3$. This limit leads to universal particle densities in pp, pA and AB–collisions. We will show that due to a rather week interaction between pomerons even at LHC energies the value of α is close to 0.9.
The problem of shadowing for inclusive spectra is not especially related to soft processes. The same interactions are relevant also for hard processes (production of jets or particles with large p_T, heavy quarks, large–mass lepton pairs, e t.c.). For hard processes, due to the QCD–factorization theorem, inclusive spectra in nucleus–nucleus collisions are given by convolutions of hard cross sections with distributions of partons in the colliding nuclei. In these cases interactions between pomerons describe shadowing effects for nuclear structure functions (i.e. distributions of quarks and gluons in nuclei). Due to a coherence condition these effects are important only in the region of very small x_i of partons ($x_i << 1/(R_A m_N)$). So these effects are important only at very high energies, when $x_i \approx M_T/\sqrt{s}$ satisfy this condition. This condition in terms of x_i of partons coincides with the condition on diffractive production of large–mass states discussed above.
These effects were calculated in ref. [39] in the same model, which was used above for description of the shadowing effects for nuclear structure functions. It was predicted that an account of extra shadowing due to interactions between pomerons will lead to a decrease of particle densities compared to the Glauber model predictions by a factor ≈ 4 at LHC and by a factor ≈ 2 at RHIC energies. Comparison with first results from RHIC [59] is given in Table 3. Experiment clearly show a large deviation from prediction of the Glauber approximation and demonstrate an importance of the shadowing effects for inclusive spectra of hadrons. This approach reproduces also dependence of particle densities at RHIC on number of participants [60].

Table 3
Densities of charged hadrons $dn/d\eta|_{\eta=0}$ in central Pb-Pb collisions at $\sqrt{s} = 130\ GeV$.

Glauber	With shadowing corrections	Experiment [59]
1200 ± 100	630 ± 120	$555 \pm 12 \pm 35$

These results show that already starting from RHIC energies interactions between pomerons (or strings) play an important role in the dynamics of heavy–ion collisions. On the other hand characteristic values of partonic $x_i \sim 10^{-2}$ at these energies and both experimental data on shadowing for nuclear structure functions [49] and their theoretical interpretation [50] show that we are very far from "saturation" in this region. Account of interactions between strings is important for the problem of equilibration in heavy ion collisions and creation of the "quark-gluon plasma". In the simplest Glauber model there are no interactions between particles from different NN–collisions and the system is not in an equilibrium. Interactions between pomerons (strings) can lead a system to an equilibrium. However due to a relative weakness of such interactions it is hardly achieved at RHIC energies.

Acknowledgements

I would like to thank organizers of Cargese School Jean-Paul Blaizot and Edmond Iancu for invitation to give this lecture course and help. I acknowledge illuminating discussions with Kostya Boreskov, Alfons Capella, Oleg Kancheli, Genya Levin, Larry McLerran and Karen Ter-Martirosyan.

References

1. V.N. Gribov, *ZhETF* **41**, 667 (1961).
2. G.F. Chew and S.C. Frautschi, *Phys. Rev. Lett.* **7**, 394 (1961).
   R. Blankenbecler and M. Goldberger, *Phys. Rev.* **126**, 766 (1962).
3. M. Froissart, *Phys. Rev.* **123**, 1053 (1959).
4. I.Ya. Pomeranchuk, *ZhETF* **34**, 725 (1958).
5. T. Regge, *Nuovo Cimento* **14**, 951 (1959).
6. D. Amati, A. Stanghellini, and S. Fubini, *Nuovo Cimento* **26**, 896 (1962).
7. A.H. Mueller, *Phys. Rev.* D **23**, 2963 (1970).
   O.V. Kancheli, *JETP Lett.* **11**, 267 (1970).
8. A.B. Kaidalov, K.A. Ter-Martirosyan, *Nucl. Phys.* B **75**, 471 (1974).
9. A.B. Kaidalov, *Phys. Rep.* **50**, 157 (1979).
   G. Alberi and G. Goggi, *Phys. Rep.* **74**, 1 (1981).
   K. Goulianos, *Phys. Rep.* **101**, 169 (1983).
10. G. t'Hooft, *Nucl. Phys.* B **72**, 461 (1974).
11. G. Veneziano, *Phys. Lett.* B **52**, 220 (1974); *Nucl. Phys.* B **117**, 519 (1976).
12. A.B. Kaidalov, *Surveys in High Energy Phys.* **13**, 265 (1999).
13. A. Dubin, A. Kaidalov, Yu. Simonov, *Phys. Lett.* B **323**, 41 (1994) ;
    Yad. Fiz. **56**, 213 (1993) .
14. Yu.A. Simonov, *Nucl. Phys.* B **307**, 512 (1988); *Yad. Fiz.* **54**, 192 (1991) .
15. F.E. Low, *Phys. Rev.* D **12**, 163 (1975).
    S. Nussinov, *Phys. Rev. Lett.* **34**, 1286 (1975).
16. L. N. Lipatov,*Sov. J. Nucl. Phys.* **23**, 338 (1976).
    E. A. Kuraev, L. N. Lipatov, V.S. Fadin, *Sov. Phys. JETP* **45**, 199 (1977).

Y. Y. Balitsky, L. N. Lipatov, *Sov. J. Nucl. Phys.* **28**, 822 (1978).
L.N. Lipatov, *Nucl. Phys.* B **365**, 614 (1991).
17. C. Morningstar, M. Peardon, *Phys. Rev.* D **60**, 034509 (1999) .
18. M. Teper, hep-th/9812187.
19. Yu.A. Simonov, *Phys. Lett.* B **249**, 514 (1990).
20. A.B. Kaidalov, Yu.A. Simonov, *Phys. Lett.* B **477**, 163 (2000).
21. L. V. Gribov, E. M. Levin and M. G. Ryskin, *Phys. Rep.* **100**, 1 (1983).
E. Laenen and E. Levin, *Ann. Rev. Nucl. Part.* **44**, 199 (1994).
A. H. Mueller, hep-ph/9911289.
22. V. S. Fadin, L. N. Lipatov, *Phys. Lett.* B **429**, 127 (1998).
G. Camici, M. Ciafaloni, *Phys. Lett.* B **430**, 349 (1998).
23. S.J.Brodsky *et al*, *JETP Lett.* **70**, 155 (1999).
24. V. N. Gribov, *ZhETF* **57**, 654 (1967).
25. V. A. Abramovsky, V. N. Gribov and O. V. Kancheli, *Sov. J. Nucl. Phys.* **18**, 308 (1974).
26. A. Kaidalov, in *"QCD at 200 TeV"*, ed. L. Cifarelli and Yu. Dokshitzer, Plenum Press (1992), p. 1.
27. A. Capella, U. Sukhatme, C.-I. Tan and J. Tran Thanh Van, *Phys. Rep.***236**, 225 (1994).
28. A. B. Kaidalov, L. A. Ponomarev and K. A. Ter-Martirosyan, *Sov. J. Nucl. Phys.* **44**, 468 (1986).
29. T. Ahmed *et al.* (H1 Collaboration), *Phys. Lett.* B **299**, 374 (1992);
C. Adloff *et al.* (H1 Collaboration), *Nucl. Phys.* B **497**, 3 (1997).
30. M. Derrick *et al.* (ZEUS Collaboration), *Phys. Lett.* B **293**, 465 (1992);
J. Breitweg *et al.* (ZEUS Collaboration), *Phys. Lett.* B **407**, 432 (1997);
J. Breitweg *et al.* (ZEUS Collaboration), Preprint DESY 00-071.
31. M. Derrick *et al.* (ZEUS Collaboration), *Z. Phys.* C **72**, 399 (1996).
32. C. Adloff *et al.* (H1 Collaboration), *Z. Phys.* C **76**, 613 (1997).
33. A. Capella, A. Kaidalov, C. Merino, J. Tran Thanh Van, *Phys. Lett.* B **337**, 358 (1994).
34. A. Kaidalov, C. Merino, *Eur. Phys. J.* **C10**, 153 (1999).
35. A. Kaidalov, C. Merino, D. Perterman, *Eur. Phys. J.* **C20**, 301 (2001) .
36. A. B. Kaidalov, *Surveys High Energy Phys.* **9**, 143 (1996).
37. A. H. Mueller, *Nucl. Phys.* B **437**, 107 (1995).
A. L. Ayala, M. B. Gay Ducati and E. M. Levin, *Phys. Lett.* B **388**, 188 (1996) ; *Nucl. Phys.* B **493**, 305 (1997).
E. Gotsman, E. Levin and U. Maor, *Nucl. Phys.* B **493**, 354 (1997); *Phys. Lett.* B **425**, 369 (1998) ; **B452**, 387 (1999).
E. Gotsman, E. Levin, U. Maor and E. Naftali, *Nucl. Phys.* B **539**, 535 (1999).
M. Mac Dermott, L. L. Frankfurt, V. Guzey and M. Strikman, *Eur. Phys. J.* **C16** ,641 (2000).
K. Golec-Biernat and M. Wüsthoff, *Phys. Rev.* D **59**, 014017 (1999); **D60**, 114023 (1999).
A. H. Mueller, *Eur. Phys. J.* **A1**, 19 (1998).
38. L. McLerran and R. Venugopalan, *Phys. Rev.* D **49**, 2233, 3352 (1994); **50**, 2225 (1994) ; **53**, 458 (1996).
J. Jalilian-Marian et al., *Phys. Rev.* D **59**, 014014; 034007 (1999).
A. Kovner, L. McLerran and H. Weigert, *Phys. Rev.* D **52**, 3809; 6231 (1995).
E. Iancu and L. McLerran, *Phys. Lett.* B **510**, 145 (2001).
Yu. V. Kovchegov and A. H. Mueller, *Nucl. Phys.* B **529**, 451 (1998).
Yu. V. Kovchegov, A. H. Mueller and S. Wallon, *Nucl. Phys.* B **507**, 367 (1997).
A. H. Mueller, *Nucl. Phys.* B **558**, 285 (1999).
39. A. Capella, A. Kaidalov, J. Tran Thanh Van, *Heavy Ion Phys.* **9**, 169 (1999).
40. A. Capella, E. Ferreiro, A.B. Kaidalov and C.A. Salgado, hep-ph/0005049, hep-ph/0006233.
41. B.L. Ioffe, *Phys. Lett.* B **30**, 123 (1969).

42. J. D. Bjorken and J. B. Kogut, *Phys. Rev.* D **8**, 1341 (1973).
43. L. L. Frankfurt and M. Strikman, *Phys. Rep.* **160**, 235 (1988).
44. N. N. Nikolaev and B. G. Zakharov, *Z. Phys.* C **49**, 607 (1990).
45. A. Capella, A. Kaidalov, C. Merino, J. Tran Than Van, *Phys. Lett.* B **343**, 403 (1995).
A. Capella, A. Kaidalov, C. Merino, D. Perterman, J. Tran Than Van, *Phys. Rev.* D **53**, 2309 (1996).
46. R.J. Glauber, Lectures in Theoretical Physics. Ed. Britten W.E. N.Y.: Int.Publ. 1959,v.1,p.315.
47. V.N. Gribov, *JETP* **56**, 892 (1969);
V.N. Gribov, *JETP* **57**, 1306 (1969).
48. A.Schwimmer, *Nucl. Phys.* B **94**, 445 (1975).
49. M.Arneodo et al.(NMC Collaboration), *Nuovo Cim.* **107A**, 2141 (1994);
P.Amaudruz et al.(NMC Collaboration), *Nucl. Phys.* B **441**, 3 (1995).
50. A. Capella, A. Kaidalov, C. Merino, D. Perterman, J. Tran Than Van, *Eur. Phys. J.* **C5**, 111 (1998).
51. A.B. Kaidalov, Proc. Quark Matter 90, Nucl. Phys. **A552** (39c) (1991).
52. I.V. Andreev, A.V. Chernov, *Yad. Fiz.* **28**, 477 (1978).
53. K.G. Boreskov, A.B. Kaidalov, *Yad. Fiz.* **48**, 575 (1988).
54. J. Formanek, *Nucl. Phys.* **B 12**, 441 (1969).
55. W.Czyz and L.C. Maximon, *Ann. Phys. (N.Y.)* **52**, 59 (1969).
56. C. Pajares and A. V. Ramallo, *Phys. Rev.* D **31**, 2800 (1985).
57. K.G. Boreskov, A.B. Kaidalov, *Acta Phys. Polon.* **B20**, 397 (1989).
58. K. Werner, *Phys. Rep.* **232**, 87 (1993).
59. B.B.Back *et al* (PHOBOS collaboration), hep-ex/0007036
60. A. Capella and D. Sousa, nucl-th/0101023.

CLASSICAL CHROMO–DYNAMICS OF RELATIVISTIC HEAVY ION COLLISIONS *

DMITRI KHARZEEV (kharzeev@bnl.gov)
Physics Department, Brookhaven National Laboratory, Upton, NY11973-5000

Abstract. Relativistic heavy ion collisions produce thousands of particles, and it is sometimes difficult to believe that these processes allow for a theoretical description directly in terms of the underlying theory – QCD. However once the parton densities are sufficiently large, an essential simplification occurs – the dynamics becomes semi–classical. As a result, a simple *ab initio* approach to the nucleus–nucleus collision dynamics may be justified. In these lectures, we describe the application of these ideas to the description of multi–particle production in relativistic heavy ion collisions. We also discuss the rôle of semi–classical fields in the QCD vacuum in hadron interactions at low and high energies.

1. What is Chromo–Dynamics?

Strong interaction is, indeed, the strongest force of Nature. It is responsible for over 80% of the baryon masses, and thus for most of the mass of everything on Earth and in the Universe. Strong interactions bind nucleons in nuclei, which, being then bound into molecules by much weaker electro-magnetic forces, give rise to the variety of the physical World. Quantum Chromo–Dynamics is *the* theory of strong interactions, and its practical importance is thus undeniable. But QCD is more than a useful tool – it is a consistent and very rich field theory, which continues to serve as a stimulus for, and testing ground of, many exciting ideas and new methods in theoretical physics.
These lectures will deal with QCD of strong color fields, which can be explored in relativistic heavy ion collisions. (See the lectures by E. Iancu, A. Leonidov, L. McLerran [1], A.H. Mueller [2], and R. Venugopalan in this volume for complementary presentation of the subject and more details.)

* Work supported by the U.S. Department of Energy under contract No. DE–AC02–98CH10886.

J.-P. Blaizot and E. Iancu (eds.), QCD Perspectives on Hot and Dense Matter, 207–236.

1.1. QCD: THE LAGRANGEAN

So what is QCD? From the early days of the accelerator experiments it has become clear that the number of hadronic resonances is very large, suggesting that all hadrons may be classified in terms of a smaller number of (more) fundamental constituents. A convenient classification was offered by the quark model, but QCD was not born until the hypothetical existence of quarks was not supplemented by the principle of local gauge invariance, previously established as the basis of electromagnetism. The resulting Lagrangian has the form

$$\mathcal{L} = -\frac{1}{4} G^a_{\mu\nu} G^a_{\mu\nu} + \sum_f \bar{q}^a_f (i\gamma_\mu D_\mu - m_f) q^a_f; \tag{1.1}$$

the sum is over different colors a and quark flavors f; the covariant derivative is $D_\mu = \partial_\mu - igA^a_\mu t^a$, where t^a is the generator of the color group $SU(3)$, A^a_μ is the gauge (gluon) field and g is the coupling constant. The gluon field strength tensor is given by

$$G^a_{\mu\nu} = \partial_\mu A^a_\nu - \partial_\nu A^a_\mu + g f^{abc} A^b_\mu A^c_\nu, \tag{1.2}$$

where f^{abc} is the structure constant of $SU(3)$: $[t^a, t^b] = i f^{abc} t^c$.

1.2. ASYMPTOTIC FREEDOM

Due to the quantum effects of vacuum polarization, the charge in field theory can vary with the distance. In electrodynamics, summation of the electron–positron loops in the photon propagator leads to the following expression for the effective charge, valid at $r \gg r_0$:

$$\alpha_{em}(r) \simeq \frac{3\pi}{2\ln(r/r_0)}. \tag{1.3}$$

This formula clearly exhibits the "zero charge" problem [3] of QED: in the local limit $r_0 \to 0$ the effective charge vanishes at any finite distance away from the bare charge due to the screening. Fortunately, because of the smallness of the physical coupling, this apparent inconsistency of the theory manifests itself only at very short distances $\sim exp\{-3\pi/[2\alpha_{em}]\}$, $\alpha_{em} \simeq 1/137$. Such short distances are (and probably will always remain) beyond the reach of experiments, and one can safely use QED as a truly effective theory.

As it has been established long time ago [4], QCD is drastically different from electrodynamics in possessing the remarkable property of "asymptotic freedom" – due to the fact that gluons carry color, the behavior of the effective charge $\alpha_s = g^2/4\pi$ changes from the familiar from QED screening to anti–screening:

$$\alpha_s(r) \simeq \frac{3\pi}{(11N_c/2 - N_f)\ln(r_0/r)}; \tag{1.4}$$

as long as the number of flavors does not exceed 16 ($N_c = 3$), the anti–screening originating from gluon loops overcomes the screening due to quark–antiquark pairs, and the theory, unlike electrodynamics, is weakly coupled at short distances: $\alpha_s(r) \to 0$ when $r \to 0$.

1.3. CHIRAL SYMMETRY

In the limit of massless quarks, QCD Lagrangian (1.1) possesses an additional symmetry $U_L(N_f) \times U_R(N_f)$ with respect to the independent transformation of left– and right–handed quark fields $q_{L,R} = \frac{1}{2}(1 \pm \gamma_5)q$:

$$q_L \to V_L q_L;\ q_R \to V_R q_R;\ V_L, V_R \in U(N_f); \tag{1.5}$$

this means that left– and right–handed quarks are not correlated. Even a brief look into the Particle Data tables, or simply in the mirror, can convince anyone that there is no symmetry between left and right in the physical World. One thus has to assume that the symmetry (1.5) is spontaneously broken in the vacuum. The flavor composition of the existing eight Goldstone bosons (3 pions, 4 kaons, and the η) suggests that the $U_A(1)$ part of $U_L(3) \times U_R(3) = SU_L(3) \times SU_R(3) \times U_V(1) \times U_A(1)$ does not exist. This constitutes the famous "$U_A(1)$ problem".

1.4. THE ORIGIN OF MASS

There is yet another problem with the chiral limit in QCD. Indeed, as the quark masses are put to zero, the Lagrangian (1.1) does not contain a single dimensionful scale – the only parameters are pure numbers N_c and N_f. The theory is thus apparently invariant with respect to scale transformations, and the corresponding scale current is conserved: $\partial_\mu s_\mu = 0$. However, the absence of a mass scale would imply that all physical states in the theory should be massless!

1.5. QUANTUM ANOMALIES AND CLASSICAL SOLUTIONS

Both apparent problems – the missing $U_A(1)$ symmetry and the origin of hadron masses – are related to quantum anomalies. Once the coupling to gluons is included, both flavor singlet axial current and the scale current cease to be conserved; their divergences become proportional to the $\alpha_s G^a_{\mu\nu} \tilde{G}^a_{\mu\nu}$ and $\alpha_s G^a_{\mu\nu} G^a_{\mu\nu}$ gluon operators, correspondingly. This fact by itself would not have dramatic consequences if the gluonic vacuum were "empty", with $G^a_{\mu\nu} = 0$. However, it appears that due to non–trivial topology of the $SU(3)$ gauge group, QCD equations of motion allow classical solutions even in the absence of external color source, i.e. in the vacuum. The well–known example of a classical solution is the instanton, corresponding to the mapping of a three–dimensional sphere S^3 into the $SU(2)$ subgroup of $SU(3)$; its existence was shown to solve the $U_A(1)$ problem.

1.6. CONFINEMENT

The list of the problems facing us in the study of QCD would not be complete without the most important problem of all – why are the colored quarks and gluons excluded from the physical spectrum of the theory? Since confinement does not appear in perturbative treatment of the theory, the solution of this problem, again, must lie in the properties of the QCD vacuum.

1.7. UNDERSTANDING THE VACUUM

As was repeatedly stated above, the most important problem facing us in the study of all aspects of QCD is understanding the structure of the vacuum, which, in a manner of saying, does not at all behave as an empty space, but as a physical entity with a complicated structure. As such, the vacuum can be excited, altered and modified in physical processes [5].

2. Strong interactions at short and large distances

In this lecture we will investigate the influence of QCD vacuum on hadron interactions at short and large distances. To make the problem treatable, we will limit ourselves to heavy quarkonia. In this lecture I will describe two recent results – one on the scattering of heavy quarkonia at very low energies, another on high-energy scattering. The common idea behind these two examples is to explore the influence of the QCD vacuum on hadron interactions. The presentation will be schematic, and I refer the interested reader to the original papers [6] and [7] for details.

2.1. THE LONG–RANGE FORCES OF QCD

2.1.1. *Perturbation theory*

Let us begin with a somewhat academic problem – the scattering of two heavy quarkonium states at very low energies. The Wilson operator product expansion allows one to write down the scattering amplitude (in the Born approximation) of two small color dipoles in the following form[8]:

$$V(R) = -i \int dt \langle 0| \mathrm{T} \left(\sum_i c_i O_i(0) \right) \left(\sum_j c_j O_j(x) \right) |0\rangle, \qquad (2.1)$$

where $x = (t, R)$, $O_i(x)$ is the set of local gauge-invariant operators expressible in terms of gluon fields, and c_i are the coefficients which reflect the structure of the color dipole. At small (compared to the binding energy of the dipole) energies, the leading operator in (2.1) is the square of the chromo-electric field $(1/2)g^2\mathbf{E}^2$ [9, 8]. Keeping only this leading operator, we can rewrite (2.1) in a simple form

$$V(R) = -i\Big(\bar{d}_2\frac{a_0^2}{\epsilon_0}\Big)^2 \int dt \langle 0|\mathrm{T}\,\frac{1}{2}g^2\mathbf{E}^2(0)\,\frac{1}{2}g^2\mathbf{E}^2(t,R)|0\rangle, \quad (2.2)$$

where $\bar{d}_2$ is the corresponding Wilson coefficient defined by

$$\bar{d}_2\frac{a_0^2}{\epsilon_0} = \frac{1}{3N}\langle\phi|r^i\frac{1}{H_a+\epsilon}r^i|\phi\rangle, \quad (2.3)$$

where we have explicitly factored out the dependence on the quarkonium Bohr radius a_0 and the Rydberg energy ϵ_0; N is the number of colors, and $|\phi\rangle$ is the quarkonium wave function, which is Coulomb in the heavy quark limit[1]. In physical terms, the structure of (2.2) is transparent: it describes elastic scattering of two dipoles which act on each other by chromo-electric dipole fields; color neutrality permits only the square of dipole interaction. It is convenient to express $g^2\mathbf{E}^2$ in terms of the gluon field strength tensor [13]:

$$g^2\mathbf{E}^2 = -\frac{1}{4}g^2G_{\alpha\beta}G^{\alpha\beta} + g^2(-G_{0\alpha}G_0^\alpha + \frac{1}{4}g_{00}G_{\alpha\beta}G^{\alpha\beta}) =$$

$$= \frac{8\pi^2}{b}\theta_\mu^\mu + g^2\theta_{00}^{(G)} \quad (2.4)$$

where

$$\theta_\mu^\mu \equiv \frac{\beta(g)}{2g}G^{\alpha\beta a}G^a_{\alpha\beta} = -\frac{bg^2}{32\pi^2}G^{\alpha\beta a}G^a_{\alpha\beta}\,. \quad (2.5)$$

Note that as a consequence of scale anomaly, θ_μ^μ is the trace of the energy-momentum tensor of QCD in the chiral limit of vanishing light quark masses.
Let us now introduce the spectral representation for the correlator of the trace of energy-momentum tensor:

$$\langle 0|\mathrm{T}\theta_\mu^\mu(0)\theta_\mu^\mu(x)|0\rangle = \int d\sigma^2\rho_\theta(\sigma^2)\Delta_F(x;\sigma^2), \quad (2.6)$$

where $\rho_\theta(\sigma^2)$ is the spectral density and $\Delta_F(x;\sigma^2)$ is the Feynman propagator of a scalar field. Using the representation (2.6) in (2.2), we get

$$V_\theta(R) = -\Big(\bar{d}_2\frac{a_0^2}{\epsilon_0}\Big)^2\Big(\frac{4\pi^2}{b}\Big)^2\int d\sigma^2\rho_\theta(\sigma^2)\frac{1}{4\pi R}e^{-\sigma R}. \quad (2.7)$$

[1] The Wilson coefficients $\bar{d}_2$, evaluated in the large N limit, are available for S [8] and P [12] quarkonium states.

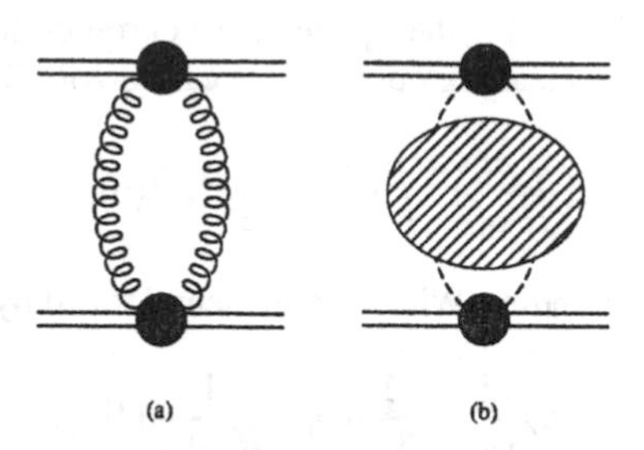

Figure 1. Contributions to the scattering amplitude from (a) two gluon exchange and (b) correlated two pion exchange.

The potential (2.7) is simply a superposition of Yukawa potentials corresponding to the exchange of scalar quanta of mass σ.
Our analysis so far has been completely general; the dynamics enters through the spectral density. In perturbation theory, for $SU(N)$, one has

$$\rho_\theta^{\rm pt}(q^2) = \left(\frac{bg^2}{32\pi^2}\right)^2 \frac{N^2-1}{4\pi^2} q^4. \tag{2.8}$$

Substituting (2.8) into (2.7) and performing the integration over invariant mass σ^2, we get, for $N = 3$

$$V_\theta(R) = -g^4 \left(\bar{d}_2 \frac{a_0^2}{\epsilon_0}\right)^2 \frac{15}{8\pi^3} \frac{1}{R^7}. \tag{2.9}$$

The $\propto R^{-7}$ dependence of the potential (2.9) is a classical result known from atomic physics [10]; as is apparent in our derivation (note the time integration in (2.7)), the extra R^{-1} as compared to the Van der Waals potential $\propto R^{-6}$ is the consequence of the fact that the dipoles we consider fluctuate in time, and the characteristic fluctuation time $\tau \sim \epsilon_0^{-1}$, is small compared to the spatial separation of the "onia" : $\tau \ll R$.
Let us note finally that the second term in (2.4) gives the contribution of the same order in g; this contribution is due to the tensor 2^{++} state of two gluons and can be evaluated in a completely analogous way. Adding this contribution to (2.9), changes the factor of 15 in (2.9) to 23, and we reproduce the result of ref. [8], which shows the equivalence of the spectral representation method used here and the functional method of ref. [8].

2.1.2. *Beyond the perturbation theory: scale anomaly and the role of pions*
At large distances, the perturbative description breaks down, because, as can be clearly seen from (2.7), the potential becomes determined by the spectral density at small q^2, where the transverse momenta of the gluons become small. At small

invariant masses, we have therefore to saturate the physical spectral density by the lightest state allowed in the scalar channel – two pions. Since, according to (2.5), θ_α^α is gluonic operator, this requires the knowledge of the coupling of gluons to pions. This looks like a hopeless non-perturbative problem, but it can nevertheless be rigorously solved, as it was shown in ref. [11] (see also [13]). The idea is the following: at small pion momenta, the energy–momentum tensor can be accurately computed using the low-energy chiral Lagrangian:

$$\theta_\mu^\mu = -\partial_\mu \pi^a \partial^\mu \pi^a + 2m_\pi^2 \pi^a \pi^a + \cdots \tag{2.10}$$

Using this expression, in the chiral limit of vanishing pion mass one gets an elegant result [11]

$$\langle 0|\frac{\beta(g)}{2g} G^{\alpha\beta a} G^a_{\alpha\beta}|\pi^+\pi^-\rangle = q^2. \tag{2.11}$$

Now that we know the coupling of gluons to the two pion state, the pion–pair contribution to the spectral density can be easily computed by performing the simple phase space integration, with the result

$$\rho_\theta^{\pi\pi}(q^2) = \frac{3}{32\pi^2} q^4, \tag{2.12}$$

which leads to the following long–distance potential [6]:

$$V^{\pi\pi}(R) \to -\left(\bar{d}_2 \frac{a_0^2}{\epsilon_0}\right)^2 \left(\frac{4\pi^2}{b}\right)^2 \frac{3}{2}(2m_\pi)^4 \frac{m_\pi^{1/2}}{(4\pi R)^{5/2}} e^{-2m_\pi R} \qquad \text{as } R \to \infty. \tag{2.13}$$

Note that, unlike the perturbative result which is manifestly $\sim g^4$, the amplitude (2.13) is $\sim g^0$ – this "anomalously" strong interaction is the consequence of scale anomaly[2].

While the shape of the potential in general depends on the spectral density, which is fixed theoretically only at relatively small invariant mass, the overall strength of the non-perturbative interactions is fixed by low energy theorems and is determined by the energy density of QCD vacuum. Indeed, in the heavy quark limit, one can derive the following sum rule [6]:

$$\int d^3\mathbf{R} \left(V(R) - V^{pt}(R)\right) = \left(\bar{d}_2 \frac{a_0^2}{\epsilon_0}\right)^2 \left(\frac{4\pi^2}{b}\right)^2 16\,|\epsilon_{vac}|\,, \tag{2.14}$$

which expresses the overall strength of the interaction between two color dipoles in terms of the energy density of the non-perturbative QCD vacuum.

[2] Of course, in the heavy quark limit the amplitude (2.13) will nevertheless vanish, since $a_0 \to 0$ and $\epsilon_0 \to \infty$.

2.1.3. *Does α_s ever get large?*

Asymptotic freedom ensures the applicability of QCD perturbation theory to the description of processes accompanied by high momentum transfer Q. However, as Q decreases, the strong coupling $\alpha_s(Q)$ grows, and the convergence of perturbative series is lost. How large can α_s get? The analyzes of many observables suggest that the QCD coupling may be "frozen" in the infrared region at the value $\langle\alpha_s\rangle_{IR} \simeq 0.5$ (see [14] and references therein). Gribov's program [15] relates the freezing of the coupling constant to the existence of massless quarks, which leads to the "decay" of the vacuum at large distances similar to the way it happens in QED in the presence of "supercritical" charge $Z > 1/\alpha$. One may try to infer the information about the behavior of the coupling constant at large distances by performing the matching of the fundamental theory onto the effective chiral Lagrangian at a scale $Q \simeq 4\pi f_\pi \simeq 1$ GeV, at which the ranges of validity of perturbative and chiral descriptions meet [6]. It is easy to see that in the chiral limit the matching of the potentials (2.13) and (2.9) yields[3] the coupling constant which freezes at the value

$$\langle\alpha_s\rangle_{IR} = \frac{6\sqrt{2}\,\pi}{11N_c - 2N_f}\sqrt{\frac{N_f^2-1}{N_c^2-1}}; \tag{2.15}$$

numerically, for QCD with $N_c = 3$ and $N_f = 2$ one finds $\langle\alpha_s\rangle_{IR} \simeq 0.56$. Note that this expression has an expected N_c dependence in the topological expansion limit of $N_c \to \infty, N_f/N_c = const.$

Since the trace of the energy momentum tensor in general relativity is linked to the curvature of space–time, the matching procedure leading to Eq. (2.15) has an interesting geometrical interpretation: it corresponds to the matching, at a relatively large distance, of curved space–time of the fundamental QCD with the flat space–time of the chiral theory.

2.2. HIGH–ENERGY SCATTERING: SCALE ANOMALY AND THE "SOFT" POMERON

In a 1972 article entitled "Zero pion mass limit in interaction at very high energies" [16], A.A. Anselm and V.N. Gribov posed an interesting question: what is the total cross section of hadron scattering in the chiral limit of $m_\pi \to 0$? On one hand, as everyone believes since the pioneering work of H. Yukawa, the range of strong interactions is determined by the mass of the lightest meson, i.e. is proportional to $\sim m_\pi^{-1}$. The total cross sections may then be expected to scale as $\sim m_\pi^{-2}$, and would tend to infinity as $m_\pi \to 0$. On the other hand, soft–pion theorems, which proved to be very useful in understanding low–energy hadronic

[3] The matching procedure of course can be performed directly for the correlation function of the energy–momentum tensor.

phenomena, state that hadronic amplitudes do not possess singularities in the limit $m_\pi \to 0$, and one expects that the theory must remain self-consistent in the limit of the vanishing pion mass. At first glance, the advent of QCD has not made this problem any easier; on the contrary, the presence of massless gluons in the theory apparently introduces another long–range interaction. Here, we will try to address this problem considering the scattering of small color dipoles.

Again, perturbation theory provides a natural starting point. In the framework of perturbative QCD, a systematic approach to high energy scattering was developed by Balitsky, Fadin, Kuraev and Lipatov [17], who demonstrated that the "leading log" terms in the scattering amplitude of type $(g^2 ln\ s)^n$ (where g is the strong coupling) can be re-summed, giving rise to the so–called "hard" Pomeron. Diagramatically, BFKL equation describes the t–channel exchange of "gluonic ladder" – a concept familiar from the old–fashioned multi–peripheral model.

It has been found, however, that at sufficiently high energies the perturbative description breaks down [18], [19]. The physical reason for this is easy to understand: the higher the energy, the larger impact parameters contribute to the scattering, and at large transverse distances the perturbation theory inevitably fails, since the virtualities of partons in the ladder diffuse to small values. At this point, the following questions arise: Does this mean that the problem becomes untreatable? Does the same difficulty appear at large distances in low–energy scattering? And, finally, what (if any) is the role played by pions?

The starting point of the approach proposed in [7] is the following: among the higher order, $O(\alpha_S^2)$ $(\alpha_S = g^2/4\pi)$, corrections to the BFKL kernel one can isolate a particular class of diagrams which include the propagation of two gluons in the scalar color singlet channel $J^{PC} = 0^{++}$. We then show that, as a consequence of scale anomaly, these, apparently $O(\alpha_S^2)$, contributions become the *dominant* ones, $O(\alpha_S^0)$. This is similar to our previous discussion in Sect. 2.1, where the interaction potential, proportional to α_S^2, at large distances turned into a "chiral" potential $\sim \alpha_S^0$ due to the scale anomaly.

One way of understanding the disappearance of the coupling constant in the spectral density of the g^2G^2 operator is to assume that the non-perturbative QCD vacuum is dominated by the semi–classical fluctuations of the gluon field. Since the strength of the classical gluon field is inversely proportional to the coupling, $G \sim 1/g$, the quark zero modes, and the spectral density of their pionic excitations, appear independent of the coupling constant.

The explicit calculation using the methods of [20] yields the power–like behavior of the total cross section:

$$\sigma_{tot} = \sum_{n=0}^{\infty} \sigma_n = \sigma^{BORN} s^{\Delta} , \tag{2.16}$$

where σ^{BORN} is the cross section due to two gluon exchange, and the non–

perturbative contribution to the intercept Δ is [7]

$$\Delta = \frac{\pi^2}{2} \times \left(\frac{8\pi}{b}\right)^2 \times \frac{18}{32\pi^2} \int \frac{dM^2}{M^6} \left(\rho_\theta^{phys}(M^2) - \rho_\theta^{pQCD}(M^2)\right) \quad (2.17)$$

Using the chiral formula (2.12) for ρ_θ^{phys} for $M^2 < M_0^2$, we obtain the following result [7]:

$$\Delta = \frac{1}{48} \ln \frac{M_0^2}{4m_\pi^2}. \quad (2.18)$$

The precise value of the matching scale M_0^2 as extracted from the low–energy theorems depends somewhat on detailed form of the spectral density, and can vary within the range of $M_0^2 = 4 \div 6\,\mathrm{GeV}^2$. Fortunately, the dependence of Eq. (2.18) on M_0 is only logarithmic, and varying it in this range leads to

$$\Delta = 0.08 \div 0.1, \quad (2.19)$$

in agreement with the phenomenological intercept of the "soft" Pomeron, $\Delta \simeq 0.08$.
At present, the language used in the description of hadron interactions at low and high energies is very different. Yet, as the two examples discussed above imply, both limits may appear to be determined by the same fundamental object – the QCD vacuum.

3. QCD in the classical regime

Most of the applications of QCD so far have been limited to the short distance regime of high momentum transfer, where the theory becomes weakly coupled and can be linearized. While this is the only domain where our theoretical tools based on perturbation theory are adequate, this is also the domain in which the beautiful non–linear structure of QCD does not yet reveal itself fully. On the other hand, as soon as we decrease the momentum transfer in a process, the dynamics rapidly becomes non–linear, but our understanding is hindered by the large coupling. Being perplexed by this problem, one is tempted to dream about an environment in which the coupling is weak, allowing a systematic theoretical treatment, but the fields are strong, revealing the full non–linear nature of QCD.
I am going to argue now that this environment can be created on Earth with the help of relativistic heavy ion colliders. Relativistic heavy ion collisions allow to probe QCD in the non–linear regime of high parton density and high color field strength.
It has been conjectured long time ago that the dynamics of QCD in the high density domain may become qualitatively different: in parton language, this is best described in terms of *parton saturation* [20, 21, 22], and in the language of color

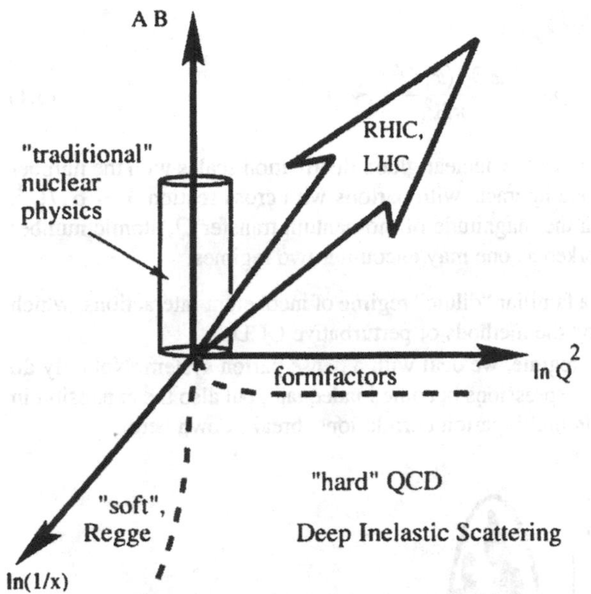

Figure 2. The place of relativistic heavy ion physics in the study of QCD; the vertical axis is the product of atomic numbers of projectile and target, and the horizontal axes are the momentum transfer Q^2 and rapidity $y = \ln(1/x)$ (x is the Bjorken scaling variable).

fields – in terms of the *classical* Chromo–Dynamics [23]; see the lectures [1] and [2] and references therein. In this high density regime, the transition amplitudes are dominated not by quantum fluctuations, but by the configurations of classical field containing large, $\sim 1/\alpha_s$, numbers of gluons. One thus uncovers new non–linear features of QCD, which cannot be investigated in the more traditional applications based on the perturbative approach. The classical color fields in the initial nuclei (the "color glass condensate" [1]) can be thought of as either perturbatively generated, or as being a topologically non–trivial superposition of the Weizsäcker-Williams radiation and the quasi–classical vacuum fields [24, 25, 26].

3.1. GEOMETRICAL ARGUMENTS

Let us consider an external probe J interacting with the nuclear target of atomic number A. At small values of Bjorken x, by uncertainty principle the interaction develops over large longitudinal distances $z \sim 1/mx$, where m is the nucleon mass. As soon as z becomes larger than the nuclear diameter, the probe cannot distinguish between the nucleons located on the front and back edges of the nucleus, and all partons within the transverse area $\sim 1/Q^2$ determined by the momentum transfer Q participate in the interaction coherently. The density of partons in the transverse plane is given by

$$\rho_A \simeq \frac{xG_A(x, Q^2)}{\pi R_A^2} \sim A^{1/3}, \tag{3.1}$$

where we have assumed that the nuclear gluon distribution scales with the number of nucleons A. The probe interacts with partons with cross section $\sigma \sim \alpha_s/Q^2$; therefore, depending on the magnitude of momentum transfer Q, atomic number A, and the value of Bjorken x, one may encounter two regimes:

- $\sigma\rho_A \ll 1$ – this is a familiar "dilute" regime of incoherent interactions, which is well described by the methods of perturbative QCD;
- $\sigma\rho_A \gg 1$ – in this regime, we deal with a dense parton system. Not only do the "leading twist" expressions become inadequate, but also the expansion in higher twists, i.e. in multi–parton correlations, breaks down here.

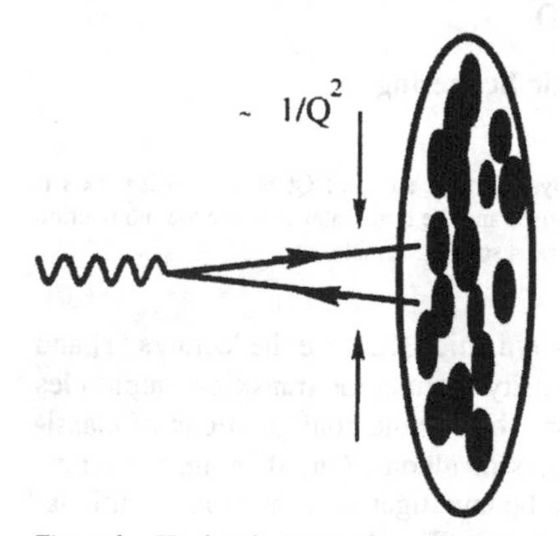

Figure 3. Hard probe interacting with the nuclear target resolves the transverse distance $\sim 1/\sqrt{Q}$ (Q^2 is the square of the momentum transfer) and, in the target rest frame, the longitudinal distance $\sim 1/(mx)$ (m is the nucleon mass and x – Bjorken variable).

The border between the two regimes can be found from the condition $\sigma\rho_A \simeq 1$; it determines the critical value of the momentum transfer ("saturation scale"[20]) at which the parton system becomes to look dense to the probe[4]:

$$Q_s^2 \sim \alpha_s \frac{xG_A(x, Q_s^2)}{\pi R_A^2}. \tag{3.2}$$

[4] Note that since $xG_A(x, Q_s^2) \sim A^{1/3}$, which is the length of the target, this expression in the target rest frame can also be understood as describing a broadening of the transverse momentum resulting from the multiple re-scattering of the probe.

In this regime, the number of gluons from (3.2) is given by

$$xG_A(x, Q_s^2) \sim \frac{\pi}{\alpha_s(Q_s^2)} Q_s^2 R_A^2, \tag{3.3}$$

where $Q_s^2 R_A^2 \sim A$. One can see that the number of gluons is proportional to the *inverse* of $\alpha_s(Q_s^2)$, and becomes large in the weak coupling regime. In this regime, as we shall now discuss, the dynamics is likely to become essentially classical.

3.2. SATURATION AS THE CLASSICAL LIMIT OF QCD

Indeed, the condition (3.2) can be derived in the following, rather general, way. As a first step, let us re-scale the gluon fields in the Lagrangian (1.1) as follows: $A_\mu^a \to \tilde{A}_\mu^a = gA_\mu^a$. In terms of new fields, $\tilde{G}_{\mu\nu}^a = gG_{\mu\nu}^a = \partial_\mu \tilde{A}_\nu^a - \partial_\nu \tilde{A}_\mu^a + f^{abc}\tilde{A}_\mu^b \tilde{A}_\nu^c$, and the dependence of the action corresponding to the Lagrangian (1.1) on the coupling constant is given by

$$S \sim \int \frac{1}{g^2} \tilde{G}_{\mu\nu}^a \tilde{G}_{\mu\nu}^a \, d^4x. \tag{3.4}$$

Let us now consider a classical configuration of gluon fields; by definition, $\tilde{G}_{\mu\nu}^a$ in such a configuration does not depend on the coupling, and the action is large, $S \gg \hbar$. The number of quanta in such a configuration is then

$$N_g \sim \frac{S}{\hbar} \sim \frac{1}{\hbar \, g^2} \rho_4 V_4, \tag{3.5}$$

where we re-wrote (3.4) as a product of four-dimensional action density ρ_4 and the four-dimensional volume V_4.

Note that since (3.5) depends only on the product of the Planck constant $\hbar$ and the coupling g^2, the classical limit $\hbar \to 0$ is indistinguishable from the weak coupling limit $g^2 \to 0$. The weak coupling limit of small $g^2 = 4\pi\alpha_s$ therefore corresponds to the semi-classical regime.

The effects of non-linear interactions among the gluons become important when $\partial_\mu \tilde{A}_\mu \sim \tilde{A}_\mu^2$ (this condition can be made explicitly gauge invariant if we derive it from the expansion of a correlation function of gauge-invariant gluon operators, e.g., $\tilde{G}^2$). In momentum space, this equality corresponds to

$$Q_s^2 \sim \tilde{A}^2 \sim (\tilde{G}^2)^{1/2} = \sqrt{\rho_4}; \tag{3.6}$$

Q_s is the typical value of the gluon momentum below which the interactions become essentially non-linear.

Consider now a nucleus A boosted to a high momentum. By uncertainty principle, the gluons with transverse momentum Q_s are extended in the longitudinal and proper time directions by $\sim 1/Q_s$; since the transverse area is πR_A^2, the four–volume is $V_4 \sim \pi R_A^2/Q_s^2$. The resulting four–density from (3.5) is then

$$\rho_4 \sim \alpha_s \frac{N_g}{V_4} \sim \alpha_s \frac{N_g Q_s^2}{\pi R_A^2} \sim Q_s^4, \tag{3.7}$$

where at the last stage we have used the non–linearity condition (3.6), $\rho_4 \sim Q_s^4$. It is easy to see that (3.7) coincides with the saturation condition (3.2), since the number of gluons in the infinite momentum frame $N_g \sim xG(x, Q_s^2)$.

In view of the significance of saturation criterion for the rest of the material in these lectures, let us present yet another argument, traditionally followed in the discussion of classical limit in electrodynamics [27]. The energy of the gluon field per unit volume is $\sim \vec{E}^{a2}$. The number of elementary "oscillators of the field", also per unit volume, is $\sim \omega^3$. To get the number of the quanta in the field we have to divide the energy of the field by the product of the number of the oscillators $\sim \omega^3$ and the average energy $\hbar\omega$ of the gluon:

$$N_{\vec{k}} \sim \frac{\vec{E}^{a2}}{\hbar \omega^4}. \tag{3.8}$$

The classical approximation holds when $N_{\vec{k}} \gg 1$. Since the energy ω of the oscillators is related to the time Δt over which the average energy is computed by $\omega \sim 1/\Delta t$, we get

$$\vec{E}^{a2} \gg \frac{\hbar}{(\Delta t)^4}. \tag{3.9}$$

Note that the quantum mechanical uncertainty principle for the energy of the field reads

$$\vec{E}^{a2}\, \omega^4 \sim \hbar, \tag{3.10}$$

so the condition (3.9) indeed defines the quasi–classical limit.
Since $\vec{E}^{a2}$ is proportional to the action density ρ_4, and the typical time is $\Delta t \sim 1/k_\perp$, using (3.7) we finally get that the classical description applies when

$$k_\perp^2 < \alpha_s \frac{N_g}{\pi R_A^2} \equiv Q_s^2. \tag{3.11}$$

3.3. THE ABSENCE OF MINI–JET CORRELATIONS

When the occupation numbers of the field become large, the matrix elements of the creation and annihilation operators of the gluon field defined by

$$\hat{A}^\mu = \sum_{\vec{k},\alpha} (\hat{c}_{\vec{k}\alpha} A^\mu_{\vec{k}\alpha} + \hat{c}^\dagger_{\vec{k}\alpha} A^{\mu *}_{\vec{k}\alpha}) \tag{3.12}$$

become very large,

$$N_{\vec{k}\alpha} = \langle \hat{c}^{\dagger}_{\vec{k}\alpha} \hat{c}_{\vec{k}\alpha} \rangle \gg 1, \tag{3.13}$$

so that one can neglect the unity on the r.h.s. of the commutation relation

$$\hat{c}_{\vec{k}\alpha} \hat{c}^{\dagger}_{\vec{k}\alpha} - \hat{c}^{\dagger}_{\vec{k}\alpha} \hat{c}_{\vec{k}\alpha} = 1 \tag{3.14}$$

and treat these operators as classical $c-$numbers.
This observation, often used in condensed matter physics, especially in the theoretical treatment of superfluidity, has important consequences for gluon production – in particular, it implies that the correlations among the gluons in the saturation region can be neglected:

$$\langle A(k_1)A(k_2)...A(k_n) \rangle \simeq \langle A(k_1) \rangle \langle A(k_2) \rangle ... \langle A(k_n) \rangle. \tag{3.15}$$

Thus, in contrast to the perturbative picture, where the produced mini-jets have strong back-to-back correlations, the gluons resulting from the decay of the classical saturated field are uncorrelated at $k_{\perp} \lesssim Q_s$.
Note that the amplitude with the factorization property (3.15) is called point–like. However, the relation (3.15) cannot be exact if we consider the correlations of final–state hadrons – the gluon mini–jets cannot transform into hadrons independently. These correlations caused by color confinement however affect mainly hadrons with close three–momenta, as opposed to the perturbative correlations among mini–jets with the opposite three–momenta.
It will be interesting to explore the consequences of the factorization property of the classical gluon field (3.15) for the HBT correlations of final–state hadrons. It is likely that the HBT radii in this case reflect the universal color correlations in the hadronization process.

Another interesting property of classical fields follows from the relation

$$\langle (\hat{c}^{\dagger}_{\vec{k}\alpha} \hat{c}_{\vec{k}\alpha})^2 \rangle - \langle \hat{c}^{\dagger}_{\vec{k}\alpha} \hat{c}_{\vec{k}\alpha} \rangle^2 = \langle \hat{c}^{\dagger}_{\vec{k}\alpha} \hat{c}_{\vec{k}\alpha} \rangle, \tag{3.16}$$

which determines the fluctuations in the number of produced gluons. We will discuss the implications of Eq. (3.16) for the multiplicity fluctuations in heavy ion collisions later.

4. Classical QCD in action

4.1. CENTRALITY DEPENDENCE OF HADRON PRODUCTION

In nuclear collisions, the saturation scale becomes a function of centrality; a generic feature of the quasi–classical approach – the proportionality of the number of gluons to the inverse of the coupling constant (3.5) – thus leads to definite predictions [28] on the centrality dependence of multiplicity.
Let us first present the argument on a qualitative level. At different centralities (determined by the impact parameter of the collision), the average density of partons (in the transverse plane) participating in the collision is very different. This density ρ is proportional to the average length of nuclear material involved in the collision, which in turn approximately scales with the power of the number N_{part} of participating nucleons, $\rho \sim N_{part}^{1/3}$. The density of partons defines the value of the saturation scale, and so we expect

$$Q_s^2 \sim N_{part}^{1/3}. \tag{4.1}$$

The gluon multiplicity is then, as we discussed above, is

$$\frac{dN_g}{d\eta} \sim \frac{S_A\, Q_s^2}{\alpha_s(Q_s^2)}, \tag{4.2}$$

where S_A is the nuclear overlap area, determined by atomic number and the centrality of collision. Since $S_A\, Q_s^2 \sim N_{part}$ by definitions of the transverse density and area, from (4.2) we get

$$\frac{dN_g}{d\eta} \sim N_{part}\, \ln N_{part}, \tag{4.3}$$

which shows that the gluon multiplicity shows a logarithmic deviation from the scaling in the number of participants.
To quantify the argument, we need to explicitly evaluate the average density of partons at a given centrality. This can be done by using Glauber theory, which allows to evaluate the differential cross section of the nucleus–nucleus interactions. The shape of the multiplicity distribution at a given (pseudo)rapidity η can then be readily obtained (see, e.g., [30]):

$$\frac{d\sigma}{dn} = \int d^2b\, \mathcal{P}(n; b)\, (1 - P_0(b)), \tag{4.4}$$

where $P_0(b)$ is the probability of no interaction among the nuclei at a given impact parameter b:

$$P_0(b) = (1 - \sigma_{NN} T_{AB}(b))^{AB}; \tag{4.5}$$

σ_{NN} is the inelastic nucleon–nucleon cross section, and $T_{AB}(b)$ is the nuclear overlap function for the collision of nuclei with atomic numbers A and B; we have used the three–parameter Woods–Saxon nuclear density distributions [29]. The correlation function $\mathcal{P}(n; b)$ is given by

$$\mathcal{P}(n;b) = \frac{1}{\sqrt{2\pi a \bar{n}(b)}} \exp\left(-\frac{(n-\bar{n}(b))^2}{2a\bar{n}(b)}\right), \tag{4.6}$$

here $\bar{n}(b)$ is the mean multiplicity at a given impact parameter b; the formulae for the number of participants and the number of binary collisions can be found in [30]. The parameter a describes the strength of fluctuations; for the classical gluon field, as follows from (3.16), $a = 1$. However, the strength of fluctuations can be changed by the subsequent evolution of the system and by hadronization process. Moreover, in a real experiment, the strength of fluctuations strongly depends on the acceptance. In describing the PHOBOS distribution [31], we have found that the value $a = 0.6$ fits the data well.

In Fig.4, we compare the resulting distributions for two different assumptions about the scaling of multiplicity with the number of participants to the PHOBOS experimental distribution, measured in the interval $3 < |\eta| < 4.5$. One can see that almost independently of theoretical assumptions about the dynamics of multiparticle production, the data are described quite well. At first this may seem surprising; the reason for this result is that at high energies, heavy nuclei are almost completely "black"; unitarity then implies that the shape of the cross section is determined almost entirely by the nuclear geometry. We can thus use experimental differential cross sections as a reliable handle on centrality. This gives us a possibility to compute the dependence of the saturation scale on centrality of the collision, and thus to predict the centrality dependence of particle multiplicities, shown in Fig. 5. (see [28] for details).

4.2. ENERGY DEPENDENCE

Let us now turn to the discussion of energy dependence of hadron production. In semi–classical scenario, it is determined by the variation of saturation scale Q_s with Bjorken $x = Q_s/\sqrt{s}$. This variation, in turn, is determined by the x–dependence of the gluon structure function. In the saturation approach, the gluon distribution is related to the saturation scale by Eq.(3.2). A good description of HERA data is obtained with saturation scale $Q_s^2 = 1 \div 2\ \mathrm{GeV}^2$ with W - dependence ($W \equiv \sqrt{s}$ is the center-of-mass energy available in the photon–nucleon system) [35]

$$Q_s^2 \propto W^\lambda, \tag{4.7}$$

where $\lambda \simeq 0.25 \div 0.3$. In spite of significant uncertainties in the determination of the gluon structure functions, perhaps even more important is the observation

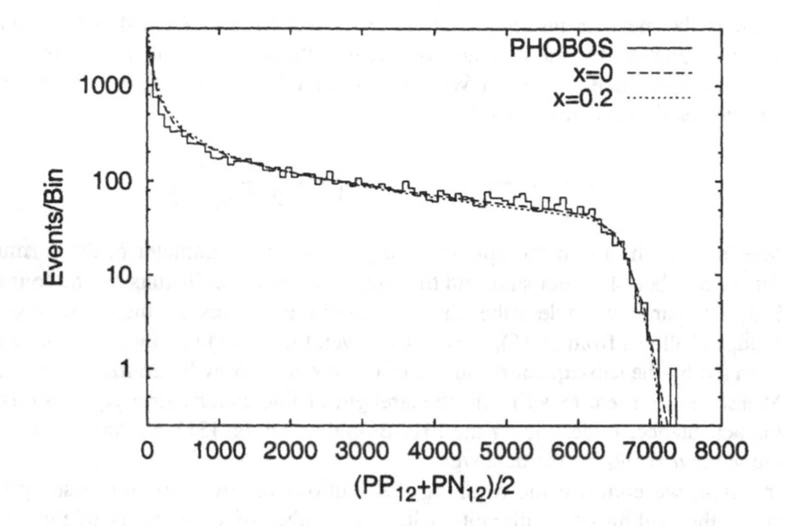

Figure 4. Charged multiplicity distribution at $\sqrt{s} = 130$ A GeV; solid line (histogram) – PHOBOS result; dashed line – distribution corresponding to participant scaling ($x = 0$); dotted line – distribution corresponding to the 37% admixture of "hard" component in the multiplicity; see text for details.

[35] that the HERA data exhibit scaling when plotted as a function of variable

$$\tau = \frac{Q^2}{Q_0^2} \left(\frac{x}{x_0} \right)^{\lambda}, \tag{4.8}$$

where the value of λ is again within the limits $\lambda \simeq 0.25 \div 0.3$. In high density QCD, this scaling is a consequence of the existence of dimensionful scale [20, 23])

$$Q_s^2(x) = Q_0^2 \, (x_0/x)^{\lambda}. \tag{4.9}$$

Using the value of $Q_s^2 \simeq 2.05$ GeV^2 extracted [28] at $\sqrt{s} = 130$ GeV and $\lambda = 0.25$ [35] used in [7], equation (4.19) leads to the following approximate formula for the energy dependence of charged multiplicity in central $Au - Au$ collisions:

$$\left\langle \frac{2}{N_{part}} \frac{dN_{ch}}{d\eta} \right\rangle_{\eta<1} \approx 0.87 \left(\frac{\sqrt{s}\,(\mathrm{GeV})}{130} \right)^{0.25} \times$$

$$\times \left[3.93 + 0.25 \ \ln \left(\frac{\sqrt{s}\,(\mathrm{GeV})}{130} \right) \right]. \tag{4.10}$$

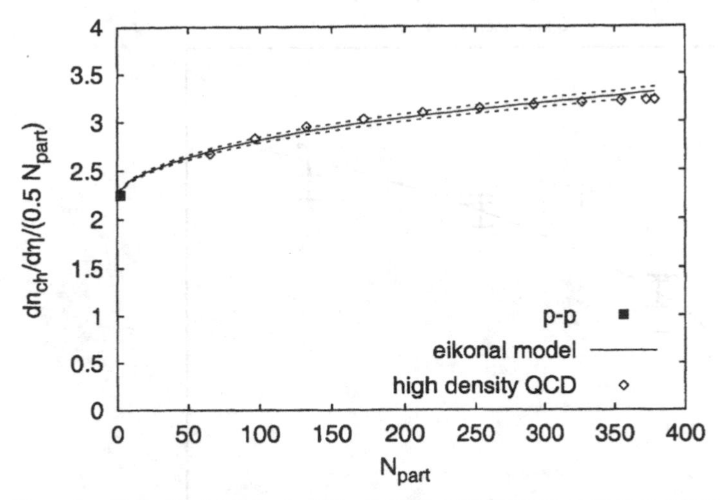

Figure 5. Centrality dependence of the charged multiplicity per participant pair near $\eta = 0$ at $\sqrt{s} = 130$ A GeV; the curves represent the prediction based on the conventional eikonal approach, while the diamonds correspond to the high density QCD prediction (see text). The square indicates the pp multiplicity.

At $\sqrt{s} = 130$ GeV, we estimate from Eq.(4.10) $2/N_{part}\ dN_{ch}/d\eta\ |_{\eta<1}= 3.42 \pm 0.15$, to be compared to the average experimental value of 3.37 ± 0.12 [31, 32, 33, 34]. At $\sqrt{s} = 200$ GeV, one gets 3.91 ± 0.15, to be compared to the PHOBOS value [31] of 3.78 ± 0.25. Finally, at $\sqrt{s} = 56$ GeV, we find 2.62 ± 0.15, to be compared to [31] 2.47 ± 0.25. It is interesting to note that formula (4.10), when extrapolated to very high energies, predicts for the LHC energy a value substantially smaller than found in other approaches:

$$\left\langle \frac{2}{N_{part}} \frac{dN_{ch}}{d\eta} \right\rangle_{\eta<1} = 10.8 \pm 0.5; \quad \sqrt{s} = 5500 \text{ GeV}, \tag{4.11}$$

corresponding only to a factor of 2.8 increase in multiplicity between the RHIC energy of $\sqrt{s} = 200$ GeV and the LHC energy of $\sqrt{s} = 5500$ GeV (numerical calculations show that when normalized to the number of participants, the multiplicity in central $Au - Au$ and $Pb - Pb$ systems is almost identical). The energy dependence of charged hadron multiplicity per participant pair is shown in Fig.6. One can also try to extract the value of the exponent λ from the energy dependence of hadron multiplicity measured by PHOBOS at $\sqrt{s} = 130$ GeV and at at $\sqrt{s} = 56$ GeV; this procedure yields $\lambda \simeq 0.37$, which is larger than the value inferred from the HERA data (and is very close to the value $\lambda \simeq 0.38$, resulting from the final–state saturation calculations [36]).

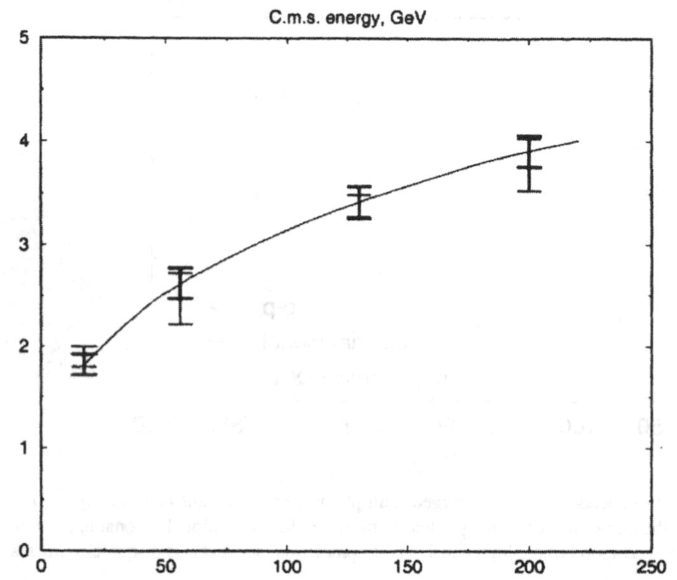

Figure 6. Energy dependence of charged multiplicity per participant pair at RHIC energies; solid line is the result (4.10).

4.3. RADIATING THE CLASSICAL GLUE

Let us now proceed to the quantitative calculation of the (pseudo-) rapidity and centrality dependences [37]. We need to evaluate the leading tree diagram describing emission of gluons on the classical level, see Fig. 7[5].
Let us introduce the unintegrated gluon distribution $\varphi_A(x, k_t^2)$ which describes the probability to find a gluon with a given x and transverse momentum k_t inside the nucleus A. As follows from this definition, the unintegrated distribution is related to the gluon structure function by

$$xG_A(x, p_t^2) = \int^{p_t^2} dk_t^2 \; \varphi_A(x, k_t^2); \tag{4.12}$$

when $p_t^2 > Q_s^2$, the unintegrated distribution corresponding to the bremsstrahlung radiation spectrum is

$$\varphi_A(x, k_t^2) \sim \frac{\alpha_s}{\pi} \frac{1}{k_t^2}. \tag{4.13}$$

[5] Note that this "mono–jet" production diagram makes obvious the absence of azimuthal correlations in the saturation regime discussed above, see eq (3.15).

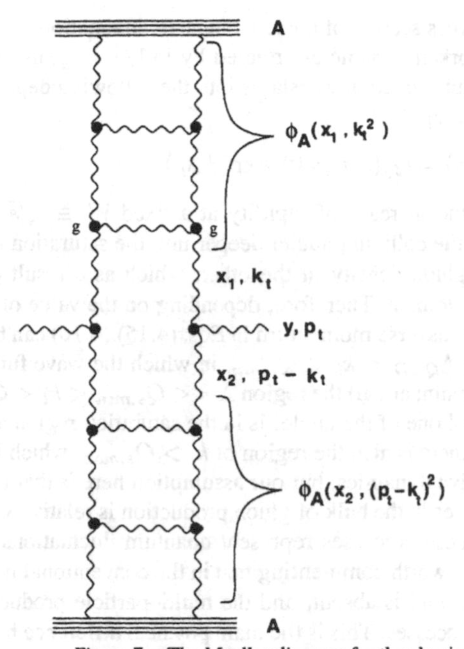

Figure 7. The Mueller diagram for the classical gluon radiation.

In the saturation region, the gluon structure function is given by (3.3); the corresponding unintegrated gluon distribution has only logarithmic dependence on the transverse momentum:

$$\varphi_A(x, k_t^2) \sim \frac{S_A}{\alpha_s}; \; k_t^2 \le Q_s^2, \tag{4.14}$$

where S_A is the nuclear overlap area, determined by the atomic numbers of the colliding nuclei and by centrality of the collision.

The differential cross section of gluon production in a AA collision can now be written down as [20, 38]

$$E\frac{d\sigma}{d^3p} = \frac{4\pi N_c}{N_c^2 - 1}\frac{1}{p_t^2}\int dk_t^2 \; \alpha_s \; \varphi_A(x_1, k_t^2) \; \varphi_A(x_2, (p-k)_t^2), \tag{4.15}$$

where $x_{1,2} = (p_t/\sqrt{s})\exp(\pm\eta)$, with η the (pseudo)rapidity of the produced gluon; the running coupling α_s has to be evaluated at the scale $Q^2 = max\{k_t^2, (p-k)_t^2\}$. The rapidity density is then evaluated from (4.15) according to

$$\frac{dN}{dy} = \frac{1}{\sigma_{AA}}\int d^2p_t \left(E\frac{d\sigma}{d^3p}\right), \tag{4.16}$$

where σ_{AA} is the inelastic cross section of nucleus–nucleus interaction.
Since the rapidity y and Bjorken variable are related by $\ln 1/x = y$, the $x-$ dependence of the gluon structure function translates into the following dependence of the saturation scale Q_s^2 on rapidity:

$$Q_s^2(s; \pm y) = Q_s^2(s; y = 0)\ \exp(\pm\lambda y). \tag{4.17}$$

As it follows from (4.17), the increase of rapidity at a fixed $W \equiv \sqrt{s}$ moves the wave function of one of the colliding nuclei deeper into the saturation region, while leading to a smaller gluon density in the other, which as a result can be pushed out of the saturation domain. Therefore, depending on the value of rapidity, the integration over the transverse momentum in Eqs. (4.15),(4.16) can be split in two regions: i) the region $\Lambda_{QCD} < k_t < Q_{s,min}$ in which the wave functions are both in the saturation domain; and ii) the region $\Lambda << Q_{s,min} < k_t < Q_{s,max}$ in which the wave function of one of the nuclei is in the saturation region and the other one is not. Of course, there is also the region of $k_t > Q_{s,max}$, which is governed by the usual perturbative dynamics, but our assumption here is that the rôle of these genuine hard processes in the bulk of gluon production is relatively small; in the saturation scenario, these processes represent quantum fluctuations above the classical background. It is worth commenting that in the conventional mini–jet picture, this classical background is absent, and the multi–particle production is dominated by perturbative processes. This is the main physical difference between the two approaches; for the production of particles with $p_t >> Q_s$ they lead to identical results.
To perform the calculation according to (4.16),(4.15) away from $y = 0$ we need also to specify the behavior of the gluon structure function at large Bjorken x (and out of the saturation region). At $x \to 1$, this behavior is governed by the QCD counting rules, $xG(x) \sim (1 - x)^4$, so we adopt the following conventional form: $xG(x) \sim x^{-\lambda}\ (1 - x)^4$.
We now have everything at hand to perform the integration over transverse momentum in (4.16), (4.15); the result is the following [37]:

$$\frac{dN}{dy} = const\ S_A\ Q_{s,min}^2\ \ln\left(\frac{Q_{s,min}^2}{\Lambda_{QCD}^2}\right) \times$$

$$\times \left[1 + \frac{1}{2}\ \ln\left(\frac{Q_{s,max}^2}{Q_{s,min}^2}\right)\ \left(1 - \frac{Q_{s,max}}{\sqrt{s}} e^{|y|}\right)^4\right], \tag{4.18}$$

where the constant is energy–independent, S_A is the nuclear overlap area, $Q_s^2 \equiv Q_s^2(s; y = 0)$, and $Q_{s,min(max)}$ are defined as the smaller (larger) values of (4.17); at $y = 0$, $Q_{s,min}^2 = Q_{s,max}^2 = Q_s^2(s) = Q_s^2(s_0) \times \times (s/s_0)^{\lambda/2}$. The first term in the brackets in (4.18) originates from the region in which both nuclear wave functions are in the saturation regime; this corresponds to the familiar $\sim (1/\alpha_s)\ Q_s^2 R_A^2$

term in the gluon multiplicity. The second term comes from the region in which only one of the wave functions is in the saturation region. The coefficient $1/2$ in front of the second term in square brackets comes from k_t ordering of gluon momenta in evaluation of the integral of Eq.(4.15).

The formula (4.18) has been derived using the form (4.14) for the unintegrated gluon distributions. We have checked numerically that the use of more sophisticated functional form of φ_A taken from the saturation model of Golec-Biernat and Wüsthoff [35] in Eq.(4.15) affects the results only at the level of about 3%.

Since $S_A Q_s^2 \sim N_{part}$ (recall that $Q_s^2 \gg \Lambda_{QCD}^2$ is defined as the density of partons in the transverse plane, which is proportional to the density of participants), we can re–write (4.18) in the following final form [37]

$$\frac{dN}{dy} = c\, N_{part} \left(\frac{s}{s_0}\right)^{\frac{\lambda}{2}} e^{-\lambda |y|} \left[\ln\left(\frac{Q_s^2}{\Lambda_{QCD}^2}\right) - \lambda |y|\right] \times$$

$$\times \left[1 + \lambda |y| \left(1 - \frac{Q_s}{\sqrt{s}}\, e^{(1+\lambda/2)|y|}\right)^4\right], \tag{4.19}$$

with $Q_s^2(s) = Q_s^2(s_0)\,(s/s_0)^{\lambda/2}$. This formula is the central result of our paper; it expresses the predictions of high density QCD for the energy, centrality, rapidity, and atomic number dependences of hadron multiplicities in nuclear collisions in terms of a single scaling function. Once the energy–independent constant $c \sim 1$ and $Q_s^2(s_0)$ are determined at some energy s_0, Eq. (4.19) contains no free parameters. At $y = 0$ the expression (4.18) coincides exactly with the one derived in [28], and extends it to describe the rapidity and energy dependences.

4.4. CONVERTING GLUONS INTO HADRONS

The distribution (4.19) refers to the radiated gluons, while what is measured in experiment is, of course, the distribution of final hadrons. We thus have to make an assumption about the transformation of gluons into hadrons. The gluon mini–jets are produced with a certain virtuality, which changes as the system evolves; the distribution in rapidity is thus not preserved. However, in the analysis of jet structure it has been found that the *angle* of the produced gluon is remembered by the resulting hadrons; this property of "local parton–hadron duality" (see [14] and references therein) is natural if one assumes that the hadronization is a soft process which cannot change the direction of the emitted radiation. Instead of the distribution in the angle θ, it is more convenient to use the distribution in pseudo–rapidity $\eta = -\ln\tan(\theta/2)$. Therefore, before we can compare (4.18) to the data, we have to convert the rapidity distribution (4.19) into the gluon distribution in pseudo–rapidity. We will then assume that the gluon and hadron distributions are dual to each other in the pseudo–rapidity space.

To take account of the difference between rapidity y and the measured pseudo-rapidity η, we have to multiply (4.18) by the Jacobian of the $y \leftrightarrow \eta$ transformation; a simple calculation yields

$$h(\eta; p_t; m) = \frac{\cosh \eta}{\sqrt{\frac{m^2+p_t^2}{p_t^2} + \sinh^2 \eta}}, \tag{4.20}$$

where m is the typical mass of the produced particle, and p_t is its typical transverse momentum. Of course, to plot the distribution (4.19) as a function of pseudo-rapidity, one also has to express rapidity y in terms of pseudo-rapidity η; this relation is given by

$$y(\eta; p_t; m) = \frac{1}{2} \ln \left[\frac{\sqrt{\frac{m^2+p_t^2}{p_t^2} + \sinh^2 \eta} + \sinh \eta}{\sqrt{\frac{m^2+p_t^2}{p_t^2} + \sinh^2 \eta} - \sinh \eta} \right]; \tag{4.21}$$

obviously, $h(\eta; p_t; m) = \partial y(\eta; p_t; m)/\partial \eta$.
We now have to make an assumption about the typical invariant mass m of the gluon mini–jet. Let us estimate it by assuming that the slowest hadron in the mini-jet decay is the ρ-resonance, with energy $E_\rho = (m_\rho^2 + p_{\rho,t}^2 + p_{\rho,z}^2)^{1/2}$, where the z axis is pointing along the mini-jet momentum. Let us also denote by x_i the fractions of the gluon energy q_0 carried by other, fast, i particles in the mini-jet decay. Since the sum of transverse (with respect to the mini-jet axis) momenta of mini-jet decay products is equal to zero, the mini-jet invariant mass m is given by

$$m_{jet}^2 \equiv m^2 = (\sum_i x_i q_0 + E_\rho)^2 - (\sum_i x_i q_z + p_{\rho,z})^2 \simeq$$

$$\simeq 2 \sum_i x_i q_z \cdot (m_{\rho,t} - p_{\rho,z}) \equiv 2 Q_s \cdot m_{eff}, \tag{4.22}$$

where $m_{\rho,t} = (m_\rho^2 + p_{\rho,t}^2)^{1/2}$. In Eq. (4.22) we used that $\sum_i x_i = 1$ and $q_0 \approx q_z = Q_s$. Taking $p_{\rho,z} \approx p_{\rho,t} \approx$ 300 MeV and ρ mass, we obtain $m_{eff} \approx 0.5$ GeV.
We thus use the mass $m^2 \simeq 2Q_s m_{eff} \simeq Q_s \cdot 1$ GeV in Eqs.(4.20,4.21). Since the typical transverse momentum of the produced gluon mini–jet is Q_s, we take $p_t = Q_s$ in (4.20). The effect of the transformation from rapidity to pseudo–rapidity is the decrease of multiplicity at small η by about $25 - 30\%$, leading to the appearance of the $\approx 10\%$ dip in the pseudo–rapidity distribution in the vicinity of $\eta = 0$. We have checked that the change in the value of the mini–jet mass by two times affects the Jacobian at central pseudo–rapidity to about $\simeq 10\%$, leading to $\sim 3\%$ effect on the final result.
The results for the $Au - Au$ collisions at $\sqrt{s} = 130$ GeV are presented in Figs 8 and 9. In the calculation, we use the results on the dependence of saturation scale

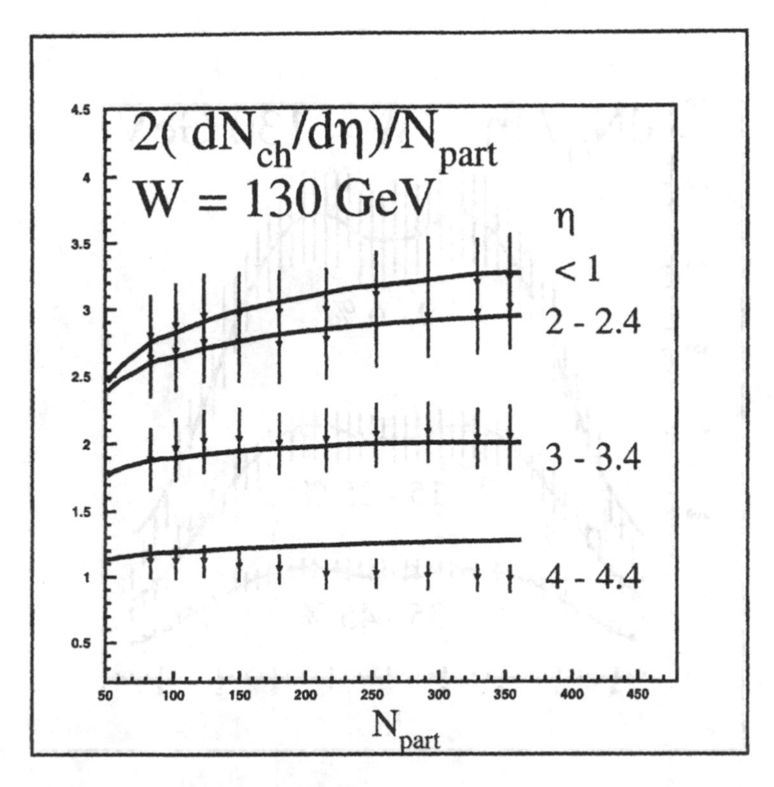

Figure 8. Centrality dependence of charged hadron production per participant at different pseudo-rapidity η intervals in $Au - Au$ collisions at $\sqrt{s} = 130$ GeV; from (Kharzeev and Levin, 2001), the data are from (PHOBOS Coll., 2000).

on the mean number of participants at $\sqrt{s} = 130$ GeV from [28], see Table 2 of that paper. The mean number of participants in a given centrality cut is taken from the PHOBOS paper [31]. One can see that both the centrality dependence and the rapidity dependence of the $\sqrt{s} = 130$ GeV PHOBOS data are well reproduced below $\eta \simeq \pm 4$. The rapidity dependence has been evaluated with $\lambda = 0.25$, which is within the range $\lambda = 0.25 \div 0.3$ inferred from the HERA data [35]. The discrepancy above $\eta \simeq \pm 4$ is not surprising since our approach does not properly take into account multi–parton correlations which are important in the fragmentation region.

Our predictions for $Au - Au$ collisions at $\sqrt{s} = 200$ GeV are presented in [37]. The only parameter which governs the energy dependence is the exponent λ, which we assume to be $\lambda \simeq 0.25$ as inferred from the HERA data. The absolute prediction for the multiplicity, as explained above, bears some uncertainty, but

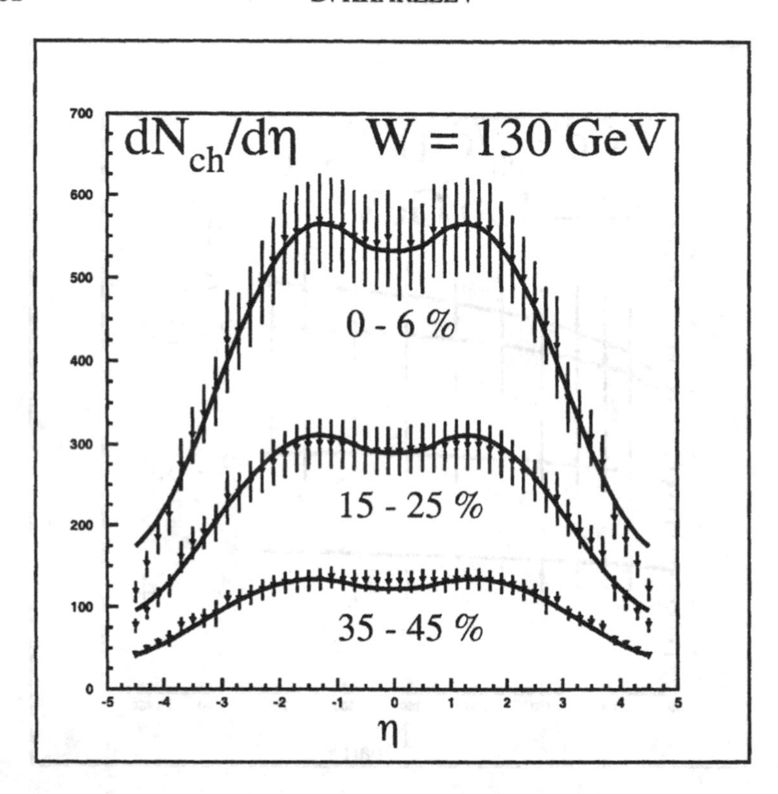

Figure 9. Pseudo–rapidity dependence of charged hadron production at different cuts on centrality in $Au - Au$ collisions at $\sqrt{s}$ = 130 GeV; from (Kharzeev and Levin, 2001), the data are from (PHOBOS Coll., 2000).

there is a definite feature of our scenario which is distinct from other approaches. It is the dependence of multiplicity on centrality, which around $\eta = 0$ is determined solely by the running of the QCD strong coupling [28]. As a result, the centrality dependence at $\sqrt{s}$ = 200 GeV is somewhat less steep than at $\sqrt{s}$ = 130. While the difference in the shape at these two energies is quite small, in the perturbative mini-jet picture this slope should increase, reflecting the growth of the mini-jet cross section with energy [39].

4.5. FURTHER TESTS

Checking the predictions of the semi–classical approach for the centrality and pseudo–rapidity dependence at $\sqrt{s}$ = 200 GeV is clearly very important. What other tests of this picture can one devise? The main feature of the classical emission is that it is coherent up to the transverse momenta of about $\sqrt{2}\, Q_s$ (about $\simeq 2$ GeV/c for central $Au - Au$ collisions). This means that if we look at the centrality dependence of particle multiplicities above a certain value of the transverse momentum, say, above 1 GeV/c, it should be very similar to the dependence without the transverse momentum cut-off. On the other hand, in the two–component "soft plus hard" model the cut on the transverse momentum would strongly enhance the contribution of hard mini–jet production processes, since soft production mechanisms presumably do not contribute to particle production at high transverse momenta. Of course, at sufficiently large value of the cutoff all of the observed particles will originate from genuine hard processes, and the centrality dependence will become steeper, reflecting the scaling with the number of collisions. It will be very interesting to explore the transition to this hard scattering regime experimentally.

Another test, already discussed above (see eq.(3.15)) is the study of azimuthal correlations between the produced high p_t particles. In the saturation scenario these correlations should be very small below $p_t \simeq 2$ GeV/c in central collisions. At higher transverse momenta, and/or for more peripheral collisions (where the saturation scale is smaller) these correlations should be much stronger.

5. Does the vacuum melt?

The approach described above allows us to estimate the initial energy density of partons achieved at RHIC. Indeed, in this approach the formation time of partons is $\tau_0 \simeq 1/Q_s$, and the transverse momenta of partons are about $k_t \simeq Q_s$. We thus can use the Bjorken formula and the set of parameters deduced above to estimate [28]

$$\epsilon \simeq \frac{< k_t >}{\tau_0} \frac{d^2 N}{d^2 b d\eta} \simeq Q_s^2 \frac{d^2 N}{d^2 b d\eta} \simeq 18 \text{ GeV/fm}^3 \tag{5.1}$$

for central $Au - Au$ collisions at $\sqrt{s} = 130$ GeV. This value is well above the energy density needed to induce the QCD phase transition according to the lattice calculations. However, the picture of gluon production considered above seems to imply that the gluons simply flow from the initial state of the incident nuclei to the final state, where they fragment into hadrons, with nothing spectacular happening on the way. In fact, one may even wonder if the presence of these gluons modifies at all the structure of the physical QCD vacuum.

To answer this question theoretically, we have to possess some knowledge about the non–perturbative vacuum properties. While in general the problem of vacuum structure still has not been solved (and this is one of the main reasons for the heavy ion research!), we do know one class of vacuum solutions – the instantons. It is thus interesting to investigate what happens to the QCD vacuum in the presence of strong external classical fields using the example of instantons [26].

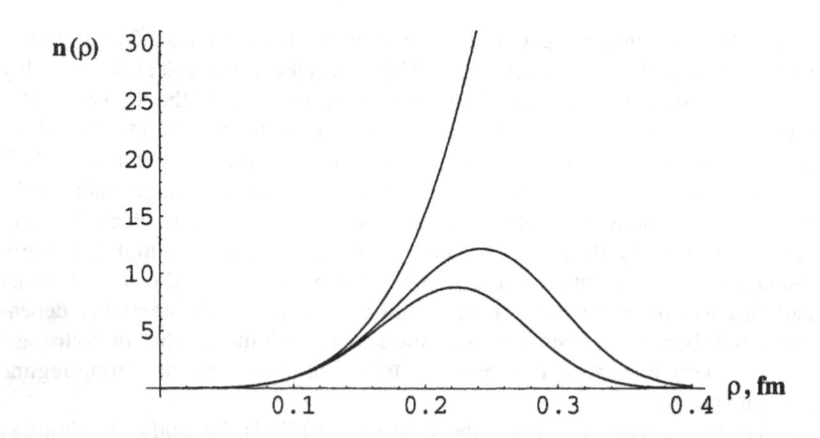

Figure 10. Distributions of instanton sizes in vacuum for QCD with three light flavors (upper curve) versus the distribution of instanton sizes in the saturation environment produced by a collision of two identical nuclei for $c = 1$ (middle curve) and $c = 2\ln 2$ (lower curve) with $Q_s^2 = 2\,\mathrm{GeV}^2$; from (Kharzeev, Kovchegov and Levin, 2002).

The problem of small instantons in a slowly varying background field was first addressed in [40, 41] by introducing the effective instanton lagrangian $L_{eff}^{I(\bar{I})}(x)$

$$L_{eff}^{I}(x_0) = \int d\rho\, n_0(\rho)\, dR \exp\left(-\frac{2\pi^2}{g}\rho^2\,\overline{\eta}_{a\mu\nu}^{M}\, R^{aa'}\, G_{\mu\nu}^{a'}(x_0)\right) \qquad (5.2)$$

in which $n_0(\rho)$ is the instanton size distribution function in the vacuum, $\overline{\eta}_{a\mu\nu}^{M}$ is the 't Hooft symbol in Minkowski space, and $R^{aa'}$ is the matrix of rotations in color space, with dR denoting the averaging over the instanton color orientations. The complete field of a single instanton solution could be reconstructed by perturbatively resumming the powers of the effective instanton lagrangian which corresponds to perturbation theory in powers of the instanton size parameter ρ^2. In our case here the background field arises due to the strong source current J_μ^a. The current can be due to a single nucleus, or resulting from the two colliding nuclei. Perturbative resummation of powers of the source current term translates itself into resummation of the powers of the classical field parameter $\alpha_s^2 A^{1/3}$ [23, 42]. Thus the problem of instantons in the background classical gluon field is described by the effective action in Minkowski space

$$S_{eff} = \int d^4x \left(-\frac{1}{4}G_{\mu\nu}^a(x)G_{\mu\nu}^a(x) + L_{eff}^{I}(x) + L_{eff}^{\bar{I}}(x) + J_\mu^a\, A_\mu^a(x)\right) \qquad (5.3)$$

The problem thus is clearly formulated; by using an explicit form for the radiated classical gluon field, it was possible to demonstrate [26] that the distribution of instantons gets modified from the original vacuum one $n_0(\rho)$ to

$$n_{sat}^{AA}(\rho) = n_0(\rho) \exp\left(-\frac{c\rho^4 Q_s^4}{8\,\alpha_s^2\, N_c\,(Q_s\tau_0)^2}\right), \tag{5.4}$$

where τ_0 is the proper time. The result Eq. (5.4) shows that large size instantons are suppressed by the strong classical fields generated in the nuclear collision (see Fig. 10)[6]. The vacuum does melt!

The results presented here were obtained together with Hirotsugu Fujii, Yuri Kovchegov, Eugene Levin, and Marzia Nardi. I am very grateful to them for the most enjoyable collaboration. I also wish to thank Jean-Paul Blaizot, Yuri Dokshitzer, Larry McLerran, Al Mueller and Raju Venugopalan for numerous illuminating discussions on the subject of these lectures.

References

1. Iancu, I., Leonidov, A., and McLerran, L. hep-ph/0202270.
2. Mueller, A.H., hep-ph/0111244.
3. Landau, L.D., and Pomeranchuk, I. Ya., Dokl.Akad.Nauk 102:489, 1955.
4. Gross, D.J., and Wilczek, F., Phys. Rev. Lett. 30:1343, 1973;
   Politzer, H.D., Phys. Rev. Lett. 30:1346, 1973.
5. Lee, T.D. and Wick, G.C., Phys. Rev. D9:2291, 1974.
6. Fujii, H. and Kharzeev, D., Phys. Rev. D60:114039, 1999; hep-ph/9807383.
7. Kharzeev, D. and Levin, E., Nucl. Phys. B578:351, 2000.
8. Peskin, M.E., Nucl. Phys. B156:365, 1977;
   Bhanot, G. and Peskin, M.E., Nucl. Phys. B156:391, 1977.
9. Voloshin, M.B., Nucl. Phys. 154:365, 1978.
10. Casimir, H.B.G. and Polder, D., Phys. Rev. 73:360, 1948.
11. Voloshin, M.B. and Zakharov, V.I., Phys.Rev.Lett. 45:688, 1980.
12. Kharzeev, D. Quarkonium Interactions in QCD. In A. DiGiacomo and D. Diakonov, editors, *Selected Topics in Nonperturbative QCD*. IOS, 1996; nucl-th/9601029.
13. Novikov, V.A. and Shifman, M.A., Z. Phys. C 8:43, 1981.
14. Dokshitzer, Yu.L., hep-ph/9812252.
15. Gribov, V.N., Eur. Phys. J. C10:71; 91, 1999.
16. Anselm, A.A. and Gribov, V.N., Physics Letters B40:487, 1972.
17. Kuraev, E.A., Lipatov, L.N. and Fadin, V.S., Sov. Phys. JETP 45:199, 1977;
    Balitsky, Ia. Ia. and Lipatov, L.N., Sov. J. Nucl. Phys. 28:822, 1978;
    Lipatov, L.N., Sov. Phys. JETP 63:904, 1986.
18. Mueller, A.H., Physics Letters B396:251, 1997.
19. Dokshitzer, Yu.L., hep-ph/9801372.
20. Gribov, L.V., Levin, E.M. and Ryskin, M.G., Phys. Rept. 100:1, 1983.
21. Mueller, A.H. and Qiu, J.-W., Nucl. Phys. B268:427, 1986.
22. Blaizot, J.P. and Mueller, A.H., Nucl. Phys. B289:847, 1987.
23. McLerran, L.D. and Venugopalan, R., Phys. Rev. D49:2233, 1994; D49:3352, 1994.

[6] Of course, at large proper times $\tau_0 \to \infty$ the vacuum "cools off", and the instanton distribution returns to the vacuum one.

24. Kharzeev, D., Kovchegov, Yu. and Levin, E., Nucl. Phys. A690:621, 2001;
25. Nowak, M., Shuryak, E., and Zahed, I., Phys.Rev. D64:034008, 2001.
26. Kharzeev, D., Kovchegov, Yu. and Levin, E., Nucl.Phys.A699:745, 2002.
27. Berestetskii, V. B., Lifshitz, E.M. and Pitaevskii, L.P., *"Quantum electrodynamics"*, Oxford New York, Pergamon Press, 1982.
28. Kharzeev, D. and Nardi, Phys. Lett. B507:121, 2001.
29. De Jager, C., De Vries, H. and De Vries, C., Atom. Nucl. Data Tabl. 14:479, 1974.
30. Kharzeev, D., Lourenço, C., Nardi, M. and Satz, H., Z.Phys. C74:307, 1997.
31. Back, B. et al., PHOBOS Coll., Phys. Rev. Lett. 85:3100, 2000; Phys. Rev. Lett. 87:102303, 2001; nucl-ex/0105011; nucl-ex/0108009.
32. Adcox, K. et al., (The PHENIX Collaboration), Phys. Rev. Lett. 86:3500, 2001; Phys. Rev. Lett. 87:052301, 2001; Milov, A. et al., (The PHENIX Collaboration), nucl-ex/0107006.
33. Adler, C. et al., (The STAR Collaboration), Phys. Rev. Lett. 87:112303, 2001; Phys. Rev. Lett. 87:082301, 2001.
34. Bearden I.G. et al., (The BRAHMS Collaboration), Phys. Rev. Lett. 87:112305, 2001; nucl-ex/0102011; nucl-ex/0108016
35. Golec-Biernat, K. and Wüsthof, M., Phys. Rev. D59:014017, 1999; Phys. Rev. D60:114023, 1999; Stasto, A., Golec-Biernat, K. and Kwiecinski, J., Phys. Rev. Lett. 86:596, 2001.
36. Eskola, K.J., Kajantie, K. and Tuominen, K., Phys. Lett. B497:39, 2001;
Eskola, K.J., Kajantie, K., Ruuskanen, P.V. and Tuominen, K., Nucl. Phys. B 570:379, 2000.
37. Kharzeev, D. and Levin, E., Phys. Lett. B523:79, 2001.
38. Gyulassy, M. and McLerran, L., Phys. Rev. C56:2219, 1997.
39. Wang, X.N. and Gyulassy, M., Phys. Rev. Lett. 86:3496, 2001.
40. Callan, C.G., Dashen, R. and Gross, D.J., Phys. Rev. D19:1826, 1979.
41. Shifman, M.A., Vainshtein, A.I., Zakharov, V.I., Nucl. Phys. B165:45, 1980.
42. Kovchegov, Yu.V., Phys. Rev. D54:5463, 1996); 55:5445, 1997.

COLLECTIVE DYNAMICS IN HEAVY ION COLLISIONS

J.-Y. OLLITRAULT (ollie@wasa.saclay.cea.fr)
Service de physique théorique, DSM, CEA Saclay

Abstract. These lectures are an elementary introduction to various experimental signatures of collective motion in ultrarelativistic heavy ion collisions: relation between transverse momenta and multiplicities, identified particle spectra, HBT correlations, and elliptic flow.

1. Introduction

One of the main goals of ultrarelativistic heavy ion collisions is to produce and study the quark-gluon plasma (QGP), a new state of matter where hadrons dissolve into their elementary constituents. One expects a QGP to be formed in these collisions for two reasons: first, a large number of particles are created and seen in detectors; second, due to the Lorentz contraction of the incoming nuclei, the volume of the system shortly after the collision is so small that the density of particles reaches values too high to be compatible with a mere gas of hadrons. Theoretical frameworks to study the quark-gluon plasma, either through perturbative calculations or lattice simulations, assume thermal equilibrium. Now, the system formed in a nucleus-nucleus collision is expanding so rapidly that it can certainly not be described as a static thermal bath. However, the interactions among its constituents may be large enough to achieve thermal equilibrium *locally*: if this were the case, the dense system formed after the nuclei have passed through each other would behave like an expanding gas, following the laws of fluid dynamics.
Looking for experimental signatures of this (local) thermal equilibrium is therefore a central problem in the phenomenology of heavy ion collisions. I consider it to be an essential preliminary step to the search for the quark-gluon plasma. These lectures are an elementary introduction to this problem. In order to put emphasis on the essential, qualitative ideas behind the various signatures, I have chosen to work out in detail a small number of oversimplified examples, rather than give a general overview of the field.
Sec. 2 is a short presentation of thermal and hydrodynamical models, based on the idea that the expanding system formed after the collision behaves as an ideal fluid. The strong Lorentz contraction at ultrarelativistic energies leads to a clear

J.-P. Blaizot and E. Iancu (eds.), QCD Perspectives on Hot and Dense Matter, 237–256.

separation between longitudinal and transverse degrees of freedom. I will show in particular that an important consequence of thermalization is the longitudinal cooling of the system. Due to this phenomenon, the relation between transverse momenta of outgoing particles and their multiplicity is directly related to the equation of state.
Sec. 3 is devoted to signatures of thermalization in the transverse expansion of the system. Local thermalization would result in a collective motion of the particles superimposed to the thermal motion, as in a flowing liquid. Various signatures of this collective motion are reviewed, based on particle spectra and on two-particle correlations.
Finally, Sec. 4 discusses a phenomenon specific to nucleus-nucleus collisions, namely the correlation between outgoing momenta and the impact direction (or reaction plane) of the two nuclei. This correlation, usually named "anisotropic flow", results solely from interactions between the particles after they have been created in the collision. Therefore, it is widely believed to be the most sensitive probe of collective motion.

2. Aspects of thermalization

The collision of two heavy nuclei produces many particles in a small volume. These particles interact. If interactions are strong enough, the mean free path of a particle becomes much smaller than the size of the system: thermalization is reached.
In this ideal limit, the evolution of the system can be described by means of a few parameters. Consider a volume V, much smaller than the total volume of the colliding system, yet large enough to contain many particles. In the local rest frame of this subsystem (i.e., the frame where the total momentum of the subsystem vanishes), the momentum ($\mathbf{p}$) distribution of the constituents are isotropic, and their energy (E) distributions are simply given by simple Bose-Einstein or Fermi-Dirac distributions, depending on their spin s:

$$\frac{dN}{d^3\mathbf{x}\,d^3\mathbf{p}} = \frac{(2s+1)}{(2\pi)^3}\frac{1}{\exp((E-\mu)/T)\pm 1}. \tag{2.1}$$

The only free parameters are the temperature T and a chemical potential μ (in practice, there is one chemical potential for each conserved quantity, baryon number, electric charge and strangeness).
Momentum distributions in the laboratory frame are then obtained by boosting the system from the local rest frame with a velocity $\beta = (\sum \mathbf{p})/(\sum E)$, where sums run over all constituents of the subsystem. The momentum distribution in the laboratory frame is then simply obtained through the following replacement in Eq. (2.1)

$$E \to p^\mu\, u_\mu, \tag{2.2}$$

where p^μ is the four-momentum and u_μ the *fluid four-velocity*,

$$u_\mu = \frac{1}{\sqrt{1-\beta^2}} \begin{pmatrix} 1 \\ -\beta \end{pmatrix} \tag{2.3}$$

2.1. CHEMICAL EQUILIBRIUM, KINETIC EQUILIBRIUM

Thermal equilibrium is a very strong assumption. Eq. (2.1) indeed constrains not only the momentum distributions, but also the total particle yields, and in particular particle ratios (i.e., relative abundances of various particle species). However, these various aspects of thermal equilibrium are achieved by means of distinct microscopic processes. While all collisions between the particles, either elastic or inelastic, tend to equilibrate momentum distribution, only inelastic collisions may change the number of particles.
Equilibrium with respect to inelastic collisions, which is in some sense the "perfect" thermal equilibrium, is usually referred to as "chemical equilibrium" and will not be studied further in these lectures. It strongly constrains the ratios of particle abundances [1] as well as phase-space densities, which are obtained by combining informations from momentum spectra and two-particle HBT correlations [2], and these constraints seem to be compatible with existing data.
Equilibrium with respect to elastic collisions constrains momentum distributions, and implies in particular that they are isotropic in the local rest frame. This is the "kinetic" equilibrium, on which I concentrate here.
Kinetic equilibrium itself has (at least) two facets. One is the equilibration between longitudinal and transverse degrees of freedom, i.e., the implication that in the local rest frame, longitudinal and transverse momenta are of the same order of magnitude. This aspect of thermalization can be discussed from first principles at the partonic level [3], and there is now a vast literature on this subject [4]. Its major consequence is longitudinal cooling, to be discussed in Sec. 2.3. In order to achieve this "longitudinal-transverse equilibrium", the mean free path of particles must be much smaller than the *longitudinal* size of the system, much smaller than the nuclear radius due to the strong Lorentz contraction.
However, most of the experimental signatures to be discussed here deal in fact rather with equilibration among the two transverse degrees of freedom. Since the typical transverse scale (the nuclear radius) is much larger than the longitudinal scale, this "transverse equilibrium" is probably easier to achieve than "longitudinal-transverse equilibrium".

2.2. HYDRODYNAMICAL MODELS

If thermalization is achieved, the evolution of the system is governed by the laws of relativistic fluid dynamics. We need not write the corresponding set of equations here. It is important, however, to understand that they completely determine the evolution of the system from three ingredients:

- An equation of state, i.e., a relation between the pressure, P, and the energy density, ϵ.
- Initial conditions, i.e., the specification of the fluid velocity and the energy density on a space-like hypersurface (i.e., at some initial time).
- A "freeze-out" temperature

Initial conditions are of course essential since they fix the geometry of the system. While transverse degrees of freedom are well controlled (the initial transverse fluid velocity is zero by symmetry, while the transverse density profile can be estimated rather accurately from the density of nucleons within the nucleus), longitudinal degrees of freedom (dependence of density on longitudinal coordinates, longitudinal fluid velocity) are to a large extent arbitrary.
Finally, the freeze-out temperature is a crude way of parameterizing the smooth transition from a thermalized state to a free-streaming regime: when the mean free path becomes as large as the system, which occurs approximately at a given temperature (the freeze-out temperature T_f), one assumes that particles follow straight lines to the detectors. One must keep in mind that this really is a rough approximation. If the mean free path at temperature T_f is equal to the system size, it is obvious that the mean free path at a temperature, say, $1.5\,T_f$ cannot be much smaller than the system size, so that hydrodynamics at temperatures not much larger than T_f cannot hold strictly.

2.3. LONGITUDINAL COOLING

Due to the strong Lorentz contraction, the primary collisions between nucleons essentially occur at $z = t = 0$, where z is the collision axis. The longitudinal size of the system then grows proportional to t, and the early stages of the evolution are dominated by this longitudinal expansion. Now, if the system is thermalized, it cools during the expansion: the energy E contained in a given fluid element decreases as its volume V increases since $dE = -PdV$. This is a nontrivial phenomenon. It is interesting to note that models of initial conditions based on minijet production do require a substantial amount of cooling in order to reproduce observables [5].
The cooling is effectively cut off by the transverse expansion of the system (whether or not the system is in equilibrium): the total energy is effectively frozen at a time of the order of half the transverse radius for a cylindrically symmetric source [6]. The energy density at this time can be estimated from the final transverse energy of particle multiplicity. On the other hand, average transverse momenta of outgoing particles are related to the energy per particle, which is given by the equation of state. Therefore, it is a nontrivial prediction of hydrodynamical models that they

can reproduce simultaneously particle multiplicities and transverse momentum spectra. Without longitudinal cooling, such agreement is much harder to reach [7].

2.4. TRANSVERSE COLLECTIVE FLOW

Initially, the fluid transverse velocity is zero by symmetry. Transverse collective motion is created by pressure gradients. For a nonrelativistic system, the fluid acceleration is given by Euler's equation:

$$\frac{d\beta}{dt} = -\frac{1}{\rho}\nabla P, \tag{2.4}$$

with ρ the mass density. The relativistic generalization of this equation is

$$\frac{d\beta}{dt} = -\frac{c^2}{\epsilon + P}\nabla P, \tag{2.5}$$

with ϵ the total energy density. Time scales associated with transverse collective flow are thus naturally of the order of the transverse size of the system, which gives rise to the gradients: any observable associated with transverse collective flow (p_T spectra, HBT correlations, elliptic flow...) probes the system on a time scale which is *essentially* of the order of the transverse radius.

3. Various one- and two-particle observables

In this section, I discuss various observables which are commonly used in order to characterize the transverse collective expansion of the system. More precisely, one tries to extract from experimental data a quantitative information about the freeze-out temperature, T, and the transverse collective velocity, $\beta_\perp$, at freeze-out. The discussion will remain at a very elementary level. The goal here is simply to grasp simple, but essential qualitative ideas.

I will first show that transverse momentum spectra of various identified particles can help to disentangle thermal and collective motion. I then turn to signatures of space-momentum correlations, which may be more specific probes of collective motion: quantum correlations between identical particles, the so-called Hanbury-Brown and Twiss (HBT) effects, yield direct information on the size of the emitting source; the production of deuterons and light nuclei, which involves the phase space density (i.e., the density in momentum *and* space) of the particles, is another probe of space-momentum correlations, which will not be studied here.

3.1. SINGLE PARTICLE P_T-SPECTRA

The four-momentum of a particle with mass m is usually parameterized in the form

$$p^\mu = \begin{pmatrix} E = m_T \cosh y \\ p_z = m_T \sinh y \\ \mathbf{p_T} \end{pmatrix}, \tag{3.1}$$

where y is the *rapidity* of the particle, $\mathbf{p_T}$ its *transverse momentum* and m_T its *transverse mass*. The mass-shell condition $p^\mu p_\mu = m^2$ gives the relation $m_T^2 = p_T^2 + m^2$.

In a two-dimensional transverse world ($p_z = 0$), the energy E coincides with the transverse mass m_T. For a thermalized fluid at rest, neglecting effects of quantum statistics, transverse momentum distributions follow a Boltzmann distribution:

$$\frac{dN}{d^2\mathbf{p_T}} \propto e^{-m_T/T}. \tag{3.2}$$

If one takes into account the longitudinal momentum, the exponential is replaced by a more complicated expression involving Bessel functions, which essentially differ from the exponential behavior by powers of m_T.

Surprisingly, it turns out that transverse momentum distributions of particles produced in heavy ion collisions, as well as in elementary processes, are rather well fit by such simple exponentials in m_T. If the system undergoes collective transverse expansion, however, the "inverse slope parameter" T obtained by fitting p_T spectra no longer corresponds to the temperature of the fluid. We therefore denote this parameter by $T_{\rm eff}$ (effective temperature).

In order to understand how the inverse slope parameter $T_{\rm eff}$ is related to the temperature T and to the fluid velocity $\beta_\perp$, we consider for simplicity a one-dimensional, non-relativistic fluid. The velocity v of a particle in the fluid is the sum of a random, thermal component, v^*, and the fluid velocity β:

$$v = v^* + \beta. \tag{3.3}$$

The average kinetic energy for a particle of mass m is therefore

$$\left\langle \frac{mv^2}{2} \right\rangle = \left\langle \frac{m(v^*)^2}{2} \right\rangle + \frac{m\beta^2}{2}, \tag{3.4}$$

where the cross term vanishes due to the random orientation of the thermal velocity v^*. Now, the thermal energy is simply related to the temperature: $\left\langle m(v*)^2 \right\rangle /2 = T/2$ for a one dimensional fluid, while the average kinetic energy is similarly related to the "effective temperature" $T_{\rm eff}$. One thus obtains

$$T_{\rm eff} = T + m\beta^2. \tag{3.5}$$

Coming back to transverse momentum spectra in nucleus-nucleus collision, one may derive a similar relation, with β replaced by the transverse fluid velocity $\beta_\perp$. [8]
Two important lessons can be drawn from this equation. First, one cannot easily disentangle effects of thermal motion and collective motion from the spectra of a single species of particles: one only sees the inverse slope (3.5) which is a combination of both. This can be seen in Fig. 1 which displays constraints on T and $\beta_\perp$ from a thermal fit to negative hadron spectra. For small $\beta_\perp$ (don't forget that Eq. (3.5) only holds for a nonrelativistic fluid), the allowed region indeed follows a curve of the type $T+m\beta_\perp^2 = \text{constant}$. Second lesson, the inverse slope parameter $T_{\rm eff}$ increases with particle mass when collective expansion is present, so that collective expansion and thermal motion may be disentangled by comparing spectra of different particles. This is also seen in Fig. 1: deuterium, which is heavier, is more sensitive to the collective fluid velocity than other negative hadrons.

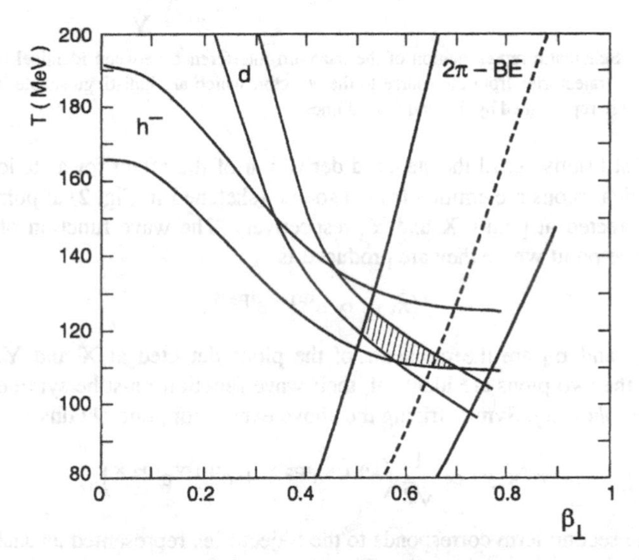

Figure 1. Thermal fits to negative hadron p_T-spectra (h^-), deuteron p_T-spectra (d) and HBT radii (2π-BE) for Pb-Pb collisions at 158 GeV per nucleon, analyzed by the NA49 Collaboration (from [9]).

To conclude this section, let us mention that other mechanisms might produce inverse slopes which increase with mass [10], so that single-particle spectra alone cannot provide conclusive evidence for transverse collective flow.

3.2. QUANTUM CORRELATIONS

Correlations of identical particles, in particular pions, provide direct information on the size of the emitting source. A huge number of papers, both experimental and theoretical (see [11] for reviews) are devoted to this rather technical issue, which I merely illustrate by means of a few simple examples.

3.2.1. *Basics of HBT*

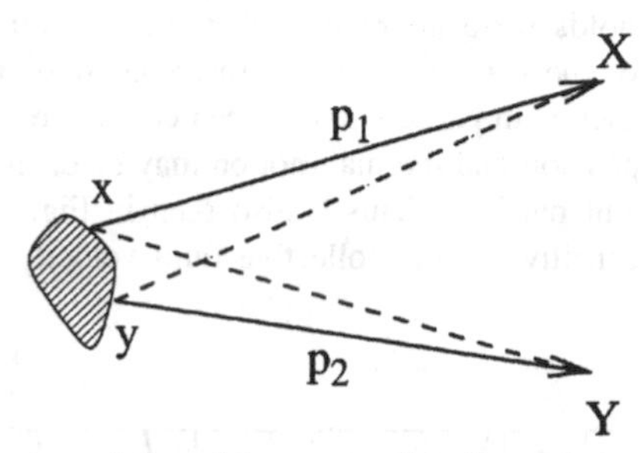

Figure 2. Schematic representation of the quantum interference between identical pions: their two possible trajectories from the source to the detector, which are indistinguishable in quantum mechanics, are represented by full and dashed lines.

Let us first briefly recall the standard derivation of the effect for a static system: two identical pions are emitted from a source (sketched in Fig. 2) at points $\mathbf{x}$ and $\mathbf{y}$, and detected at points $\mathbf{X}$ and $\mathbf{Y}$, respectively. The wave function of the two pions at the point where they are produced is

$$\psi(\mathbf{x},\mathbf{y}) \propto e^{i\mathbf{p_1}\cdot\mathbf{x}} e^{i\mathbf{p_2}\cdot\mathbf{y}}, \tag{3.6}$$

where $\mathbf{p_1}$ and $\mathbf{p_2}$ are the momenta of the pions detected at $\mathbf{X}$ and $\mathbf{Y}$, respectively. If the two pions are identical, their wave function must be symmetric, i.e., $\psi(\mathbf{x},\mathbf{y}) = \psi(\mathbf{y},\mathbf{x})$. Symmetrizing the above expression, one obtains

$$\psi(\mathbf{x},\mathbf{y}) \propto \frac{1}{\sqrt{2}} \left(e^{i\mathbf{p_1}\cdot\mathbf{x}} e^{i\mathbf{p_2}\cdot\mathbf{y}} + e^{i\mathbf{p_1}\cdot\mathbf{y}} e^{i\mathbf{p_2}\cdot\mathbf{x}} \right), \tag{3.7}$$

where the second term corresponds to the trajectories represented as dashed lines in Fig. 2, obtained by exchanging $\mathbf{X}$ and $\mathbf{Y}$. (Note that the source size is much, much smaller than the distance to the detector, so that the direction of the momentum of a particle seen at a given in the detector does not depend on the point where it was emitted).

The probability is obtained by squaring the above amplitude, which yields an interference term:

$$|\psi(\mathbf{x},\mathbf{y})|^2 \propto 1 + \mathrm{Re}\left(e^{i(\mathbf{p_1}-\mathbf{p_2})\cdot\mathbf{x}} e^{-i(\mathbf{p_1}-\mathbf{p_2})\cdot\mathbf{y}} \right). \tag{3.8}$$

If $\rho(\mathbf{x})$ denotes the density of the source normalized to unity, averaging the above expression over $\mathbf{x}$ and $\mathbf{y}$ with the weight $\rho(\mathbf{x})\rho(\mathbf{y})$, one obtains the two particle correlation function

$$C_2(\mathbf{p_1},\mathbf{p_2}) \equiv \frac{f(\mathbf{p_1},\mathbf{p_2})}{f(\mathbf{p_1})f(\mathbf{p_2})} = 1 + \left| \int d^3\mathbf{x}\, \rho(\mathbf{x})\, e^{i(\mathbf{p_1}-\mathbf{p_2})\cdot\mathbf{x}} \right|^2, \tag{3.9}$$

where $f(\mathbf{p_1},\mathbf{p_2})$ denotes the two-body momentum-space density.
The scaled correlation $C_2(\mathbf{p_1},\mathbf{p_2})$ is therefore directly related to the Fourier transform of the source density at the point $\mathbf{p_1}-\mathbf{p_2}$. The Fourier transform typically has a bell shape, and decrease from 1 for $\mathbf{p_1}-\mathbf{p_2}=0$, down to 0 for large $|\mathbf{p_1}-\mathbf{p_2}|$, with a typical width of $1/R$, where R is the source size, the so-called "HBT radius".
The final question is: what is the "source" in the case of a heavy ion collision? Here, the system is no longer static, and its expansion must be taken into account. It is believed that HBT radii reflect the position of the pions when they "freeze-out", i.e., when they scatter for the last time. Typical HBT radii are of order $R \sim 4$ fm, thus the correlation due to the quantum interference is sizable only for relative momenta less than $1/R$ ~50 MeV/c, much smaller than the average transverse momentum of a pion, roughly 400 MeV/c. Therefore, quantum correlations are *short-range* correlations in momentum space.

3.2.2. *Collective motion reduces HBT radii*

We have assumed above that the density of the source depends on the position $\mathbf{x}$ only. Within a flowing system, however, there are space-momentum correlations, so that the density also depends on momenta. In practice, the above analysis can be carried out for a fixed value of the average momentum $(\mathbf{p_1}+\mathbf{p_2})/2$. Since the relative momenta $\mathbf{p_1}-\mathbf{p_2}$ of interest are small, this amounts to fixing the value of $\mathbf{p_1}$. Therefore, the HBT radius given by the analysis is the size of the source that emits pions *with a given momentum* $\mathbf{p_1}$.

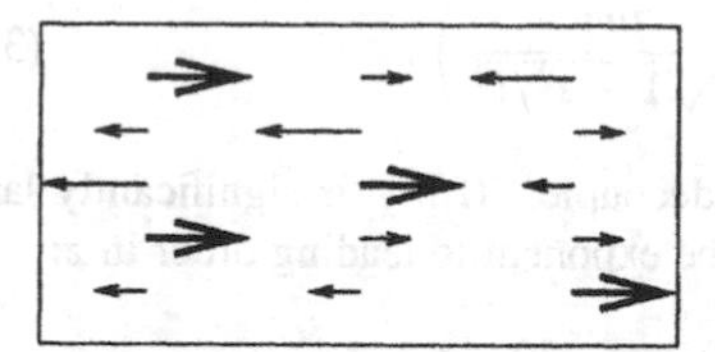
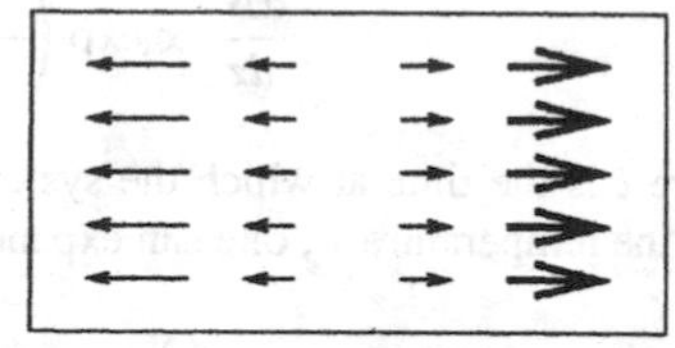

Figure 3. Schematic picture of two one-dimensional sources, of equal length R. Left: disordered source; right: perfectly ordered source.

One then easily understands how HBT radii may reflect collective motion in the system. For this purpose, we consider a simplified, one-dimensional world depicted in Fig. 3. If there is no collective motion, space and momentum coordinates are uncorrelated, and the HBT radius is the true width of the source, R (Fig. 3, left). In the case when the motion is purely collective (no thermal dispersion), on the other hand, there is a one-to-one correspondence between momentum and space coordinates (Fig. 3, right), so that all particles with a given momentum originate from the same point. In this extreme case, the HBT radius vanishes. One understands from this simple picture that collective motion generally results in smaller HBT radii. Indeed, HBT radii measured in Pb-Pb collisions are not much larger than in collisions with lighter projectiles, and significantly smaller than the radius of the Pb nucleus.

3.2.3. *Longitudinal expansion and m_T scaling of longitudinal radii*

Sizable space-momentum correlations are naturally expected in the longitudinal direction, whether or not thermal equilibrium is achieved. Indeed, we have seen that due to the strong Lorentz contraction, primary nucleon-nucleon collisions essentially occur at the origin $z = t = 0$. If a particle is issued from this point without subsequent rescattering, its longitudinal velocity is exactly given by $v_z = z/t$: space and momentum are strictly correlated for a free-streaming system, as in Fig. 3, right! If thermalization occurs, a random thermal motion will destroy part of this correlation, but it is reasonable to assume that the average (fluid) longitudinal velocity is indeed given by $\beta_z = z/t$ [12].

Let us now compute the source density for particles with a given momentum in this longitudinally expanding fluid. For simplicity, we approximate Eq. (2.1) by a Boltzmann distribution. For a fixed momentum p^μ, the spatial density is given by

$$\frac{dN}{d^3\mathbf{x}} \propto \exp\left(-\frac{p^\mu u_\mu}{T}\right). \tag{3.10}$$

For a particle emitted at zero rapidity, Eq. (3.1) gives $p^\mu = (m_T, 0, \mathbf{p_T})$. Neglecting the transverse fluid velocity, Eq. (2.3) gives $u_\mu = (1, -z/t, 0)/\sqrt{1 - z^2/t^2}$. One thus obtains the source density:

$$\frac{dN}{dz} \propto \exp\left(-\frac{m_T}{T\sqrt{1 - z^2/t^2}}\right), \tag{3.11}$$

where t is the time at which the system decouples. If m_T is significantly larger than the temperature T, one can expand the exponent to leading order in z:

$$\frac{dN}{dz} \propto \exp\left(-\frac{m_T}{2T}\frac{z^2}{t^2}\right). \tag{3.12}$$

This is a gaussian profile, with a longitudinal width

$$R_{\text{long}} = t\sqrt{\frac{T}{m_T}}. \tag{3.13}$$

This leads to the prediction that longitudinal HBT radii scale like $1/\sqrt{m_T}$ in a thermalized fluid [13], a prediction indeed supported by data. Whether or not this m_T scaling constitutes a strong evidence for thermalization is left to the appreciation of the reader.

3.2.4. *Transverse radii and transverse collective flow*

Similarly, the m_T dependence of the source transverse size can be used to constrain parameters of thermal models, namely, the temperature T and the fluid velocity $\beta_\perp$. For simplicity, we again consider a one-dimensional, nonrelativistic model. We assume that the source has a gaussian density profile with radius R:

$$\rho(x) \propto \exp\left(-\frac{x^2}{2R^2}\right). \tag{3.14}$$

In order to mimic the transverse expansion in a heavy ion collision, we further assume that the source is expanding symmetrically away from the origin, with a simple fluid velocity profile:

$$\beta(x) = \beta_\perp \frac{x}{R}, \tag{3.15}$$

where $\beta_\perp$ is a constant, which represents the typical velocity of the fluid. The density in coordinate (x) and velocity (v) space is then given by

$$\frac{dN}{dx\,dv} \propto \exp\left(-\frac{E^*}{T}\right) \rho(x), \tag{3.16}$$

where E^* is the kinetic energy in the rest frame of the fluid, i.e.,

$$E^* = m(v - \beta(x))^2/2, \tag{3.17}$$

where $\beta(x)$ is given by Eq. (3.15) Inserting Eqs. (3.14) and Eq. (3.17) into Eq. (3.16), one obtains for fixed v the following density profile

$$\frac{dN}{dx} \propto \exp\left(-\frac{x^2}{2R^2}\left(\frac{m\beta_\perp^2}{T} + 1\right)\right). \tag{3.18}$$

This is again a gaussian, with radius

$$R_{\mathrm{HBT}} = \frac{R}{\sqrt{1 + m\beta_\perp^2/T}}. \tag{3.19}$$

As expected, collective expansion reduces the HBT radius.

For a relativistic fluid, one obtains a similar expression, where m is replaced by the transverse mass m_T: transverse HBT radii also decrease with transverse mass.

From the observed decrease, one may constrain the parameter $\beta_\perp^2/T$. This is illustrated in Fig. 1: the constraint from HBT correlations indeed follows a curve of the type $\beta_\perp^2/T = \text{constant}$. Remarkably, the values of $\beta_\perp$ and T obtained from single-particle spectra automatically satisfy this new constraint from HBT correlations.

4. Elliptic flow and multiparticle correlations

The phenomenon of elliptic flow is widely considered as the most direct evidence that final state interactions, driving the colliding system towards local thermal equilibrium, are indeed large in nucleus-nucleus collisions. I first give a general definition of azimuthal anisotropies, to which elliptic flow belongs; then I show that elliptic flow is a sensitive probe of thermalization; finally, I discuss how elliptic flow is obtained experimentally: this subtle issue indeed deserves some interest.

4.1. DIRECTED AND ELLIPTIC FLOW: DEFINITIONS

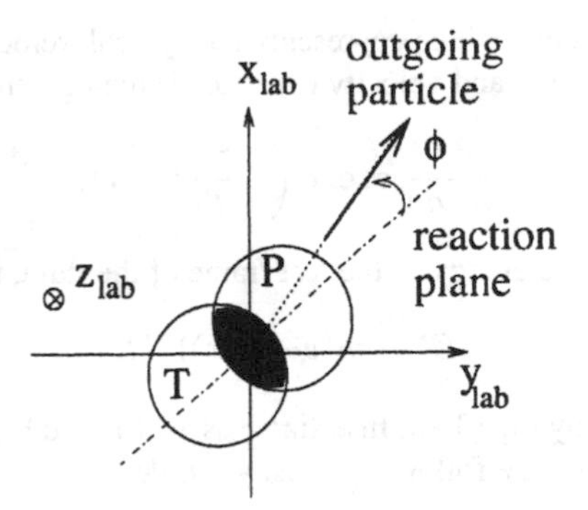

Figure 4. Nucleus-nucleus collisions viewed in the transverse plane. $x_{\rm lab}$ and $y_{\rm lab}$ are fixed directions in the laboratory system. The dash-dotted line is the direction of impact parameter, or reaction plane.

In a non-central nucleus-nucleus collision, the impact parameter defines a reference direction in the transverse plane. One usually calls "reaction plane" the plane spanned by the impact parameter and the collision axis. It turns out that azimuthal angles of outgoing particles are most often correlated to this reference direction. This is the phenomenon of *anisotropic flow*. If ϕ denotes the azimuthal angle of a particle with respect to the reaction plane (see Fig. 4), such a correlation means that the ϕ distribution is not flat. The latter is usually expanded in Fourier series [14]

$$\frac{dN}{d\phi} \propto 1 + 2\sum_{n=1}^{\infty} v_n \cos(n\phi), \tag{4.1}$$

where terms proportional to $\sin(n\phi)$ vanish due to the $\phi \to -\phi$ symmetry. The Fourier coefficients v_n characterize the strength of anisotropic flow:

$$v_n = \left\langle e^{in\phi} \right\rangle, \tag{4.2}$$

where brackets denote a statistical average. The first two Fourier coefficients v_1 and v_2 are usually called "directed flow" and "elliptic flow", and have been measured at various colliding energies, from below 50 MeV per nucleon up to RHIC energies. [15]

This phenomenon is of crucial importance for the following reason: if the nucleus-nucleus collision was a mere superposition of independent nucleon-nucleon collisions, the ϕ distribution would be flat: a pair of colliding nucleons does not see the impact parameter of the whole nucleus-nucleus collision. For this reason, anisotropies in the ϕ distribution must result from final state interactions of the produced particles. This is illustrated below, where we discuss a mechanism that produces elliptic flow at ultrarelativistic energies.

4.2. THE PHYSICS OF ELLIPTIC FLOW

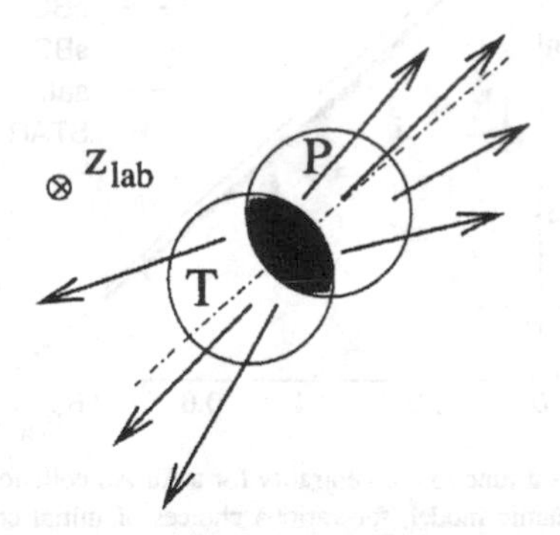

Figure 5. Typical directions of outgoing particles at ultrarelativistic energies.

At ultrarelativistic energies, the azimuthal anisotropy is dominated by elliptic flow, and v_2 is positive. This can be easily understood. A large number of particles are created in an almond-shaped region, represented by the shaded area in Fig. 5. Interactions between these particles result in a pressure which is highest at the center of the almond, and zero outside. At a given point, the resulting force per unit volume is opposite to the pressure gradient, see Eq. (2.5). Now, the gradient is larger along the smaller direction of the almond, which is precisely the direction of the reaction plane. Thus one expects stronger collective motion in the direction of the reaction plane than in the perpendicular direction. This results in a positive value of v_2, which was predicted in [16] and later observed at the top AGS energy [17] and at SPS [9].

4.3. CENTRALITY DEPENDENCE OF ELLIPTIC FLOW

Hydrodynamical models are able to provide stable, quantitative predictions for this effect. Indeed, what matters here is essentially the shape of the almond, i.e., the distribution of energy in the transverse plane, which is well controlled theoretically. On the other hand, v_2 depends weakly on the scenario chosen for the longitudinal expansion, which is to a large extent arbitrary as discussed in Sec. 2. It turns out that the resulting v_2 always decreases linearly [16] with centrality (as estimated from the total charged multiplicity or transverse energy produced in the collision). This is illustrated in Fig. 6. The absolute magnitude on v_2, however, depends on the equation of state. More precisely, it increases with the velocity of sound $c_s \equiv \sqrt{dP/d\epsilon}$, as one would expect intuitively: a higher pressure produces more v_2.

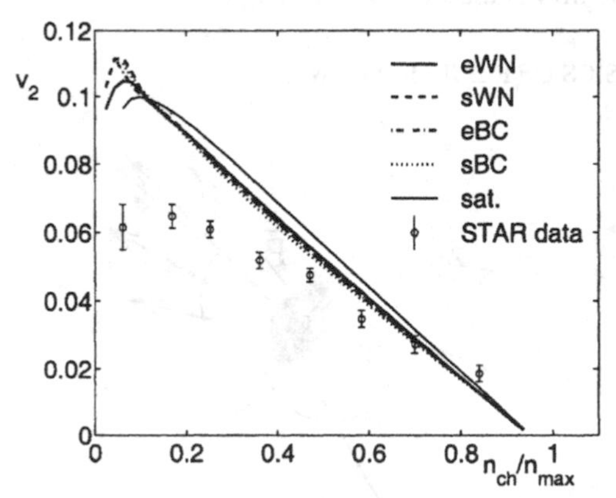

Figure 6. Elliptic flow as a function of centrality for a Au-Au collision at RHIC. The curves are predictions of a hydrodynamic model, for various choices of initial conditions. From P. Kolb et al. [19].

The centrality dependence of elliptic flow therefore yields valuable information about the degree of thermalization achieved in the system[20]. If thermalization is only partial, departures from thermalization (which yield smaller values of elliptic flow) are expected to be more significant for the more peripheral collisions: the size of the system is smaller, so that particles undergo fewer collisions. Then, the maximum of v_2 occurs at less peripheral collisions than if thermalization if fully achieved. This is indeed observed in all calculations done with transport models [21, 22] which contain final state interactions but do not assume perfect thermalization. This is illustrated in Fig. 7.

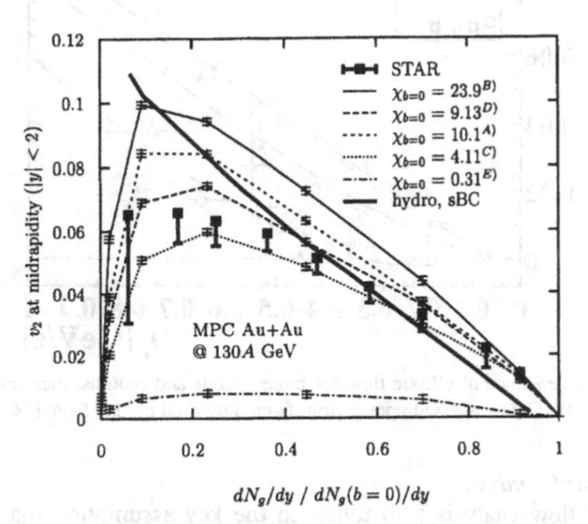

Figure 7. Predictions from a transport calculation. The strength of final state interactions increases from bottom to top. As interactions increase, one gets closer to the hydrodynamical limit, and the maximum of v_2 correspondingly shifts towards more peripheral collisions, From D. Molnar and M. Gyulassy [22].

4.4. P_T DEPENDENCE OF ELLIPTIC FLOW

A great success of thermal models is that they also naturally reproduce the non-trivial p_T dependence of elliptic flow for various identified particles [23]. This is illustrated in Fig. 8. However, a saturation of elliptic flow is observed for p_T above 2 GeV (see Fig. 11), which is not reproduced by hydrodynamical calculations. This phenomenon has been proposed as a possible signature of jet quenching [25].

4.5. FLOW ANALYSIS

Measuring elliptic flow, and more generally anisotropic flow, is far from obvious. Indeed, the orientation of the reaction plane is unknown experimentally, so that the azimuthal angle ϕ defined in Fig. 4 is not a measurable quantity, and elliptic flow defined by Eq. (4.2) is also not directly measurable. Only *relative* azimuthal angles can be measured experimentally.

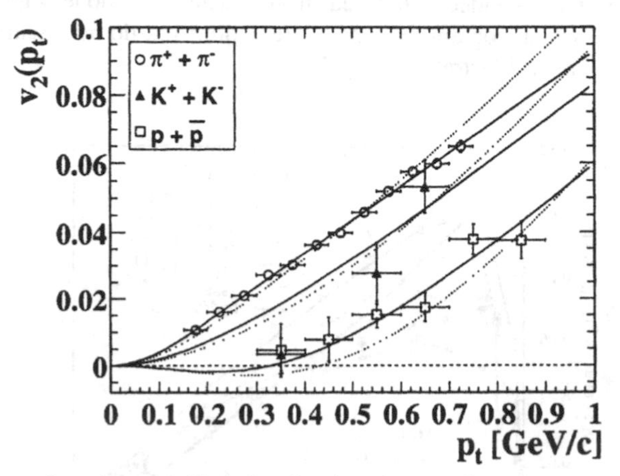

Figure 8. p_T dependence of elliptic flow for pions, kaons and protons, measured by the STAR Collaboration at RHIC, together with predictions from a thermal model, from [24].

4.5.1. *Standard analysis*

The standard flow analysis [26] relies on the key assumption that particles are independent. This allows one to write [27]:

$$\left\langle e^{2i(\phi_1-\phi_2)}\right\rangle = \left\langle e^{2i\phi_1}\right\rangle \left\langle e^{-2i\phi_2}\right\rangle = (v_2)^2, \tag{4.3}$$

where brackets denote an average over pairs of particles belonging to the same event. From the measured two-particle average in the left-hand side, one thus obtains the elliptic flow v_2, up to a sign.

4.5.2. *Beware of nonflow correlations!*

However, the above equation is not quite correct, since it neglects correlations between particles. Such correlations generally produce an additional term in the right-hand side of Eq. (4.3). The magnitude of a two-particle correlation generally scales with the total number of particles N like $1/N$, and this is generally the order of magnitude of the additional, "nonflow" term:

$$\left\langle e^{2i(\phi_1-\phi_2)}\right\rangle = (v_2)^2 + \mathcal{O}(1/N). \tag{4.4}$$

At SPS energies, $N \simeq 2500$ for a central Pb-Pb collision, while v_2 is of the order of 3%: both terms in the right-hand side of Eq. (4.4) are of the same order, and one may no longer ignore nonflow correlations. Similar arguments apply to directed flow v_1.

Several sources of correlations between particles are well known, and their contribution to the last term of Eq. (4.4) can be estimated. For example, the quantum

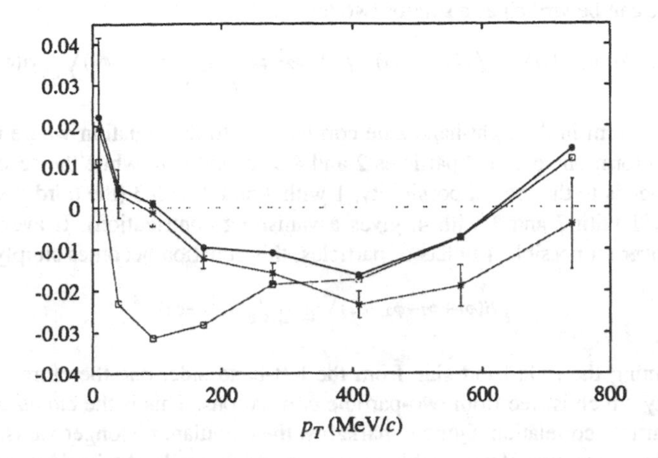

Figure 9. Directed flow of pions in Pb-Pb collisions at 158 GeV per nucleon. Open squares: result of the standard flow analysis performed by NA49 [9]; full squares: after subtraction of HBT correlations; stars: after subtraction of correlations from global momentum conservation (from [28]).

correlations already studied in Sec. 3.2 do produce sizable azimuthal correlations between particles with low relative momenta. Taking this effect into account, one is led to revise significantly the values of the flow given by the standard analysis. This is illustrated in Fig. 9 which shows the corresponding modification in the case of pion directed flow. Correlations due to global momentum conservation also lead to significant corrections.

4.5.3. *Cumulants of multiparticle correlations*

The contribution of nonflow correlations can be greatly reduced by going beyond two-particle methods. One may for instance construct a four-particle average such as $\left\langle e^{2i(\phi_1+\phi_2-\phi_3-\phi_4)}\right\rangle$, where $\phi_1, \cdots, \phi_4$ are azimuthal angles of four particles belonging to the same events. Repeating the argument leading to Eq. (4.3), this quantity is also related to elliptic flow:

$$\left\langle e^{2i(\phi_1+\phi_2-\phi_3-\phi_4)}\right\rangle = (v_2)^4, \tag{4.5}$$

but the corresponding estimate of v_2 is also biased by nonflow correlations. However, the contribution of nonflow correlations can be greatly reduced by combining the informations from two- and four-particle averages, Eqs. (4.3) and (4.5). Indeed, let us assume that particles are pairwise correlated. Then, the four-particle average can be written as a sum of two terms:

$$\left\langle e^{2i(\phi_1+\phi_2-\phi_3-\phi_4)}\right\rangle = \left\langle e^{2i(\phi_1-\phi_3)}\right\rangle \left\langle e^{2i(\phi_2-\phi_4)}\right\rangle + \left\langle e^{2i(\phi_1-\phi_4)}\right\rangle \left\langle e^{2i(\phi_2-\phi_3)}\right\rangle . \tag{4.6}$$

The first term in the right-hand side corresponds to the situation where particles 1 and 3 form one pair and particles 2 and 4 a second pair, while the second term corresponds to the second possibility, 1 with 4 and 2 with 3 (the third possibility, namely 1 with 2 and 3 with 4, gives a vanishing contribution). If averages are taken over all possible 4-uplets of particles, this equation becomes simply

$$\left\langle e^{2i(\phi_1+\phi_2-\phi_3-\phi_4)}\right\rangle = 2\left\langle e^{2i(\phi_1-\phi_3)}\right\rangle^2 . \tag{4.7}$$

Subtracting the right-hand side from the left-hand side, one therefore obtains a quantity which is free from two-particle correlations. This is the *cumulant* of the four-particle correlation. Quite remarkably, the cumulant no longer vanishes if elliptic flow is present. It thus yields an estimate of v_2, easily obtained by combining Eqs. (4.3) and (4.5):

$$\left\langle e^{2i(\phi_1+\phi_2-\phi_3-\phi_4)}\right\rangle - 2\left\langle e^{2i(\phi_1-\phi_2)}\right\rangle^2 = -(v_2)^4, \tag{4.8}$$

and this estimate is essentially *free from nonflow correlations.* [29]

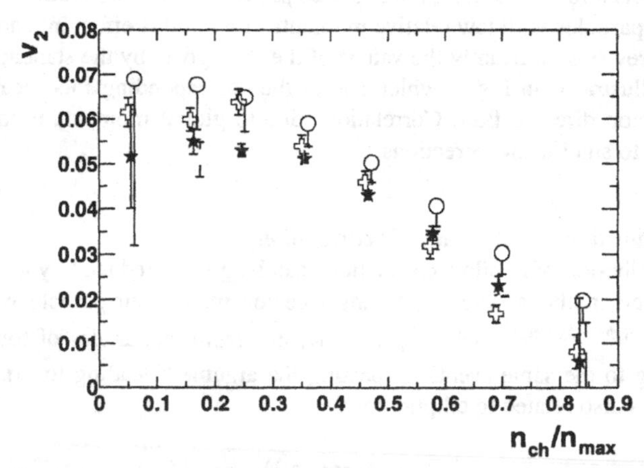

Figure 10. Elliptic flow versus centrality. Circles: from the standard, two-particle analysis; stars: from the cumulant of four-particle correlations (from [31]).

This method was recently applied to STAR data [30, 31]. The corresponding centrality dependence of elliptic flow is displayed in Fig. 10. The values of v_2 from cumulants of four-particle correlations are significantly smaller than those obtained with the standard flow analysis, in particular for the most peripheral collisions. this is precisely where nonflow effects are expected to give the largest contribution since the multiplicity N is smaller (see Eq. (4.4). The centrality dependence obtained with this method suggests that departures from thermalization at RHIC may be larger than was previously thought.

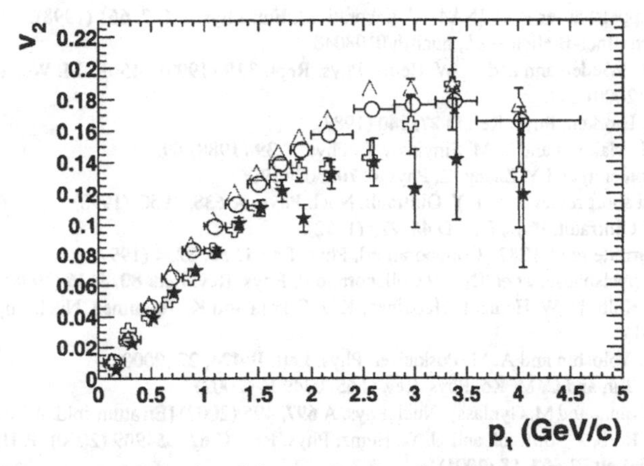

Figure 11. Elliptic flow of charged particles as a function of transverse momentum, analyzed by the STAR Collaboration at RHIC. Triangles and circles: from two-particle analyses; stars: from the cumulant of four-particle correlations (from [31]).

The p_T dependence of elliptic flow has also be analyzed with the four-particle method. The corresponding results are displayed in Fig. 11. Deviations from the standard method seem to be larger at high p_T. The corresponding nonflow correlations may be due to minijets, which produce high p_T particles with azimuthal angles strongly correlated[32]

This cumulant expansion can be worked out to arbitrary orders, and allows one to extract the genuine 4-, 6-particle correlations and beyond.[29] Flow, which is essentially a collective phenomenon, contributes to all orders, while the relative contribution of nonflow correlations decreases as the order increases. Higher order cumulants are therefore a unique tool to check quantitatively that azimuthal correlations are indeed of collective origin.

References

1. F. Becattini, M. Gaździcki and J. Sollfrank, Eur. Phys. J. C **5**, 143 (1998); P. Braun-Munzinger, I. Heppe and J. Stachel, Phys. Lett. B **465**, 15 (1999).
2. G. F. Bertsch, Phys. Rev. Lett. **72**, 2349 (1994) [Erratum-ibid. **77**, 789 (1996)]; D. Ferenc *et al.*, Phys. Lett. B **457**, 347 (1999).
3. E. V. Shuryak, Phys. Rev. Lett. **68**, 3270 (1992).

4. See for instance R. Baier, A. H. Mueller, D. Schiff and D. T. Son, Phys. Lett. B **502**, 51 (2001) and references therein.
5. K. J. Eskola, H. Niemi, P. V. Ruuskanen and S. S. Räsänen, hep-ph/0206230.
6. J.-Y. Ollitrault, Phys. Lett. B **273**, 32 (1991).
7. U. W. Heinz and S. M. Wong, Phys. Rev. C **66** (2002) 014907
8. E. Schnedermann, J. Sollfrank and U. W. Heinz, Phys. Rev. C **48**, 2462 (1993); I. G. Bearden *et al.* [NA44 Collaboration], Phys. Rev. Lett. **78**, 2080 (1997).
9. H. Appelshäuser *et al.* [NA49 Collaboration], Eur. Phys. J. C **2**, 661 (1998).
10. J. Schaffner-Bielich *et al.*, nucl-th/0108048.
11. U. A. Wiedemann and U. W. Heinz, Phys. Rept. **319** (1999) 145; R. M. Weiner, Phys. Rept. **327** (2000) 249.
12. J. D. Bjorken, Phys. Rev. D **27**, 140 (1983).
13. A. N. Makhlin and Y. M. Sinyukov, Z. Phys. C **39** (1988) 69.
14. S. Voloshin and Y. Zhang, Z. Phys. C **70**, 665 (1996).
15. For a short review, see J. Y. Ollitrault, Nucl. Phys. A **638**, 195C (1998).
16. J.-Y. Ollitrault, Phys. Rev. D **46**, 229 (1992).
17. J. Barrette *et al.* [E877 Collaboration], Phys. Rev. C **56**, 3254 (1997).
18. H. Appelshäuser *et al.* [NA49 Collaboration], Phys. Rev. Lett. **80**, 4136 (1998).
19. P. F. Kolb, U. W. Heinz, P. Huovinen, K. J. Eskola and K. Tuominen, Nucl. Phys. A **696**, 197 (2001).
20. S. A. Voloshin and A. M. Poskanzer, Phys. Lett. B **474**, 27 (2000).
21. Z. w. Lin and C. M. Ko, Phys. Rev. C **65**, 034904 (2002).
22. D. Molnar and M. Gyulassy, Nucl. Phys. A **697**, 495 (2002) [Erratum-ibid. A **703**, 893 (2002)].
23. P. F. Kolb, J. Sollfrank and U. W. Heinz, Phys. Rev. C **62**, 054909 (2000); P. Huovinen *et al.*, Phys. Lett. B **503**, 58 (2001).
24. C. Adler *et al.* [STAR Collaboration], Phys. Rev. Lett. **87**, 182301 (2001).
25. X. N. Wang, Phys. Rev. C **63**, 054902 (2001).
26. P. Danielewicz and G. Odyniec, Phys. Lett. B **157**, 146 (1985); A. M. Poskanzer and S. A. Voloshin, Phys. Rev. C **58**, 1671 (1998).
27. S. Wang *et al.*, Phys. Rev. C **44**, 1091 (1991).
28. P. M. Dinh, N. Borghini and J.-Y. Ollitrault, Phys. Lett. B **477**, 51 (2000); Phys. Rev. C **62**, 034902 (2000).
29. N. Borghini, P. M. Dinh and J.-Y. Ollitrault, Phys. Rev. C **63**, 054906 (2001); Phys. Rev. C **64**, 054901 (2001).
30. A. H. Tang (STAR Collaboration), hep-ex/0108029.
31. C. Adler *et al.* [STAR Collaboration], nucl-ex/0206001.
32. Y. V. Kovchegov and K. L. Tuchin, hep-ph/0203213.

EXPERIMENTALISTS ARE FROM MARS, THEORISTS ARE FROM VENUS

J.L. NAGLE (nagle@nevis.columbia.edu)
Columbia University, New York, NY 10027, USA

T. ULLRICH (ullrich@bnl.gov)
Brookhaven National Laboratory, Upton, NY 11973, USA

Abstract. We present a written version of four lectures given at the NATO Advanced Study Institute on "QCD Perspectives on Hot and Dense Matter" in Cargese, Corsica during August, 2001. Over the last year the first exciting results from the Relativistic Heavy Ion Collider (RHIC) and the four experiments BRAHMS, PHENIX, PHOBOS, and STAR have been presented. In these lectures we review the state of RHIC and the experiments and the most exciting current results from Run I which took place in 2000. A complete review is not possible yet with many key results still preliminary or to be measured in Run II, which is currently underway, and thus the emphasis will be on the approach experimentalists have taken to address the fundamental physics issues of the field. We have not attempted to update the RHIC results for this proceedings, but rather present it as a snapshot of what was discussed in the workshop. The field is developing very quickly, and benefits greatly from contact and discussions between the different approaches of experimentalists and theorists.

1. Introduction

A good way to proceed in understanding the construction, operation, and physics output from current high-energy heavy ion experiments is to start with the original physics motivations for the experimental program. The Relativistic Heavy Ion Collider project was started in the 1980's with a list of experimental observables for characterizing hot and dense quark and gluonic matter and the expected restoration of approximate chiral symmetry and screening of the long range confining potential of QCD. We give a brief and certainly not complete review of these physics signals. Then we discuss some of the cost, schedule and technology constraints that impacted the design and construction of the RHIC experiments. Finally we present a sample of the first results from the RHIC experimental program from data taken during Run I in the summer of 2000.

2. Physics Motivations

There are three main categories of observables that were originally proposed to study the matter produced in relativistic heavy ion collisions: (1) Thermodynamic

J.-P. Blaizot and E. Iancu (eds.), QCD Perspectives on Hot and Dense Matter, 257–304.

properties of the system and indications of a possible first or second order phase transition between hadronic matter and quark-gluon plasma, (2) Signatures for the restoration of approximate chiral symmetry transition, and (3) Signatures for deconfinement and the screening of the long range confining potential between color charges. In the following list, there will be specific channels that need to be observed that essentially specify the types of detectors and experiments necessary.

2.1. THERMODYNAMIC PROPERTIES

One of the most important question in the physics of heavy-ion collisions is thermalization. We want to describe the system in terms of a few thermodynamic properties, otherwise it is not possible to discuss an equation-of-state and a true order to any associated phase transitions. The use of thermodynamic concepts to multi-particle production has a long history. One of the first to apply them in elementary collisions was Hagedorn in the early 1960's [1]. The concept of a *temperature* applies strictly speaking only to systems in at least local thermal equilibrium. Thermalization is normally only thought to occur in the transverse degrees of freedom as reflected in the Lorentz invariant distributions of particles. The measured hadron spectra contain two pieces of information: *(i)* their normalization, *i.e.* their yields ratios, provide the chemical composition of the fireball at the chemical freeze-out point where the hadron abundances freeze-out and *(ii)* their transverse momentum spectra which provides information about thermalization of the momentum distributions and collective flow. The latter is caused by thermodynamic pressure and reflects the integrated equation of state of the fireball matter. It is obvious that the observed particle spectra do not reflect earlier conditions, *i.e.* the hot and dense deconfined phase, where chemical and thermal equilibrium may have been established, since re-scattering erases most traces from the dense phase. Nevertheless, those which are accumulative during the expansion, such as flow, remain. Only direct photons, either real photons or virtual photons that split into lepton pairs, escape the system without re-scattering. Thus these electromagnetic probes yield information on the earliest thermodynamic state which may be dominated by intense quark-quark scattering.
The assumption of a locally thermalized source in chemical equilibrium can be tested by using statistical thermal models to describe the ratios of various emitted particles. This yields a baryon chemical potential μ_B, a strangeness saturation factor γ_s, and the temperature T_{ch} at chemical freeze-out.
So far these models are very successful in describing particle ratios at SPS [2] and now also at RHIC [3]. At RHIC the derived chemical freeze-out temperature are found to be around 175 MeV (165 MeV at SPS) and a baryon chemical potential of around 45 MeV (270 MeV at SPS). It should be stressed that these models assume thermal equilibration but their success together with the large collective flow (radial and elliptic) measured at RHIC is a strong hint that this picture indeed applies.

2.2. CHIRAL SYMMETRY RESTORATION

The phase transition to a quark-gluon plasma is expected to be associated with a strong change in the chiral condensate, often referred to as the restoration of approximate chiral symmetry in relativistic heavy ion collisions is discussed here. Note that although many in the field refer to the restoration of chiral symmetry, the system always breaks chiral symmetry at a small scale due to the non-zero neutral current masses of the up and down quarks.

There are multiple signatures of this transition, including disoriented chiral condensates (DCC), strangeness enhancement, and many others. However, the most promising signature is the in medium modification to the mass and width of the low mass vector mesons. Nature has provided an excellent set of probes in the various low mass vector meson states (ρ, ω, ϕ) whose mass poles and spectra are dynamically determined via the collisions of hadrons or partons. If the hot and dense state produced in heavy ion collisions is composed of nearly massless partons, the ρ(770) meson mass distribution is expected to broaden significantly and shift to lower values of invariant mass. The ρ meson has a lifetime that is $\tau \approx 1$ fm$/c$ and the plasma state created in RHIC collisions has a lifetime of order 10 fm/c. Thus, the ρ meson is created and decays many times during the entire time evolution of the collision.

The ρ has a dominant decay (nearly 100%) into two pions. However, in this decay mode if the pions suffer re-scattering with other hadrons after the decay, one cannot experimentally reconstruct the ρ meson and information is lost. Given the dense, either partonic or hadronic environment, the probability for pion re-scattering is very large, unless the ρ is created and decays at the latest stages of the time evolution. This time period is often referred to as thermal freeze-out, when elastic collisions cease. Thus a measurement of the ρ as reconstructed via its pion decay channel gives interesting information on the final hadronic stage, but not on the dense phase where chiral symmetry may be restored.

There is an additional decay channel into electron pairs and muon pairs, though with small branching ratios of $4.5 \cdot 10^{-5}$ and $4.6 \cdot 10^{-5}$, respectively. Since the leptons do not interact strongly, after they are produced, they exit the dense system essentially unaffected carrying out crucial information from the core of the system. There is a good analogy in understanding the processes in the center of the sun via neutrino emission, since only neutrinos have a small enough interaction cross-section to pass out of the sun's core largely unaffected. In the case of neutrinos, the more interesting physics of possible neutrino oscillations complicates matters, but that is not a concern in the case of our electron and muons measurements.

One other additional point of interest is that the apparent branching ratio of the ρ into pions and leptons should be modified as observed by experiment. The ρ mesons that decay in medium into pions are not reconstructed, but the ones decaying into electrons are. If one can reconstruct the ρ in both channels one can

gauge the number of lifetimes of the ρ the dense medium survives for.
The lifetime of the ρ meson in the rest frame of the plasma depends on its gamma boost in this frame, and thus to study the earliest stages, a measurement of low transverse momentum ρ mesons is desirable. If the ρ decays at rest in the plasma frame, the maximum transverse momentum for the electron or positron is $p_\perp \approx$ 385 MeV/c. These electrons are considered low $p_\perp$ and present an experimental challenge to measure for two reasons. First, there are a large number of low momentum charged pions created in these collisions, that results in a charged pion to electron ratio in this $p_\perp$ range of 1000 : 1. Thus one needs detectors that can cleanly identify electrons with good momentum resolution, while rejecting the copiously produced pions. The second challenge is that most of the electrons come from pion Dalitz decays ($\pi^0 \rightarrow e^+e^-\gamma$), η Dalitz decays ($\eta \rightarrow e^+e^-\gamma$), and conversions of photons ($\gamma \rightarrow e^+e^-$) mostly resulting from π^0 decays. Ideally one wants to reject these other electrons to enhance the signal contribution from the low mass vector mesons. Conversions are reduced by reducing the amount of material in the path of produced photons. This restriction is often at odds with the desire to have substantial inner tracking detectors and these needs must be balanced.
The $\phi(1020)$ meson spectral function is also sensitive to in-medium chiral symmetry restoration; however its substantially longer lifetime $\tau \approx 40$ fm/c means that most ϕ decays occur outside the medium. However, measuring the ϕ in its various decay modes (kaon pairs, electron pairs, muon pairs) remains an interesting signal at low transverse momentum.
These low mass vector mesons also decay into muons pairs. These muons have low momentum and are an real experimental challenge to measure as detailed later in these proceedings.

2.3. DECONFINEMENT

There are many signatures that result from the deconfinement of color charges over an extended volume, often referred to as the quark-gluon plasma. Two are detailed in this proceedings: (1) Suppression of heavy quarkonium states and (2) Parton energy loss via gluon emission, also referred to as jet quenching.

2.3.1. *Quarkonium Suppression*

The suppression of heavy quarkonium states was originally proposed by Matsui and Satz [4] in the late 1980's as a signature for color deconfinement. The Debye screening in a QED plasma is a reasonable analogue for the scenario in our QCD plasma. A charm-anticharm ($c\bar{c}$) quark pair produced via gluon fusion in the initial phase of the heavy ion collision can form a J/ψ if the pair has low relative momentum. The total production of such states in proton-proton collisions relative to the total charm production is less than a few percent. If the Debye screening

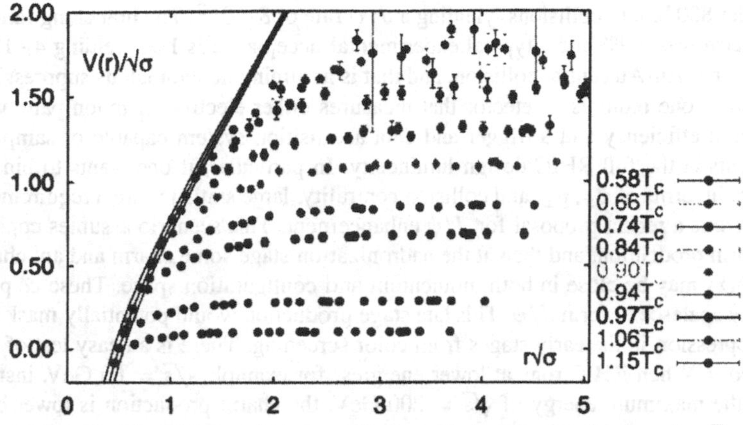

Figure 1. The QCD potential between heavy quarks in shown as a function of $r\sqrt{\sigma}$

length is of order the same size as the quarkonium state, then the pair is screened. The charm and anticharm quark scatter away from each other and, eventually at the hadronization point, pair with surrounding light quarks and antiquarks to form D mesons. This color screening is displayed in recent lattice QCD calculations described at this workshop in terms of a modification in the linear rise at large distances of the QCD potential. This change in the QCD potential as a function of temperature is shown in Fig. 1.

There are a variety of heavy vector mesons with a large range in binding energy (and associated hadronic size). The J/ψ, χ_c, and ψ' have binding energies of 0.64, 0.20, and 0.05 GeV respectively. The $\Upsilon(1s)$, χ_b, $\Upsilon(2s)$, χ_b', and $\Upsilon(3s)$ have binding energies of 1.10, 0.67, 0.54, 0.31, 0.20 GeV respectively. Since the suppression of these states is determined by the relative plasma temperature and the binding energy (or by the quarkonium hadronic size and the Debye screening length), measuring the sequential disappearance of these states acts as a QCD thermometer.

The $J/\psi(1s)$ state decays into almost anything hadronic with a large branching ratio of 87.7%. However, the experimentally accessible decay channels are 5.93% to e^+e^- and 5.88% to $\mu^+\mu^-$. Similar decay channels are available for the ψ' and the Υ states. The accessible decay channel of the χ_c for heavy ion experiments is $\chi_c \to \gamma + J/\psi$ with a branching ratio $6.6 \cdot 10^{-3}$. Since the decay γ is quite soft, the χ_c represents an experimental challenge.

Similar to the low mass vector mesons there is interest in these states at low transverse momentum, where they reside in the plasma state longer. At rest the J/ψ decays into electrons or muons with a characteristic $p_\perp \approx 1.5$ GeV. A rough rate estimate (good to a factor of 2-3) is that the production of J/ψ is approximately $1 \cdot 10^{-4}$ per proton-proton collision. In a central Au-Au collision, there are of order 800 binary collisions, yielding a J/ψ rate of $8 \cdot 10^{-2}$. The branching ratio to electrons is 5.9% and a typical experimental acceptance is 1%, yielding $4 \cdot 10^{-5}$ J/ψ per Au-Au central collision, and that is assuming no anomalous suppression! Hence, one requires a detector that measures either electron or muon pairs with a high efficiency and a trigger and data acquisition system capable of sampling events at the full RHIC design luminosity. In particular, if one wants to bin the data in terms of $x_F, p_\perp$, and collision centrality, large statistics are a requirement. There is a recent proposal for J/ψ enhancement. This scenario assumes copious charm prodcution, and then at the hadronization stage some charm and anticharm quarks may be close in both momentum and configuration space. These $c\bar{c}$ pairs may coalesce to form J/ψ. This late stage production would potentially mask any suppression in the early stages from color screening. There is an easy test of this theory. When RHIC runs at lower energies, for example $\sqrt{s} = 60$ GeV, instead of the maximum energy of $\sqrt{s} = 200$ GeV, the charm production is lower by a factor of approximately three and the effect of recombination should be reduced substantially. In addition, we expect different $p_\perp$ dependence of J/ψ production from original hard processes compared with late stage $c\bar{c}$ coalescence.

2.3.2. *Parton Energy Loss*

An ideal experiment would be to contain the quark-gluon plasma, and send well calibrated probes through it, and measure the resulting transparency or opacity of the system. There is no experimental way of aiming a third beam of particle at the collision. Therefore any probes of the system must be generated in the collision itself. These probes must have calculable production rates in order to be considered calibrated. An excellent example of such a probe is a hard scattered parton. A parton traversing a color confined medium of hadrons sees a relatively transparent system. However, a parton passing through a hot colored deconfined medium will lose substantial energy via gluon radiation[5, 6].

The source of these partons is from hard scattering processes producing back-to-back parton jets. In a deconfined medium the parton will lose energy before escaping the system and fragmenting into a jet cone of hadrons. The total energy of the initial parton jet is conserved since eventually the radiated gluons will also hadronize. It is likely that the radiated gluons will have a larger angular dispersion than the normally measured jet cone. Thus one might be able to measure a modification in the apparent jet shape. When the parton fragments into hadrons it has less energy, and hence the fragmentation will result in a much reduced energy for the leading hadron. A measurement of high transverse momentum hadrons ($\pi^0, \pi^{+/-}, K^{+/-}, h^{+/-}$) is a strong indicator of the opacity of the medium. An exciting additional observable was mentioned at the workshop in the context of the high $p_\perp$ spectra of charm D mesons. There is a reduction in the induced gluon radiation for charm and bottom quarks relative to light quarks due to their slower velocity through the medium.

3. The RHIC Complex

The scope of the RHIC program is to operate a colliding beam facility which allows studies of phenomena in ultra-relativistic heavy-ion collisions and in collisions of polarized protons. The collider is located in the northwest section of the Brookhaven National Laboratory (BNL) in Upton, New York. Its construction began in 1991 and the completion of the complex was accomplished in Spring 2000.

The collider, which consists of two concentric rings of 1740 super-conducting magnets, was constructed in an already existing ring tunnel of $\sim$ 3.8 km circumference. This tunnel was originally constructed for the proposed ISABELLE project. It offers an extraordinary combination of energy, luminosity and polarization. A schematic diagram of the whole RHIC complex, including the various facilities used to produce and pre-accelerate the beams of particles is displayed in Figure 2.

RHIC is able to accelerate and store counter-rotating beams of ions ranging from those of gold to protons at the top energy of 100 GeV/nucleon for gold and 250 GeV for protons. The stored beam lifetime for gold in the energy range of 30 to 100 GeV/nucleon is expected to be approximately 10 hours. The major performance parameters are summarized in Figure 3.

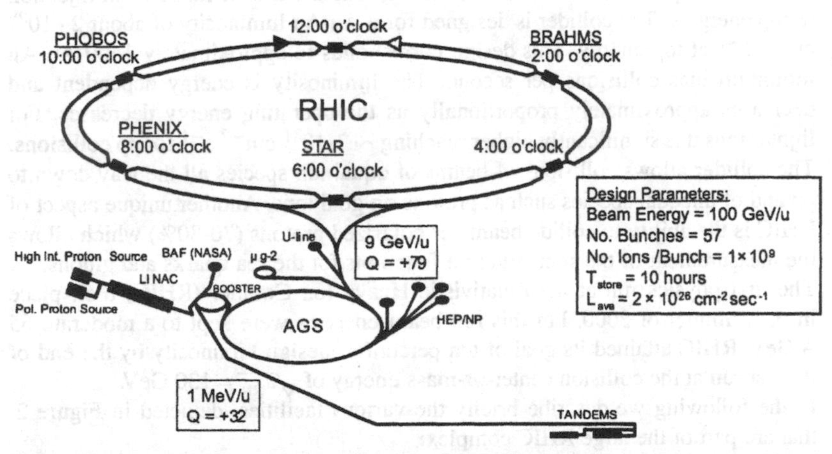

Figure 2. The Relativistic Heavy Ion Collider (RHIC) accelerator complex at Brookhaven National Laboratory. Nuclear beams are accelerated from the tandem Van de Graaff, through the transfer line into the AGS Booster and AGS prior to injection into RHIC. Details of the characteristics of proton and Au beams are also indicated after acceleration in each phase.

The layout of the tunnel and the magnet configuration allow the two rings to intersect at six locations along their circumference. The top kinetic energy is 100+100

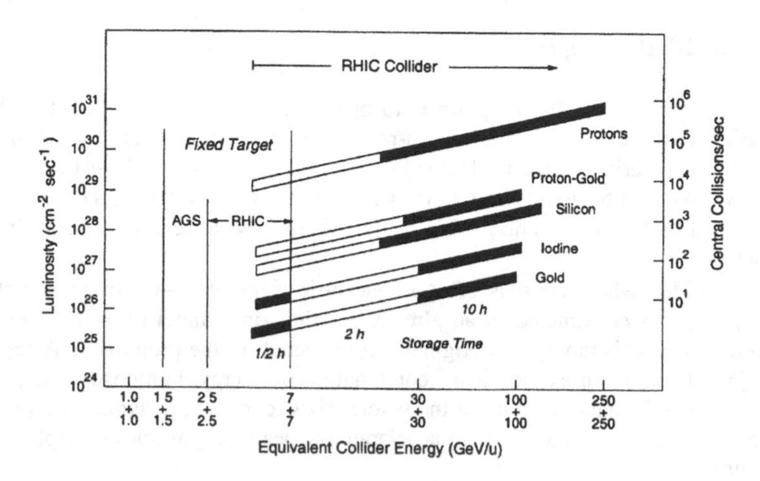

Figure 3. RHIC performance parameters.

GeV/nucleon for gold ions. The operational momentum increases with the charge-to-mass ratio, resulting in kinetic energy of 125 GeV/nucleon for lighter ions and 250 GeV for protons. The collider is able to operate a wide range from injection to top energies. The collider is designed for a Au-Au luminosity of about $2 \cdot 10^{26}$ $\mathrm{cm}^{-2}\ \mathrm{s}^{-1}$ at top energy. This design corresponds to approximately 1400 Au-Au minimum bias collisions per second. The luminosity is energy dependent and decreases approximately proportionally as the operating energy decreases. For lighter ions it is significantly higher reaching $\sim 1 \cdot 10^{31}\ \mathrm{cm}^{-2}\ \mathrm{s}^{-1}$ for pp collisions. The collider allows collisions of beams of equal ion species all the way down to pp and of unequal species such as protons on gold ions. Another unique aspect of RHIC is the ability to collide beams of polarized protons (70-80%) which allows the measurement of the spin structure functions for the sea quarks and gluons.
The first physics run at the Relativistic Heavy Ion Collider (RHIC) took place in the Summer of 2000. For this run beam energies were kept to a moderate 65 A GeV. RHIC attained its goal of ten percent of design luminosity by the end of its first run at the collision center-of-mass energy of $\sqrt{s_{\mathrm{NN}}} = 130$ GeV.
In the following we describe briefly the various facilities, depicted in Figure 2, that are part of the large RHIC complex:

Tandem Van de Graaff Completed in 1970, the Tandem Van de Graaff facility was for many years the world's largest electrostatic accelerator facility. It can provide beams of more than 40 different types of ions ranging from hydrogen to uranium. The facility consists of two 15 MV electrostatic accelerators, each about 24 meters long, aligned end-to-end.In the Tandem the atoms are stripped of some of their electrons (e.g. Au to Q = +32) and accelerated to a kinetic energy of 1 MeV/nucleon.

Heavy Ion Transfer Line (HITL) To study heavy ion collisions at high energies, a 700 meter-long tunnel and beam transport system called the Heavy Ion Transport Line were completed in 1986, allowing the delivery of heavy ions from the Tandem to the Booster for further acceleration. The HITL makes it possible for the Tandem to serve as the Relativistic Heavy Ion Collider's ions source.

Linear Accelerator (Linac) For the study of pp or pA collisions at the experiments, energetic protons are supplied by an Linear Accelerator (Linac). The Brookhaven Linear Accelerator was designed and built in the late 1960's as a major upgrade to the Alternating Gradient Synchroton (AGS) complex. The basic components of the Linac include ion sources, a radiofrequency quadrapole pre-injector, and nine accelerator radiofrequency cavities spanning the length of a 150 m tunnel. The Linac is capable of producing up to a 35 milliampere proton beam at energies up to 200 MeV for injection into the AGS Booster.

Booster The Alternating Gradient Synchrotron Booster is less than one quarter the size of the AGS. It is used to preaccelerate particles entering the AGS ring and plays an important role in the operation of the Relatavistic Heavy Ion Collider (RHIC) by accepting heavy ions from the Tandem Van de Graaff facility via the Heavy Ion Transfer Line (HITL) and protons from the Linac. It then feeds them to the AGS for further acceleration and delivery to RHIC. After the installation of the HITL in 1986, the AGS was capable of accelerating ions up to silicon with its atomic mass of 28. However, due to its superior vacuum, the Booster makes it possible for the AGS to accelerate and deliver heavy ions up to gold with its atomic mass of 197.

AGS Since 1960, the Alternating Gradient Synchrotron (AGS) has been one of the world's premiere particle accelerators and played a major role in the study of relativistic heavy ion collisions in the last decade. The AGS name is derived from the concept of alternating gradient focusing, in which the field gradients of the accelerator's 240 magnets are successively alternated inward and outward, permitting particles to be propelled and focused in both the horizontal and vertical plane at the same time. Among its other duties, the AGS is now used as an injector for the Relativistic Heavy Ion Collider. For RHIC operation the fully stripped ions are accelerated in the AGS to 9 GeV/nucleon before ejection.

ATR The AGS sends the ions (or protons) down another beamline called the AGS-to-RHIC Transfer Line (ATR). At the end of this line, there's a "fork in the road", where sorting magnets separate the ion bunches. From here, the counter-rotating beams circulate in the RHIC where they are collided at one of four intersecting points.

4. Experimental Program

4.1. LETTERS OF INTENT

In July 1991 there were a set of experimental Letters of Intent that were put forward to an advisory committee. The proposals are listed below, including the lead institution and in parenthesis the physics observable focus.

1. LBL-TPC (inclusive charged hadrons)
2. BNL-TPC (inclusive charged hadrons)
3. TOYKO-TALES (electron pairs, hadrons)
4. SUNY-SB (direct photons)
5. Columbia-OASIS (electron pairs, hadrons, high $p_\perp$)
6. ORNL Di-Muon (muon pairs)
7. BNL Forward Angle Spectrometer (hadrons at large rapidity)
8. MIT MARS (hadrons and particle correlations)

The experiments span the range of hadronic, leptonic and photonic capabilities to cover the broad spectrum of physics topics listed above. At the time, only the LBL-TPC proposal was approved and became the STAR experiment. Eventually the MARS proposal evolved into the PHOBOS experiment (note that Phobos is a moon of the planet Mars) and PHENIX (should be spelled Phoenix) rose from the ashes of OASIS, Di-Muon, TALES and the other lepton focussed experiments.
Eventually there were four approved experiments which have now been constructed and operated during the first year of RHIC running. BRAHMS, PHENIX, PHOBOS and STAR are briefly described below. These experiments have various approaches to study the deconfinement phase transition to the quark gluon plasma. The STAR experiment [7] concentrates on measurements of hadron production over a large solid angle in order to measure single- and multi-particle spectra and to study global observables on an event-by-event basis. The PHENIX experiment [8] focuses on measurements of lepton and photon production and has the capability of measuring hadrons in a limited range of azimuth and pseudo-rapidity. The two smaller experiments BRAHMS (a forward and mid-rapidity hadron spectrometer) [9] and PHOBOS (a compact multiparticle spectrometer) [10] focus on single- and multi-particle spectra. The collaborations, which have constructed these detector systems and which will exploit their physics capabilities, consist of approximately 900 scientists from over 80 institutions internationally. In addition to colliding heavy ion beams, RHIC will collide polarized protons to study the

spin content of the proton [11]. STAR and PHENIX are actively involved in the spin physics program planned for RHIC.

4.2. BRAHMS

The **BR**oad **R**ange **H**adron **M**agnetic **S**pectrometer BRAHMS experiment is designed to measure and identify charged hadrons ($\pi^{\pm}$, $K^{\pm}$, $^{(}\overline{p}^{)}$) over a wide range of rapidity and transverse momentum for all beams and energies available at RHIC. Because the conditions and thus the detector requirements at mid-rapidity and forward angles are different, the experiment uses two movable spectrometers for the two regions.

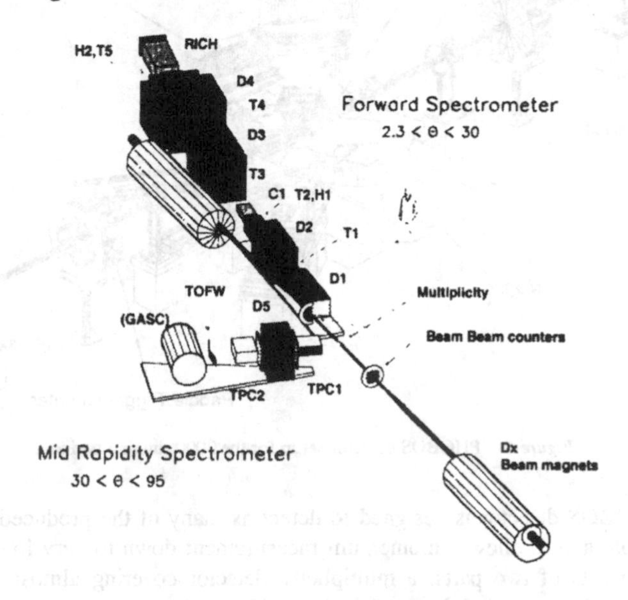

Figure 4. Layout of the BRAHMS detector.

As shown in Figure 4, there is a mid-rapidity spectrometer to cover the pseudo-rapidity range $0 \leq \eta \leq 1.3$ and a forward spectrometer to cover $1.3 \leq \eta \leq 4.0$. The latter employs four dipole magnets, three time projection chambers (TPC), and drift chambers. Particle identification is achieved with time-of-flight hodoscopes, a threshold Cherenkov counter, and one ring-imaging Cherenkov counter (RICH). The solid angle acceptance of the forward arm is 0.8 mstr. The mid-rapidity spectrometer has been designed for charged particle measurements for $p \leq 5\,\mathrm{GeV}/c$. The spectrometer has two TPCs for tracking, a magnet for momentum measurement, and a time-of-flight wall and segmented gas Cherenkov counter (GASC) for particle identification. It has a solid angle acceptance of 7 mstr. A set of beam counters and a silicon multiplicity array provide the experiment with trigger information and vertex determination.

4.3. PHOBOS

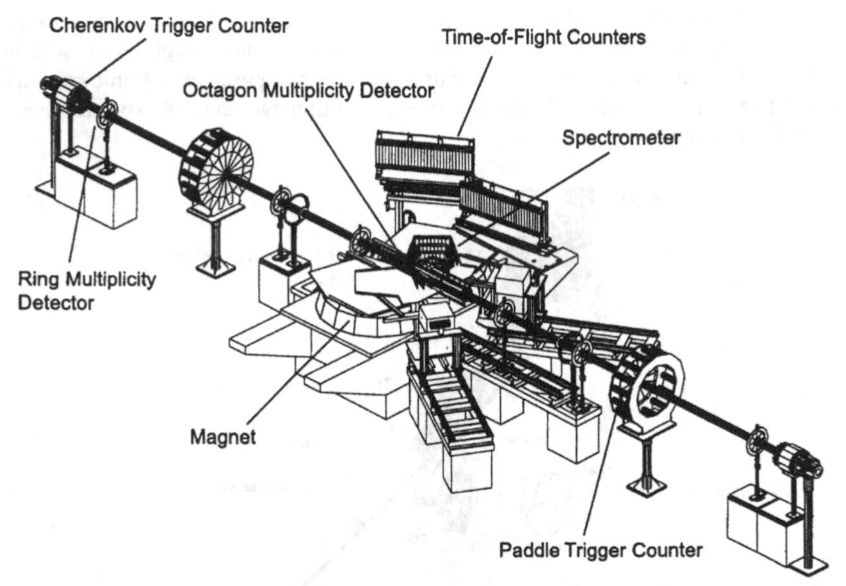

Figure 5. PHOBOS detector setup for the 2000 running period.

The PHOBOS detector is designed to detect as many of the produced particles as possible and to allow a momentum measurement down to very low $p_{\perp}$. The setup consists of two parts: a multiplicity detector covering almost the entire pseudo-rapidity range of the produced particles and a two arm spectrometer at mid-rapidity. Figure 5 shows the detector, including the spectrometer arms, the multiplicity and vertex array, and the lower half of the magnet.

One aspect of the design is that all detectors are produced using a common technology, namely as silicon pad or strip detectors. The multiplicity detector covers the range $-5.4 < \eta < 5.4$, measuring total charged multiplicity $dN_{ch}/d\eta$ over almost the entire phase space. For approximately 1% of the produced particles, information on momentum and particle identification will be provided by a two arm spectrometer located on either side of the interaction volume (only one arm was installed for the 2000 run). Each arm covers about 0.4 rad in azimuth and one unit of pseudo-rapidity in the range $0 < \eta < 2$, depending on the interaction vertex, allowing the measurement of $p_{\perp}$ down to 40 MeV/c. Both detectors are capable of handling the 600 Hz minimum bias rate expected for all collisions at the nominal luminosity.

4.4. PHENIX

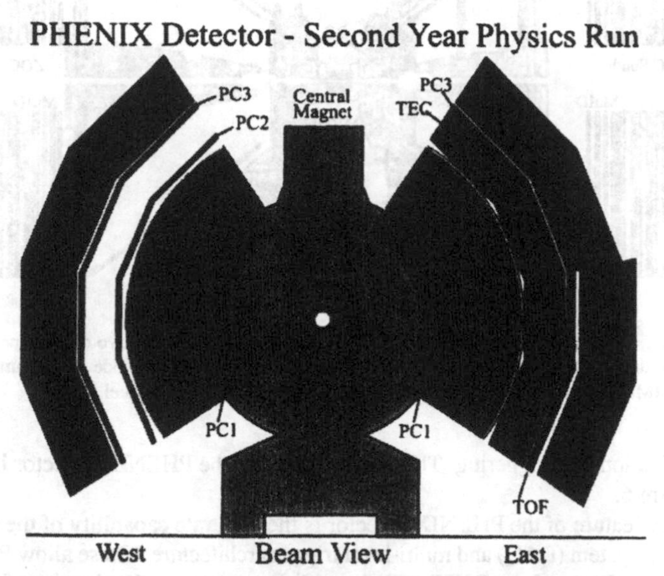

Figure 6. Shown is a beam view of the PHENIX two central spectrometer arms. Their is a axial field magnet in the middle. The detectors from the inner radius out are the multiplicity and vertex detector (MVD), beam-beam counters (BBC), drift chambers (DC), pad chambers (PC1-3), ring imaging cherenkov counter (RICH), time-expansion chamber (TEC), time-of-flight (TOF), and a Lead Glass and Lead Scintillator electro-magnetic calorimeter (PbSc, PbGl).

The PHENIX experiment is specifically designed to measure electrons, muons, hadrons and photons. The experiment is capable of handling high event rates, up to ten times RHIC design luminosity, in order to sample rare signals such as the J/ψ decaying into muons and electrons, high transverse momentum π^0's, direct photons, and others. The detector consists of four spectrometer arms. Two central arms have a small angular coverage around central rapidity and consist of a silicon vertex detector, drift chamber, pixel pad chamber, ring imaging Cerenkov counter, a time-expansion chamber, time-of-flight and an electromagnetic calorimeter. These detectors allow for electron identification over a broad range of momenta in order to measure both low mass and high mass vector mesons. Two forward spectrometers are used for the detection of muons. They employ cathode strip chambers in a magnetic field and interleaved layers of Iarocci tubes and steel for muon

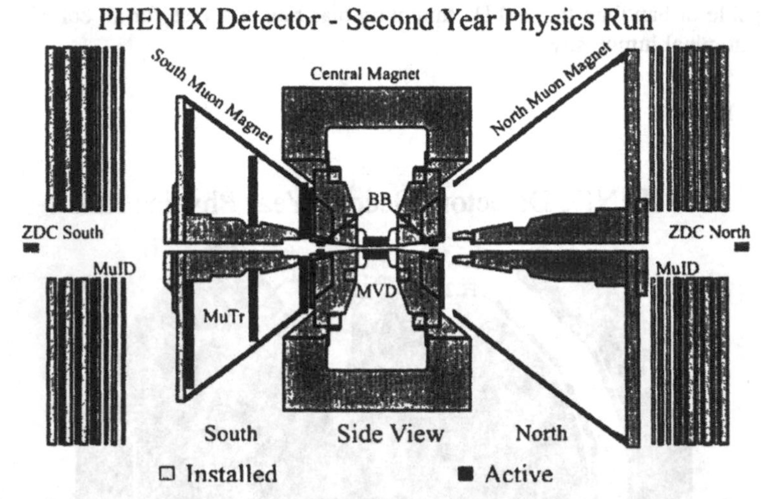

Figure 7. Shown is a side view of the PHENIX detector including the two central spectrometer arms and the two muon spectrometers. The muons systems consist of cathode strip chamber muon trackers (MuTr) and muon identifiers (MuID) interleaved with layers of steel.

identification and triggering. The overall layout of the PHENIX detector is shown in Figure 6.
One key feature of the PHENIX detector is the high rate capability of the data acquisition system (DAQ) and multi-level trigger architecture. These allow PHENIX to sample physics from RHIC collisions above the design luminosity of the machine. This high rate is crucial for studying rare leptonic, photonic and high $p_\perp$ processes.

4.5. STAR

The **S**olenoidal **T**racker **A**t **R**HIC (STAR) is a large acceptance detector capable of tracking charged particles and measuring their momenta in the expected high multiplicity environment. It is also designed for the measurement and correlations of global observables on an event-by-event basis and the study of hard parton scattering processes. The layout of the STAR experiment is shown in Figure 4.5. The initial configuration of STAR in 2000 consists of a large time projection chamber (TPC) covering $|\eta| < 2$, a ring imaging Cherenkov detector covering $|\eta| < 0.3$ and $\Delta\phi = 0.1\pi$, and trigger detectors inside a solenoidal magnet with 0.25 T magnetic field. The solenoid provides a uniform magnetic field of maximum strength 0.5 T for tracking, momentum analysis and particle identification via ionization energy loss measurements in the TPC. Measurements in the TPC were carried out at mid-rapidity with full azimuthal coverage ($\Delta\phi = 2\pi$) and

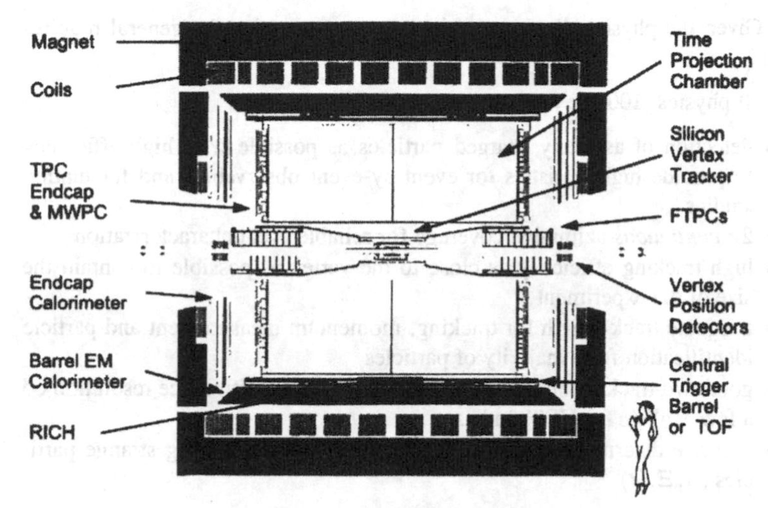

Figure 8. Schematic view of the STAR detector.

symmetry. A total of 1M minimum bias and 1M central events were recorded during the summer run 2000.

Additional tracking detectors will be added for the run in 2001. These are a silicon vertex tracker (SVT) covering $|\eta| < 1$ and two Forward TPCs (FTPC) covering $2.5 < |\eta| < 4$. The electromagnetic calorimeter (EMC) will reach approximately 20% of its eventual $-1 < \eta < 2$ and $\Delta\phi = 2\pi$ coverage and will allow the measurement of high transverse momentum photons and particles. The endcap EMC will be constructed and installed over the next 2 – 3 years.

4.6. RHIC SPIN PROGRAM

The design of both the STAR and PHENIX experiments includes a polarized proton program to conduct studies of the spin structure of the proton. Critical to this measurement is the identification of high transverse momentum photons and leptons. The STAR experiment is phasing in an electromagnetic calorimeter that will be crucial for such observations. In particular the RHIC experiments are ideally suited for measuring the gluon contribution of the proton spin.

5. Experimental Techniques

5.1. TRACKING CHARGED PARTICLES (THE STAR EXAMPLE)

The STAR experiment aims at the observation of hadronic observables and their correlations, global observables on an event-by-event base, and the measurement of hard scattering processes. The physics goals dictate the design of an experiment. Given the physics directions it is easy to summarize the general requirements:

– Soft physics (100 MeV/c $< p_\perp < 1.5$ GeV$/c$)

  - detection of as many charged particles as possible with high efficiency to provide high statistics for event-by-event observables and fluctuation studies
  - 2π *continuous* azimuthal coverage for reliable event characterization
  - high tracking efficiency as close to the vertex as possible to contain the size of the experiment
  - adequate track length for tracking, momentum measurement and particle identification for a majority of particles
  - good two-track resolution providing a momentum difference resolution od a few MeV$/c$ for HBT studies
  - accurate determination of secondary vertices for detecting strange particles (Λ, Ξ, Ω)

– Hard physics ($>$ 1.5 GeV$/c$ and jets)

  - large uniform acceptance to maximize rates and minimize edge effects in jet reconstruction
  - accurate determination of the primary vertex in order to achieve high momentum resolution for primary particles
  - electromagnetic calorimetry combined with tracking and good momentum resolution up to $p_\perp = 12$ GeV$/c$ to trigger on jets
  - segmentation of electromagnetic calorimeters which is considerable finer than the typical jet size, *i.e.* jet radius $r = \sqrt{d\eta^2 + d\phi^2} \sim 1$.

The challenge is to find a detector concept which meets all these requirements with minimal costs. The detector of choice to solve the main tracking tasks was a large Time Projection Chamber (TPC) operated in a homogeneous magnetic field for continuous tracking, good momentum resolution and particle identification (PID) for tracks below 1 GeV$/c$. The requirements not met by the TPC needed to be covered by more specialized detectors such as a large acceptance electromagnetic calorimeter (hard processes), two Forward-TPCs (coverage of $\eta > 1.7$), and an inner silicon vertex tracker (better primary and secondary vertex measurement). For particle identification of high-$p_\perp$ particles detectors a small Ring Imaging Detector (RICH) and a Time-of-Flight (ToF) patch were added. Both detectors are not usable for event-by-event physics due to their small acceptance but allow to extend the PID capabilities for *inclusive* distributions.
The need for an homogeneous field along the beam direction puts an stringent constraint on the design of the whole experiment. Only large solenoidal magnets are able to provide uniform fields of considerable strength (0.5 T). To keep down the costs the magnet cannot be too large which limits significantly the amount of

"real estate" (*i.e.* detectors) it can house inside for tracking, PID, and calorimetry. The final magnet has coils with an inner radius of 2.32 m and a yoke radius of 2.87 m. The total length is 6.9 m. Note, that this concept is very different from the design of PHENIX where the axial-field magnet does not, or only weakly, constrain the dimensions of the required detectors.

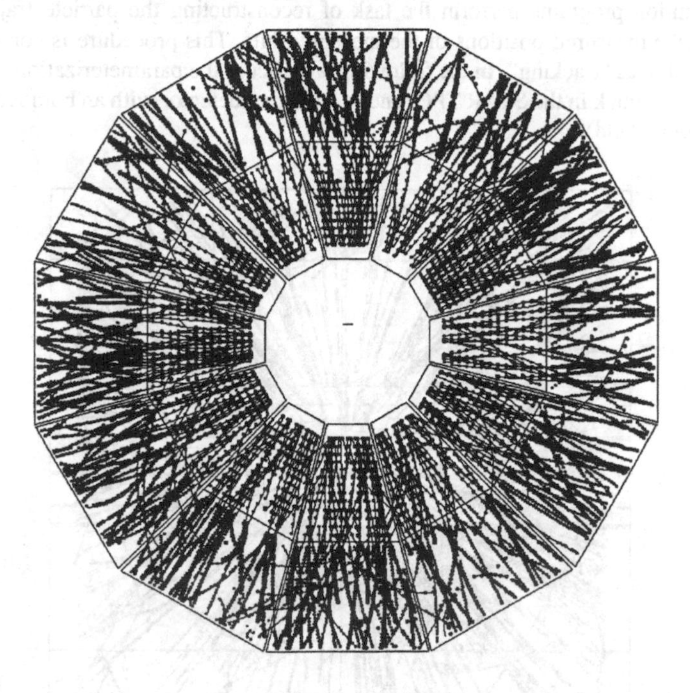

Figure 9. Peripheral Au+Au event recorded in the STAR TPC. Shown is the projection of all hits (points) and reconstructed tracks (solid lines) in the event onto the xy plane perpendicular to the beamline.

In the following we focus on the TPC, STARs main tracking detector, that essentially performs the role of a 3D camera with around 70 million pixel resolution. The TPC is divided into two longitudinal drift regions, each 2.1 m long. Electrons created from track ionization drift in the longitudinal direction, along the TPC electric field lines, to the end-caps of the TPC. Each end-cap is instrumented with 70,000 pads. Each pad reads out 512 time samples. The position of the ionization charge in the readout plane provides the x and y coordinates of a space point along the particle trajectory while the arrival *time* of the charge allows to determine the original z position. The ionization pattern (chain of hits) of a traversing charged particle curved in the magnetic field allows the complete reconstruction of the particle trajectory and its 3-momentum. Fig. 9 shows the xy-projection (front view) of a low multiplicity event recorded in the STAR TPC. Each point represents one reconstructed hit, i.e. the local ionization charge created by one particle. The lines represent the reconstructed trajectories of the particles. Together with the timing information it is possible to also reconstruct the z-position of the hits and such the polar angle of the tracks as shown in Fig. 10. Sophisticated pattern recognition programs perform the task of reconstructing the particle trajectory using the measured positions of the measured hits. This procedure is commonly referred to as "tracking". In the following we discuss the parameterization used to describe a track in the STAR TPC (and in any other detector with an homogeneous solenoidal field).

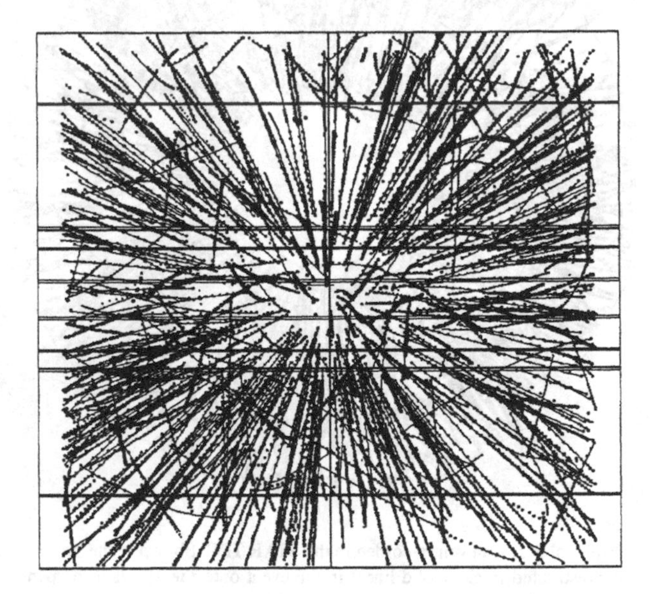

Figure 10. Same event as shown in Fig. 9 but viewed from the side. The beamline runs horizontally from the left to the right.

5.1.1. *Track Parameterization Momentum Determination*

The trajectory of a charged particle in a static uniform magnetic field with $\vec{B} = (0, 0, B_z)$ is a helix. In principle five parameters are needed to define a helix. From the various possible parameterizations we describe here the version which is most suited for the geometry of a collider experiment and therefore used in STAR. This parameterization describes the helix in Cartesian coordinates, where x, y and z are expressed as functions of the track length s.

$$x(s) = x_0 + \frac{1}{\kappa}[\cos(\Phi_0 + h\, s\, \kappa\, \cos\lambda) - \cos\Phi_0] \quad (5.1)$$

$$y(s) = y_0 + \frac{1}{\kappa}[\sin(\Phi_0 + h\, s\, \kappa\, \cos\lambda) - \sin\Phi_0] \quad (5.2)$$

$$z(s) = z_0 + s\, \sin\lambda \quad (5.3)$$

where: s is the path length along the helix
$\mathbf{x_0, y_0, z_0}$ is the starting point at $s = s_0 = 0$
λ is the dip angle
κ is the curvature, i.e. $\kappa = 1/R$
B is the z component of the homogeneous magnetic field ($B = (0, 0, B_z)$)
q is charge of the particle in units of positron charge
h is the sense of rotation of the projected helix in the xy-plane, *i.e.* $h = -\text{sign}(qB) = \pm 1$
$\mathbf{\Phi_0}$ is the azimuth angle of the starting point (in cylindrical coordinates) with respect to the helix axis ($\Phi_0 = \Psi - h\pi/2$)
$\mathbf{\Psi}$ is the $\arctan(dy/dx)_{s=0}$, *i.e.* the azimuthal angle of the track direction at the starting point.
The meaning of the different parameters is visualized in Fig. 11.

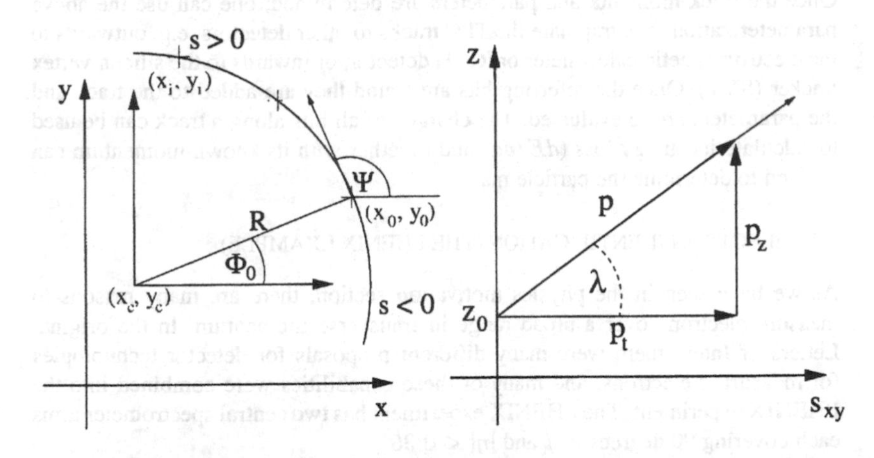

Figure 11. Helix parameterization: shown on the left is the projection of a helix on the xy plane. The crosses mark possible data points. The right plot depicts the projection of a helix on the sz plane. For the meaning of the various parameters see text.

The circle fit in the xy-plane gives the center of the fitted circle (x_c, y_c) and the curvature $\kappa = 1/R$ while the linear fit gives z_0 and $\tan\lambda$. The phase of the helix (see Fig. 11) is defined as follows:

$$\Phi_0 = \arctan\left(\frac{y_0 - y_c}{x_0 - x_c}\right) \tag{5.4}$$

The reference point (x_0, y_0) is then calculated as follows:

$$x_0 = x_c + \frac{\cos\Phi_0}{\kappa} \tag{5.5}$$

$$y_0 = y_c + \frac{\sin\Phi_0}{\kappa} \tag{5.6}$$

and the helix parameters can be evaluated as:

$$\Psi = \Phi_0 + h\pi/2 \tag{5.7}$$

$$p_\perp = c\,q\,B/\kappa \tag{5.8}$$

$$p_z = p_\perp \tan\lambda \tag{5.9}$$

$$p = \sqrt{p_\perp^2 + p_z^2} \tag{5.10}$$

where κ is the curvature in [m^{-1}], B the value of the magnetic field in [Tesla], c the speed of light in [m/ns] (≈ 0.3) and $p_\perp$ and p_z are the transverse and longitudinal momentum in [GeV$/c$].
Once the track momenta and parameters are determined, one can use the above parameterization to extrapolate the TPC tracks to other detectors, e.g. outwards to the electromagnetic calorimeter or RICH detector, or inwards to the silicon vertex tracker (SVT). Once the referring hits are found they are added to the track and the parameters are re-evaluated. The charges of all hits along a track can be used to calculate its energy loss (dE/dx) and together with its known momentum can be used to determine the particle mass.

5.2. ELECTRON IDENTIFICATION (THE PHENIX EXAMPLE)

As we have seen in the physics motivation section, there are many reasons to measure electrons over a broad range in transverse momentum. In the original Letters of Intent there were many different proposals for detector technologies for measuring electrons, and many of these capabilities were combined into the PHENIX experiment. The PHENIX experiment has two central spectrometer arms each covering 90 degrees in ϕ and $|\eta| < 0.35$.
Electrons and also charged hadrons that are produced near mid-rapidity are bent in an axial magnetic field over a radial distance of approximately two meters after which the aperture is relatively field free. There is a large aperture drift chamber that measures the projective trajectory of the charged particle tracks in the field free region with multiple wire layers oriented in the x direction (giving maximum resolution in the bend plane of the field) and also in u and v direction (stereo planes

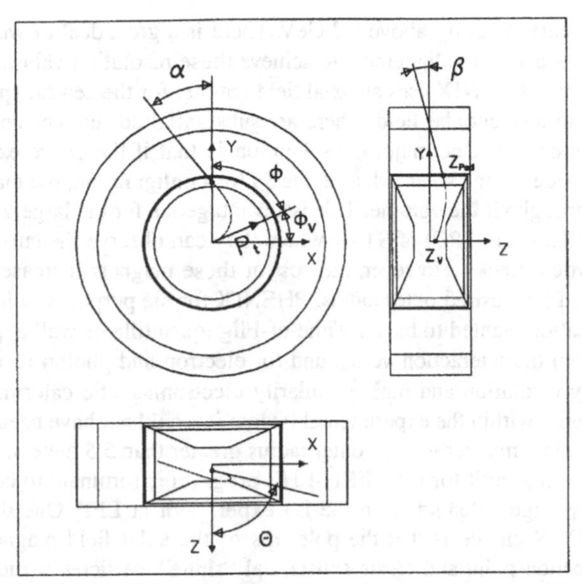

Figure 12. **A schematic diagram of the PHENIX central spectrometer magnetic reconstruction technique is shown.**

for pattern recognition). The drift chamber is augmented by a series of moderate resolution pad chambers, which yield three dimensional space point along the particle's track and aid significantly in pattern recognition and non-vertex background rejection. As shown in Figure 12 the particle track is characterized by an angle (α) in the bend plane of the magnetic field. From this vector and the assumption that the track originated at the Au-Au collision vertex, the rigidity of the track is determined. The rigidity is the momentum divided by the particle charge, and with the assumption of a $Z = \pm 1$ particle, the momentum is known. Typically the momentum resolution has two components as shown below.

$$\delta p/p = (\approx 1\%) + (\approx 1\%) \times p[\mathrm{GeV}/c] \qquad (5.11)$$

The first term, typically of order 1% is due to the multiple scattering of the charge particle in material before and in the tracking devices. The second term which scales with the particle's momentum is related to the finite spatial resolution of the detector. The PHENIX detector has been designed with a minimum of inner region material including a Beryllium beam pipe, which has a low Z value to minimize multiple scattering while maintaining integrity for the vacuum. All four experiments at RHIC have Be beam-pipes in the interaction region. The PHENIX design resolution from the drift chamber is of order 0.5% at $p_\perp \approx 0.2$ GeV and

increases linearly with $p_\perp$ above 0.7 GeV. There is a great deal of work involved in calibrations and wire alignement to achieve these resolution values.
It is notable that PHENIX uses an axial field magnet for the central spectrometers as opposed to a solenoidal field. There are substantial advantages and disadvantages to this choice. One major consideration is that if the entire experiment is contained inside a large solenoid, as in the STAR configuration, the magnetic field is present throughout the volume. This is advantageous for the large volume Time Projection Chamber (TPC) of STAR where they can observe the curvature of the charged particle tracks. However, the cost of these magnets increases steeply as one increased the desired outer radius. PHENIX for the purposes of hadron particle identification wanted to have a Time-of-Flight scintillator wall approximately 5 meters from the interaction vertex and for electron and photon identification a good energy resolution and high granularity electromagnetic calorimeter behind that. In keeping within the experimental budget it would not have been possible to have a solenoidal magnet with an outer radius greater than 5.5 meters. The ALICE experiment being built for the CERN-LHC program is fortunate to be able to reuse the very large solenoid from the L3 experiment at LEP. One disadvantage of the PHENIX choice is that the pole tips of the axial field magnet are close to the interaction point and create substantial "shine", particles scattering off the poles into the detector aperture. The overall choice for the field configuration is an important starting point for many detector designs.
Now that we have characterized the particle's momentum vector, we must discriminate all of the charged pions from our interesting electrons. The first detector in PHENIX that is employed is a Ring Imaging Cherenkov detector (RICH). It is a large gas volume detector with a thin mirror plane for reflecting Cherenkov light onto an array of photo-multiplier tubes (PMT) that are situated off to the side of the spectrometer acceptance. The radiator gas used is either ethane with an index of refraction n=1.00082 or methane with n=1.00044. Requiring more than three PMT hits yields almost 100% efficiency for electrons and rejects pions with $p_\perp \approx< 4$ GeV at the level of $< 10^3$ in a single track environment. The detector, requiring only three PMT's, does not reconstruct a ring radius for further particle characterization, but rather is used as a threshold detector only.
In a multiple track collisions (remember of order 5000 charged particles are being produced) one can incorrectly match a track to a RICH signal. Further electron identification is provided by the electromagnetic calorimeter (EMCal). The calorimeter is composed of both Lead Scintillator (PbSc) and Lead Glass (PbGl) modules, with the later being originally used in the WA98 experiment at the CERN-SPS heavy ion program. The calorimeter has a radiation length of $\approx 18X_0$ and $\approx 16X_0$ for the PbSc and PbGl respectively, but does not fully interact and contain hadronic showers. Thus, an electron or photon incident on the calorimeter deposits most of its energy, while a hadron (eg. charged pion) has a large probability to pass through the module depositing a small dE/dx minimum ionizing

radiation energy. Even when the pion suffers an inelastic collision in the calorimeter, only a fraction of its energy is contained and measured in the PMT at the back of the module. The excellent energy resolution $\approx 5 - 8\%/\sqrt{E(\text{GeV})}$ and high granularity give precise electron identification and pion rejection by requiring the energy match the measured momentum ($E/p \approx 1$). However, the background rejection degrades when the particle momentum is low and hadronic showers have a higher probability to match the measured momentum.

Therefore, an additional detector is necessary to help identify low $p_\perp$ electrons for the crucial low mass vector meson physics. For this purpose, PHENIX uses a Time Expansion Chamber (TEC) that samples energy loss in a gas radiator. The TEC determines the particle species using dE/dx information. It has a rejection of $e/\pi \approx 5\%$ for particles with a momentum p = 500 MeV$/c$ with P10 gas and 2% with Xe gas, which is much more expensive. It is the combination of all these detectors that allow for efficient electron measurements with a minimum of pion contamination. Thus, the challenge of electron identification over a broad range in $p_\perp$ is met.

5.3. MUON IDENTIFICATION (THE PHENIX EXAMPLE)

Direct muons and hadrons decaying into muons require a rather different experimental approach to measuring electrons, photons and other hadrons. Muons interact with a low cross section in material, and are easiest to identify by placing steel or other material in the particle path and removing all other particles through interaction. Then a detector placed after the steel should measure a clean muon sample.

However, the muons must have a large enough energy to penetrate the steel without stopping due to ionization energy loss dE/dx. In the central rapidity region at a collider, the muons from low $p_\perp$ J/ψ decays make it impossible to have enough steel to range out other hadrons effectively, while allowing the muons to pass through. In particle experiments that focus on high $p_\perp$ muons, the detector can consist of a similar one to that described above for electrons. After the calorimeter, one can have some steel absorber (often the return in the magnet steel) and then have a muon identifier detector outside of that. This design does not work well for heavy ion physics. First, it restricts the measurable $p_\perp$ range at too high a value and also has a large background from low $p_\perp$ pions decaying into muons.

Another option is to measure muons are forward rapidity where they have a substantial momentum in the longitudinal direction. PHENIX measures muons in the forward and backward pseudo-rapidity regions (at angles of 10-35 degrees from the beam line). The detector consists of a brass and steel absorber to range out hadrons followed by a cathode strip chamber muon tracking device that measures the particle's bend in a magnetic field. There is a delicate balance in the amount and type of absorber material used. Too much material and the multiple scattering reduces the resolution which is important for cleanly separating states such as

J/ψ and ψ', and too little material in which case the particle occupancy in the tracking device is too large. After the muon tracking device there is more steel absorber with Iarocci tube muon identifiers interspersed. The coverage is from $1.1 \leq |\eta| \leq 2.4$ and the muons must have $E \geq 2.1$ GeV in order to penetrate the absorber material. The identifier detectors also provide the necessary trigger information to sample the muons from the high luminosity RHIC collisions.
The most substantial background in measuring J/ψ and D mesons are muons from pions and kaons that decay before they hit the brass nose-cone absorbers. There is a competing requirement in PHENIX in that one wants the absorber as close to the interaction vertex as possible to reduce this decay contribution, but they need to be far enough apart for there to be a good acceptance for electrons, hadrons and photons in the two central arm spectrometers. There are many benefits to the comprehensive design of PHENIX, but there are definite drawbacks as well. Of the two muon arms, the south muon arm was completed for running in Run II at RHIC, and the north arm will be complete for Run III.

6. Physics Results

In this chapter we present the physics results from the Run I data and a preview of expected results from Run II.

6.1. GLOBAL OBSERVABLES

The first result with a measurement from all four RHIC experiments is the charged particle multiplicity [12, 13, 14, 15]. The four experiments' results for central (small impact parameter) collisions are in excellent agreement and are shown in Figure 13. The multiplicity rises more sharply as a function of center-of-mass energy in heavy ion collisions than in $p+p$ and $p+\overline{p}$ collisions, which is attributed to the increased probability for hard parton scattering in the thick nuclear target seen by each parton.
We expect the charge particle yield to increase for collisions of larger nuclei. However, at low x values, the high density of gluons may in fact saturate due to gluon fusion processes. The contribution to the yield from hard processes should exhibit point-like scaling (scaling with the number of binary collisions) and would thus scale as $A^{4/3}$. However, parton saturation depends upon the nuclear size and would limit the growth of the number of produced partons as $A^{1/3}$. If present, this initial parton saturation would limit the hard process contribution to the total charged particle multiplicity.
Only one nuclear species (Au) was accelerated in Run I at RHIC. Thus, rather that changing the mass number A directly we control the collision volume by varying the centrality or the number of participating nucleons for Au-Au collisions. Shown in Figure 14 are the published results from the PHENIX [13] and PHOBOS [16]

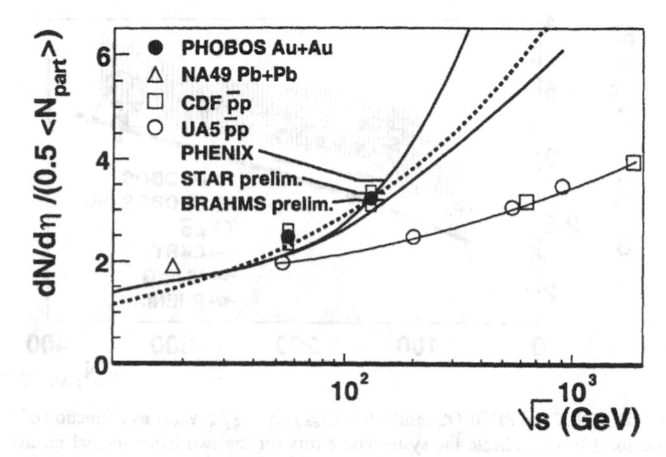

Figure 13. Charged particle multiplicity measurement from all four RHIC experiments is shown for Au-Au collisions at $\sqrt{s_{NN}} = 130$ GeV. Also shown are data for $p + p$ and $p + \overline{p}$ collisions. A model of heavy ion collisions HIJING is shown for comparison.

experiments for the number of charged particles per participant nucleon pair as a function of the number of participating nucleons. The number of participating nucleons is determined in a slightly different manner by the different experiments. However, the general method is to calibrate the number of spectator nucleons (= $2 \times A$ - participant nucleons) using a measurement of spectator neutrons in a set of zero degree calorimeters that are common to all experiments. By correlating the number of forward neutrons to the number of charged particle produced in the large pseudo-rapidity region, the event geometry can be understood.
In Figure 14 one can also see theory comparisons that indicate that a model including parton saturation (EKRT [17]) fails to agree with the more peripheral data. Results from the HIJING model [18] are also shown which does not include parton saturation and thus has a more continuous rise in the particle multiplicity. Since saturation phenomena are only likely to have observable consequences for large collision volumes, it is not possible with present systematics to rule out the saturation picture for the most central collisions.
In order to better test the saturation picture lighter ion, smaller A, collisions will be studied in Run II. In addition, heavy flavor (charm and bottom) and Drell-Yan production should be a sensitive probe to the initial parton density. Another proposal is that by varying the collision energy and keeping the nuclear geometry the same one can get a better handle on systematics and test scenarios dependent on the coupling constant and the saturation scale. The physics of parton saturation and color glass condensates is at the forefront of theoretical development in the field. Many recent developments were discussed in this workshop and will be

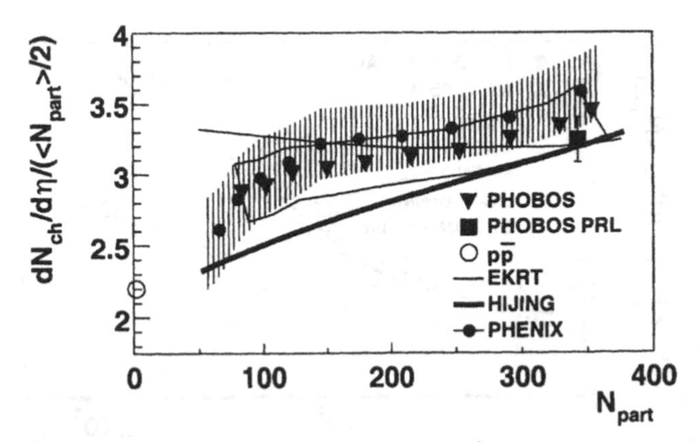

Figure 14. PHENIX and PHOBOS results for $dN_{ch}/d\eta|_{\eta=0}/\frac{1}{2}N_{part}$ as a function of N_{part}. The hashed and solid bands indicate the systematic errors for the two experimental results. The data point for $p\overline{p}$ with two participants is shown for comparison. Also theoretical predictions from the HIJING and EKRT models are shown.

described in other contributions.

In addition to the initial parton density, the energy density is of great interest. There are published results estimating the initial thermalized energy density achieved in these collisions. Bjorken originally derived a formula, shown in Eqn. 6.12, relating the measured transverse energy per unit rapidity to the thermal energy density [19].

$$\epsilon_{Bj} = \frac{1}{\pi R^2}\frac{1}{c\tau}\frac{dE_\perp}{dy} \tag{6.12}$$

It should be noted that there is a trivial factor of two error in the original reference that is corrected here. This formulation assumes a boost invariant expanding cylinder of dense nuclear matter and a thermalization time τ. There are two important assumptions in this particular formulation. The first is the boost invariant nature of the collision. There are recent preliminary measurements from STAR and PHOBOS that indicate the distribution of particles is relatively flat over ± 2 units of pseudo-rapidity. However, shown in Figure 15 is the measured distribution of $\overline{p}/p$ from the BRAHMS experiment [20]. This indicates the the system is already changing at $y \approx 2$, though it is not clear that this is enough to invalidate the energy density formulation. The second question is what is the relevant thermalization time τ.

The PHENIX experiment has published [21] the transverse energy distribution for minimum bias Au-Au collisions. For the 5% most central events, the extracted transverse energy $< dE_\perp/d\eta > |_{\eta=0} = 503 \pm 2$ GeV. Shown in Figure 16 is $dE_\perp/d\eta/(0.5N_p)$ versus the number of participating nucleons. One sees a similar increase in transverse energy as was seen in the charged particle multiplicity yield.

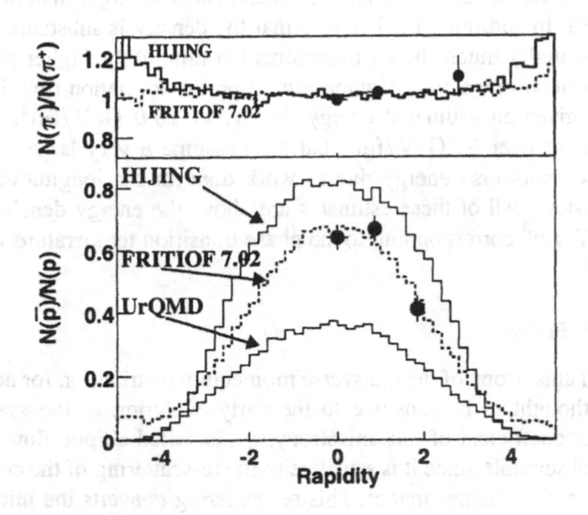

Figure 15. Plotted is the π^-/π^+ and $\overline{p}/p$ ratio as a function of rapidity from the BRAHMS experiment for central collisions.

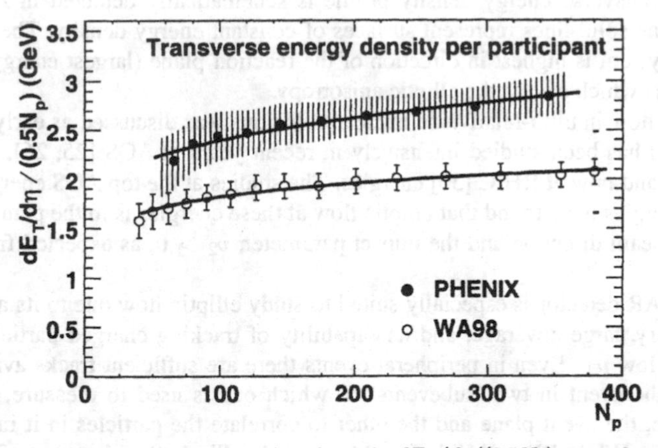

Figure 16. The PHENIX experiment result is shown for $dE_\perp/d\eta/(0.5N_p)$ at $\eta = 0$ as a function of the number of participating nucleons. Also shown in the result from experiment WA98 at the lower energy CERN-SPS.

The canonical thermalization time used in most calculations is $\tau = 1$ fm/c, that yields an energy density of 4.6 GeV/fm^3, which is is 60% larger than measured at the CERN-SPS. In addition, it is believed that the density is substantially higher due to the potentially much shorter thermalization time in the higher parton density environment. If one achieves gluon saturation the formation time is of order 0.2 fm/c and gives an estimated energy density of 23.0 $\mathrm{GeV/fm^3}$. There are even estimates of over 50 $\mathrm{GeV/fm^3}$, but they assume a very large drop in the final measured transverse energy due to work done in the longitudinal expansion of the system. All of these estimates are above the energy density of order $0.6 - 1.8\ \mathrm{GeV/fm^3}$ corresponding to the phase transition temperature 150 – 200 MeV.

6.2. ELLIPTIC FLOW

The azimuthal anisotropy of the transverse momentum distribution for non-central collisions is thought to be sensitive to the early evolution of the system. The second Fourier coefficient of this anisotropy, v_2, is called elliptic flow [22]. It is an important observable since it is sensitive to the re-scattering of the constituents in the created hot and dense matter. This re-scattering converts the initial spatial anisotropy, due to the almond shape of the overlap region of non-central collisions, into momentum anisotropy. The spatial anisotropy is largest early in the evolution of the collision, but as the system expands and becomes more spherical, this driving force quenches itself. Therefore, the magnitude of the observed elliptic flow reflects the extent of the re-scattering at early time [23]. The time evolution of the transverse energy density profile is schematically depicted in Figure 17 where the solid lines represent surfaces of constant energy density. The pressure in the system is highest in direction of the reaction plane (largest energy density gradient) which causes the elliptic anisotropy.

Elliptic flow in ultra-relativistic nuclear collisions was discussed as early as 1992 [24] and has been studied intensively in recent years at AGS [25, 26], SPS [27, 28, 29] and now at RHIC [30] energies. The studies at the top AGS energy and at SPS energies have found that elliptic flow at these energies is in the plane defined by the beam direction and the impact parameter, $v_2 > 0$, as expected from most models.

The STAR detector is especially suited to study elliptic flow due to its azimuthal symmetry, large coverage, and its capability of tracking charged particles down to very low $p_\perp$. Even in peripheral events there are sufficient tracks available to divide the event in two 'subevents' of which one is used to measure, or better estimate, the event plane and the other to correlate the particles in it in order to derive v_2. While BRAHMS is not able to study elliptic flow because of its small acceptance, PHOBOS has some capabilities to measure v_2 but only integrated over all $p_\perp$. Because of its restricted azimuthal coverage the PHENIX collaboration follows a different approach to reconstruct v_2 by studying the $\Delta\phi$ correlation

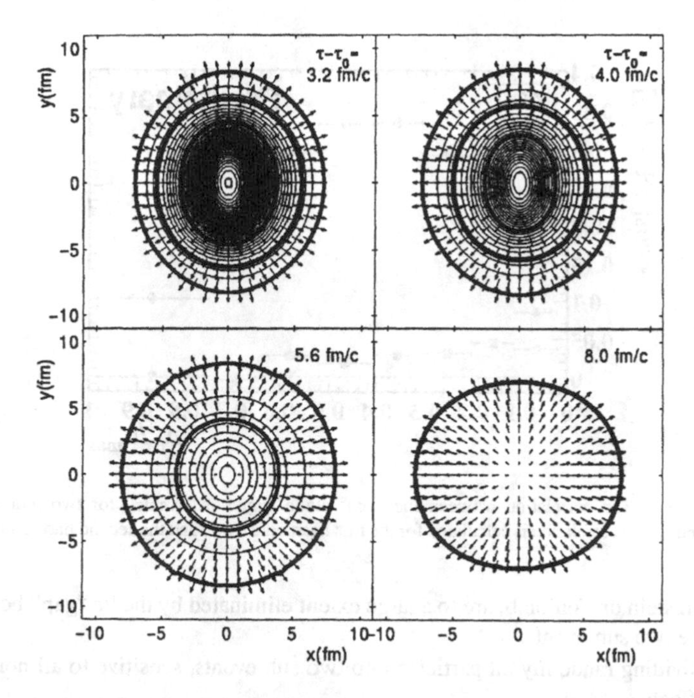

Figure 17. Schematic view of a evolution of the transverse en ergy density profile (indicated by constant energy density con tours spaced by) and of the flow velocity field (indicated by arrows) for Pb+Pb collisions at impact parameter b=7.0 fm. The four panels show snapshots at times $\tau - \tau_0 =$ 3.2, 4.0, 5.6, and 8.0 fm/c. At these times the maximal energy densities in the center are 5.63, 3.62, 1.31 and 0.21 GeV/fm^3, respectively. The figure is taken from [32].

between particles thus circumventing the event plane determination. However, work is still in progress and in the following we concentrate on the more direct event plane method used in STAR.

The flow analysis method involves the calculation of the event plane angle, which is an experimental estimator of the real reaction-plane angle. The second harmonic event plane angle, Ψ_2, is calculated for two sub-events, which are independent subsets of all tracks in each event. Figure 18 shows the results for the correlation between the sub-events for the first and second harmonic as a function of centrality [30]. The peaked shape of the centrality dependence of $\langle\cos[2(\Psi_a - \Psi_b)]\rangle$ is a signature of anisotropic flow. However, the correlation between the sub-events may not be due entirely to anisotropic flow. To estimate the magnitude of non-flow effects one can use the sub-events in three different ways:

1. Assigning particles with pseudo-rapidity $\eta < 0 + \epsilon$ to one sub-event and particles with $\eta > 0 + \epsilon$ to the other. Short range correlations, such as Bose-

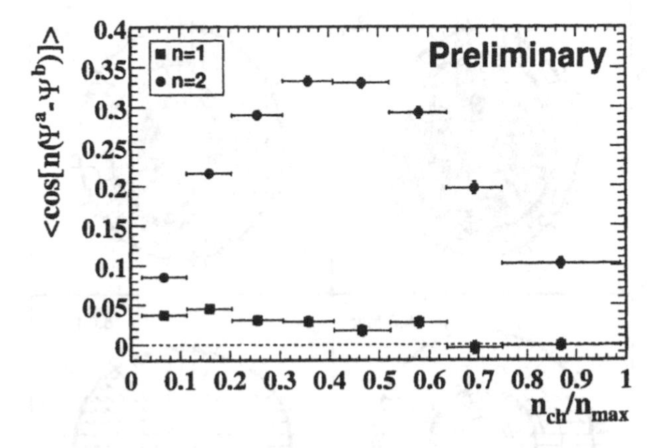

Figure 18. The correlation between the event plane angles determined for two independent sub-events. The correlation is calculated for the first harmonic (n=1) and the second harmonic (n=2).

Einstein or Coulomb, are to a large extent eliminated by the 2ϵ "gap" between the two sub-events.

2. Dividing randomly all particles into two sub-events, sensitive to all non-flow effects.
3. Assigning positive particles to one sub-event and negative particles to the other, allowing an estimation of the contribution from resonance decays.

Studies have shown that the results from all three methods are for the central and mid-peripheral events very similar. For the most peripheral events the results vary among the methods by about 0.005. Not all non-flow contributions might be known and the effects of others, such as jets, are difficult to estimate because of their long-range correlation. In order to estimate the systematic uncertainty due to the effects of jets, one can assume that jets contribute at the same level to both the first and second order correlations. Taking the maximum observed positive first order correlation, as being completely due to non-flow will reduce the calculated v_2 values.

Figure 19 shows the final v_2 integrated over all $p_\perp$ as a function of centrality. The statistical uncertainties are smaller than the markers and the uncertainties shown are the systematic uncertainties due to this estimated non-flow effect.

Figure 20 shows the maximum v_2 value as a function of collision energy. It rises monotonically from about 0.02 at the top AGS energy [25], 0.035 at the SPS [28] to about 0.06 at RHIC energies [30]. This increasing magnitude of the integrated elliptic flow indicates that the degree of thermalization, which is associated with the amount of re-scattering, is higher at the higher beam energies. However, in-

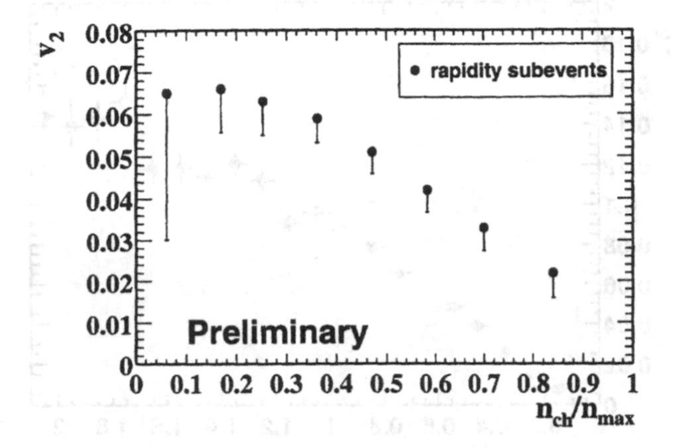

Figure 19. The integrated elliptic flow signal, v_2, with the estimated systematic uncertainties as measured by STAR.

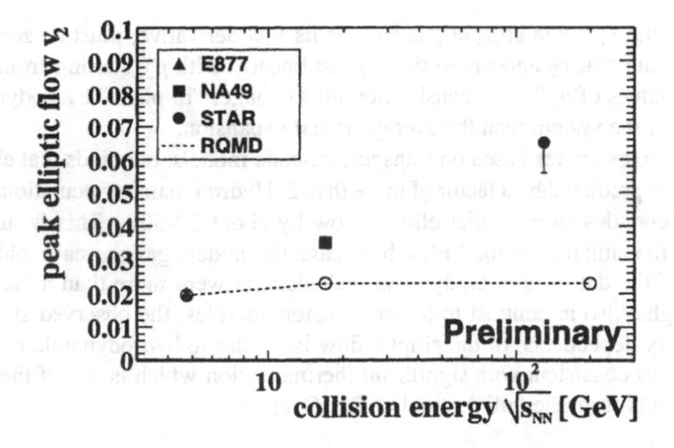

Figure 20. Excitation function of v_2 from top AGS to RHIC energies.

terpretation of the excitation function has to be done with care. The v_2 values used here are the maximum values as a function of centrality for each energy. The centrality where v_2 peaks can change as a function of beam energy, indicating different physics [31].

The differential anisotropic flow is a function of η and $p_\perp$. Figure 21 shows v_2 for charged particles as a function of $p_\perp$ for a minimum bias event sample. Math-

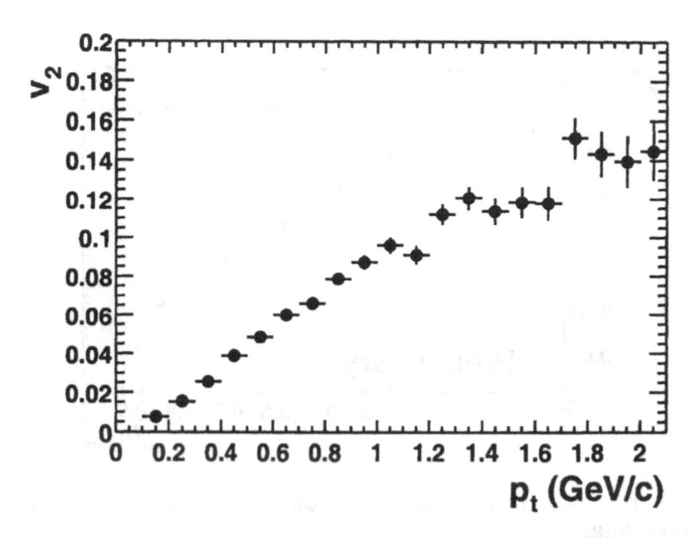

Figure 21. v_2 as a function of $p_\perp$, as measured by STAR in $\sqrt{s_{NN}}$ = 130 GeV Au-Au collisions.

ematically, the v_2 value at $p_\perp$=0, as well as its first derivative, must be zero, but it is interesting that v_2 appears to rise almost linearly with $p_\perp$ starting from relatively low values of $p_\perp$. This is consistent with a stronger "in-plane" hydrodynamic expansion of the system than the average radial expansion.
Comparing to estimates based on transport cascade models, one finds that elliptic flow is under predicted by a factor of more than 2. Hydrodynamic calculations [32] for RHIC energies over predict elliptic flow by about 20-50%. This is just the reverse of the situation at the SPS where cascade models gave a reasonable description of the data and hydrodynamic calculations were more than a factor of two too high. Also in contrast to lower collision energies, the observed shape of the centrality dependence of the elliptic flow is similar to hydrodynamic calculations and thus consistent with significant thermalization which is one of the most striking results from the initial round of RHIC results.

6.3. TWO-PARTICLE INTERFEROMETRY (HBT)

The study of small relative momentum correlations, a technique also known as HBT [33] interferometry, is one of the most powerful tools to study complicated space-time dynamics of heavy ion collisions [34]. It provides crucial information which helps to improve our understanding of the reaction mechanisms and to constrain theoretical models of the heavy ion collisions. Interpretation of the extracted HBT parameters in terms of source sizes and lifetime is more or less straightforward for the case of chaotic static sources. In the case of expanding sources with strong space-momentum correlations (due to flow, etc.) the situation is more difficult, but the concept of length of homogeneity [35] provides a useful framework for the interpretation of data.

The dependence of the pion-emitting source parameters on the transverse momentum of the particle pairs (K_T) and on centrality can in principle be measured by all RHIC experiments with high statistics. For more detailed analysis as for example event-by-event HBT, HBT radii versus reaction plane, and the correlation of HBT results with other observables can only be performed by STAR due to its large acceptance and azimuthal coverage. These studies, however, are still in progress and it is by far too early to discuss them here. In the following we show results from the STAR experiment [36] that performed a multi-dimensional analysis using the standard Pratt-Bertsch decomposition [37] into outward, sideward, and longitudinal momentum differences and radius parameters. The data are analyzed in the longitudinally co-moving source frame, in which the total longitudinal momentum of the pair (collinear with the colliding beams) is zero.

As expected, larger sizes of the pion-emitting source are found for the more central (*i.e.* decreasing impact parameter) events, which in turn have higher pion multiplicities. This source size is observed to decrease with increasing transverse momentum of the pion pair. This dependence is similar to what has been observed at lower energies and is understood to be an effect of collective transverse flow. Shown in Figure 22 is the coherence parameter λ and the radius parameters R_{out}, R_{side}, and R_{long} obtained in the analysis. Also shown are values of these parameters extracted from similar analyses at lower energies. All analyses are for low transverse momentum ($\sim$ 170 MeV/c) negative pion pairs at mid-rapidity for central collisions of Au + Au or Pb + Pb. From Figure 22 the values of λ, R_{out}, R_{side}, and R_{long} extend smoothly from the dependence at lower energies and do not reflect significant changes in the source from those observed at the CERN SPS energy. One of the biggest surprises is that the anomalously large source sizes or source lifetimes predicted for a long-lived mixed phase [38] have not been observed in this study. Preliminary results of the HBT analysis by the PHENIX Collaboration [39] agree with the STAR results within error bars.

One of the big puzzles, however, is the magnitude and the tranverse momentum (K_T) dependence of the ratio of R_{out}/R_{side} which contradicts *all* model predictions [38, 40]. These model calculations predict the ratio to be greater than unity due to system lifetime effects which cause R_{out} to be larger than R_{side}. They also predict that the ratio increases with K_T. Such an increase seems to be a generic feature of the models based on the Bjorken-type, boost-invariant expansion scenario. Hence, it was surprising to see that the experimentally observed ratio is less than unity and is decreasing as a function of K_T. Currently, it is far from clear what kind of scenario can lead to such a puzzling K_T dependence.

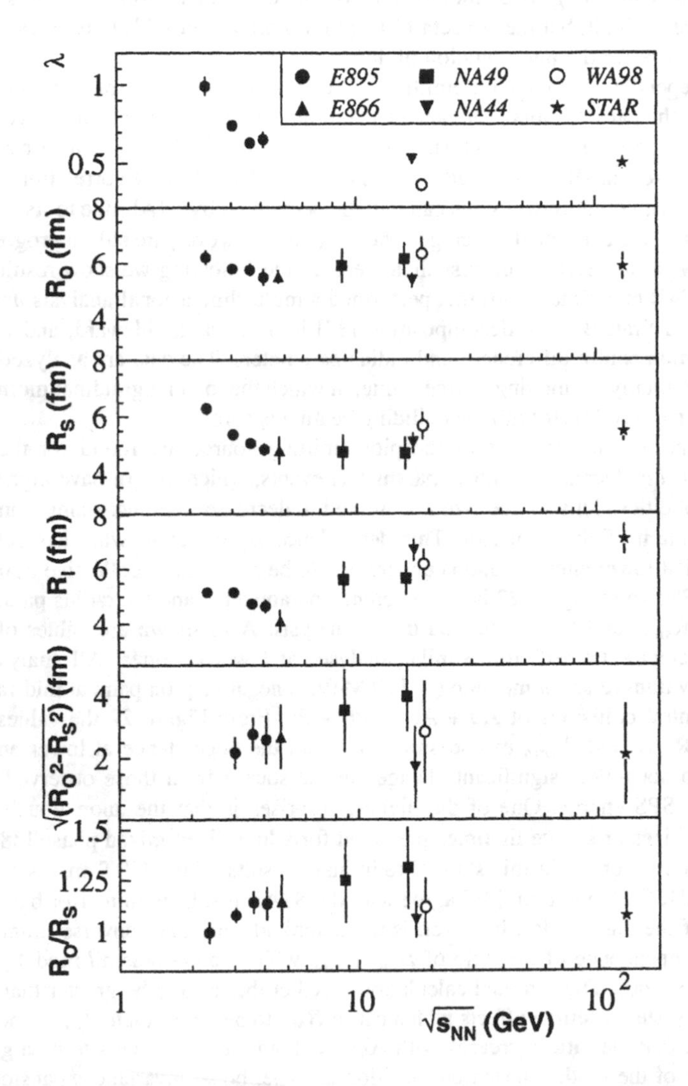

Figure 22. Compilation of results on two-particle correlation (HBT) parameters from measurements using central collisions of Au + Au at the BNL-AGS, Pb + Pb at the CERN-SPS and Au + Au data from the STAR experiment at RHIC. Plotted are the coherence parameter λ, R_{out}, R_{side}, and $R_{longitudinal}$.

6.4. HARD PROCESS PROBES OF THE PLASMA

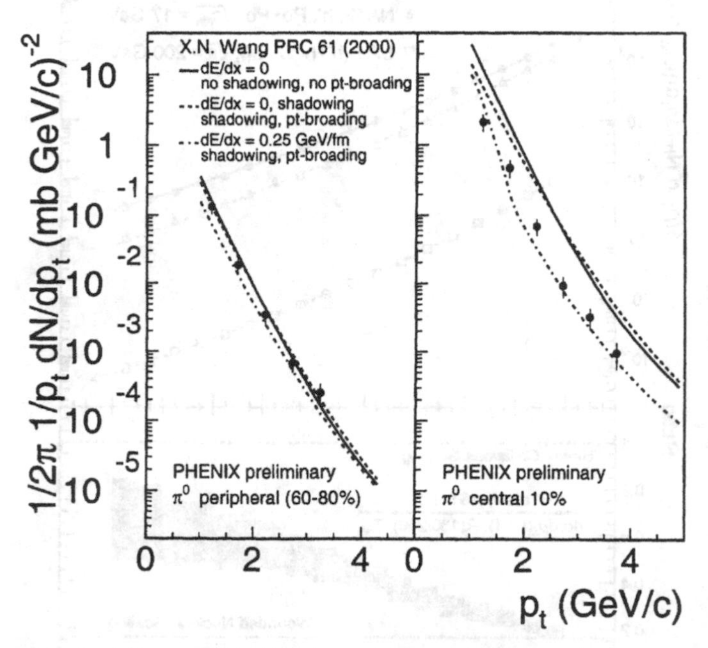

Figure 23. Preliminary PHENIX invariant multiplicity of identified π^0 as a function of transverse momentum are shown for peripheral and central collisions. Comparison with theoretical calculations with and without parton energy loss are also shown.

Jet processes and their associated hadronic fragmentation provide one of the most exciting probes of the color deconfined plasma. The PHENIX experiment has measured the distribution of identified π^0 for both central and peripheral Au-Au collisions as shown in Figure 23 [41]. The peripheral results appear to be in good agreement within systematic errors of an extrapolation from *pp* collisions scaled up by the number of binary collisions expected in this centrality class. However, the central collision results show a significant suppression in the π^0 yield relative to this point like scaling expected for large momentum transfer parton-parton interactions. If the created fireball in RHIC collisions is transparent to quark jets, then we expect the yield of high $p_\perp$ hadrons to obey point-like scaling and equal the *pp* (or equivalently $p\bar{p}$) distribution scaled up by the number of binary NN collisions, or equivalently by the nuclear thickness function T_{AA}. This is not what is observed. A more sophisticated calculation [42] yields the same qualitative conclusion.

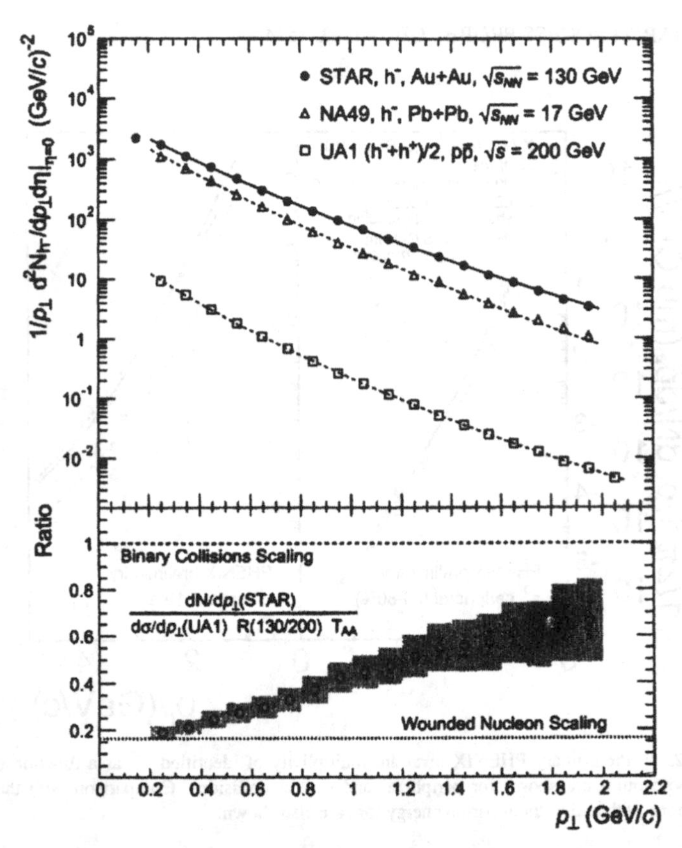

Figure 24. Invariant multiplicity of charged hadrons as a function of transverse momentum (top). Ratio of unidentified charge hadrons per calculated binary collision from Au + Au central collisions to those from p + p($\bar{p}$) collisions extrapolated to $\sqrt{s_{NN}}$ =130 GeV as a function of transverse momentum (bottom).

The STAR experiment has recently submitted for publication [43] the $p_\perp$ spectra for unidentified negatively charged hadrons in central Au + Au collisions as shown in Figure 24. Also shown are the equivalent spectra from experiment NA49 at the CERN-SPS at $\sqrt{s}$ = 17 GeV and from UA1 in $p\bar{p}$ at $\sqrt{s_{NN}}$ = 200 GeV. The STAR spectra is then divided by the spectra from $p\bar{p}$ scaled by the number of binary collisions, and the result is shown in the lower panel of Fig 24. At low transverse momentum the particle production is dominated by soft interactions which scale with the number of wounded nucleons as indicated by the line at 0.2. The rise from 0.2 as a function of $p_\perp$ certainly has a large contribution from hydrodynamic flow that will push particles to higher transverse momentum in

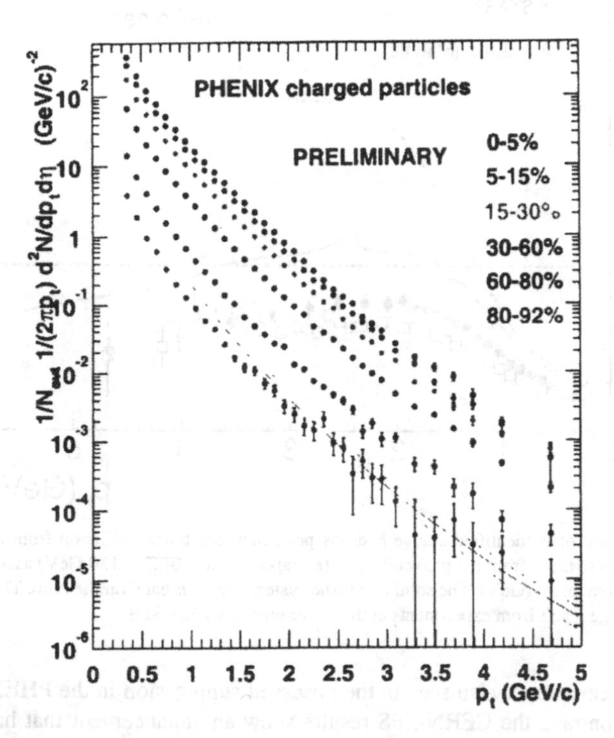

Figure 25. PHENIX preliminary results for unidentified charged hadron invariant multiplicity as a function of transverse momentum.

central Au-Au collisions.

The PHENIX experiment has shown preliminary results extending out further in transverse momentum. Preliminary results for six centrality classes are shown from PHENIX in Fig 25. STAR also has preliminary results for central collisions extending out to $p_{\perp} > 5$ GeV that are in reasonable agreement with the PHENIX results. If one takes the ratio of the central spectra to the unidentified spectra in *pp* collisions scaled up by the number of binary collisions one gets a ratio R_{AA} as shown in Fig 26. It needs to be noted that there is no *pp* data at $\sqrt{s_{\rm NN}}$ = 130 GeV and thus an extrapolation to that energy is done to calculate R_{AA}. This extrapolation is included in the systematic error band, and should be reduced when both experiments measure the spectra in *pp* in Run II.

There are many important physics points to understand in these results. The ratio appears to stay below one, although that is a marginal conclusion with the present systematic errors. However, this is certainly in qualitative agreement with a parton

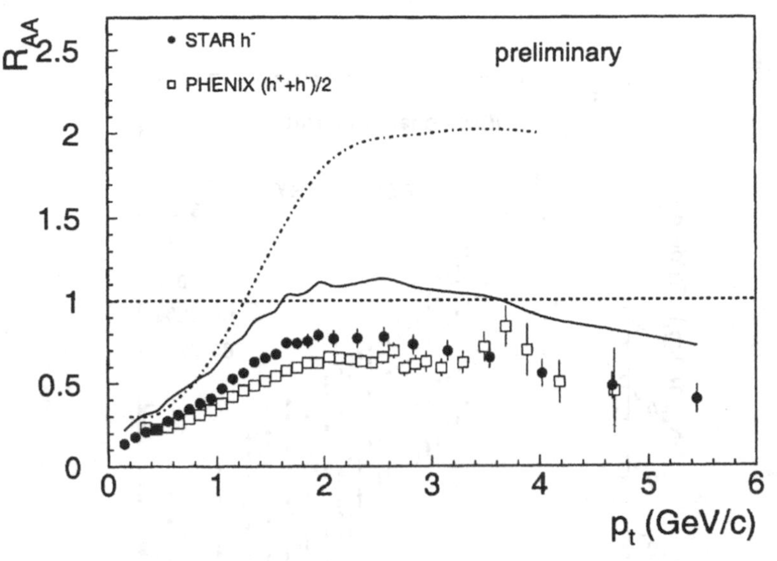

Figure 26. Ratio of unidentified charge hadrons per calculated binary collision from Au + Au central collisions to those from p + p($\overline{p}$) collisions (extrapolated to $\sqrt{s_{NN}}$ = 130 GeV) as a function of transverse momentum (GeV). The solid line is the systematic error band on the ratio. The dashed line is the average result from experiments at the lower energy CERN-SPS.

energy loss scenario, as also seen in the observed suppression in the PHENIX π^0 spectra. In contrast, the CERN-SPS results show an enhancement that has been attributed to the Cronin effect, or initial state parton scattering that gives a k_T kick to the final transverse momentum distribution. This expected enhancement makes the suppression seen at RHIC all the more striking.

There are a number of open questions that must be considered before drawing any conclusions. The most basic is that these $p_\perp$ values are low relative to where one might have confidence in the applicability of perturbative QCD calculations. In addition, the separation between soft and hard scale physics is blurred in this $p_\perp$ range, and in fact the CERN-SPS ratio R_{AA} has also been explained in terms of hydrodynamic boosting of the soft physics to higher $p_\perp$. The preliminary results from PHENIX on the ratios of $\pi/K/p(\overline{p})$ in the middle of this $p_\perp$ range look more like soft physics than a parton fragmentation function in vacuum. One additional point of concern is that these models of energy loss assume that the parton exits the collision region before finally fragmenting into a jet of forward hadrons. Thus the final hadronization takes place in vacuum. In the $p_\perp$ range of these early measurements, that conclusion is not so clear. The parton is traveling through the medium with various k_T scatters, and if it hadronizes inside a bath of other particles, the leading hadrons may be slowed down by inelastic collisions with co-moving pions. Lastly, the point-like scaling is known to be violated due to the nuclear shadowing of parton distribution functions. These nuclear modifications are known to reduce the pdf for quarks of order 20% for $x \approx 10^{-2}$; however, the shadowing for gluons is not currently measured. The calculations of [42] have included modeling of this shadowing, but must be viewed with caution at this time. These points need further theoretical investigation. In addition, as will be discussed in the next section, many of these concerns are reduced when the measurements extend to much higher transverse momentum.

7. Future Measurements

In the following chapter, we discuss a few select topics within our areas interest that have exciting results expected in the near future.

7.1. CHARM AND BOTTOM

The measurement of open charm and bottom in relativistic heavy ion collions is both an extreme experimental challenge and rich with physics information. First, the measurement of quarkonium states such as the J/ψ require a comparison measure of the original $c\bar{c}$ production to determine the effect of color screening. Also, the total charm production is sensitive to the initial gluon density in the incoming nuclei and is thus sensitive to any shadowing of the gluon distribution function and may even comment on the possible color glass condensate postulated to describe the phase space saturated gluon distributions in the highly Lorentz contracted nuclei. Lastly, recent predictions of charm quark energy loss in tranversing a hot partonic medium have generated much interest. Now for the difficult part. The best was to measure charm via D mesons is either via direct reconstruction from its $\pi + K$ decay mode or via the semi-leptonic decay $K + e + \nu_e$. The combinatoric background in the purely hadronic channel are close to overwhelming and the semi-leptonic decay cannot be completely reconstructed. In particle physics experiments measuring the decay products with a few micron displaced vertex from the collision vertex allows for a dramatic reduction in the combinatoric background. However, the level of silicon detector technology was not advanced enough for this very high multiplicity environment at the time of the RHIC detector designs. This is now being discussed as a possible upgrade to the experiments.
One promising way of determining the charm production is through the measurement of single electrons. Shown in Figure 27 is a simulation of single electrons as a function of transverse momentum. The top curve in the sum of all contributions. The next two curves that dominate the spectra at low $p_\perp < 1.0$ GeV and from the Dalitz decays of pions and η and from photon conversions. The next two contributions are from charm D meson and beauty B meson decays. Charm yields

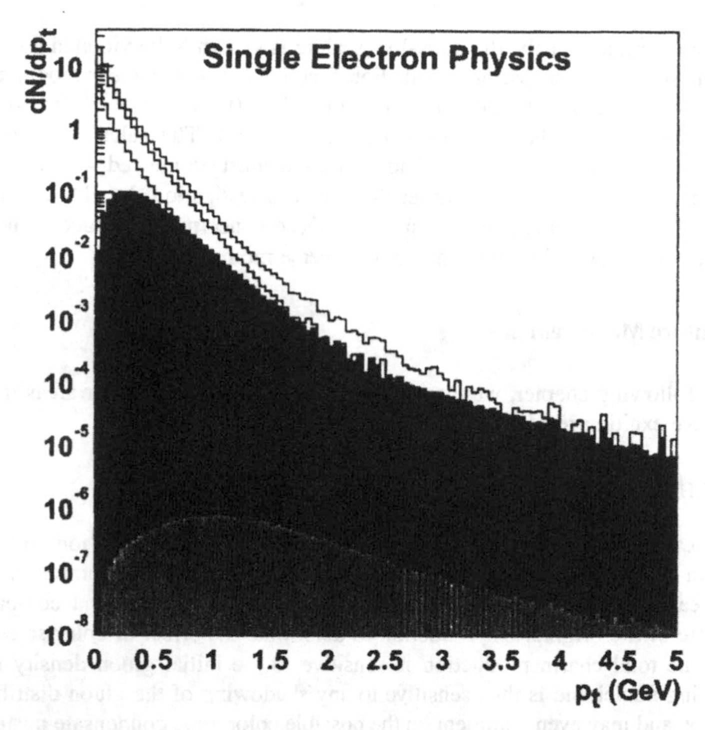

Figure 27. Simulation of the transverse momentum spectrum for mid-rapidity single electrons and positrons. The top curve is the sum of all contributions. the two lower curves are from Dalitz decays and photon conversions. The subsequent three grey shaded areas are from charm D mesons, beauty B mesons, and Drell-Yan.

$\approx$ 50% of the counts at $p_\perp$ = GeV, and beauty yields > 50% of the counts above $p_\perp > 3.5$ GeV. The lowest curve is the contribution from Drell-Yan which never has a major contribution to the single electrons.

PHENIX has made a preliminary measurement of the single electron transverse momentum spectra from the limited statistics in Run I as shown in Figure 28. The analysis of these results is proceeding and implications on charm production are forthcoming.

Additional handles on heavy flavor production can be had with the measurement of correlated leptons. For example, electron-muon pairs at large relative momentum (Q^2 or M_{inv}) have a substantial contribution from $c\bar{c}$ and $b\bar{b}$ pairs. For example, the $c\bar{c}$ can fragment into $D\overline{D}$ followed by the decays $D \longrightarrow K+e+\nu_e$ and $\overline{D} \longrightarrow K + \mu + \nu_\mu$. Although the initial Q^2 of the $c\bar{c}$ pair is significantly modified when measured as a Q^2 of the $e\mu$ pair, there is enough information to attempt to extract a total charm cross section and maybe something about the

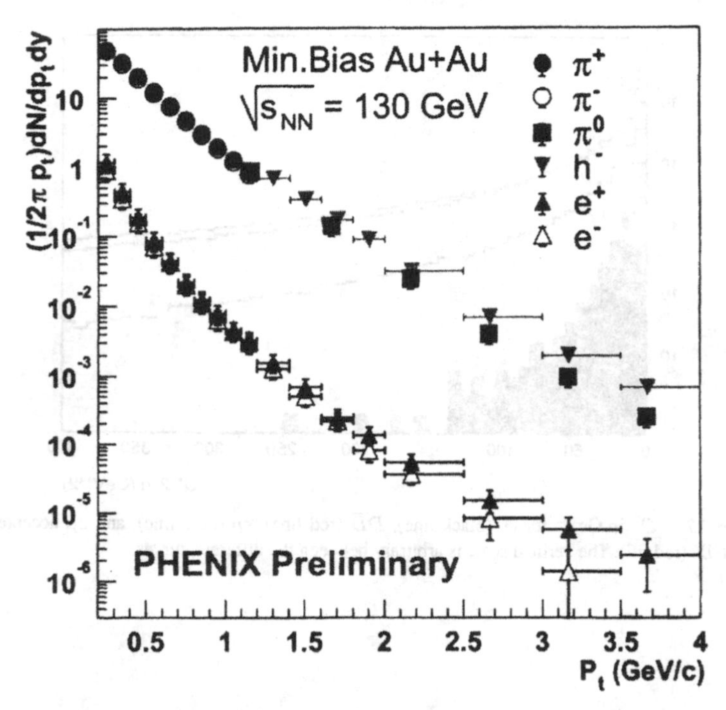

Figure 28. PHENIX preliminary results for unidentified charged hadron invariant multiplicity as a function of transverse momentum.

initial Q^2 of the $c\bar{c}$. This measurement has the advantage over e^+e^- and $\mu^+\mu^-$ pairs in that $e\mu$ pairs are free from Drell-Yan and thermal contributions.

One of us (J.N.) has used PYTHIA 6.0 to estimate the rate of $e\mu$ pairs into the PHENIX acceptance (Central arms for the electron and South muon arm only for the μ). PYTHIA has been run with a charm quark mass of 1.5 GeV/c^2 and $< \langle k_T \rangle$ = 1.5 GeV/c.

We show in Figure 29 the distribution of Q^2 for all $c\bar{c}$, $D\overline{D}$, $e\mu$ and $e\mu$ pairs accepted by PHENIX. One can see that the charm mesons carry most of the information from the $c\bar{c}$ pair. For the $e\mu$ pair the correlation with the $c\bar{c}$ pair Q^2 is substantially washed out and of much lower slope since the kaon in the D decay takes away a large fraction of the original charm momentum that is not measured. It should be noted that the modeling of the fragmentation of the charm quark will be a source of systematic error, whereas the blurring of the Q^2 from the $e\mu$ decay kinematics can be modeled exactly.

We show the invariant mass distribution of the $e\mu$ pairs into the PHENIX acceptance in Figure 30. Also shows are three mass distribution with electron energy

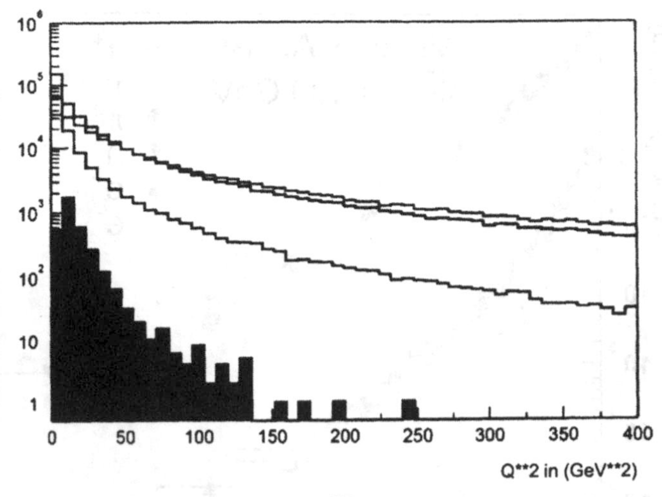

Figure 29. Q^2 in GeV^2 for $c\bar{c}$ (black line), $D\overline{D}$ (red line), $e\mu$ (blue line), and $e\mu$ accepted by PHENIX (red fill). The vertical scale is arbitrary between the different curves.

cuts of 1.0, 1.5, and 2.0 GeV. The cut on the electron energy is strongly correlated with the invariant mass selection on the $e\mu$ pair. Since lower masses (in particular $m < 4.0$ GeV) are thought to have large background contamination, the loss of these pairs with higher electron energy threshold are not as worrisome as they might otherwise be. It may be necessary to use the electron energy in order to selectively trigger on these events.

Measuring charm by this method requires an accurate model of the $c\bar{c}$ distribution in Q^2, and the corresponding Δy (rapidity gap) and $\Delta p_{\perp}$. In Figure 31 PYTHIA is compared to some of the only data on the rapidity gap between D and $\overline{D}$ mesons [44]. The agreement is not bad, but it the comparison is not a great confidence builder in the ability to model and then test by data checking this input. More experimental data are needed and theoretical work to model charm production.

In order to draw a full picture of charm production, multiple measurements must be done in the single lepton (electrons and muons) and correlated leptons (electron pairs, muon pairs, and electron-muon pairs). It would be extremely useful for theorists interested in total charm or beauty production, and also high $p_{\perp}$ energy loss of heavy flavor partons, to make some predictions for the leptonic signatures that will be measured in the next year. Future upgrades to the detectors for tagging displaced vertices from D meson decays are probably more than five years away.

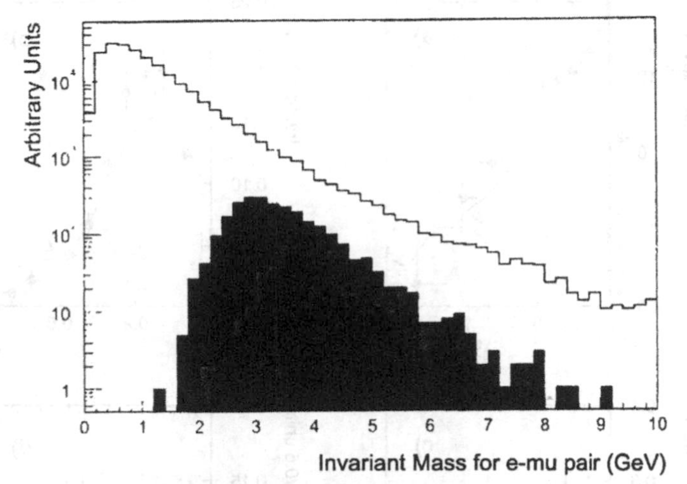

Figure 30. $e\mu$ Invariant Mass Distribution in GeV. Shown are all $e\mu$ pairs (black line), $e\mu$ pairs with the muon penetrating to gap 5 and $E(\text{electron}) > 1.0$ GeV (black fill), with $E(\text{electron}) > 1.5$ GeV (red fill), $E(\text{electron}) > 2.0$ GeV (blue fill).

7.2. QUARKONIA (PHENIX)

There are no early measurement results on quarkonia J/ψ states from Run I due to the low luminosity and short running period. The PHENIX experiment will make a measurement of J/ψ and other states in both the muon and electron channels in Run II. PHENIX has a large acceptance in x_F and $p_\perp$ that we be crucial to constraint models of coloring screening absorbtion and test theories with recoalescence at the hadronization phase. Shown in Figure 32 is a simulation of the type of measurement that could be made in the PHENIX muons arms with 37 weeks at 10% of design luminosity, or equivalently in less than four weeks with the luminosity averaging the the RHIC design specification. The first measurements are being made now in Run II, and high statistics should be available in Run III.

7.3. QUARKONIA (STAR)

The STAR detector system is unique among the RHIC experiments in its capabilities to simultaneously measure many experimental observables on a event-by-event basis such as energy-density, entropy, baryochemical potential, strangeness content, temperature, and flow. The measurement of J/Ψ production as a function of these quantities allows the study of the suppression mechanism in great detail. This advantage becomes immediately apparent when it comes to the study of the *onset* of the anomalous suppression which reflects the point where the system

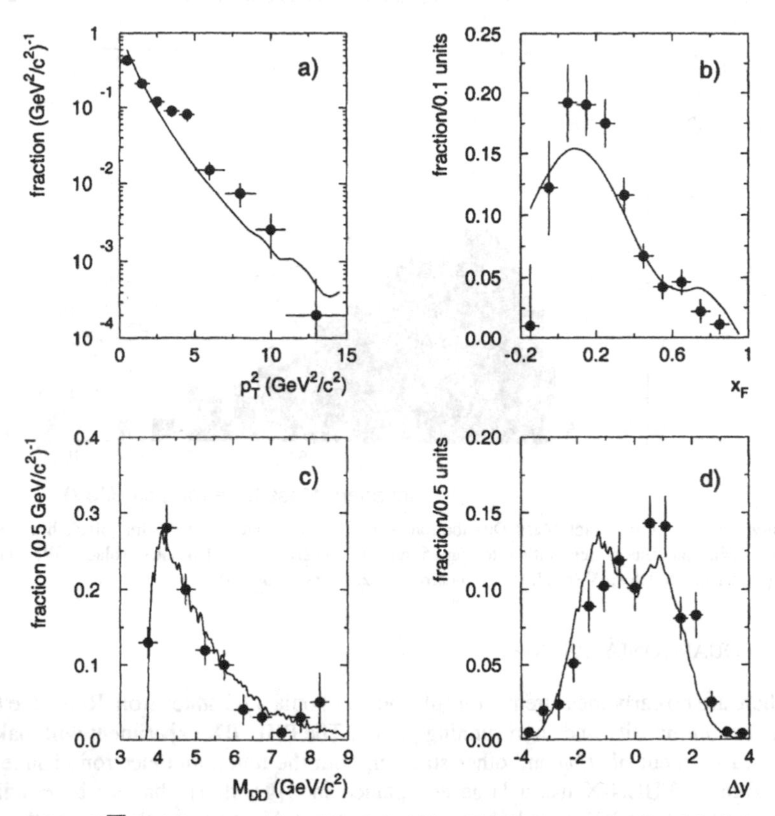

Figure 31. $D\overline{D}$ data from $\pi^- - Cu$ collisions at $\sqrt{s} = 26$ GeV. In particular, panel d) shows the $D\overline{D}$ rapidity difference $\Delta y = y_D - y_{\overline{D}}$.

reaches critical conditions and, at least partially, undergoes a phase transition. Thus, the correlation of the $c\bar{c}$ break-up with the many single-event variables provides a new promising analysis tool.

The golden J/Ψ decay mode for STAR is $J/\Psi \rightarrow e^+e^-$. Prima facie this poses a problem since the STAR detector has been designed to focus primarily on hadronic observables over a large phase-space and thus lacks two essential features of a dedicated lepton experiment: hadron-blind detectors and fast event recording rates. The two essential components which help to overcome these shortcomings are *(i)* the EMC barrel which allows to suppress hadrons to a level sufficient to achieve signal-to-background ratios around 1:3 or better, and *(ii)* a fast level-3 trigger which is designed to efficiently trigger on electron-positron pairs with a given invariant mass at rates in the order of 100 Hz, thus improving STARs bandwidth for recording J/Ψ decays by almost two orders of magnitude.

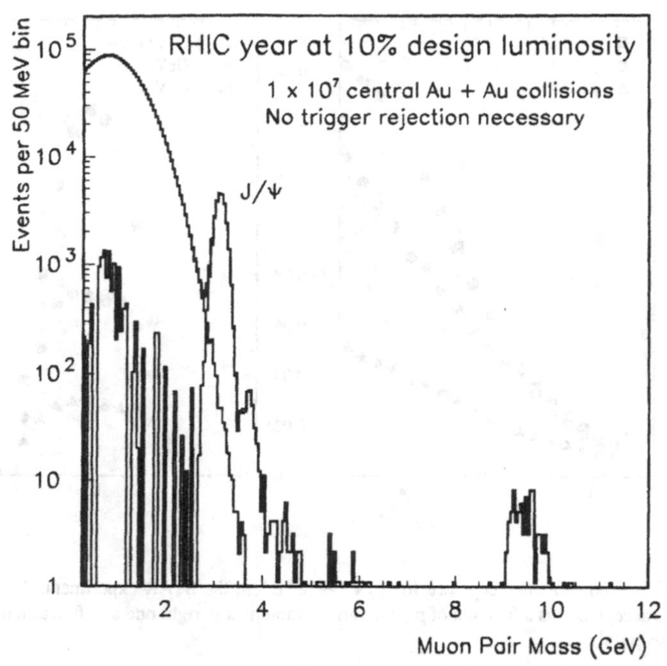

Figure 32. Simulation study of dimuons in the PHENIX muon spectrometers.

The geometric acceptance for the decay channel $J/\Psi \to e^+e^-$ in STAR is shown in Fig. 33. A J/Ψ is accepted if both electrons carry momenta p > 1.5 GeV$/c$ and fall into the EMC acceptance. Both requirements imply that the electron tracks cross all layers of inner tracking detectors (SVT+SSD) and all TPC padrows. This ensures maximum momentum and dE/dx resolution and therefore maximum additional electron identification from detectors other than the EMC. As is depicted in the left plot, the full coverage of the EMC $|\eta| < 1$ ensures relatively high efficiencies ($\sim$ 8%) at very low $p_\perp$. At these low values the two leptons run essentially back-to-back, a decay topology that requires symmetric coverage around $y = 0$. This region of phasespace is of great interest since here the J/Ψ remains longest in the hot dense medium and its breakup probability is maximal. Up to $p_\perp = 1$ GeV$/c$ the acceptance then drops significantly since, still at large opening angles, it becomes more likely to loose one electron because it either carries too low momentum and/or falls outside our acceptance. At larger $p_\perp$ the acceptance raises dramatically due to the decreasing opening angle of the pair an the higher average momenta of the electrons. However, the J/Ψ cross-section drops exponentially towards larger $p_\perp$ resulting in little net benefit in terms of total yields. In this region the acceptance scales approximately *linear* with the

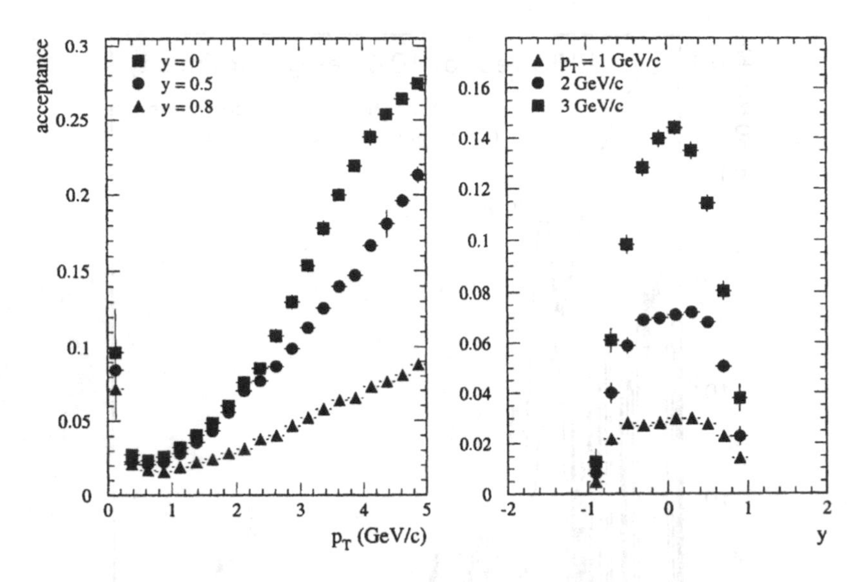

Figure 33. Geometrical acceptance for $J/\Psi \rightarrow e^+e^-$ in the STAR experiment. The left plot shows the acceptance as a function of $p_\perp$ for various rapidities y, right one as a function of rapidity for various $p_\perp$ slices.

EMC coverage while at lower $p_\perp$ it scales almost *quadratically*.
The dominant background source is the 'combinatorial background' due to pions misidentified as electrons. Other sources as electrons from π^0- and η-Dalitz decays, photo-conversions, and decays of light vector-mesons (ρ, ϕ) turned out to be neglectable, since *(i)* their average $p_\perp$ is too low and *(ii)* their rate is small compared to the abundance of misidentified hadrons. In addition, most electrons from photo-conversion can easily be rejected by requiring that all tracks point back to the primary event vertex.
The most crucial factor in the overall background rejection is the hadron rejection power (e/h) of the EMC. However, for $p < 2$ GeV$/c$, a region where the hadron rejection capabilities of the EMC are degraded, the combined dE/dx information of SVT and TPC helps to augment the S/B ratio considerably. It is important to note, that the hadron rejection enters quadratically into the background.
The magnitude of the background depends strongly on the momentum cut applied to the electron candidates. In order to access the low-$p_\perp$ J/Ψ region STAR has chosen $m_{J/\psi}/2 \simeq 1.5$ GeV$/c$ as the lowest value to study. The higher the cut the better the e/h rejection from the EMC and the lower the charged pion yield.
Before any measurement can be addressed it is important to estimate the achievable yields and the statistical significance of the signal under realistic assumptions.

TABLE I. Current estimates on J/Ψ yields, signal-to-background ratio, and statistical significance of the signal in the STAR experiment.

cut	J/Ψ-yield	S/B	σ after 10^7 sec	σ after 10^6 sec
$p_e > 1.5$ GeV/c	40 k	1:3	76	24
$p_e > 2.0$ GeV/c	10 k	3:1	77	24

Table 7.3 summarizes the resulting yields, the signal-to-background ratio, and the statistical significance after 10^7 and 10^6 sec. Note, that a nominal RHIC run has 10^7sec. Here one assumes 100% level-3 trigger efficiency running at the design rate of 100 Hz, and full EMC coverage including pre-shower detectors. The estimate includes the reconstruction efficiencies of the various detectors. Known cross-section from elementary collisions were used to exptrapolate to Au+Au at $\sqrt{s_{NN}}$ = 200 GeV. Two different sets of cuts on the electron momenta are shown in order to demonstrate its effect on the S/B ratio. Interestingly, the higher S/B ratio for the larger $p_\perp$ cut balances the decreased signal strength, thus resulting in the same statistical significance for both cases.
For the actual measurement one has to study J/Ψ production not only in the most central collisions, but has to vary the centrality and possibly the colliding systems. Although the J/Ψ yield will decrease for semi-central and peripheral collisions the signal-to-background ratio will decrease even stronger (almost quadratically) which will ease the extraction of the signal significantly.

8. Summary and Conclusions

The workshop was a wonderful forum for learning and exchanging new ideas about the physics relevant at RHIC. There is lots of exciting physics in a field with much potential for discovery. We wish to thank the organizers and all the students for their active participation and thought provoking questions.

References

1. R. Hagedorn, Suppl. A. Nuovo Cimento VolIII, No.2 (1965) 150.
2. P. Braun-Munzinger, I. Heppe, and J. Stachel, Phys. Lett. B465 (1999) 15.
3. P. Braun-Munzinger *et al.*, Phys.Lett. B518 (2001) 41-46.
4. T. Matsui and H. Satz, Phys. Lett. B178, 416 (1986).
5. R. Baier, Y.L. Dokshitzer, A.H. Mueller and D. Schiff, Phys. Rev. C58, 1706 (1998); hep-ph/9803473.
6. R. Baier, Y.L. Dokshitzer, A.H. Mueller, S. Peigne and D. Schiff, Nucl. Phys. B483, 291 (1997); hep-ph/9607355.

7. Conceptual Design Report for the Solenoidal Tracker At RHIC, The STAR Collaboration, PUB-5347 (1992); J.W. Harris *et al.*, Nucl. Phys. A 566, 277c (1994).
8. PHENIX Experiment at RHIC - Preliminary Conceptual Design Report, PHENIX Collaboration Report (1992).
9. Interim Design Report for the BRAHMS Experiment at RHIC, BNL Report (1994).
10. RHIC Letter of Intent to Study Very Low pt Phenomena at RHIC, PHOBOS Collaboration (1991).
11. Proposal on Spin Physics Using the RHIC Polarized Collider, RHIC Spin Collaboration (1992).
12. B.B. Back *et al.* (PHOBOS), Phys. Rev. Lett. 85, 3100 (2000); hep-ex/0007036.
13. K. Adcox *et al.* (PHENIX), Phys. Rev. Lett. 86, 3500 (2001); nucl-ex/0012008.
14. F. Videbaek *et al.* (BRAHMS), Proceedings to the 15th International Conference on Ultrarelativistic Nucleus-Nucleus Collisions 2001.
15. J. Harris *et al.* (STAR), Proceedings to the 15th International Conference on Ultrarelativistic Nucleus-Nucleus Collisions 2001.
16. B.B. Back *et al.* (PHOBOS); nucl-ex/0105011.
17. K.J. Eskola, K. Kajantie, and K. Tuominen, Phys. Lett. B497, 39 (2001); hep-ph/0009246.
18. X.N. Wang and M. Gyulassy, Phys. Rev. Lett. 86, 3496 (2001); nucl-th/0008014.
19. J.D. Bjorken, Phys. Rev. D, Vol. 27, page 140 (1983).
20. I.G. Bearden *et al.* (BRAHMS); nucl-ex/0106011.
21. K. Adcox *et al.* (PHENIX), Phys. Rev. Lett. 87, 052301 (2001); nucl-ex/0104015.
22. See contribution of J.-Y. Ollitrault in these proceedings.
23. H. Sorge, Phys. Lett. B402, 251 (1997).
24. J.-Y. Ollitrault, Phys. Rev. D **46**, 229 (1992).
25. E877 Collaboration, J. Barrette *et al.*, Phys. Rev. C **55**, 1420 (1997).
26. E895 Collaboration, C. Pinkenburg *et al.*, Phys. Lett. **83**, 1295 (1999).
27. NA49 Collaboration, H. Appelshäuser *et al.*, Phys. Lett. **80**, 4136 (1998).
28. A.M. Poskanzer and S.A. Voloshin for the NA49 Collaboration, Nucl. Phys. **A661**, 341c (1999).
29. WA98 Collaboration, M.M. Aggarwal et al., Phys. Lett. **B**403, 390 (1997); M.M. Aggarwal *et al.*, Nucl. Phys. **A638**, 459 (1998).
30. STAR Collaboration, K.H. Ackermann *et al.*, Phys. Lett. **86**, 402 (2001).
31. S.A. Voloshin and A.M. Poskanzer, Phys. Lett. **B**474, 27 (2000).
32. P.F. Kolb, J. Sollfrank, and U. Heinz, Phys. Rev. C **62**, 054909 (2000).
33. R. Hanbury Brown and R.Q. Twiss, Phil. Mag. **45** (1954) 663.
34. U.A. Wiedemann and U. Heinz, Rhys. Rept. **319** (1999) 145.,
35. A. Makhlin and Y. Sinyukov, Z. Phys. **C39** (1998) 69.
36. C. Adler *et al.*, Phys. Rev. Lett. 87, 082301 (2001).
37. S. Pratt, Phys. Rev. D **33**, 1314 (1986); G. Bertsch, M. Gong and M. Tohyama, Phys. Rev. C**37**, 1896 (1988); and G. Bertsch, Nucl. Phys. A**498**, 151c (1989).
38. D. H. Rischke, Nucl. Phys. A610 (1996) 88c; D.H. Rischke and M. Gyulassy, Nucl. Phys. A608 (1996) 479.
39. S.C. Johnson, nucl-ex/0104020.
40. S. Soff, S.A. Bass and A. Dumitru, nucl-th/0012085.
41. W.A. Zajc *et al.* (PHENIX), Proceedings to the 15th International Conference on Ultrarelativistic Nucleus-Nucleus Collisions 2001; nucl-ex/0106001.
42. X.N. Wang, Phys. Rev. C61, 064910 (2000); nucl-th/9812021.
43. C. Adler *et al.* (STAR); nucl-ex/0106004.
44. P. Braun-Munzinger *et al.*, Eur. Phys. J. C1 (1998) 123-130; nucl-ex/9704011.

QUANTUM FIELDS AT FINITE TEMPERATURE: A BRIEF INTRODUCTION

J.-P. BLAIZOT (blaizot@wasa.saclay.cea.fr)
Service de Physique Théorique
CEA/DSM/SPhT
CEA Saclay
91191 Gif-sur-Yvette cedex, France

1. Introduction

These lectures are a brief introduction to the methods of "finite temperature field theory". By this one refers to a collection of techniques which have been developed, for a large part of them, long ago in the context of quantum statistical mechanics, the so-called "many-body" problem. These were typically applied to non relativistic systems such as the electron gas, liquid Helium, nuclear matter, etc. More recently, these methods have been extended to systems of relativistic particles interacting with gauge fields, for which new interesting technical and conceptual issues appear.
The interest for such questions was originally triggered by the developments which took place in particle physics in the early seventies, when it was understood that elementary particle interactions could be unified in the framework of gauge theories. Two essential concepts emerged then: *spontaneous symmetry* breaking providing a mechanism for the generation of particle masses, and *asymptotic freedom* according to which the interactions become weak at short distances.
These developments changed our ideas about matter at high temperature or density. Because thermal fluctuations tend to suppress the expectation value of the fields responsible for symmetry breaking giving massses to the particles, one expects all particles to become massless at high temperature. Furthermore, because of asymptotic freedom, theses massless particles interact weakly. Thus at high enough temperature, matter becomes simple and turns essentially into an ultrelativistic plasma of weakly interacting particles.

J.-P. Blaizot and E. Iancu (eds.), QCD Perspectives on Hot and Dense Matter, 305–326.

Several phase transitions are expected in the early universe, where the appropriate conditions for the creation of such plasmas are found. We are concerned in this school with the transition which takes place at a temperature $T \gtrsim 150$ MeV. At this point chiral symmetry is restored and the hadrons dissolve into a gas of their elementary constituents, the quarks and the gluons. This is called the quark-gluon plasma. Much of the present interest in this transition is coming from the hope to observe the quark-gluon plasma in laboratory experiments, by colliding heavy nuclei at high energies. An important experimental program is underway, both in the USA (RHIC at Brookhaven), and in Europe at CERN (see the contributions by Nagle, Ullrich and Schukraft in this volume).
These lectures are meant to be an elementary introduction to some of the techniques and concepts which form the background for other lectures, in particular those of A. Rebhan, F. Karsch or R. Pisarski. Part of the material presented at the beginning is text book material [1, 2, 3, 4, 5, 6]. My goal is to arrive at general considerations on the physics of the quark-gluon plasma at high temperature [7]. But first, I shall recall the essential features and difficulties of weak coupling calculations at finite temperature using the simple example of the scalar field. (Pedagogical complements may be found in Refs. [8, 9].)

2. Statistical mechanics of fermions and bosons

We shall start by introducing the method of the imaginary time at the heart of most perturbative calculations. We discuss first the operator formalism which often leads to more transparent physical interpretations. Then we turn to the path integral formalism, more convenient for dealing with quantum fields, in particular with gauge fields.

2.1. GENERALITIES

We begin by general considerations which are not specific to field theory. We consider a physical system whose dynamics is governed by a hamiltonian H and such that a charge Q (e.g. the particle number for non relativistic particles, or more generally the electric charge, the baryonic charge, etc.) is conserved, i.e., $[H, Q] = 0$. In the grand canonical ensemble, the equilibrium state of the system at temperature $T = 1/\beta$ is described by the density operator:

$$\mathcal{D} = \frac{1}{\mathcal{Z}} \exp\{-\beta(H - \mu Q)\} = \sum_n |\Psi_n\rangle p_n \langle\Psi_n|, \tag{2.1}$$

where the $|\psi_n\rangle$'s are the common eigenstates of H and Q, and $\mathcal{Z}$ is the partition function:

$$\mathcal{Z} \equiv \mathrm{Tr} \exp\{-\beta(H - \mu Q)\} = \sum_n \exp\{-\beta(E_n - \mu Q_n)\}. \tag{2.2}$$

All useful thermodynamical functions can be obtained from $\mathcal{Z}$. For example, the thermodynamical potential Ω is given by

$$\Omega = -kT \ln \mathcal{Z} = E - TS - \mu N, \tag{2.3}$$

where

$$S = -k\mathrm{Tr}\mathcal{D} \ln \mathcal{D} = -k \sum_n p_n \ln p_n \tag{2.4}$$

is the entropy of the system, $E = \mathrm{Tr}\, H\mathcal{D} = \sum_n E_n p_n$ is the average energy, and $N = \mathrm{Tr}\, Q\mathcal{D} = \sum_n Q_n p_n$.

2.2. MANY-PARTICLE STATES

We denote the single particle states by $|\alpha\rangle$ (α represents a set of quantum numbers required to specify completely the state of a single particle, e.g. the momentum $\mathbf{k}$ and the spin). These are assumed to form a complete orthonormal basis:

$$\sum_\alpha |\alpha\rangle\langle\alpha| = 1, \qquad \langle\alpha|\beta\rangle = \delta_{\alpha\beta}. \tag{2.5}$$

A basis of the space of states with arbitrary numbers of identical bosons or fermions, the so-called *Fock space*, is conveniently written in terms of creation operators. Thus the single particle state α is written as

$$|\alpha\rangle = a_\alpha^\dagger|0\rangle \tag{2.6}$$

where the state $|0\rangle$, which contains no particles, is called the *vacuum* state. A state with two particles is of the form $a_\alpha^\dagger a_\beta^\dagger|0\rangle$, etc. Annihilation operators reduce the number of particles in a state: $a_\alpha|\alpha\rangle = |0\rangle$, etc.
The symmetry or antisymmetry of the many-particle states are guaranteed by the following commutation (-) or anticommutation relations (+) of the creation and annihilation operators:

$$\left[a_\alpha^\dagger, a_\beta^\dagger\right]_\pm = [a_\alpha, a_\beta]_\pm = 0 \qquad \left[a_\alpha, a_\beta^\dagger\right]_\pm = \langle\alpha|\beta\rangle = \delta_{\alpha\beta}. \tag{2.7}$$

We can associate to the state $|x\rangle$ a creation operator $\Psi^\dagger(x)$ and a destruction operator $\Psi(x)$. These operators, commonly referred to as field operators, are related to the operators $a_\alpha^\dagger$ and a_α by a linear transformation:

$$\Psi^\dagger(x) = \sum_\alpha a_\alpha^\dagger \langle\alpha|x\rangle \qquad \Psi(x) = \sum_\alpha \langle x|\alpha\rangle\, a_\alpha \tag{2.8}$$

Operators acting in Fock space can be given a simple representation in terms of creation or annihilation operators. For instance the operator which measures the density of particles at point x may be written:

$$\rho(x) = \Psi^\dagger(x)\Psi(x). \tag{2.9}$$

2.3. THE PARTITION FUNCTION OF FREE PARTICLES

Consider non interacting fermions or bosons with the hamiltonian

$$H_0 = \sum_\alpha \epsilon_\alpha a^\dagger_\alpha a_\alpha. \tag{2.10}$$

The corresponding partition function is:

$$Z_0 = \mathrm{Tr} e^{-\beta H_0} = \mathrm{Tr} e^{-\beta \sum_\alpha \varepsilon_\alpha \hat{n}_\alpha} \tag{2.11}$$

where we have set $\hat{n}_\alpha \equiv a^\dagger_\alpha a_\alpha$. The operator $\hat{n}_\alpha$ counts the number of particles in the state α. The trace in (2.11) is over all the states in Fock space. A simple calculation gives:

$$Z_0 = \prod_\alpha \left(1 + e^{-\beta\varepsilon_\alpha}\right) \qquad \text{(fermions)}$$

$$Z_0 = \prod_\alpha \left(\frac{1}{1 - e^{-\beta\varepsilon_\alpha}}\right) \qquad \text{(bosons)} \tag{2.12}$$

The average value of $\hat{n}_\alpha$, which we shall denote temporarily by $\langle \hat{n}_\alpha \rangle$ is easily deduced from these partitions functions:

$$\langle \hat{n}_\alpha \rangle = -\frac{\partial \ln Z_0}{\partial \beta \varepsilon_\alpha} = \frac{1}{e^{\beta\varepsilon_\alpha} \pm 1}, \tag{2.13}$$

where the $+$ sign is for fermions, the $-$ sign for bosons.

2.4. THERMODYNAMICS OF RELATIVISTIC PARTICLES

Systems of relativistic particles will be our main concern in these lectures. In the absence of interactions, the energy of a particle with momentum p is and mass m is $\epsilon_p = \sqrt{p^2 + m^2}$. The distribution functions of these particles are then of the form:

$$\langle \hat{n}_p \rangle \equiv f(p) = \frac{1}{e^{(\epsilon_p - \mu)/T} \pm 1} \tag{2.14}$$

with the $+$ sign for fermions, the $-$ sign for bosons, and μ is the chemical potential (the chemical potential of an antiparticle is opposite to that of the corresponding particle). For one particle species, the *number density* n, the *energy density* ε and the *pressure* P are given respectively by:

$$n = \int \frac{d^3p}{(2\pi)^3} f(p), \quad \varepsilon = \int \frac{d^3p}{(2\pi)^3} \epsilon_p f(p), \quad P = \frac{1}{3} \int \frac{d^3p}{(2\pi)^3} \frac{p^2}{\epsilon_p} f(p). \tag{2.15}$$

For massless particles, and vanishing chemical potential, $\epsilon_p = p$, and the pressure is simply $P = \varepsilon/3$, while the entropy density is related to ε and P by $Ts = \varepsilon + P$. Note also the temperature dependences

$$n \sim T^3, \qquad \varepsilon \sim T^4, \qquad P \sim T^4, \tag{2.16}$$

which follow immediately from dimensional considerations.

3. Perturbative evaluation of the partition function

The direct evaluation of the partition function (2.2) is rarely possible as this requires a complete knowledge of the spectrum of the hamiltonian H. Various approximation schemes have therefore been devised to calculate $\mathcal{Z}$. We briefly describe one of them, the perturbative expansion, widely used in field theory. We assume that we can split the hamiltonian into $H = H_0 + H_1$ with $H_1 \ll H_0$, and that the spectrum of H_0 is known: $H_0|\psi_n^0\rangle = E_n^0|\psi_n^0\rangle$.

3.1. THE IMAGINARY TIME FORMALISM

We define the following "evolution operator":

$$U(\tau) = \exp(-\tau H) \equiv U_0(\tau)U_I(\tau), \tag{3.1}$$

where $U_0(\tau) \equiv \exp(-H_0\tau)$. The operator $U(\tau)$ is analogous to the familiar evolution operator of quantum mechanics, $\exp(-iHt)$. It differs from it solely by the replacement of the time t by $-i\tau$. Because of this analogy, we refer to τ as the "imaginary time" (τ is real). It has no direct physical interpretation: its role is to properly keep track of ordering of operators in the perturbative expansion (indeed in a "classical" approximation, where the operators are allowed to commute, that time dependence disappears). The operator $U_I(\tau) = \exp(\tau H_0)\exp(-\tau H)$ is called the *interaction representation* of U. We also define the interaction representation of the perturbation H_1:

$$H_1(\tau) = e^{\tau H_0} H_1 e^{-\tau H_0}, \tag{3.2}$$

and similarly for other operators. The following formula can be established for $\mathcal{Z}$:

$$\mathcal{Z} = \mathcal{Z}_0 \langle \mathrm{T} \exp\left\{ -\int_0^\beta d\tau H_1(\tau) \right\}\rangle_0, \tag{3.3}$$

where, for any operator $\mathcal{O}$,

$$\langle \mathcal{O} \rangle_0 \equiv \mathrm{Tr}\left(\frac{e^{-\beta H_0}}{\mathcal{Z}_0} \mathcal{O} \right), \tag{3.4}$$

and the symbol T implies an ordering of the operators on its right, from left to right in decreasing order of their time arguments. This formula provides a way to calculate $\mathcal{Z}$ as an expansion in powers of H_1. The operator H_1 is usually expressed in terms of creation and annihilation operators a and $a^\dagger$. Then the calculation of $\mathcal{Z}$ reduces to that of the expectation values of time ordered products of such operators. Whenever H_0 is a quadratic function of these operators, Wick's theorem applies, and all expectation values can be expressed in terms of single particle propagators. A diagrammatic expansion can be worked out following standard techniques [1, 3, 4]. The partition function can be written as $\mathcal{Z} = \mathcal{Z}_0 \exp(\Gamma_c)$ where Γ_c represents the sum of all connected diagrams.

3.2. FREE PROPAGATORS

The study of the free propagators will give us the opportunity to add a few remarks on the structure of perturbation theory at finite temperature. Let us consider a system whose unperturbed hamiltonian is the hamiltonian (2.10). We define time dependent creation and annihilation operators in the interaction picture:

$$\begin{aligned} a_\alpha^\dagger(\tau) &\equiv e^{\tau H_0} a_\alpha^\dagger e^{-\tau H_0} = e^{\epsilon_\alpha \tau} a_\alpha^\dagger \\ a_\alpha(\tau) &\equiv e^{\tau H_0} a_\alpha e^{-\tau H_0} = e^{-\epsilon_\alpha \tau} a_\alpha. \end{aligned} \tag{3.5}$$

The last equalities follow (for example) from the commutation relations:

$$[H_0, a_\alpha^\dagger] = \epsilon_\alpha a_\alpha^\dagger \qquad [H_0, a_\alpha] = -\epsilon_\alpha a_\alpha \tag{3.6}$$

which hold for bosons and fermions. The single particle propagator can then be obtained by a direct calculation:

$$\begin{aligned} G_\alpha(\tau_1 - \tau_2) &= \langle \mathrm{T} a_\alpha(\tau_1) a_\alpha^\dagger(\tau_2) \rangle_0 \\ &= e^{-\epsilon_\alpha(\tau_1 - \tau_2)} \left[\theta(\tau_1 - \tau_2)(1 \pm n_\alpha) \pm n_\alpha \theta(\tau_2 - \tau_1) \right], \end{aligned} \tag{3.7}$$

where:

$$n_\alpha \equiv \langle a_\alpha^\dagger a_\alpha \rangle_0 = \frac{1}{e^{\beta \epsilon_\alpha} \mp 1}, \tag{3.8}$$

and the upper (lower) sign is for bosons (fermions). One can verify on the expression (3.7) that, in the interval $[-\beta, \beta]$ where it is defined, $G_\alpha(\tau)$ is a periodic (boson) or antiperiodic (fermion) function of τ:

$$G_\alpha(\tau - \beta) = \pm G_\alpha(\tau) \qquad (0 \leq \tau \leq \beta). \tag{3.9}$$

To show this, the relation

$$e^{\beta \epsilon_\alpha} n_\alpha = 1 \pm n_\alpha \tag{3.10}$$

is useful. Because of its periodicity, the propagator can be represented by a Fourier series

$$G_\alpha(\tau) = \frac{1}{\beta}\sum_\nu e^{-i\omega_\nu \tau} G_\alpha(i\omega_\nu), \tag{3.11}$$

where the ω_ν's are called the Matsubara frequencies:

$$\begin{aligned} \omega_\nu &= 2\nu\pi/\beta && \text{bosons,} \\ \omega_\nu &= (2\nu+1)\pi/\beta && \text{fermions.} \end{aligned} \tag{3.12}$$

The inverse transform is given by

$$G(i\omega_\nu) = \int_0^\beta \mathrm{d}\tau\, e^{i\omega_\nu \tau} G(\tau) = \frac{1}{H_0 - i\omega_\nu}. \tag{3.13}$$

Using the property

$$\delta(\tau) = \frac{1}{\beta}\sum_\nu e^{-i\omega_\nu \tau} \qquad -\beta < \tau < \beta \tag{3.14}$$

and eq. (3.11), it is easily seen that $G(\tau)$ satisfies the differential equation

$$(\partial_\tau + H_0)\, G(\tau) = \delta(\tau), \tag{3.15}$$

which may be also verified directly from eq. (3.7). Alternatively, the single propagator at finite temperature may be obtained as the solution of this equation with periodic (bosons) or antiperiodic (fermions) boundary conditions.

Remark 1. The periodicity or antiperiodicity that we have uncovered on the explicit form of the unperturbed propagator is, in fact, a general property of the propagators of a many-body system in thermal equilibrium. It is a consequence of the commutation relations of the creation and annihilation operators and the cyclic invariance of the trace.

Remark 2. The statistical factor n_k can be obtained from the relation $n_k = \pm G(\tau = 0^-, k)$. In the approximation in which the sum over Matsubara frequencies is limited to the term $\omega_n = 0$, one gets from (3.11): $n_k \approx \frac{T}{\epsilon_k}$. Such an approximation corresponds to a "classical field" approximation valid when the occupation factors are large. This approximation, typically valid for long wavelength (small k), should not be confused with the classical limit reached when the thermal wavelength of the particles becomes small compared to their average separation distances. In this limit, the occupation of the single particle states becomes small, and the statistical factors can be approximated by their Boltzmann form:

$$\frac{1}{\mathrm{e}^{\beta(\epsilon_p - \mu)} \pm 1} \approx \mathrm{e}^{-\beta(\epsilon_p - \mu)} \ll 1 \qquad \mathrm{e}^{-\beta\mu} \gg 1. \tag{3.16}$$

3.3. PATH INTEGRALS

In field theory, it is often more convenient to use the formalism of path integrals. Let us recall that for one particle in one dimension the matrix element of the evolution operator can be written as

$$\langle q_2|e^{-iHt}|q_1\rangle = \int_{q(0)=q_1}^{q(t)=q_2} \mathcal{D}\left(q\left(t\right)\right) \quad e^{i\int_{t_1}^{t_2}\left(\frac{1}{2}m\dot{q}^2 - V(q)\right)dt}, \tag{3.17}$$

where q_1 and q_2 denote the positions of the particle at times 0 and t respectively. Changing $t \to -i\tau$, and taking the trace, one obtains the following formula for the partition function:

$$\mathcal{Z} = \mathrm{tr}\; e^{-\beta H} = \int_{q(\beta)=q(0)} \mathcal{D}(q) \exp\left\{-\int_0^\beta \left(\frac{1}{2}m\dot{q}^2 + V(q)\right)\right\}. \tag{3.18}$$

This expression immediately generalizes to the case of a scalar field, for which the Lagrangian is of the form:

$$\begin{aligned} \mathcal{L} &= \frac{1}{2}\partial_\mu\phi\partial^\mu\phi - \frac{m^2}{2}\phi^2 - V(\phi) \\ &= \frac{1}{2}(\partial_0\phi)^2 - \frac{1}{2}(\nabla\phi)^2 - \frac{m^2}{2}\phi^2 - V(\phi). \end{aligned} \tag{3.19}$$

Again, we replace t by $-i\tau$, $\partial_0 = \partial_t$ by $i\partial_\tau$, so that $(\partial_0\phi)^2 \to -(\partial_\tau\phi)^2$, and finally iS by $-S_E$. The partition function becomes then:

$$\mathcal{Z} = \int \mathcal{D}(\phi) \exp\left\{-\int_0^\beta d\tau \int d^3r \left(\frac{1}{2}(\partial_\tau\phi)^2 + \frac{1}{2}(\nabla\phi)^2 + \frac{m^2}{2}\phi^2 + V(\phi)\right)\right\}, \tag{3.20}$$

where the integral is over periodic fields: $\phi(0) = \phi(\beta)$. One recovers here, for the integration variables of the path integral, the periodicity properties which in the previous section were discussed for the propagator.
The perturbative expansion involves the propagator

$$\Delta(1,2) = \frac{\int \mathcal{D}\phi e^{-S_0}\phi(\tau_1)\phi(\tau_2)}{\int \mathcal{D}\phi e^{-S_0}} = \langle \mathrm{T}\phi(\tau_1)\phi(\tau_2)\rangle_0, \tag{3.21}$$

where S_0 is the action corresponding to the free Lagrangian, i.e. to the Lagrangian (3.19) without the interaction term $V(\phi)$. The propagator $\Delta(1,2)$ satisfies the differential equation

$$\left[-\partial_{\tau_1}^2 - \nabla_1^2 + m^2\right] \Delta(\tau_1 r_1; \tau_2 r_2) = \delta(\tau_1 - \tau_2)\delta(r_1 - r_2) \tag{3.22}$$

with *periodic* boundary conditions. The Fourier transform of Δ is given by (cp. with (3.13)):

$$\Delta(i\omega_\nu, k) = \frac{1}{\omega_k^2 - (i\omega_\nu)^2} = \frac{1}{2\omega_k}\left(\frac{1}{\omega_k - i\omega_n} + \frac{1}{\omega_k + i\omega_n}\right) \tag{3.23}$$

where $\omega_k^2 = k^2 + m^2$ and $\omega_\nu = 2\pi\nu/\beta$. As a simple exercise, one can show that:

$$\Delta(\tau) = \frac{1}{2\omega_k}\left\{(1 + N_k)\mathrm{e}^{-\omega_k|\tau|} + N_k \mathrm{e}^{\omega_k|\tau|}\right\}, \tag{3.24}$$

and verify that it is a periodic function of τ.

Remark 1. The partition function (3.20) may be viewed formally as a sum over classical field configurations in four dimensions, with particular boundary conditions in the (imaginary) time direction.

Remark 2. At high temperature, $\beta \to 0$, the time dependence of the fields play no role. The partition function becomes that of a classical field theory in three dimensions:

$$\mathcal{Z} = \int \mathcal{D}(\phi) \exp\left\{-\beta \int \mathrm{d}^3r \left(\frac{1}{2}(\nabla\phi)^2 + \frac{m^2}{2}\phi^2 + V(\phi)\right)\right\}. \tag{3.25}$$

Ignoring the time dependence of the fields amounts to take into account only the Matsubara frequency $i\omega_\nu = 0$, and this "dimensional reduction" is related to the classical field approximation mentioned earlier.

Remark 3. Note the Euclidean metric in (3.20). Since the integrand is the exponential of a negative definite quantity, it is well suited to numerical evaluations, using for instance Monte Carlo techniques.

4. Thermodynamics of the scalar field

In this section, as a warm up for the more complicated case of QCD, we shall discuss some typical problems encountered in the perturbative calculations of the thermodynamic potential.

We consider a scalar field ϕ whose dynamics is described by the hamiltonian

$$H = \frac{\pi^2}{2} + \tfrac{1}{2}(\nabla\phi)^2 + \tfrac{1}{2}m^2\phi^2 + \frac{\lambda}{4!}\phi^4 \equiv H_0 + H_1, \tag{4.1}$$

where $H_1 \equiv \lambda\phi^4/4!$. We want to calculate the thermodynamic potential as a power series of the coupling λ and we write:

$$\begin{aligned}\Omega &= -\frac{1}{\beta}\ln Z \\ &= \Omega_0 + \Omega_1 + \Omega_2 + \cdots\end{aligned} \tag{4.2}$$

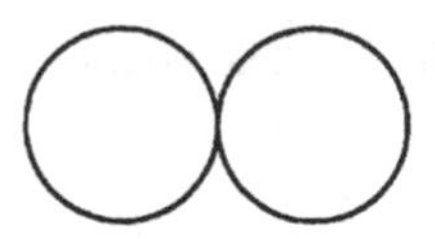

Figure 1. The lowest order correction to the thermodynamical potential in ϕ^4 theory.

where Ω_0 is the thermodynamic potential for the free field. This may be obtained directly in the operator formalism, starting with the unperturbed hamiltonian written in terms of the normal modes, $H_0 = \sum_k \omega_k (a_k^\dagger a_k + \frac{1}{2})$, where $\omega_k = \sqrt{k^2+m^2}$ is the energy of the "scalar particle" with momentum k. One gets:

$$\Omega_0 = \tfrac{1}{2}\sum_k \omega_k + \frac{1}{\beta}\sum_k \ln\left(1 - e^{-\beta\omega_k}\right). \tag{4.3}$$

4.1. CANCELLATION OF ULTRAVIOLET DIVERGENCES

One of the difficulties of field theoretical calculations is the occurence of ultraviolet divergences in the sums over modes with arbitrarily small wavelengths. A first example of such divergences is provided by the expresssion (4.3) of Ω_0, in which the zero temperature contribution

$$\sum_k \omega_k = \int^\Lambda \frac{d^3k}{(2\pi)^3}\sqrt{k^2+m^2} \tag{4.4}$$

has no limit when $\Lambda \to \infty$. In contrast, the finite temperature correction is finite. At this level however, this divergence is harmless: it can be absorbed in the redefinition of the vacuum energy, and simply discarded.
Consider now the first order correction to Ω, given by the diagram in fig. 4.1. We have:

$$\begin{aligned}\Omega_1 &= \frac{\lambda}{4!}\langle\phi^4\rangle_0 = \frac{\lambda}{4!}\cdot 3\cdot[\Delta(x=\tau=0)]^2 \\ &= \frac{\lambda}{8}\left\{\left(\sum_k \frac{1}{2\omega_k}\right)^2 + \left(\sum_k \frac{n_k}{\omega_k}\right)^2 + 2\left(\sum_k \frac{1}{2\omega_k}\right)\left(\sum_k \frac{n_k}{\omega_k}\right)\right\},\end{aligned} \tag{4.5}$$

where again we have separated the zero temperature contribution, from finite temperature corrections (note that the possibility to achieve such a separation is limited to lowest orders). The zero temperature piece in Ω_1 (the first term in eq. (4.5)) is diverging. The second term is a temperature dependent correction and is finite. But the last term is potentially dangerous, as it is diverging *and* temperature dependent. However, it is easy to show that this divergence is eliminated

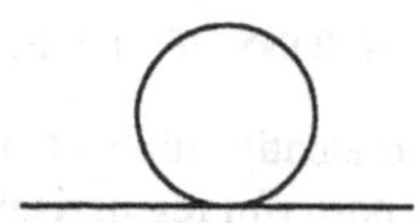

Figure 2. Lowest order correction to the self-energy in scalar ϕ^4 theory.

by a standard mass renormalization. To see this explicitly, let us write the Dyson equation for the propagator

$$D^{-1} = \Delta^{-1} + \Sigma, \tag{4.6}$$

where the self-energy Σ is given by the simple "tadpole" diagram of fig. 4.1:

$$\Sigma = \frac{\lambda}{2}\langle\phi^2\rangle = \frac{\lambda}{2}\Delta(0) = \frac{\lambda}{2}\sum_k \frac{1+2n_k}{2\omega_k}. \tag{4.7}$$

To cancel the divergence in Σ, one adds a counter-term $\frac{1}{2}\delta m^2\phi^2$ in the hamiltonian, and chooses δm^2 so that:

$$\frac{\lambda}{2}\sum_k \frac{1}{2\omega_k} + \delta m^2 = 0. \tag{4.8}$$

With this choice, there is no correction to the mass of the scalar particle at this order of approximation (and at $T = 0$).
We note now that the mass counter-term should be also added to the hamiltonian in the calculation of the thermodynamic potential to order λ (δm^2 is formally of order λ). Its contribution to Ω is:

$$\frac{1}{2}\delta m^2\Delta(0) = \frac{1}{2}\left(\frac{-\lambda}{2}\sum_k \frac{1}{2\omega_k}\right)\left(\sum_k \frac{1+2n_k}{2\omega_k}\right). \tag{4.9}$$

Adding this contribution to Ω_1 obtained in eq. (4.5), one finally gets:

$$\Omega_1 = \frac{\lambda}{8}\left(\sum_k \frac{n_k}{\omega_k}\right)^2 - \frac{\lambda}{8}\left(\sum_k \frac{1}{2\omega_k}\right)^2, \tag{4.10}$$

so that the finite temperature correction is indeed finite. The vacuum part is still diverging; it takes therefore a further subtraction to make the thermodynamic potential finite. The final result for the pressure, to first order in λ may then be written [6]:

$$\begin{aligned} P = -\frac{\Omega}{V} &= -\frac{1}{\beta}\int \frac{d^3k}{(2\pi)^3}\ln(1-e^{-\beta\omega_k}) - \frac{\lambda}{8}\left(\int \frac{d^3k}{(2\pi)^3}\frac{n_k}{\omega_k}\right)^2 \\ &\underset{m=0}{=} T^4\left(\frac{\pi^2}{90} - \frac{\lambda}{1152}\right). \end{aligned} \tag{4.11}$$

4.2. HIGHER ORDER CALCULATIONS OF THE PRESSURE

In higher order calculations the identification of ultraviolet divergences can become quite involved, and more powerful regularisations of the diverging integrals become necessary. We shall now introduce briefly dimensional regularisation, by working out a simple example. This is a powerful technique which can also be used in gauge theories [10].
Consider then the zero temperature contribution of self-energy (4.7):

$$\Sigma = \frac{\lambda_0}{2} \int \frac{d^d p}{(2\pi)^d} \frac{1}{p^2 + m_0^2}, \tag{4.12}$$

where we consider the dimension d as a free parameter (the physical result corresponds to $d = 4$). The calculation of the integral proceeds then as follows:

$$\int \frac{d^d p}{(2\pi)^d} \frac{1}{p^2 + m_0^2} = \int \frac{d^d p}{(2\pi)^d} \int_0^\infty ds\, e^{-s(p^2+m_0^2)} \tag{4.13}$$

$$= \int_0^\infty ds\, e^{-s\, m_0^2} \frac{1}{(2\pi)^d} \left(\frac{\pi}{s}\right)^{d/2} \tag{4.14}$$

$$= \frac{(m_0^2)^{\frac{d}{2}-1}}{(4\pi)^{d/2}} \Gamma(1 - d/2). \tag{4.15}$$

The ultraviolet divergence at $d = 4$ is reflected in the divergence of the Γ function, which appears as a pole in $\epsilon = 4 - d$:

$$\Gamma(\epsilon - 1) = -\frac{1}{\epsilon} - 1 + \gamma, \tag{4.16}$$

where $\gamma \approx 0.5772$ is Euler's constant.
We note now that when $d = 4$, the integral has dimension m^2, and so does the self-energy Σ because λ_0 is then dimensionless. When $d \neq 4$, λ_0 acquires mass dimension $4-d$. It is then convenient to introduced a dimensionless coupling constant λ, such that $\lambda_0 = \lambda\, \mu^{4-d}$, where μ, referred to as the renormalisation scale, is an arbitrary parameter with the dimension of a mass. After this substitution, the self-energy becomes

$$\Sigma = \frac{\lambda\, \mu^{4-d}}{2} \frac{(m^2)^{\frac{d}{2}-1}}{(4\pi)^{d/2}} \Gamma\left(1 - \frac{d}{2}\right), \tag{4.17}$$

and its expansion for small ε reads:

$$\Sigma = -\frac{\lambda}{2} \frac{m_0^2}{16\pi^2} \left(\frac{2}{\epsilon} + \ln \frac{4\pi\mu^2 e^{-\gamma}}{m_0^2} + 1\right) \tag{4.18}$$

The quantity $m_0^2 + \Sigma$, to be identified later with the physical mass, can be made finite by defining a renormalized mass m^2 by:

$$m_0^2 = m^2 \left(1 + \frac{\lambda}{16\pi^2} \frac{1}{2\epsilon}\right) \tag{4.19}$$

One obtains then:

$$m_0^2 + \Sigma = m^2 + \frac{\lambda}{2} \frac{m^2}{16\pi^2} \left(\ln \frac{m^2}{4\pi\mu^2 e^{-\gamma}} - 1\right) \tag{4.20}$$

The procedure by which one subtract from the bare parameters, here the bare mass m_0, the divergent contribution in $1/\epsilon$ is called minimal subtraction.
We turn now to the finite temperature contribution, whose expansion for $m \ll T$ reads:

$$\int \frac{d^3k}{(2\pi)^3} \frac{n(\varepsilon_k)}{\varepsilon_k} \approx \frac{T^2}{12} - \frac{mT}{4\pi} - \frac{m^2}{16\pi^2} \left(\ln \frac{m^2}{T^2} + 2\gamma - 2\ln 4\pi - 1\right) \tag{4.21}$$

By combining the two contributions, one observe that the term in $\log m^2$ disappears, leaving a contribution in $\log \mu/T$. The appearance of such logarithms is a general feature in finite temperature calculations in high order.
In dimensional regularisation, the mass counterterm vanishes if $m^2 = 0$. In such a scheme, the ultraviolet divergences associated to mass renormalisation are automatically cancelled and we can work directly in the massless limit. For the present purpose which is to illustrate the cancellation of temperature dependent ultraviolet divergences, this is a useful simplification. We shall deal later with the infrared divergences which occur in the massless case.
We shall now discuss the results of a calculation of the pressure of a massless scalar field up to 3-loop order. Our goal is to illustrate the role of the renormalization scale μ, and explain why a natural choice is $\mu \simeq 2\pi T$ which makes the effective coupling temperature dependent.
The final result for the pressure after renormalisation reads [11]:

$$P = \frac{\pi^2 T^4}{9} \left\{ \frac{1}{10} - \frac{1}{8} \frac{\lambda}{16\pi^2} + \frac{1}{8} \left[3 \ln \frac{\mu}{4\pi T} + \frac{31}{15} + C \right] \left(\frac{\lambda}{16\pi^2} \right)^2 \right\}, \tag{4.22}$$

where C is a numerical constant.
Note the apparent dependence on μ. In fact to order λ^2, P is *formally* independent of μ. This is verified using the renormalization group equation satisfied by the renormalized coupling $g(\mu)$:

$$\mu \frac{d\lambda}{d\mu} = 3 \frac{\lambda^2}{16\pi^2} + \mathcal{O}\left(\lambda^3\right). \tag{4.23}$$

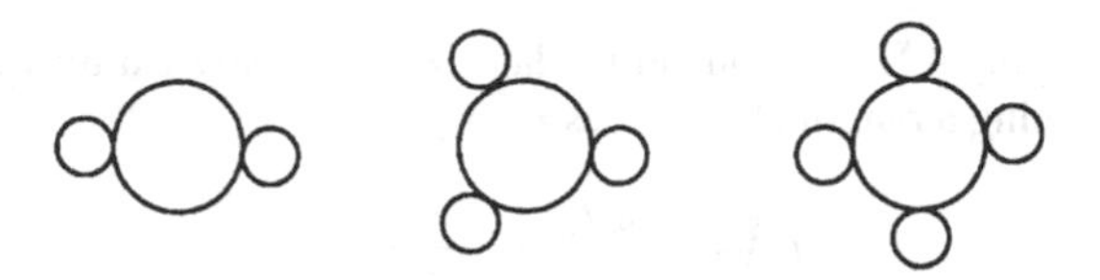

Figure 3. Simple "ring" diagrams for ϕ^4 theory.

We have

$$\mu\frac{dP}{d\mu}=\frac{\pi^2T^4}{72}\mu\frac{d}{d\mu}\left(\frac{\lambda}{16\pi^2}\right)+\frac{\pi^2T^4}{72}3\left(\frac{\lambda}{16\pi^2}\right)^2+\cdots=\mathcal{O}\left(\lambda^3\right) \quad (4.24)$$

Note also the occurence of terms in $\ln\frac{\mu}{4\pi T}$. These terms can be large if μ is very different from T. It is then "natural" to choose $\mu\simeq 2\pi T$. With this choice $\lambda(\mu)$ becomes effectively a function of the temperature. In the case of QCD where the β-function is negative, this leads to the expectation that the effective coupling becomes small at large temperature.

4.3. INFRARED DIVERGENCES. A SIMPLE RESUMMATION

The terms $\propto\lambda$ and λ^2 do not account completely for the weak coupling expansion of the pressure, which reads rather

$$P=P_0+\lambda P_2+\lambda^{3/2}P_3+\lambda^2P_4+\ldots \quad (4.25)$$

The presence of the term $\lambda^{3/2}$ which signals a breakdown of the naive perturbative expansion (which would lead to a series in powers of λ) finds its origin in the resummation of an infinite class of Feynman diagrams. Resummation is made necessary by the occurence of infrared divergences which appear in the calculation of individual diagrams.

It is not difficult to isolate diagrams in which such divergences occur. Consider the diagram of order λ^r in the family of diagrams displayed in Fig.4.3 (often referred to as ring diagrams). This contribution contains the integral

$$\frac{1}{\beta}\sum_n\int\frac{d^3p}{(2\pi)^3}\left(\frac{1}{\omega_p^2-(i\omega_n)^2}\right)^r. \quad (4.26)$$

The term with $n=0$ diverges as $1/m^r$ when $m\to 0$.

The contribution to the thermodynamic potential of the sum of all the ring diagrams is however finite. We have:

$$\Omega_R=\frac{1}{2\beta}\sum_n\int\frac{d^3p}{(2\pi)^3}\left[\ln\left(1+\Sigma\Delta\left(i\omega_n,\mathbf{p}\right)\right)-\Sigma\Delta\left(i\omega_n,\mathbf{p}\right)\right], \quad (4.27)$$

where the self energy Σ is independent of the loop momentum:

$$\Sigma = \frac{g^2}{2} \int \frac{\mathrm{d}^3k}{(2\pi)^3} \frac{n_k}{\omega_k} = \frac{g^2T^2}{24} \equiv m_T^2, \tag{4.28}$$

and $\omega_k = \sqrt{k^2 + m^2}$. Note that this expression for the self-energy includes the contribution of the mass counter-term whose role here is simply to cancel the vacuum contribution (see eqs. (4.7) and (4.8)). By inserting this result in the formula (4), and keeping only the dominant contribution to Ω_R, i.e. that of the mode $\omega_n = 0$, or in other words the most divergent individual contributions, one finds:

$$\begin{aligned}
\Omega_R^{\nu=0} &= \frac{1}{2\beta} \int \frac{\mathrm{d}^3k}{(2\pi)^3} \left\{ \ln\left(1 + \frac{m_T^2}{k^2}\right) - \frac{m_T^2}{k^2} \right\} \\
&= \frac{Tm_T^3}{2} \int \frac{\mathrm{d}^3x}{(2\pi)^3} \left\{ \ln\left(1 + \frac{1}{x^2}\right) - \frac{1}{x^2} \right\} \\
&= -\frac{Tm_T^3}{12\pi}.
\end{aligned} \tag{4.29}$$

a finite result indeed.

Hidden in this result, is the fact that the thermal fluctuations have changed the propagator of the field fluctuations into that of a massive particle, i.e., the propagator of the massless scalar field at $T = 0$, $1/k^2$, is changed into $1/\left(k^2 + m_T^2\right)$, with $m_T^2 = \lambda T^2/24$. The physical mechanism at work here, namely the generation of mass by thermal fluctuations, is very much similar to the phenomenon of Debye screening of electric charges in the electron gas. The occurence of infrared divergences is the manifestation, at the computational level, of collective phenomena which are not properly accounted for by strict perturbation theory: even though the coupling is weak, many particles are involved. Such non-perturbative phenomena can, however, be taken into account via appropriate resummations.

The dominant contribution at order $\lambda^{3/2}$ is coming entirely from the contributions of diagrams in which only $\omega_n = 0$ is kept. This may be summarized in terms of an effective theory for the static fields whose Lagrangian is of the form

$$\mathcal{L} = \frac{1}{2} (\Delta\phi)^2 + \frac{1}{2} m_T^2 \phi^2 + \frac{\lambda}{4} \phi^2 + \delta\mathcal{L}. \tag{4.30}$$

One may view the Lagrangian above as that obtained after "integrating out" modes with $n \neq 0$. The effects of these non static fields enter the parameters of the effective Lagrangian (for instance non static modes determine the value of m_T), and vertices included in $\delta\mathcal{L}$. The Lagrangian (4.30) is appropriate for the calculation of long wavelength properties $(k \lesssim m_T)$. For large k, for instance $k \sim T$, the mass term can be treated as a small perturbative correction. However for momenta $k \sim m_T$ the mass correction must be consistently included in the propagator.

As a preparation to the more elaborate treatment for QCD that will be presented in the lecture by A. Rebhan, we show how we can accomodate the effect of thermal fluctuations to all orders in the coupling by using a simple self-consistent approximation which is analogous to the Hartree approximation of the non relativistic many-body problem, of the large N approximation of field theory [12]. This will also provide us with an explicit treatment of the sum over zero point energies in a case where the mass depends on the temperature, so that the zero point energies cannot be simply subtracted as we did earlier.

5. A simple self-consistent approximation

The calculation of the scalar self-energy that we have performed earlier may be viewed as the beginning of the expansion of the complete solution of a self-consistent equation (gap equation) of the form:

$$m^2 = \frac{\lambda}{2} I(m), \tag{5.1}$$

where

$$I(m) = \sum_n \int \frac{d^3k}{(2\pi)^3} \frac{1}{\omega_n^2 + k^2 + m^2}. \tag{5.2}$$

The loop integral $I(m)$ can be written as the sum of a vacuum piece $I_0(m)$ and a finite temperature piece $I_T(m)$ such that, at fixed m, $I_T(m) \to 0$ as $T \to 0$. We use dimensional regularisation to control the ultraviolet divergences present in I_0, from which it follows in particular that $I_0(0) = 0$. Explicitly, one has:

$$\mu^\epsilon I(m^2) = -\frac{m^2}{16\pi^2}\left(\frac{2}{\epsilon} + \ln\frac{\bar{\mu}^2}{m^2} + 1\right) + I_T(m^2) + O(\epsilon), \tag{5.3}$$

with

$$I_T(m^2) = \int \frac{d^3k}{(2\pi)^3} \frac{n(\varepsilon_k)}{\varepsilon_k}, \tag{5.4}$$

and $\varepsilon_k \equiv (k^2 + m^2)^{1/2}$. In Eq. (5.3), μ is the scale of dimensional regularisation, $\epsilon = 4 - d$, with d the number of space-time dimensions, and $\bar{\mu}^2 = 4\pi e^{-\gamma}\mu^2$. We use the modified minimal subtraction scheme ($\overline{\text{MS}}$) and define a dimensionless renormalized coupling λ by:

$$\frac{1}{\lambda} = \frac{1}{\lambda_0 \mu^{-\epsilon}} + \frac{1}{16\pi^2 \epsilon}. \tag{5.5}$$

Note that the coefficient of the $1/\epsilon$ term differs by a factor 3 from that of conventional perturbative renormalization (cp. Eq. (4.23)). The same renormalized

coupling constant is obtained in the large N expansion for a N-component scalar field, where one ignores, at a given order in λ, diagrams which are formally down by powers of $1/N$.
When expressed in terms of the renormalized coupling, the gap equation becomes free of ultraviolet divergences. It reads:

$$m^2 = \frac{\lambda}{2}\int \frac{d^3k}{(2\pi)^3}\frac{n(\varepsilon_k)}{\varepsilon_k} + \frac{\lambda m^2}{32\pi^2}\left(\log\frac{m^2}{\bar{\mu}^2} - 1\right), \tag{5.6}$$

and its solution is independent of μ.
The thermodynamic potential, or equivalently the pressure, can be written:

$$-P = \frac{\Omega}{V} = \frac{1}{2}\int \frac{d^3k}{(2\pi)^3}\varepsilon_k + \frac{1}{\beta}\int \frac{d^3k}{(2\pi)^3}\ln(1 - e^{-\beta\varepsilon_k}) - \frac{m^4}{2\lambda}, \tag{5.7}$$

where $\varepsilon_k^2 \equiv k^2 + m^2$. This differs from the pressure of an ideal gas of massive bosons, given by:

$$P = -\int \frac{d^3k}{(2\pi)^3}\left\{T\log(1 - e^{-\epsilon_k/T}) + \frac{\epsilon_k}{2}\right\}, \tag{5.8}$$

by the term m^4/λ which corrects for the double counting of the interactions included in the thermal mass. In the present approximation, the renormalization (5.5) of the coupling constant is sufficient to also make the pressure (5.7) finite.
As a final remark we note that the entropy in this approximation takes the form:

$$\mathcal{S} = \int \frac{d^3k}{(2\pi)^3}\left\{(1 + n_k)\log(1 + n_k) - n_k \log n_k\right\}. \tag{5.9}$$

This formula shows that, in the present approximation, the entropy of the interacting scalar gas is formally identical to the entropy of an ideal gas of massive bosons, with mass m. (The pressure does not have this simple property.) Note that in relating the pressure to the entropy, $\mathcal{S} = dP/dT$, one should take into account that m depends on the temperature.

6. Effective theories for the quark-gluon plasma

We turn now to the quark-gluon plasma and analyze the various scales and degrees of freedom which are relevant in the weak coupling regime. We show that there is a hierarchy of scales controlled by powers of the gauge coupling g. We focus in these lectures on two particular momentum scales, the 'hard' one which is that of the plasma particles with momenta $k \sim T$, and the 'soft' one with $k \sim gT$ at which collective phenomena develop. We shall be in particular interested in the effective theory obtained when the hard degrees of freedom are 'integrated out'. This effective theory describes long wavelength, low frequency collective phenomena [7]. As can be seen in the lecture by A. Rebhan, getting a complete description of the dynamics of the collective excitations turns out to be important also for the calculation of the equilibrium properties of the quark-gluon plasma.

6.1. SCALES AND DEGREES OF FREEDOM IN ULTRARELATIVISTIC PLASMAS

A property of QCD which is essential in the present discussion is that of asymptotic freedom, according to which the coupling constant depends on the renormalization scale $\bar{\mu}$ as

$$\alpha_s(\bar{\mu}) \equiv \frac{g^2}{4\pi} \propto \frac{1}{\ln(\bar{\mu}/\Lambda_{QCD})}. \tag{6.1}$$

At high temperature, the natural scale is $\bar{\mu} = 2\pi T$, so that the coupling becomes weak when $2\pi T \gg \Lambda_{QCD}$. At extremely high temperature the interactions become negligible and hadronic matter turns into an ideal gas of quarks and gluons: this is the quark-gluon plasma.
In the absence of interactions, the plasma particles are distributed in momentum space according to the Bose-Einstein or Fermi-Dirac distributions:

$$N_k = \frac{1}{e^{\beta\varepsilon_k} - 1}, \qquad n_k = \frac{1}{e^{\beta\varepsilon_k} + 1}, \tag{6.2}$$

where $\varepsilon_k = k \equiv |\mathbf{k}|$ (massless particles), $\beta \equiv 1/T$, and chemical potentials are assumed to vanish. In such a system, the particle density n is determined by the temperature: $n \propto T^3$ (see eq. (2.16)). Accordingly, the mean interparticle distance $n^{-1/3} \sim 1/T$ is of the same order as the thermal wavelength $\lambda_T = 1/k$ of a typical particle in the thermal bath for which $k \sim T$. Thus the particles of an ultrarelativistic plasma are quantum degrees of freedom for which in particular the Pauli principle can never be ignored.
In the weak coupling regime ($g \ll 1$), the interactions do not alter significantly the picture. The hard degrees of freedom, i.e. the plasma particles, remain the dominant degrees of freedom and since the coupling to gauge fields occurs typically through covariant derivatives, $D_x = \partial_x + igA(x)$, the effect of interactions on particle motion is a small perturbation unless the fields are very large, i.e., unless $A \sim T/g$, where g is the gauge coupling: only then do we have $\partial_X \sim T \sim gA$, where ∂_X is a space-time gradient. We should note here that we rely on considerations, based on the magnitude of the gauge fields, which depend on the choice of a gauge. What is meant is that there exists a large class of gauge choices for which they are valid. And one can verify a posteriori that within such a class, the final results are gauge invariant.
Considering now more generally the effects of the interactions, we note that these depend both on the strength of the gauge fields and on the wavelength of the

modes under study. A measure of the strength of the gauge fields in typical situations is obtained from the magnitude of their thermal fluctuations, that is $\bar{A} \equiv \sqrt{\langle A^2(t,\mathbf{x})\rangle}$. In equilibrium $\langle A^2(t,\mathbf{x})\rangle$ is independent of t and $\mathbf{x}$ and given by $\langle A^2\rangle = G(t=0,\mathbf{x}=\mathbf{0})$ where $G(t,\mathbf{x})$ is the gauge field propagator. In the non interacting case we have (with $\varepsilon_k = k$):

$$\langle A^2\rangle = \int \frac{\mathrm{d}^3k}{(2\pi)^3}\frac{1}{2\varepsilon_k}(1+2N_k). \tag{6.3}$$

Here we shall use this formula also in the interacting case, assuming that the effects of the interactions can be accounted for simply by a change of ε_k. We shall also ignore the (divergent) contribution of the vacuum fluctuations (the term independent of the temperature in (6.3)).

For the plasma particles $\varepsilon_k = k \sim T$ and $\langle A^2\rangle_T \sim T^2$. The associated electric (or magnetic) field fluctuations are $\langle E^2\rangle_T \sim \langle(\partial A)^2\rangle_T \sim k^2\langle A^2\rangle_T \sim T^4$ and are a dominant contribution to the plasma energy density. As already mentioned, these short wavelength, or *hard*, gauge field fluctuations produce a small perturbation on the motion of a plasma particle. However, this is not so for an excitation at the momentum scale $k \sim gT$, since then the two terms in the covariant derivative ∂_X and $g\bar{A}_T$ become comparable. That is, the properties of an excitation with momentum gT are expected to be non perturbatively renormalized by the hard thermal fluctuations. And indeed, the scale gT is that at which collective phenomena develop. The emergence of the Debye screening mass $m_D \sim gT$ is one of the simplest examples of such phenomena.

Let us now consider the fluctuations at the *soft* scale $gT \ll T$. These fluctuations can be accurately described by classical fields. In fact the associated occupation numbers N_k are large, and accordingly one can replace N_k by T/ε_k in (6.3). Introducing an upper cut-off gT in the momentum integral, one then gets:

$$\langle A^2\rangle_{gT} \sim \int^{gT} \mathrm{d}^3k\, \frac{T}{k^2} \sim gT^2. \tag{6.4}$$

Thus $\bar{A}_{gT} \sim \sqrt{g}T$ so that $g\bar{A}_{gT} \sim g^{3/2}T$ is still of higher order than the kinetic term $\partial_X \sim gT$. In that sense the soft modes with $k \sim gT$ are still perturbative, i.e. their self-interactions can be ignored in a first approximation (if $g\bar{A} \ll \partial_X$, $gA^2 \ll \partial_X A$ and the non linear terms in the field strength tensor are negligible). Note however that they generate contributions to physical observables which are not analytic in g^2, as shown by the example of the order g^3 contribution to the energy density of the plasma (see the corresponding calculation for the scalar case in section 4.3):

$$\epsilon^{(3)} \sim \int_0^{\omega_{pl}} \mathrm{d}^3k\, \omega_{pl}\, \frac{1}{\mathrm{e}^{\omega_{pl}/T}-1} \sim \omega_{pl}^3\, \omega_{pl}\, \frac{T}{\omega_{pl}} \sim g^3T^4, \tag{6.5}$$

where $\omega_{pl} \sim gT$ is the typical frequency of a collective mode.

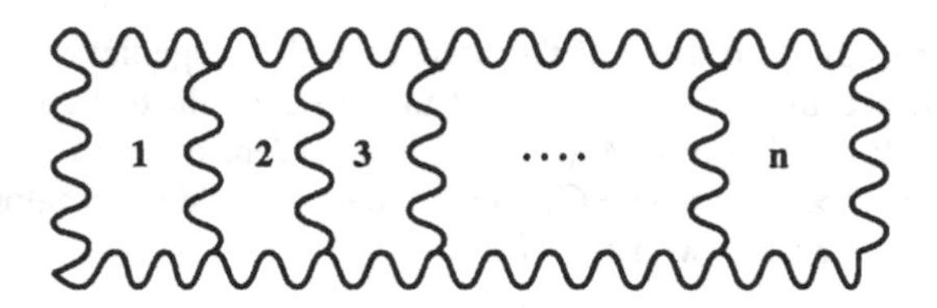

Figure 4. Example of a multiloop diagram which is infrared divergent

Moving down to a lower momentum scale, one meets the contribution of the unscreened magnetic fluctuations which play a dominant role for $k \sim g^2T$. At that scale, to be referred to as the *ultrasoft* scale, it becomes necessary to distinguish the electric and the magnetic sectors (which provide comparable contributions at the scale gT). The electric fluctuations are damped by the Debye screening mass ($\varepsilon_k^2 = k^2 + m_D^2 \approx m_D^2$ when $k \sim g^2T$) and their contribution is negligible, of order g^4T^2. However, because of the absence of static screening in the magnetic sector, we have here $\varepsilon_k \sim k$ and

$$\langle A^2 \rangle_{g^2T} \sim T \int_0^{g^2T} \mathrm{d}^3k \frac{1}{k^2} \sim g^2T^2, \tag{6.6}$$

so that $g\bar{A}_{g^2T} \sim g^2T$ is now of the same order as the ultrasoft derivative $\partial_X \sim g^2T$: the fluctuations are no longer perturbative. This is the origin of the breakdown of perturbation theory in high temperature QCD.
To appreciate the difficulty from another perspective, let us observe that the dominant contribution to the fluctuations at scale g^2T comes from the zero Matsubara frequency:

$$\langle A^2 \rangle_{g^2T} = T \sum_n \int_0^{g^2T} \mathrm{d}^3k \, \frac{1}{\omega_n^2 + k^2} \sim T \int_0^{g^2T} \mathrm{d}^3k \, \frac{1}{k^2}. \tag{6.7}$$

Thus the fluctuations that we are discussing are those of a three dimensional theory of static fields. Simple power counting arguments [13] reveal that the most divergent n-loop diagrams contributing to the pressure are of order $g^6T^4 \ln(T/\mu)$ if $n = 4$ and of the order $g^6T^4 \left(g^2T/\mu\right)^{n-4}$ if $n > 4$, where μ is an infrared cut-off. Accordingly, if $\mu \sim g^2T$, all the diagrams with $n \geq 4$ loops contribute to the same order, namely to $\mathcal{O}(g^6)$. In other words, the correction of $\mathcal{O}(g^6)$ to the pressure cannot be computed in perturbation theory. In order to calculate the effects of the magnetic fluctuations, one needs to use non perturbative techniques such as those based on dimensional reduction and effective theory [14, 15].

6.2. EFFECTIVE THEORY AT SCALE gT

Having identified the main scales and degrees of freedom, one may proceed and construct appropriate effective theories at the various scales, obtained by eliminating the degrees of freedom at higher scales. We shall consider here the effective theory at the scale gT obtained by eliminating the hard degrees of freedom with momenta $k \sim T$.

The soft excitations at the scale gT can be described in terms of *average fields* [7]. Such average fields develop for example when the system is exposed to an external perturbation, such as an external electromagnetic current. In QED, we can summarize the effective theory for the soft modes by the equations of motion:

$$\partial_\mu F^{\mu\nu} = j^\nu_{ind} + j^\nu_{ext} \tag{6.8}$$

that is, Maxwell equations with a source term composed of the external perturbation j^ν_{ext}, and an extra contribution j^ν_{ind} which we refer to as the *induced current*. The induced current is generated by the collective motion of the charged particles, i.e. the hard degrees of freedom. It may be regarded itself as a functional of the average gauge fields and, once this functional is known, the equations above constitute a closed system of equations for the soft fields.

The main problem is to calculate j_{ind}. This is done by considering the dynamics of the hard particles in the background of the soft fields. For QED, the induced current can be obtained using linear response theory. To be more specific, consider as an example a system of charged particles on which is acting a perturbation of the form $\int \mathrm{d}x\, j_\mu(x) A^\mu(x)$, where $j_\mu(x)$ is the current operator and $A^\mu(x)$ some applied gauge potential. Linear response theory leads to the following relation for the induced current:

$$j_\mu^{ind} = \int \mathrm{d}^4y\, \Pi^R_{\mu\nu}(x-y) A^\nu(y), \tag{6.9}$$

where the (retarded) response function $\Pi^R_{\mu\nu}(x-y)$ is also referred to as the polarization operator. Note that in (6.9), the expectation value is taken in the equilibrium state. Thus, within linear response, the task of calculating the basic ingredients of the effective theory for soft modes reduces to that of calculating appropriate equilibrium correlation functions.

In fact we shall need the response function only in the weak coupling regime, and for particular kinematic conditions which allow for important simplifications. In leading order in weak coupling, the polarization tensor is given by the one-loop approximation. In the kinematic regime of interest, where the incoming momentum is soft while the loop momentum is hard, we can write $\Pi(\omega, p) = g^2T^2 f(\omega/p, p/T)$ with f a dimensionless function, and in leading order in $p/T \sim g$, Π is of the form $g^2T^2 f(\omega/p)$. This particular contribution of the one-loop polarization tensor is an example of what has been called a "hard thermal loop"

[16, 17, 18, 19]; for photons in QED, this is the only one. It turns out that this hard thermal loop can be obtained from simple *kinetic theory*.
In non Abelian theory, linear response is not sufficient: constraints due to gauge symmetry force us to take into account specific non linear effects and a more complicated formalism needs to be worked out. Still, simple kinetic equations can be obtained in this case also, but in contrast to QED, the resulting induced current is a non linear functional of the gauge fields [7]. As a result, it generates an infinite number of "hard thermal loops".

References

1. A.A. Abrikosov, L.P. Gorkov and I.E. Dzyaloshinskii, *Methods of Quantum Field Theory in Statistical Physics*, (Dover, New-York, 1963).
2. L. Kadanov and G. Baym, *Quantum Statistical mechanics*, (Benjamin/Cummings, London,1962).
3. A. Fetter and J.D. Walecka, *Quantum Theory of Many Particle Systems*, (McGraw Hill, New-York, 1971).
4. J.P. Blaizot and G. Ripka, *Quantum Theory of Finite Systems*, (MIT Press, Cambridge, 1986).
5. M. Le Bellac, *Thermal field theory*, Cambridge Monographs in Mathematical Physics, (Cambridge University Press, 1996).
6. J.I. Kapusta, *Finite temperature field theory*, Cambridge Monographs in Mathematical Physics, (Cambridge University Press, 1989).
7. J. Blaizot and E. Iancu, "The quark-gluon plasma: Collective dynamics and hard thermal loops," Phys. Repts.**359** (2002) 355 , hep-ph/0101103.
8. J.P. Blaizot, *"QCD at finite temperature"*, in "Probing the Standard Model of Particle Interactions", Les Houches, Session LXVIII, 1997, ed. by R. Gupta et al. (Elsevier, Amsterdam, 1999).
9. J. P. Blaizot, "The quark-gluon plasma and nuclear collisions at high energy," *Lecture given at Les Houches Summer School on Theoretical Physics, Session 66: Trends in Nuclear Physics, 100 Years Later, Les Houches, France, 30 Jul - 30 Aug 1996.*
10. J. Zinn-Justin, *Quantum Field Theory and Critical Phenomena*, Clarendon Press (Oxford 1989, third ed. 1996).
11. P. Arnold and C. Zhai, Phys. Rev. **D50** (1994) 7603.
12. J. P. Blaizot, E. Iancu and A. Rebhan, "Approximately self-consistent resummations for the thermodynamics of the quark-gluon plasma. I: Entropy and density," Phys. Rev. D **63** (2001) 065003 [hep-ph/0005003].
13. A. Linde, Phys. Lett. **B96** (1980) 289
14. E. Braaten and A. Nieto, Phys. Rev. **D51** (1995) 6990.
15. K. Kajantie, M. Laine, K. Rummukainen and Y. Schroder, Phys. Rev. Lett. **86** (2001) 10 [arXiv:hep-ph/0007109].
16. R.D. Pisarski, Phys. Rev. Lett.**63** (1989) 1129
17. E. Braaten and R.D. Pisarski, Phys. Rev. Lett.**64** (1990) 1338; Phys. Rev. **D42** (1990) 2156; Nucl. Phys. **B337** (1990) 569.
18. J. Frenkel and J.C. Taylor, Nucl. Phys. **B334** (1990) 199; J.C. Taylor and S.M.H. Wong, *ibid.* **B346** (1990) 115.
19. R. Efraty and V.P. Nair, Phys. Rev. Lett.**68** (1992) 2891; R. Jackiw and V.P. Nair, Phys. Rev. **D48** (1993) 4991.

HTL PERTURBATION THEORY AND QCD THERMODYNAMICS

ANTON REBHAN (rebhana@hep.itp.tuwien.ac.at)
Institut für Theoretische Physik
Technische Universität Wien
Wiedner Hauptstr. 8-10, A-1040 Vienna, Austria

1. Introduction

At sufficiently high temperature and/or density, quantum chromodynamics (QCD) should become accessible by perturbative methods due to asymptotic freedom. Thermal perturbation theory is however a surprisingly intricate subject, and it was only during the last decade that the necessary methodology has been developed, and this development is not yet completely finished. An important milestone was the hard-thermal-loop (HTL) resummation programme introduced in particular by Braaten and Pisarski [1] which generalized the static ring resummations that were known to be required in the calculation of thermo-*static* quantities since long [2, 3] to dynamic quantities such as quasiparticle properties as well as production and loss rates.

There are, however, well-known limitations of a fundamental nature which present an impenetrable barrier to perturbation theory at a certain order of the coupling (depending on the quantity under consideration), caused by the inherently nonperturbative chromo-magnetostatic sector of nonabelian gauge theories [4, 5, 6]. This does not mean that thermal perturbation theory is completely futile, though, but rather that at particular points certain nonperturbative input is required in addition [7], for example from lattice calculations of the inherently nonperturbative 3-d Yang-Mills theory describing the self-interactions of chromo-magnetostatic fields. There are in fact even more (less well-known) limitations from collinear singularities that occur in real-time quantities involving external light-like momenta such as the production rate of real photons from a QCD plasma [8, 9], which require improved resummations [74] and/or nonperturbative input.

But even in those cases where thermal perturbation theory has not yet run into one of those barriers, such as the landmark three-loop calculation of the free

J.-P. Blaizot and E. Iancu (eds.), QCD Perspectives on Hot and Dense Matter, 327–351.

energy of QCD to order $\alpha_s^{5/2}$ by Arnold, Zhai, and others [11, 12], there are severe problems caused by extremely poor convergence and strong renormalizaton scheme dependences which seem to render quantitative predictions possible only beyond preposterously high temperatures. While some improvement seemed possible through tricks like Padé approximations [13, 14], the sad conclusion appeared to be that thermal perturbation theory was not applicable in QCD at temperatures of practical interest [10].
Recently, however, it became clear that the problem of poor convergence is not specific to QCD, but arises already in such simple theories as massless scalar φ^4 theory [15] and that alternative resummations can be found that greatly improve the apparent convergence [16, 17]. In these lectures, after an introduction to hard thermal loops, I will present an approach to the problem of calculating thermodynamical quantities in QCD in an *approximately self-consistent* resummation of hard thermal loops and their next-to-leading order corrections [18, 19, 20, 21] that appears to work well down to temperatures a few times the deconfinement transition temperature and that suggests that at such temperatures the still strongly interacting QCD may (at least in certain cases) be adequately described by weakly interacting HTL quasiparticles after all.

2. Thermal field theory

The Feynman rules of thermal field theory [22, 23] are most easily formulated in the imaginary-time (ITF) or Matsubara formalism. The statistical density operator $e^{-\beta H}$ is then equivalent to a time evolution operator over an imaginary time interval of length β, the inverse temperature T^{-1}, and the traces in $\langle A\rangle = \mathrm{Tr}[e^{-\beta H}A]$ require (anti-)periodic boundary conditions at the ends of the imaginary time interval for bosons (fermions). This gives rise to discrete imaginary (Matsubara) frequencies when going to momentum space, so that the only change in the Feynman rules is in the replacement

$$\int \frac{d^4k}{i(2\pi)^4} \to \beta^{-1}\sum_\nu \int \frac{d^3k}{(2\pi)^3}, \qquad i(2\pi)^4\delta^4(k) \to \beta(2\pi)^3\delta_{\nu,0}\delta^3(k). \quad (2.1)$$

with

$$k_0 \to 2\pi i T\nu, \qquad \nu \in \begin{cases} \mathbb{Z} & \text{bosons} \\ \mathbb{Z}-\frac{1}{2} & \text{fermions} \end{cases}. \quad (2.2)$$

Since it is usually hard to evaluate the resulting sums directly, they are best turned into integrals again by writing, in the bosonic case,

$$T\sum_{n=-\infty}^{\infty} f(k_0 = 2\pi i n T) = \frac{T}{2\pi i}\oint_C dk_0 f(k_0)\frac{\beta}{2}\coth\frac{\beta k_0}{2}$$

$$= \underbrace{\frac{1}{2\pi i}\int_{-i\infty}^{i\infty} f(k_0)}_{\text{vacuum contribution}} + \underbrace{\frac{1}{2\pi i}\int_{-i\infty+\epsilon}^{i\infty+\epsilon} [f(k_0)+f(-k_0)]\frac{1}{e^{\beta k_0}-1}}_{\text{thermal contribution}} \quad (2.3)$$

for a function $f(k_0)$ that is regular for $k_0 \in i\mathbb{R}$ and where C is a contour encircling only the poles of coth at the location of the Matsubara frequencies. (In the fermionic case, an analogous formula can be easily obtained by using tanh in place of coth, which in the right-hand side leads to minus the Fermi-Dirac distribution in place of the Bose-Einstein one.)

This evaluation through contour integrals leads to a nice separation into a (Wick rotated) $T = 0$ contribution, and a purely thermal one, which vanishes for $\beta \to \infty$, i.e. $T \to 0$. There exists also a formulation directly in real time (Schwinger-Keldysh and variants thereof) [24] where this separation is already conspicuous in the Feynman rules. However, this requires a doubling of fields, a 2×2 matrix structure for propagators and even more components for n-point vertex functions, which in fact correspond to the many possibilities of analytic continuation to real frequencies (in particular if there are several independent external frequencies) [25].

A simple case that leads to a thermal contribution in (2.3) is the frequently occurring one that there is a simple pole in $f(k_0)$ at $k_0 = \pm E$ with $E > 0$. Then one has

$$T\sum_n \frac{1}{k_0 \pm E} = \pm(-1)^\sigma n_\sigma(E) + \text{vacuum contributions}, \quad (2.4)$$

where $n_\sigma(E) = [e^{\beta E} - \sigma]^{-1}$ and $\sigma = +1$ for bosons, -1 for fermions.

3. Hard thermal loops

The simplest example of a hard thermal loop is given by the one-loop self-energy diagram in a scalar $g\varphi^4$ theory. This is a tadpole diagram, independent of external momenta:

$$\Pi = \frac{4!g^2}{2}\sum_K \frac{-1}{K^2 - m^2} \quad (3.1)$$

where we have introduced the notation $\sum_K = T\sum_n \int \frac{d^3k}{(2\pi)^3}$, $K^\mu = (k^0, k^m)$. With $E = \sqrt{k^2 + m^2}$, its thermal contribution is easily evaluated as

$$\sum_K \frac{-1}{K^2 - m^2} = \sum_K \frac{1}{2E}\left(\frac{1}{k_0 - E} - \frac{1}{k_0 + E}\right)$$

$$= \int \frac{d^3k}{(2\pi)^3}\frac{1}{2E}2n(E) = \int_0^\infty \frac{dk\, k^2}{2\pi^2 E} n(E). \quad (3.2)$$

(The vacuum contribution is removed by standard (T=0) renormalization.) Without the Bose-Einstein factor, this integral would be quadratically divergent. Thanks to the former, it is finite, but dominated by momenta $k \sim T$. If $T \gg m$, the mass terms of the T=0 theory can be neglected, and the leading contribution to the self-energy is given by a self-energy contribution proportional to T^2, which is called a *hard thermal loop* (HTL):

$$\hat{\Pi} = \frac{4!g^2}{2} \sum_K \frac{-1}{K^2} = g^2 \frac{6}{\pi^2} \int_0^\infty dk\, k\, n(k) = g^2 T^2 \,, \qquad (3.3)$$

where the hat is a reminder of the HTL approximation – in this case the neglect of the bare mass m.
$\hat{\Pi}$ clearly corresponds to a mass term for scalar excitations generated by interactions with the particles in the heat bath. It should be understood, however, that this *thermal mass* is qualitatively different from ordinary masses. In particular, one can show that unlike ordinary masses it does not spoil conformal invariance (if any) as it does not contribute to the trace of the energy momentum tensor [26].
This example for a HTL is, however, deceptively simple. In general, thermal masses are not constant but depend on momentum, that is, they correspond to nonlocal terms in an effective action.
The thermal masses generated for gauge fields are of this form. Indeed, a constant mass term would violate gauge invariance. Let us consider as a simple gauge theory example the case of scalar electrodynamics with covariant gauge fixing,

$$\mathcal{L} = (D_\mu \phi)^* D^\mu \phi - \frac{1}{4} F_{\mu\nu} F^{\mu\nu} - \frac{1}{2\alpha} (\partial_\mu A^\mu)^2 \qquad (3.4)$$

where $D_\mu = \partial_\mu + ieA_\mu$. The tree-level propagator reads $G^0_{\mu\nu} = g_{\mu\nu}/K^2 + (\alpha - 1) K_\mu K_\nu / K^4$ and this will receive thermal corrections through the photon polarization tensor, which in momentum space reads

$$\Pi_{\mu\nu}(K) = e^2 \sum_P \left[\frac{(2P-K)_\mu (2P-K)_\nu}{P^2 (P-K)^2} - \frac{2 g_{\mu\nu}}{P^2} \right] . \qquad (3.5)$$

The last term is from the seagull diagram, which is essentially the same as the tadpole diagram before. The first term is however dependent on the external momentum, and there is no reason to expect a Lorentz invariant form, because the heat bath is singling out a preferred frame of reference.
There are in fact four symmetric tensors that can be built from the available quantities $g_{\mu\nu}$, K_μ, and the four-velocity of the heat bath which we have tacitly chosen as $U_\mu = \delta^0_\mu$.
In electrodynamics, the polarization tensor has to be transverse, $K^\mu \Pi_{\mu\nu} \equiv 0$. This additional requirement still leaves two possible tensors, for from U and K one can build the transverse vector $\tilde{U}_\mu = K^2 U_\mu - (U \cdot K) K_\mu$. We choose them as

$$A_{\mu\nu} = g_{\mu\nu} - K_\mu K_\nu / K^2 - B_{\mu\nu}, \quad B_{\mu\nu} = \tilde{U}_\mu \tilde{U}_\nu / \tilde{U}^2 . \qquad (3.6)$$

$A_{\mu\nu}$ can easily be shown to have vanishing components $A_{0\nu} = 0 = A_{\mu 0}$, whereas $A_{ij} = -\delta_{ij} + k_i k_j / k^2$, so this is a projection onto spatially transverse momenta; $B_{\mu\nu}$ is a projector orthogonal to $A_{\mu\nu}$.

In nonabelian gauge theories like QCD, one generally has a more complicated structure. Then one needs two more, nontransverse, tensors, which one may choose as

$$C_{\mu\nu} = \frac{\tilde{U}_\mu K_\nu + K_\mu \tilde{U}_\nu}{\sqrt{2K^2}k}, \quad D_{\mu\nu} = \frac{K_\mu K_\nu}{K^2}. \tag{3.7}$$

Decomposing $\Pi_{\mu\nu} = -\Pi_t A_{\mu\nu} - \Pi_\ell B_{\mu\nu} - \Pi_c C_{\mu\nu} - \Pi_d D_{\mu\nu}$ and

$$G^{\mu\nu} \equiv [G_0^{-1} + \Pi]^{-1\mu\nu} = \Delta_t A_{\mu\nu} + \Delta_\ell B_{\mu\nu} + \Delta_c C_{\mu\nu} + \Delta_d D_{\mu\nu} \tag{3.8}$$

one has

$$\Delta_t = [K^2 - \Pi_t]^{-1}, \quad \Delta_\ell = [K^2 - \Pi_\ell + \alpha \frac{\Pi_c^2}{K^2 - \alpha \Pi_d}]^{-1}, \tag{3.9}$$

$$\Delta_c = \alpha \Pi_c [K^2 - \alpha \Pi_d]^{-1}, \quad \Delta_d = \frac{\alpha (K^2 - \Pi_\ell)}{K^2 - \alpha \Pi_d} \Delta_\ell. \tag{3.10}$$

In view of the explicit gauge parameter dependences (there are in fact more hidden within the structure functions of Π), it is remarkable that one can prove that the singularities of Δ_t and Δ_ℓ are gauge-fixing independent [27, 28, 29].

In electrodynamics this situation is much simpler (unless one introduces nonlinear gauge fixing): $\Pi_{\mu\nu}$ is both transverse and completely gauge parameter independent.

Because of transversality, which is easily verified for (3.5), there are only two independent components, e.g. $\Pi^\mu{}_\mu$ and Π_{00}. The former reads

$$\Pi^\mu{}_\mu(K) = e^2 \sum_P \left[\frac{-4}{P^2} - \frac{K^2}{P^2 (P-K)^2} \right] = 4e^2 \sum_P \frac{-1}{P^2} + O(T) \approx \frac{e^2 T^2}{3} \tag{3.11}$$

where the term proportional to K^2 does not constitute a hard thermal loop, because its integrand does not involve a quadratic divergence in its vacuum piece. For $k_0, k \ll T$, $\Pi^\mu_\mu(K) \sim \hat{\Pi}^\mu_\mu = e^2 T^2/3$. Notice, however, that this 'HTL approximation' remains valid even for $k_0, k \sim T$ as long as $K^2 \ll T^2$.

The other component, Π_{00} is more complicated to evaluate. Its HTL piece, which is contained in

$$\Pi_{00}(K) = e^2 \sum_P \left[\frac{4p_0(p_0 - k_0)}{P^2 (P-K)^2} - \frac{2}{P^2} \right], \tag{3.12}$$

can however be extracted rather easily using

$$\frac{p_0}{P^2} = \frac{1}{2} \left(\frac{1}{p_0 - p} + \frac{1}{p_0 + p} \right), \quad \frac{1}{p_0 + X} \frac{1}{p_0 + Y} = \frac{1}{X - Y} \left(\frac{1}{p_0 + Y} - \frac{1}{p_0 + X} \right)$$

which gives a number of terms of the form $\frac{1}{X-Y}(n(Y) - n(X))$ after summing over the Matsubara frequencies. Now, HTL contributions arise from $p \sim T$, and for $k_0, k \ll T$ one can approximate the energies X and Y by $\pm p$, except when two hard energies form a soft difference. This gives

$$\Pi_{00}(K) \sim e^2 \int \frac{d^3p}{(2\pi)^3} \left\{ \frac{1}{k_0 + p - |p-k|}[n(|p-.k|) - n(p)] \right. \\ \left. + \frac{1}{k_0 - p + |p-k|}[-n(|p-k|) + n(p)] \right\} \quad (3.13)$$

where the tadpole-like last term in (3.12) has cancelled against terms where the energy denominators contain the sum of two hard energies.
Since $k \ll p \sim T$, we have $n(|p-k|) - n(p) \sim n'(p)[|p-k| - p]$ and $|p-k| - p \sim \vec{p} \cdot \vec{k}/p \equiv zk$, so the HTL piece of Π_{00} is finally given by

$$\begin{aligned} \hat{\Pi}_{00}(K) &= 2e^2 \int \frac{p^2\, dp}{(2\pi)^2} n'(p) \int_{-1}^{1} dz \left[-1 + \frac{k_0}{k_0 - zk} \right] \\ &= \frac{e^2 T^2}{3} \left[1 - \frac{k_0}{2k} \ln \frac{k_0 + k}{k_0 - k} \right]. \end{aligned} \quad (3.14)$$

Originally, k_0 was restricted to Matsubara frequencies. In order to allow for soft $k_0 \ll T$ without being restricted to the zero mode, we in fact need analytic continuation, e.g. $k_0 \to \omega + i\epsilon$ for retarded boundary conditions, and this defines the cut of the logarithm in (3.14).
The above results for $\hat{\Pi}^{\mu}_{\mu}$ and $\hat{\Pi}_{00}(K)$ are actually universal. They have the same form in nonabelian gauge theories, only the overall coefficient differs. In SU(N) gauge theories with N_f quark flavors one just needs to replace [30, 31] $e^2 \to g^2(N + N_f/2)$. They also retain their form in the presence of a nonvanishing chemical potential μ_f, which leads to the replacement of $T^2 \to T^2 + 3\mu_f^2/\pi^2$ in the contributions $\propto N_f$. In the HTL approximation, there are moreover no nontransverse contributions[1] so the HTL gauge boson propagator involves two independent branches determined by $\hat{\Pi}_\ell = -\hat{\Pi}_{00} K^2/k^2$ and $\hat{\Pi}_t = (\hat{\Pi}^{\mu}_{\mu} - \hat{\Pi}_\ell)/2$. The poles of the propagators Δ_t and Δ_ℓ determine the dispersion laws of two sorts of quasiparticles, which in contrast to the scalar φ^4 example are not given by simple mass hyperboloids. These are displayed in Fig. 1 in a plot of $\omega_{t,\ell}(k)$ in quadratic scales (where a relativistic mass hyperboloid would show up as a straight line parallel to the light-cone).
Above a common plasma frequency $\omega_{\rm pl.} = eT/3$, there are propagating modes, which for large momenta in the transverse branch tend to a mass hyperboloid with asymptotic mass $m_\infty^2 = \frac{3}{2}\omega_{\rm pl.}^2$, and in branch ℓ approach the light-cone

[1] At order kT one has however $\Pi_c \neq 0$ for $\alpha \neq 1$ in the nonabelian case, and $\Pi_d \neq 0$ at two-loop order whenever $\Pi_c \neq 0$.

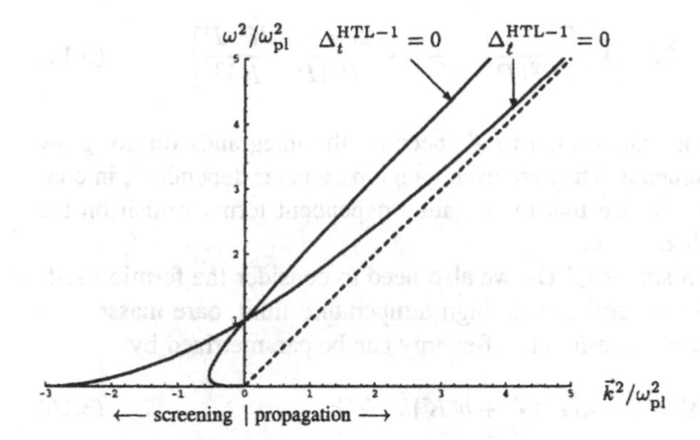

Figure 1. The location of the zeros of $\Delta_t^{\rm HTL-1}$ (spatially transverse gauge bosons) and of $\Delta_\ell^{\rm HTL-1}$ (longitudinal plasmons) in quadratic scales such as to show propagating modes and screening phenomena on one plot.

exponentially with exponentially vanishing residue. Indeed, this mode does not have an analogue in the T=0 theory but is a purely collective phenomenon, so it has to disappear from the spectrum as $k \to \infty$. The spatially transverse mode, on the other hand, represents quasiparticles that are in-medium versions of the physical polarisations of gauge bosons.

For $\omega < \omega_{\rm pl.}$, $|\mathbf{k}|$ is the inverse screening length, which in the static limit vanishes for mode t (absence of magnetostatic screening), but reaches the Debye mass, $\hat{m}_D^2 = 3\omega_{\rm pl.}^2$, for mode ℓ (electrostatic screening). A vanishing magnetic screening mass is required by gauge invariance in abelian gauge theories [32, 33], but not in the nonabelian case. In fact, lattice simulations of gauge fixed propagators in nonabelian theories do find a screening behaviour in the transverse sector, however the corresponding singularity is certainly quite different from a simple pole [34].

For $\omega^2 < k^2$, there is a large imaginary part $\sim e^2T^2$ from (3.14) which prevents the appearance of poles in this region. This imaginary part corresponds to the possibility of Landau damping, which is the transfer of energy from soft fields to hard plasma constituents moving in phase with the field [35, 36] and is an important part of the spectral density of HTL propagators. At higher, subleading orders of perturbation theory, it is, however, not protected against gauge dependences in nonabelian gauge theories.

To complete the discussion of HTL's in scalar electrodynamics, let us also consider briefly the scalar self-energy. In contrast to φ^4 theory, this is now a nonlocal quantity. Nevertheless, the HTL part is still a constant thermal mass as given by the first term in

$$\Xi = \frac{e^2T^2}{4} + e^2K^2 \sum_P \left[\frac{3-\alpha}{P^2(P-K)^2} + \frac{2(\alpha-1)K\cdot P}{P^4(P-K)^2} \right]. \tag{3.15}$$

The other terms are not proportional to T^2 because the integrands do not grow sufficiently at large momenta. They are even gauge parameter dependent, in contrast to the HTL piece. Notice that these gauge dependent terms vanish on the lowest-order mass shell $K^2 = 0$.

In QCD (and already in spinor QED), we also need to consider the fermion self-energies [37, 38]. In the ultrarelativistic high-temperature limit, bare masses can be neglected. In this case the fermion self-energy can be parametrized by

$$\Sigma(K) = a(K)\gamma^0 + b(K)\vec{k}\cdot\vec{\gamma}/k\,. \tag{3.16}$$

Using the same methods as above one easily computes the HTL contributions in

$$\frac{1}{4}\mathrm{tr}\, K\!\!\!/\Sigma = \omega a + kb = \frac{e^2T^2}{8} \equiv \hat{M}^2, \tag{3.17}$$

$$\frac{1}{4}\mathrm{tr}\gamma^0\Sigma = a = \frac{e^2T^2}{16k}\ln\frac{\omega+k}{\omega-k}, \tag{3.18}$$

where in nonabelian gauge theories now $e^2 \to g^2(N^2-1)/(2N)$.

The structure of the HTL fermion propagator is

$$S(K) = \frac{1}{2}\left(\gamma^0 + \frac{\vec{k}\cdot\vec{\gamma}}{k}\right)\Delta_+ + \frac{1}{2}\left(\gamma^0 - \frac{\vec{k}\cdot\vec{\gamma}}{k}\right)\Delta_- \tag{3.19}$$

with $\Delta_\pm^{-1} = -[\omega \mp (k+\Sigma_\pm)]$ and $\Sigma_\pm \equiv b \pm a$. The two branches correspond to spinors whose chirality is equal (+) or opposite (−) to their helicity.

The additional collective modes of branch (−) ("plasminos") have a curious minimum of ω at $\omega/\hat{M} \approx 0.93$ and $|\vec{k}|/\hat{M} \approx 0.41$ and approach the light-cone for large momenta, but with exponentially vanishing residue. The regular branch approaches a mass hyperboloid (in Fig. 2 a straight line parallel to the diagonal) with asymptotic mass $\sqrt{2}\hat{M}$. Again, for space-like momenta, $K^2 < 0$, there is a large imaginary part corresponding to Landau damping, which now corresponds to the transmutation of soft fermionic fields together with hard fermionic (bosonic) plasma constituents into hard bosonic (fermionic) ones.

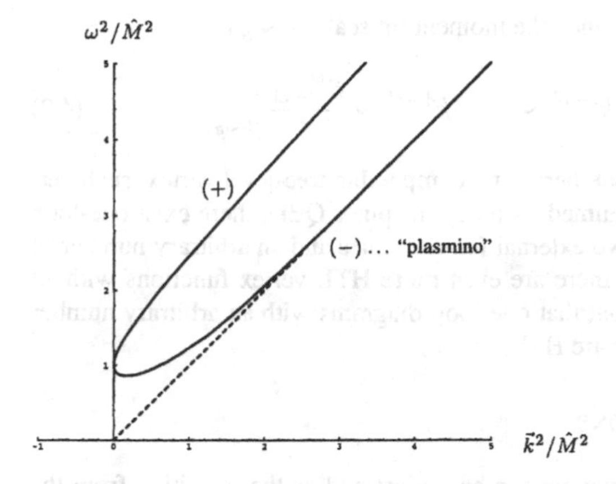

Figure 2. The location of the zeros of $\Delta_{\pm}^{-1}$ in the HTL approximation in quadratic scales.

4. Standard HTL perturbation theory

When writing down dressed propagators, as we have done above in the discussion of the spectrum of thermal quasiparticles, we have already performed a resummation of infinitely many loops—the geometric series of self-energy insertions according to Dyson's equation, which in the scalar φ^4 theory is simply

$$\frac{-1}{K^2-(gT)^2}=\frac{-1}{K^2}\left(1+g^2\frac{T^2}{K^2}+g^4\frac{T^4}{K^4}+\dots\right). \tag{4.1}$$

Clearly, the perturbative version is useful only when $K^2 \gg (gT)^2$. In particular, it fails for $k_0, k \sim gT$, the scale where collective phenomena transform the familiar quanta into quasiparticles.
The appearance of thermal masses presents unavoidable problems at some order of perturbation theory, namely when these higher orders probe the energy/momentum scale $k_0, k \sim gT$, and, in particular in the bosonic sector, the Bose-Einstein distribution factors enhance the sensitivity to the infrared.
It is therefore mandatory to switch from bare perturbation theory to one that uses resummed (dressed) propagators. In simple cases like scalar φ^4 theory, where the only HTL is a constant mass term, this resummation has been studied already long ago [39]. Conceptually (although not practically) equally simple is the example of scalar electrodynamics, where the only HTL's are the self energy diagrams considered above. However, already in spinor QED and to a larger extent in QCD, it turns out that there are HTL vertex functions [40] which have to be treated on a par with the HTL self energies to achieve a systematic resummed perturbation theory [1]. Indeed, if N-point vertex functions give rise to HTL's $\propto T^2$, they are

as important as bare vertices when the momentum scale is $\sim gT$:

$$\Gamma^{\rm HTL}_{,N} \sim g^N T^2 k^{2-N} \sim g^{N-2} k^{4-N} \sim \frac{\partial^N \mathcal{L}_{\rm cl.}}{\partial A^N}\Big|_{k\sim gT}. \quad (4.2)$$

In fact, even for N so high that there is no comparable tree-level vertex, such vertex functions have to be resummed. Already in spinor QED, there exist one-loop HTL vertex functions with two external fermion lines and an arbitrary number of gauge boson lines. In QCD, there are even more HTL vertex functions without tree-level analogue. It turns out that one-loop diagrams with an arbitrary number of external gauge boson lines are HTL.

4.1. HTL EFFECTIVE ACTIONS

The HTL resummation programme can be understood as the transition from the fundamental Lagrangian to an effective one generated by 'integrating out' the hard momentum modes $k_0, k \sim T$ in one-loop order.
In scalar φ^4 theory, this effective theory differs from the bare one only in a simple thermal mass term, $\mathcal{L}^{\rm HTL}_{\rm scalar} = -\frac{1}{2}\mu^2\varphi^2$.
For gauge bosons and fermions, the effective theory is necessarily a non-local one as gauge invariance forbids a simple local thermal mass term and in the case of fermions it is due to the fact that HTL's do not spoil chiral symmetries. Remarkably, the infinitely many non-local vertex functions can be summarized by a comparatively simple and manifestly gauge-invariant integral representation [41, 42, 43]

$$\begin{aligned}\mathcal{L}^{\rm HTL} &= \hat{M}^2 \int \frac{d\Omega_v}{4\pi} \bar{\psi}\gamma^\mu \frac{v_\mu}{v\cdot D(A)}\psi \\ &\quad -\frac{3}{2}\omega_{pl.}^2 {\rm tr} \int \frac{d\Omega_v}{4\pi} F^{\mu\alpha} \frac{v_\alpha v^\beta}{(v\cdot D_{adj.}(A))^2} F_{\mu\beta}\end{aligned} \quad (4.3)$$

$v = (1, \vec{v})$ is a light-like 4-vector, i.e. with $\vec{v}^2 = 1$, and its spatial components are averaged over by $d\Omega_v$. v is the remnant of the hard plasma constituents' momenta $p^\mu \sim Tv^\mu$, namely their light-like 4-velocity, and the overall scale T has combined with the coupling constant to form the scale of thermal masses, $\hat{M}, \omega_{pl.} \sim gT$.
The covariant derivatives in the denominators of (4.3) are responsible for the fact that there are infinitely many HTL's involving external fermions and an arbitrary number of gauge bosons, even in QED, where only the pure gauge-field sector becomes bilinear because of $D_{adj.}(A) \to \partial$.
Technically, resummed perturbation theory amounts to the replacement

$$\mathcal{L}_{\rm cl.} \to \mathcal{L}_{\rm cl.} + \mathcal{L}^{\rm HTL} - \ell\mathcal{L}^{\rm HTL} \quad (4.4)$$

where ℓ is a loop counting parameter that is sent to 1 in the end, after the last term has been treated as a 'thermal counterterm'.
Because $\mathcal{L}^{\rm HTL}$ has been derived under the assumption of soft external momenta, this prescription is in fact only to be followed for soft propagators and vertices [1]. Those involving hard momenta (if present) do not require this resummation, and they can be excluded from this resummation by the introduction of some intermediate scale Λ with $gT \ll \Lambda \ll T$. Complete results have to come out independent of Λ, of course.

4.2. EXAMPLE: NLO TERMS IN SCALAR ELECTRODYNAMICS

Massless scalar electrodynamics [44] is a particularly simple toy model as its HTL effective action (4.3) is bilinear in all fields. The scalar HTL self-energy (3.15) is moreover a simple mass term, and in order to consider the one-loop corrections to the photon polarization tensor, nothing more is needed. Let us consider two limiting cases of this to illustrate some important points of the HTL resummation programme.

4.2.1. *Debye mass*

The Debye mass, i.e. the inverse screening length of electrostatic fields, is determined by the zero of $\Delta_\ell(k_0 = 0, k)$ at imaginary k leading to $m_D^2 = \Pi_{00}(k_0 = 0, \vec{k}^2 = -m_D^2)$. Because at leading order $\hat\Pi_{00}(k_0 = 0, k)$ turns out to be independent of k (see (3.14)), the frequently found definition [22] of m_D^2 as $\hat\Pi_{00}(k_0 = 0, k \to 0)$ happens to be correct, but becomes unphysical in general [45]: beyond LO, only the former, self-consistent definition is renormalization-group invariant and (in nonabelian theories) gauge invariant.
The resummation programme sketched above makes the LO result $\hat m_D^2 = e^2T^2/3$ part of the new lowest-order Lagrangian $\mathcal{L}_{\rm cl.} + \mathcal{L}^{\rm HTL}$. The NLO correction therefore is given by one-loop diagrams using this Lagrangian. These are scalar loops which now have massive propagators with thermal mass $\mu^2 = \hat\Xi = e^2T^2/4$ from (3.15), from which the HTL result $\hat m_D^2$ has to be subtracted as thermal counterterm. Indeed, without this subtraction, the LO result would be generated a second time, since for large loop momenta the thermal mass of the scalar is negligible. Because of the subtraction, the one-loop integrals are now receiving their leading contributions from soft loop momenta $k \sim eT$:

$$\delta\Pi_{00}(0,q) = e^2 \sum_K \frac{4k_0^2}{(K^2-\mu^2)[(K-Q)^2-\mu^2]} - e^2\sum_K \frac{2}{K^2-\mu^2} - \hat m_D^2$$
$$= \frac{e^2}{\pi^2}\int_0^\infty dk\, k^2 \left\{ \frac{n(\sqrt{k^2+\mu^2})}{\sqrt{k^2+\mu^2}} \left[1 + \frac{k^2+\mu^2}{kq} \ln\left|\frac{2k+q}{2k-q}\right|\right] - (\mu \to 0)\right\} . \quad (4.5)$$

To obtain the leading contribution, we can replace $n(E) \to T/E$ which gives

$$\delta\Pi_{00}(0,q) = -\frac{e^2T}{\pi^2}\int_0^\infty dk \left\{ \frac{\mu^2}{k^2+\mu^2} + 1 - \frac{k}{q}\ln\left|\frac{2k+q}{2k-q}\right| \right\} + O(e^2q^2\ln(T)). \tag{4.6}$$

The NLO correction to the Debye mass is now given by evaluating this at $q = i\hat{m}_D$. Incidentially, the above integral is q-independent, as can be seen from an integration by parts, which finally gives

$$m_D^2 = \hat{m}_D^2 - \frac{e^2T\mu}{2\pi} = \frac{e^2T^2}{3}\left(1 - \frac{3}{4\pi}e\right). \tag{4.7}$$

Notice that the perturbative result at NLO involves a single power of e and so is *non-analytic* in $\alpha = e^2/(4\pi)$.
The above calculation can actually be simplified by noting [46] that the only terms capable of producing odd powers in e are the $n = 0$ terms in the sums over Matsubara frequencies in (4.5). For $n \neq 0$, one has $-k_0^2 = (2\pi n)^2T^2 \gg \mu^2$ so that the thermal masses can be expanded out, leading to powers of e^2 only. Keeping only the $n = 0$ contributions, the first integral in (4.5) vanishes, and we have

$$\delta\Pi_{00}(0,q) = 2e^2T\sigma^{2\epsilon}\int \frac{d^{3-2\epsilon}k}{(2\pi)^{3-2\epsilon}}\frac{1}{k^2+\mu^2} \overset{\epsilon\to 0}{\longrightarrow} -\frac{e^2T\mu}{2\pi} \tag{4.8}$$

in agreement with (4.7). Here we have introduced dimensional regularization (with mass scale σ) to render the integral finite as in Ref. [46].
As we shall see presently, this simplified resummation by ***dimensional reduction*** is only possible in the static case ($q_0 = 0$), and not for dynamical quantities.

4.2.2. *Plasma frequency*

Propagating modes exist for frequencies $q_0 \geq \omega_{\text{pl.}}$, and this plasma frequency is the same for the transverse and longitudinal modes, because it corresponds to the long-wavelength limit $q \to 0$, and with $\vec{q} = 0$ there is no way to distinguish the polarizations. In the HTL approximation, $\omega_{\text{pl.}}^2 = \hat{m}_D^2/3 = e^2T^2/9$.
The NLO correction in the case of scalar electrodynamics can be calculated in full analogy to (4.5), but now with $q_0 = \omega_{pl.}$ and $\vec{q} \to 0$. Because of $\vec{q} = 0$, the angular integrals are now trivial and one finds

$$\delta\Pi_\ell(q_0,0) = \delta\Pi_t(q_0,0) = -\frac{e^2T}{2\pi}\left\{\mu + \frac{4}{3q_0^2}\left([\mu^2 - q_0^2/4]^{3/2} - \mu^3\right)\right\}. \tag{4.9}$$

Evaluated at $q_0 = \omega_{\text{pl.}}$ this gives $\delta\omega_{\text{pl.}}^2/\omega_{\text{pl.}}^2 = -e(8\sqrt{2}-9)/(2\pi) \approx -0.37e$. Notice that without resumming μ, the result would have been completely misleading:

evaluating the unresummed result, i.e. (4.9) with $\mu = 0$, at $q_0 = \omega_{pl.}$ would have given a purely imaginary result that one would have wrongly identified with a damping constant.[2]

Furthermore, the correct result (4.9) is now only obtained if the nonstatic modes are resummed along with the static ones. If one keeps only the zero modes and ignors that in the imaginary time formalism the external frequency q_0 has to be a multiple of $2\pi iT$ (and so cannot be soft and nonzero), but immediately 'continues' to $q_0 = \omega_{pl.} \sim eT$, one would find

$$\delta\Pi_{\ell,t}(q_0,0)|_{\substack{\text{0-mode}\\ \text{contr.}}} = -\frac{e^2T}{6\pi}\left\{\frac{2}{q_0}\left[\frac{\mu}{q_0} - \sqrt{\frac{\mu^2}{q_0^2} - 1}\right](q_0^2 - \mu^2) + \mu\right\} \quad (4.10)$$

which clearly differs from (4.9). The resulting $\delta\omega_{pl.}^2$ would in fact be only about a quarter of the true result. So dynamic quantities require the full HTL resummation method; resumming only the zero modes is not sufficient (see also [47]).

4.3. NLO CORRECTIONS FOR QCD QUASIPARTICLES

The calculation of NLO corrections to the long-wavelength plasmons in QCD was in fact one of the first applications of the HTL resummation programme. In particular, the damping constant of order $g^2T \sim g\omega_{pl.}$ was the subject of a long controversy (in particular with regard to its gauge-fixing (in)dependence) before it was calculated in Ref. [48] with the gauge independent[3] result

$$\gamma \approx 0.264\sqrt{N}\, g\, \omega_{pl.} \quad (4.11)$$

(for pure-glue QCD). The analogous calculation for the damping constant of long-wavelength fermionic quasiparticles was carried out in Refs. [52, 53]. The significance of these results is that gluonic and fermionic quasiparticles, which in the HTL approximation appear to be stable, experience damping. For $g \ll 1$, they are weakly damped, whereas for $g \sim 1$ which is more relevant for experimentally accessible quark-gluon plasmas, the damping is significant: $\gamma \sim \frac{1}{2}\omega_{pl.}$.

The NLO correction to the gluonic plasma frequency has also been calculated [54] with the result $\delta\omega_{pl.}^2/\omega_{pl.}^2 \approx -0.18\sqrt{N}\, g$.

The NLO correction to the Debye mass, however, runs into IR problems. Naively one would expect problems from the masslessness of magnetostatic gluons only

[2] In scalar electrodynamics, unlike QCD, the plasmon damping is a higher order effect because the scalar HTL quasiparticles are heavier than plasmons and moreover do not have Landau damping cuts in the HTL approximation.

[3] In fact, later investigations using covariant gauges encountered again gauge dependences [49, 50], which are removed, however, when an infrared cut-off is retained while taking the on-mass-shell limit [51]. This means that there is an infrared singularity in the residue of the pole, which is gauge dependent, but the position of the singularity (no longer a simple pole) is gauge independent in accordance with the theorems of Refs. [27, 28].

at two-loop order resummed perturbation theory, when their self-interactions become relevant. However, because gauge independence requires evaluation on mass-shell (which in the case of the Debye mass means $q_0 = 0$, $\vec{q}^2 = -\hat{m}_D^2$), there appear 'mass-shell singularities' caused by the massless magnetostatic modes. Because these singularities are only logarithmic, the leading log is perturbatively calculable and reads [45, 55]

$$\delta m_D^2/\hat{m}_D^2 = \frac{N}{2\pi}\sqrt{\frac{6}{2N+N_f}}\,g\log\frac{1}{g} + \mathcal{O}(g). \tag{4.12}$$

For small coupling, the logarithm dominates over the non-perturbative constant behind the logarithm, and thus the perturbative prediction is that of a *positive* correction to the screening mass. Indeed, lattice simulations of both (gauge-fixed) chromo-electrostatic propagators [34] and gauge-invariant lattice definitions of the nonabelian Debye mass [56, 57] give significant positive corrections to the HTL value.[4]

Such a logarithmic sensitivity to the nonperturbative physics of the chromo-magnetostatic sector has in fact been encountered earlier in the damping of hard excitations [58, 59, 60]. More generally, it arises for all propagating modes [61] as well as for all finite screening lengths [62]. For nonzero wave-vector $\vec{q}$, only the real corrections to the dispersion law $\omega = \omega(q)$ turn out to be IR safe in one-loop resummed perturbation theory.

5. HTL-resummed thermodynamics

In the previous section we have seen that dynamic quantities cannot be treated by the simplified resummation scheme that resums only static modes. For static quantities like the thermodynamic potential, however, such a resummation works in the sense that it gives a scheme to systematically compute the series expansion in powers (and log's) of the coupling. This calculation has been performed to order $\alpha_s^{5/2}$ in QCD [11] with the result (for pure glue)

$$P = -\Omega/V = \frac{8\pi^2T^4}{45}\Big[1 - \frac{15\alpha_s}{4\pi} + 30(\frac{\alpha}{\pi})^{3/2} + \frac{135}{2}(\frac{\alpha}{\pi})^2\log\frac{\alpha}{\pi} \tag{5.1}$$
$$- \frac{165}{8}\left(\log\frac{\bar{\mu}}{2\pi T} - 11.49\right)(\frac{\alpha}{\pi})^2 + \frac{495}{2}\left(\log\frac{\bar{\mu}}{2\pi T} - 3.23\right)(\frac{\alpha}{\pi})^{5/2} + \dots\Big].$$

[4] These positive corrections are so large in fact that the nonperturbative contributions at order g dominate over the HTL result in the gauge-invariant lattice definitions. When extracted from lattice propagators, which also give gauge-independent results [34], these nonperturbative contributions are considerably smaller and about 1/3 of them can be accounted for by one-loop resummed perturbation theory if one introduces a simple phenomenological magnetic screening mass taken from the lattice [55, 29].

Unfortunately, this is very poorly convergent: only when $\alpha_s < 0.05$ one has apparent convergence, but this corresponds to temperatures higher than $10^5 T_c$!
In what follows we shall attempt a different route that resums also the nonstatic modes, and tries to keep resummation effects even when they are formally of higher order than that achievable at a given loop order.

5.1. SCREENED PERTURBATION THEORY

In scalar φ^4 theory, it has been shown [16, 17] that the convergence of thermal perturbation theory can be improved if the thermal mass of the scalar quasiparticles is kept within thermal integrals and not treated as proportional to a coupling constant when setting up the perturbation series. Technically, this is just as in (4.4), but without the requirement that the resummation has to take place in soft quantities only. Because this changes the UV structure at any finite order of perturbation theory, this introduces new UV divergences and associated renormalization scheme dependences, which in principle can become arbitrarily large. But starting from two-loop order, these can be minimized if the thermal mass used in screened perturbation theory is determined by a variational principle, i.e. a prinicple of 'minimal sensitivity' to the mass parameter used in this reorganization of perturbation theory [17, 63].
In Refs. [64, 65], this approach has been adapted to a one-loop calculation of the thermodynamic potential of QCD where in place of a simple mass term the gauge-invariant HTL effective action is used. While the leading-order interaction term $\propto g^2$ is incomplete (in fact, it is over-included), it does contain the plasmon term $\propto g^3$ without leading to the disastrous result of a thermodynamic pressure in excess of the Stefan-Boltzmann limit. However, it remains to be seen if this is still true in the (technically very difficult) two-loop approximation, which presumably has to contribute with positive sign to make up for the over-included negative leading-order interaction term of the one-loop approximation.
In the following, I shall present instead the result of a different strategy. Instead of using the HTL effective action as a gauge invariant mass term for an optimization of perturbation theory, which does not (have to) care whether the HTL effective action remains accurate for hard momenta (which it does not), the formalism of self-consistent "Φ-derivable" [66] approximations will be invoked to find expressions that keep resummation effects for both soft and hard momenta. As we shall see, the leading-order effects will arise exclusively from kinematical regimes where the HTL approximation remains justifiable. Moreover, it will be possible to avoid the spurious UV problems of screened perturbation theory.

5.2. APPROXIMATELY SELF-CONSISTENT RESUMMATIONS

In the Luttinger-Ward representation of the thermodynamic potential $\Omega = -PV$ [67] (to particle physicists often more familiar as the composite operator effective potential [68]) is expressed as a functional of full propagators D and two-particle irreducible diagrams. Considering for simplicity a scalar field theory for the moment, $\Omega[D]$ has the form

$$\begin{aligned}\Omega[D] &= -T\log Z = \tfrac{1}{2}T\ \mathrm{Tr}\log D^{-1} - \tfrac{1}{2}T\ \mathrm{Tr}\ \Pi D + T\Phi[D] \\ &= \int \frac{d^4k}{(2\pi)^4} n(\omega)\mathrm{Im}\left[\log D^{-1}(\omega,k) - \Pi(\omega,k)D(\omega,k)\right] + T\Phi[D],\end{aligned} \tag{5.2}$$

where Tr denotes the trace in configuration space, and $\Phi[D]$ is the sum of the 2-particle-irreducible "skeleton" diagrams

$$-\Phi[D] = \ 1/12 \ \ldots + 1/8 \ \ldots + 1/48 \ \ldots + \ldots \tag{5.3}$$

The self energy $\Pi = D^{-1} - D_0^{-1}$, where D_0 is the bare propagator, is related to $\Phi[D]$ by

$$\delta\Phi[D]/\delta D = \tfrac{1}{2}\Pi. \tag{5.4}$$

An important property of the functional $\Omega[D]$, which is easily verified using (5.4), is that it is stationary under variations of D:

$$\delta\Omega[D]/\delta D = 0. \tag{5.5}$$

Self-consistent ("Φ-derivable") [66] approximations are obtained by selecting a class of skeletons in $\Phi[D]$ and calculating Π from Eq. (5.4) above, preserving the stationarity condition.

The stationarity of $\Omega[D]$ has an interesting consequence for the entropy $\mathcal{S} = -\partial(\Omega/V)/\partial T$. Because of Eq. (5.5), the temperature derivative of the spectral density in the dressed propagator cancels out and only the explicit Bose-Einstein factors need to be differentiated in (5.2), yielding [69, 70, 18, 20]

$$\begin{aligned}\mathcal{S} = &-\int \frac{d^4k}{(2\pi)^4} \frac{\partial n(\omega)}{\partial T} \mathrm{Im}\ \log D^{-1}(\omega,k) \\ &+\int \frac{d^4k}{(2\pi)^4} \frac{\partial n(\omega)}{\partial T} \mathrm{Im}\,\Pi(\omega,k)\,\mathrm{Re}\,D(\omega,k) + \mathcal{S}'\end{aligned} \tag{5.6}$$

with

$$\mathcal{S}' = -\left.\frac{\partial(T\Phi)}{\partial T}\right|_D + \int \frac{d^4k}{(2\pi)^4} \frac{\partial n(\omega)}{\partial T}\,\mathrm{Re}\,\Pi\,\mathrm{Im}\,D = 0 \tag{5.7}$$

up to terms that are of loop-order 3 or higher. Thus, in contrast to Ω, where Φ contributes already to order g^2 in perturbation theory, Eq. (5) with $\mathcal{S}' = 0$ is perturbatively correct to order g^3. The first two terms in Eq. (5.6) represent essentially the entropy of "independent quasiparticles", while $\mathcal{S}'$ may be viewed as the residual interactions among these quasiparticles [70].

The same simplification holds true in the presence of fermions and, with nonzero chemical potential, extends to the fermion density $\mathcal{N} = -\partial(\Omega/V)/\partial\mu$ [19, 20]. In a *self-consistent* two-loop approximation one thus has the remarkably simple formulae, now for general theories

$$\mathcal{S} = -\,\mathrm{tr}\int\frac{d^4k}{(2\pi)^4}\frac{\partial n(\omega)}{\partial T}\left[\mathrm{Im}\,\log D^{-1}(\omega,k) - \mathrm{Im}\,\Pi(\omega,k)\,\mathrm{Re}\,D(\omega,k)\right]$$
$$-2\,\mathrm{tr}\int\frac{d^4k}{(2\pi)^4}\frac{\partial f(\omega)}{\partial T}\left[\mathrm{Im}\,\log S^{-1}(\omega,k) - \mathrm{Im}\,\Sigma(\omega,k)\,\mathrm{Re}\,S(\omega,k)\right], \quad (5.8)$$
$$\mathcal{N} = -2\,\mathrm{tr}\int\frac{d^4k}{(2\pi)^4}\frac{\partial f(\omega)}{\partial\mu}\left[\mathrm{Im}\,\log S^{-1}(\omega,k) - \mathrm{Im}\,\Sigma(\omega,k)\,\mathrm{Re}\,S(\omega,k)\right], \quad (5.9)$$

where $n(\omega) = (e^{\beta\omega}-1)^{-1}$, $f(\omega) = (e^{\beta(\omega-\mu)}+1)^{-1}$, and "tr" refers to all discrete labels, including spin, color and flavor when applicable.
In gauge theories, the above expressions have to be augmented by Faddeev-Popov ghost contributions which enter like bosonic fields but with opposite over-all sign, unless a gauge is used where the ghosts do not propagate such as in axial gauges. But because Φ-derivable approximations do not generally respect gauge invariance,[5] the self-consistent two-loop approximation will not be gauge-fixing independent. It is in fact not even clear that the corresponding gap equations (5.4) have solutions at all or that one can renormalize these (nonperturbative) equations. For this reason, we shall construct *approximately self-consistent* solutions which are gauge invariant and which maintain equivalence with conventional perturbation theory up to and including order g^3, the maximum (perturbative) accuracy of the two-loop approximation for Φ. For these approximations it will be sufficient to keep only the two transverse structure functions of the gluon propagator and to neglect ghosts.
For soft momenta, we know that the leading order contribution is given by the HTL's, and indeed there is no HTL ghost self-energy. For hard momenta, one can identify the contributions to (5.8) below order g^4 as those linear in the self-energies,

$$\mathcal{S}^{\mathrm{hard}} = \mathcal{S}_0 + 2N_g\int\frac{d^4k}{(2\pi)^4}\frac{\partial n}{\partial T}\,\mathrm{Re}\,\Pi_t\,\mathrm{Im}\,\frac{1}{\omega^2-k^2}$$
$$-\,4NN_f\int\frac{d^4k}{(2\pi)^4}\frac{\partial f}{\partial T}\left\{\mathrm{Re}\,\Sigma_+\mathrm{Im}\,\frac{-1}{\omega-k} - \mathrm{Re}\,\Sigma_-\mathrm{Im}\,\frac{-1}{\omega+k}\right\} \quad (5.10)$$

considering now a gauge theory with N_g gluons and N_f fermion flavors. Because the imaginary parts of the free propagators restrict their contribution to the light-cone, only the light-cone projections of the self-energies enter. At order g^2 this is

[5] For this, one would have to treat vertices on an equal footing with self-energies, which is in principle possible using the formalism developed in Ref. [71].

exactly given by the HTL results, without having to assume soft ω, k [73, 74] (as in the above example of scalar electrodynamics, see Eqs. (3.11), (3.15))

$$\mathrm{Re}\,\Pi_t^{(2)}(\omega^2 = k^2) = \hat{\Pi}_t(\omega^2 = k^2) = \tfrac{1}{2}\hat{m}_D^2 \equiv m_\infty^2, \tag{5.11}$$

$$2k\ \mathrm{Re}\,\Sigma_\pm^{(2)}(\omega = \pm k) = 2k\,\hat{\Sigma}_\pm(\omega = \pm k) = 2\hat{M}^2 \equiv M_\infty^2, \tag{5.12}$$

and without contributions from the other components of $\Pi_{\mu\nu}$ and the Faddeev-Popov self-energy.
There is no contribution $\propto g^2$ from soft momenta in (5.8) and (5.9) so that one is left with remarkably simple general formulae for the leading-order interaction contributions to the thermodynamic potentials expressed through the asymptotic thermal masses of the bosonic and fermionic quasiparticles:

$$\mathcal{S}^{(2)} = -T\left\{\sum_B \frac{m_{\infty B}^2}{12} + \sum_F \frac{M_{\infty F}^2}{24}\right\}, \quad \mathcal{N}^{(2)} = -\frac{1}{8\pi^2}\sum_F \mu_F M_{\infty F}^2. \tag{5.13}$$

Here the sums run over all the bosonic (B) and fermionic (F) degrees of freedom (e.g. 4 for each Dirac fermion), which are allowed to have different asymptotic masses and, in the case of fermions, different chemical potentials.
Turning now to the next order, g^3, let us first recapitulate how this ususally arises in the thermodynamic potential. Since this is a static quantity, we can use the imaginary-time formalism and concentrate on the zero-modes as in Sect. 4.2.1. For $\omega = 0$ the leading contributions to the self-energies are $\hat{\Pi}_\ell(0,k) = \hat{m}_D^2 \propto (gT)^2$ and $\hat{\Pi}_t(0,k) = 0$. This gives the "plasmon-effect"[6] term

$$P^{(3)} = -N_g T \int \frac{d^3k}{(2\pi)^3}\left[\log\left(1 + \frac{\hat{m}_D^2}{k^2}\right) - \frac{\hat{m}_D^2}{k^2}\right] = N_g \frac{\hat{m}_D^3 T}{12\pi}. \tag{5.14}$$

In deriving the self-consistent expressions for entropy and density, Eqs. (5.8) and (5.9), we can no longer use this argument to extract the order g^3 term, for we have first rewritten the thermodynamic potential in terms of real-time propagators and self-energies and then used stationarity to drop T and μ derivatives of the spectral densities hidden in the full propagators. In fact, from the second line of (5.2), one can still isolate the zero-mode contribution (5.14) by $\int \frac{d\omega}{\pi\omega}\mathrm{Im}\,[\ldots] = [\ldots](\omega = 0)$, but in (5.8) and (5.9) we have products of imaginary and real parts times statistical distribution functions.
Indeed, the order g^3 contributions now arise from both soft and hard momentum regimes. In $\mathcal{S}^{\mathrm{soft}}$ such contributions are due to the singular behaviour of $\partial n/\partial T \sim 1/\omega$ which does not allow us to expand out the self-energy insertions perturbatively, but on dimensional grounds gives a contribution $\sim \Pi^{3/2}$. In $\mathcal{S}^{\mathrm{hard}}$,

[6] Obviously a misnomer. It is caused exclusively by the Debye screening mass, not the plasmon mass = plasma frequency $\neq \hat{m}_D$.

on the other hand, where we could expand out the self-energies as in Eq. (5.10), g^3 contributions arise from NLO order contributions to Π and Σ themselves. (In the case of the pressure, this did not happen because of the stationarity property of the pressure [20].)

With this insight, we can now formulate an *approximately self-consistent* dressing of the propagators that is in line with the maximum perturbative accuracy of the Φ-derivable two-loop approximation: for soft momenta, we take the (gauge-invariant and gauge-independent) HTL expressions for self-energies and propagators; for hard momenta, where according to (5.10) only the light-cone limit of the self-energies contribute below order g^4, the correct leading-order contribution is still given by the HTL expressions (5.11), (5.12), but in order to include the g^3 contributions completely, we also require the NLO corrections to the on-light-cone self-energies. The latter can be calculated by standard HTL perturbation theory, and the theorems of Ref. [28] ensure their gauge independence.

5.2.1. *HTL approximation*

As a first approximation let us consider one which only uses the HTL expressions without NLO corrections thereof. We have seen that this gives the correct leading-order interaction term $\propto g^2$, some part of the g^3 contribution, and infinitely many formally higher-order terms as well, since we are going to use (5.8) and (5.9) "non-perturbatively", i.e. without expanding out in powers of g and truncating. We can do so because expressions (5.8) and (5.9) are manifestly UV finite, for they involve only the derivatives of the statistical distribution functions—in (5.2), the $T = 0$ UV-divergences are contained in the integration domain $\omega \to -\infty$, where the undifferentiated n's and f's do not fall off exponentially.

Using the HTL expressions in (5.8) and considering for simplicity the pure-glue case, we obtain two physically distinct contributions. The first corresponds to the transverse and longitudinal gluonic quasiparticle poles,

$$\mathcal{S}^{\rm QP}_{\rm HTL} = -N_g \int \frac{k^2\,dk}{2\pi^2} \frac{\partial}{\partial T} \left[2T \log(1 - e^{-\omega_t(k)/T}) + T \log \frac{1 - e^{-\omega_\ell(k)/T}}{1 - e^{-k/T}} \right] \quad (5.15)$$

where only the explicit T dependences are to be differentiated, and not those implicit in the HTL dispersion laws $\omega_t(k)$ and $\omega_\ell(k)$. Secondly, there are the Landau-damping contributions which read

$$\mathcal{S}^{\rm LD}_{\rm HTL} = -N_g \int_0^\infty \frac{k^2 dk}{2\pi^3} \int_0^k d\omega \frac{\partial n(\omega)}{\partial T} \Big\{ 2 \arg[k^2 - \omega^2 + \hat\Pi_t]$$
$$-2 {\rm Im}\, \hat\Pi_t \, {\rm Re}[\omega^2 - k^2 - \hat\Pi_t]^{-1} + \arg[k^2 + \hat\Pi_\ell] - {\rm Im}\, \hat\Pi_\ell \, {\rm Re}[k^2 + \hat\Pi_\ell]^{-1} \Big\} \quad (5.16)$$

The usual perturbative g^2-contribution (5.13) is contained in the first term of Eq. (5.15); all the other terms in Eqs. (5.15),(5.16) are of order g^3 in a small-g expansion.

In the HTL approximation, only the soft plasmon effect $\sim g^3$ contained in $\mathcal{S}^{\rm soft}$ is present, which turns out to equal

$$\mathcal{S}_3^{\rm soft} = \frac{\partial P^{(3)}}{\partial T}\Big|_{\hat{m}_D}, \tag{5.17}$$

which is $\frac{1}{4}\mathcal{S}_3$ in the case of pure glue. However, this identification requires a peculiar sum rule

$$\begin{aligned}\Delta\mathcal{S}_3 \equiv N_g \int \frac{d^4k}{(2\pi)^4}\frac{1}{\omega}\Big\{2\,\mathrm{Im}\,\hat{\Pi}_t\,\mathrm{Re}\Big(\hat{D}_t - D_t^{(0)}\Big) - \mathrm{Im}\,\hat{\Pi}_\ell\,\mathrm{Re}\Big(\hat{D}_\ell - D_\ell^{(0)}\Big)\Big\}\\ \equiv \Delta\mathcal{S}_t^{(3)} + \Delta\mathcal{S}_\ell^{(3)} = 0\,,\end{aligned} \tag{5.18}$$

which we found to hold numerically by cancellations in more than 8 significant digits. Rather unusually (cp. Ref. [23]), this does not hold separately for the longitudinal and transverse sector and moreover holds only after carrying out both, the frequency and the momentum integrations in (5.18).
Although the g^3 term in $\mathcal{S}_{\rm HTL}$ is only a fraction of the full one, it would make similar troubles when expanded out perturbatively, throwing away all higher-order terms in g: for large enough g, the perturbative approximation would lead to an entropy in excess of the Stefan-Boltzmann limit. $\mathcal{S}_{\rm HTL}$, on the other hand, is a monotonically decreasing function of $\hat{m}_D/T \sim g$.

5.2.2. *Next-to-leading approximation*

The plasmon term $\sim g^3$ becomes complete only upon inclusion of the next-to-leading correction to the asymptotic thermal masses m_∞ and M_∞. These are determined in standard HTL perturbation theory through

$$\begin{aligned}\delta m_\infty^2(k) &= \mathrm{Re}\,\delta\Pi_T(\omega = k)\\ &= \mathrm{Re}(\ \cdots + \cdots + \cdots + \cdots\ )|_{\omega=k}\end{aligned} \tag{5.19}$$

where thick dashed and wiggly lines with a blob represent HTL propagators for longitudinal and transverse polarizations, respectively. Similarly,

$$\frac{1}{2k}\delta M_\infty^2(k) = \delta\Sigma_+(\omega = k) = \mathrm{Re}(\ \cdots + \cdots\ )|_{\omega=k}\,. \tag{5.20}$$

The explicit proof that these contributions indeed restore the correct plasmon term is given in Ref. [20].
These corrections to the asymptotic thermal masses are, in contrast to the latter, nontrivial functions of the momentum, which can be evaluated only numerically. However, as far as the generation of the plasmon term is concerned, these functions contribute in the averaged form

$$\bar{\delta} m_\infty^2 = \frac{\int dk\, k\, n_{\rm BE}'(k)\,\mathrm{Re}\,\delta\Pi_T(\omega = k)}{\int dk\, k\, n_{\rm BE}'(k)} \tag{5.21}$$

(cp. Eq. (5.13)) and similarly

$$\bar{\delta}M_\infty^2 = \frac{\int dk\, k\, n'_{\rm FD}(k)\, {\rm Re}\, 2k\delta\Sigma_+(\omega = k)}{\int dk\, k\, n'_{\rm FD}(k)} . \tag{5.22}$$

These averaged asymptotic thermal masses turn out to be given by the remarkably simple expressions [20]

$$\bar{\delta}m_\infty^2 = -\frac{1}{2\pi}g^2 NT\hat{m}_D, \quad \bar{\delta}M_\infty^2 = -\frac{1}{2\pi}g^2 C_f T\hat{m}_D, \tag{5.23}$$

where $C_f = N_g/(2N)$. Since the integrals in (5.21) and (5.22) are dominated by hard momenta, these thermal mass corrections only pertain to hard excitations. Indeed, in Sect. 4.3 we have seen that e.g. the plasmon mass at $k = 0$ receives a different, namely smaller correction, whereas the NLO contribution to the Debye mass is even positive and logarithmically enhanced.

Pending the full evaluation of the NLO corrections to ${\rm Re}\,\delta\Pi$ and ${\rm Re}\,\delta\Sigma$, it has therefore been proposed in Refs. [18, 19, 20] to define a next-to-leading approximation through (for gluons)

$$\mathcal{S}_{NLA} = \mathcal{S}_{HTL}\Big|_{\rm soft} + \mathcal{S}_{\bar{m}_\infty^2}\Big|_{\rm hard}, \tag{5.24}$$

where $\bar{m}_\infty^2$ includes (5.23). To separate soft ($k \sim \hat{m}_D$) and hard ($k \sim 2\pi T$) momentum scales, we introduce the intermediate scale $\Lambda = \sqrt{2\pi T\hat{m}_D c_\Lambda}$ and consider a variation of $c_\Lambda = \frac{1}{2}\ldots 2$ as part of our theoretical uncertainty.

Another crucial issue concerns the definition of the corrected asymptotic mass $\bar{m}_\infty$. For the range of coupling constants of interest ($g \gtrsim 1$), the correction $|\bar{\delta}m_\infty^2|$ is greater than the LO value m_∞^2, leading to tachyonic masses if included in a strictly perturbative manner.

However, this problem is not at all specific to QCD. In the simple $g^2\varphi^4$ model, one-loop resummed perturbation theory gives

$$m^2 = g^2T^2(1 - \frac{3}{\pi}g) \tag{5.25}$$

which also turns tachyonic for $g \gtrsim 1$. On the other hand, the self-consistent solution of the corresponding one-loop gap equation [39, 15]

$$m^2 = \Pi(m) = 12g^2 \sum_K \frac{-1}{K^2 - m^2} \tag{5.26}$$

(properly renormalized), whose perturbative expansion begins exactly like (5.25), is a monotonic function in g. In fact, it turns out that the first two terms in a (m/T)-expansion of this gap equation,

$$m^2 = g^2T^2 - \frac{3}{\pi}g^2Tm, \tag{5.27}$$

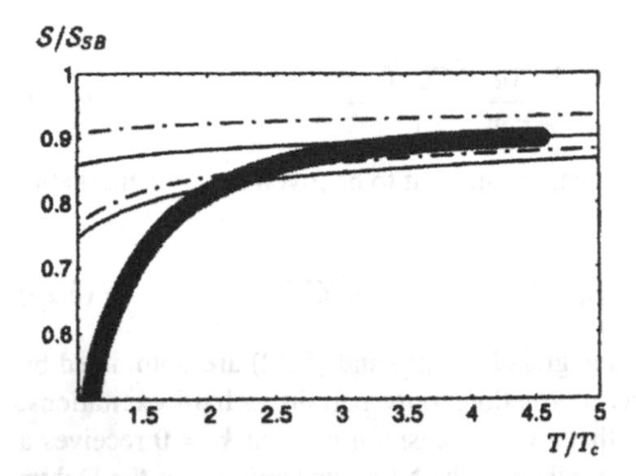

Figure 3. Comparison of the lattice data for the entropy of pure-glue SU(3) gauge theory of Ref. [72] (gray band) with the range of $\mathcal{S}_{HTL}$ (solid lines) and $\mathcal{S}_{NLA}$ (dash-dotted lines) for $\bar{\mu} = \pi T \ldots 4\pi T$ and $c_\Lambda = 1/2 \ldots 2$.

which is perturbatively equivalent to (5.25), has a solution that is extremely close to that of the full gap equation (for $\overline{\text{MS}}$ renormalization scales $\bar{\mu} \approx 2\pi T$) [20].

In QCD, the (non-local) gap equations are way too complicated to be attacked directly. We instead consider perturbatively equivalent expressions for the corrected $\bar{m}_\infty$ which are monotonic functions in g. Besides the solution to a quadratic equation analogous to (5.27) we have tried the simplest Padé approximant $m^2 = g^2T^2/(1+\frac{3}{\pi}g)$, which also gives a greatly improved approximation to the solution of scalar gap equations. In QCD, our final results do not depend too much on whether we use the Padé approximant [18, 19] or a quadratic gap equation [20].

The main uncertainty rather comes from the choice of the renormalization scale which determines the magnitude of the strong coupling constant when this is taken as determined by the renormalization group equation (2-loop in the following).

In Fig. 3, the numerical results for the HTL entropy and the NLA one are given as a function of T/T_c with T_c chosen as $T_c = 1.14\Lambda_{\overline{MS}}$. The full lines show the range of results for $\mathcal{S}_{HTL}$ when the renormalization scale $\bar{\mu}$ is varied from πT to $4\pi T$; the dash-dotted lines mark the corresponding results for $\mathcal{S}_{NLA}$ with the additional variation of c_Λ from $1/2$ to 2. The dark-gray band are lattice data from Ref. [72]. Evidently, there is very good agreement for $T \gtrsim 2.5T_c$.

From the above results for the entropy density, one can recover the thermodynamic pressure by simple integration, $P(T) - P(T_0) = \int_{T_0}^{T} dT' \mathcal{S}(T')$. The integration constant $P(T_0)$, however, is a strictly nonperturbative input. It cannot be fixed by requiring $P(T = 0) = 0$, as this is in the confinement regime. It is also not sufficient to know that $\lim_{T\to\infty} P = P_{\text{free}}$ by asymptotic freedom. In fact, the undetermined integration constant in $P(T)/P_{\text{free}}(T)$ when expressed as a function of $\alpha_s(T)$ corresponds to a term [19] $C \exp\{-\alpha_s^{-1}[4\beta_0^{-1} + O(\alpha_s)]\}$, which

vanishes for $\alpha_s \to 0$ with all derivatives and thus is not fixed by any order of perturbation theory. It is, in essence, the nonperturbative bag constant, which can be added on to standard perturbative results, too. However, in $P(T)/P_{\text{free}}(T)$ this term becomes rapidly unimportant as the temperature is increased, as it decays like T^{-4}. Fixing it by $P(T_c) = 0$, which is a good approximation in particular for the pure-glue case because glue balls are rather heavy, one finds again good agreement with lattice data for $T \gtrsim 2.5T_c$.
This approach can be generalized [19, 20] also to nonzero chemical potentials μ_f, where lattice data are not available[7]. Simpler quasiparticle models [79, 78] have already been used to extrapolate lattice data to finite chemical potential [80]. The HTL approach offers a possible refinement, but that has still to be worked out.

6. Conclusion

Hard thermal loops, the leading-order contributions to self-energies and vertices at high temperature and/or density, form a gauge-invariant basis for a systematic perturbative expansion, as long as one does not run into the perturbative barrier formed by the inherently nonperturbative sector of self-interacting chromomagnetostatic modes. But in QCD one faces the additional problem that corrections to leading-order results are so large for almost all values of the coupling of interest that they lead to a complete loss of (apparent) convergence. However, we have seen that further resummations which keep as much as possible of the effects of HTL resummation without expanding in a power series in the coupling may lead to results that remain valid down to temperatures a few times the deconfinement phase transition temperature.

References

1. Braaten, E. and R. D. Pisarski: 1990a. *Nucl. Phys.* **B337**, 569.
2. Gell-Mann, M. and K. A. Brueckner: 1957. *Phys.Rev.* **106**, 364.
3. Akhiezer, I. A. and S. V. Peletminskii: 1960. *Sov. Phys. JETP* **11**, 1316.
4. Polyakov, A. M.: 1978. *Phys. Lett.* **B72**, 477.
5. Linde, A. D.: 1980. *Phys. Lett.* **B96**, 289.
6. Gross, D. J., R. D. Pisarski, and L. G. Yaffe: 1981. *Rev. Mod. Phys.* **53**, 43.
7. Braaten, E.: 1995. *Phys. Rev. Lett.* **74**, 2164.
8. Baier, R., S. Peigné, and D. Schiff: 1994. *Z. Phys.* **C62**, 337.
9. Aurenche, P., F. Gelis, R. Kobes, and E. Petitgirard: 1997. *Z. Phys.* **C75**, 315.
10. Braaten, E. and A. Nieto: 1996b. *Phys. Rev. Lett.* **76**, 1417.
11. Arnold, P. and C.-X. Zhai: 1995. *Phys. Rev.* **D51**, 1906.
12. Braaten, E. and A. Nieto: 1996a. *Phys. Rev.* **D53**, 3421.

[7] Lattice results exist however for the response of thermodynamic quantities to infinitesimal chemical potentials, namely for quark number susceptibilities [75, 76, 77], and the above approach has been applied to those, too, by now [21].

13. Kastening, B.: 1997. *Phys. Rev.* **D56**, 8107.
14. Hatsuda, T.: 1997. *Phys. Rev.* **D56**, 8111.
15. Drummond, I. T., R. R. Horgan, P. V. Landshoff, and A. Rebhan: 1998. *Nucl. Phys.* **B524**, 579.
16. Karsch, F., A. Patkós, and P. Petreczky: 1997. *Phys. Lett.* **B401**, 69.
17. Andersen, J. O., E. Braaten, and M. Strickland: 2001. *Phys. Rev.* **D63**, 105008.
18. Blaizot, J.-P., E. Iancu, and A. Rebhan: 1999a. *Phys. Rev. Lett.* **83**, 2906.
19. Blaizot, J.-P., E. Iancu, and A. Rebhan: 1999b. *Phys. Lett.* **B470**, 181.
20. Blaizot, J.-P., E. Iancu, and A. Rebhan: 2001a. *Phys. Rev.* **D63**, 065003.
21. Blaizot, J.-P., E. Iancu, and A. Rebhan: 2001b, 'Quark number susceptibilities from HTL-resummed thermodynamics', hep-ph/0110369.
22. Kapusta, J. I.: 1989, *Finite-temperature field theory*. Cambridge, UK: Cambridge University Press.
23. Le Bellac, M.: 1996, *Thermal Field Theory*. Cambridge, UK: Cambridge University Press.
24. Landsman, N. P. and C. G. van Weert: 1987. *Phys. Rept.* **145**, 141.
25. Kobes, R.: 1990. *Phys. Rev.* **D42**, 562.
26. Nachbagauer, H., A. K. Rebhan, and D. J. Schwarz: 1996. *Phys. Rev.* **D53**, 882.
27. Kobes, R., G. Kunstatter, and A. Rebhan: 1990. *Phys. Rev. Lett.* **64**, 2992.
28. Kobes, R., G. Kunstatter, and A. Rebhan: 1991. *Nucl. Phys.* **B355**, 1.
29. Rebhan, A.: 2001, 'Thermal gauge field theories', hep-ph/0105183.
30. Kalashnikov, O. K. and V. V. Klimov: 1980. *Sov. J. Nucl. Phys.* **31**, 699.
31. Weldon, H. A.: 1982. *Phys. Rev.* **D26**, 1394.
32. Fradkin, E. S.: 1965. *Proc. Lebedev Inst.* **29**, 1.
33. Blaizot, J.-P., E. Iancu, and R. R. Parwani: 1995. *Phys. Rev.* **D52**, 2543.
34. Cucchieri, A., F. Karsch, and P. Petreczky: 2001. *Phys. Rev.* **D64**, 036001.
35. Lifshitz, E. M. and L. P. Pitaevsky: 1981, *Physical Kinetics*. Oxford, UK: Pergamon Press.
36. Blaizot, J.-P. and E. Iancu: 2002, 'The quark-gluon plasma: Collective dynamics and hard thermal loops', *Phys. Rept.* **359**, 355.
37. Klimov, V. V.: 1981. *Sov. J. Nucl. Phys.* **33**, 934.
38. Weldon, H. A.: 1989. *Phys. Rev.* **D40**, 2410.
39. Dolan, L. and R. Jackiw: 1974. *Phys. Rev.* **D9**, 3320.
40. Frenkel, J. and J. C. Taylor: 1990. *Nucl. Phys.* **B334**, 199.
41. Taylor, J. C. and S. M. H. Wong: 1990. *Nucl. Phys.* **B346**, 115.
42. Braaten, E. and R. D. Pisarski: 1992a. *Phys. Rev.* **D45**, 1827.
43. Frenkel, J. and J. C. Taylor: 1992. *Nucl. Phys.* **B374**, 156.
44. Kraemmer, U., A. K. Rebhan, and H. Schulz: 1995. *Ann. Phys.* **238**, 286.
45. Rebhan, A. K.: 1993. *Phys. Rev.* **D48**, 3967.
46. Arnold, P. and O. Espinosa: 1993. *Phys. Rev.* **D47**, 3546.
47. Elze, H. T., U. Heinz, K. Kajantie, and T. Toimela: 1988. *Z. Phys.* **C37**, 305.
48. Braaten, E. and R. D. Pisarski: 1990b. *Phys. Rev.* **D42**, 2156.
49. Baier, R., G. Kunstatter, and D. Schiff: 1992a. *Phys. Rev.* **D45**, 4381.
50. Baier, R., G. Kunstatter, and D. Schiff: 1992b. *Nucl. Phys.* **B388**, 287.
51. Rebhan, A.: 1992b. *Phys. Rev.* **D46**, 4779.
52. Kobes, R., G. Kunstatter, and K. Mak: 1992. *Phys. Rev.* **D45**, 4632.
53. Braaten, E. and R. D. Pisarski: 1992b. *Phys. Rev.* **D46**, 1829.
54. Schulz, H.: 1994. *Nucl. Phys.* **B413**, 353.
55. Rebhan, A. K.: 1994. *Nucl. Phys.* **B430**, 319.
56. Arnold, P. and L. G. Yaffe: 1995. *Phys. Rev.* **D52**, 7208.
57. Laine, M. and O. Philipsen: 1999. *Phys. Lett.* **B459**, 259.
58. Pisarski, R. D.: 1989. *Phys. Rev. Lett.* **63**, 1129.

59. Lebedev, V. V. and A. V. Smilga: 1991. *Phys. Lett.* **B253**, 231.
60. Rebhan, A.: 1992a. *Phys. Rev.* **D46**, 482.
61. Pisarski, R. D.: 1993. *Phys. Rev.* **D47**, 5589.
62. Flechsig, F., A. K. Rebhan, and H. Schulz: 1995. *Phys. Rev.* **D52**, 2994.
63. Rebhan, A.: 2000, 'Improved resummations for the thermodynamics of the quark gluon plasma'. In: C. P. Korthals Altes (ed.): *Strong and Electroweak Matter 2000*. Singapore: World Scientific Publ., p. 199.
64. Andersen, J. O., E. Braaten, and M. Strickland: 1999. *Phys. Rev. Lett.* **83**, 2139.
65. Andersen, J. O., E. Braaten, and M. Strickland: 2000. *Phys. Rev.* **D61**, 014017.
66. Baym, G.: 1962. *Phys. Rev.* **127**, 1391.
67. Luttinger, J. M. and J. C. Ward: 1960. *Phys. Rev.* **118**, 1417.
68. Cornwall, J. M., R. Jackiw, and E. Tomboulis: 1974. *Phys. Rev.* **D10**, 2428.
69. Riedel, E.: 1968. *Z. Phys.* **210**, 403.
70. Vanderheyden, B. and G. Baym: 1998. *J. Stat. Phys.* **93**, 843.
71. Freedman, B. A. and L. D. McLerran: 1977. *Phys. Rev.* **D16**, 1130.
72. Boyd, G., J. Engels, F. Karsch, E. Laermann, C. Legeland, M. Lütgemeier, and B. Petersson: 1996. *Nucl. Phys.* **B469**, 419.
73. Kraemmer, U., M. Kreuzer, and A. Rebhan: 1990. *Ann. Phys.* **201**, 223 [Appendix].
74. Flechsig, F. and A. K. Rebhan: 1996. *Nucl. Phys.* **B464**, 279.
75. Gottlieb, S., W. Liu, D. Toussaint, R. L. Renken, and R. L. Sugar: 1987. *Phys. Rev. Lett.* **59**, 2247.
76. Gavai, R. V. and S. Gupta: 2001. *Phys. Rev.* **D64**, 074506.
77. Gavai, R. V., S. Gupta, and P. Majumdar: 2001, 'Susceptibilities and screening masses in two flavor QCD', hep-lat/0110032.
78. Levai, P. and U. Heinz: 1998. *Phys. Rev.* **C57**, 1879.
79. Peshier, A., B. Kämpfer, O. P. Pavlenko, and G. Soff: 1996. *Phys. Rev.* **D54**, 2399.
80. Peshier, A., B. Kämpfer, and G. Soff: 2000. *Phys. Rev.* **C61**, 045203.

NOTES ON THE DECONFINING PHASE TRANSITION

ROBERT D. PISARSKI (pisarski@bnl.gov)
Department of Physics
Brookhaven National Laboratory
Upton, NY 11973 USA

Abstract. I review the deconfining phase transition in an $SU(N)$ gauge theory without quarks. After computing the interface tension between $Z(N)$ degenerate vacua deep in the deconfined phase, I follow Giovannangeli and Korthals Altes, and suggest a new model for (discrete) Polyakov loop spins. Effective theories for (continuous) Polyakov loop spins are constructed, including those with $Z(N)$ charge greater than one, and compared with Lattice data. About the deconfining transition, the expectation values of $Z(N)$ singlet fields ("quarkless baryons") may change markedly. Speculations include: a possible duality between Polyakov loop and ordinary spins in four dimensions, and how $Z(N)$ bubbles might be guaranteed to have positive pressure.

1. Overview

In these lectures I review the deconfining phase transition in a "pure" $SU(N)$ gauge theory, without dynamical quarks. Gauge theories are ubiquitous in physics, so their phase transitions are manifestly of fundamental importance. Two examples may include the collisions of large nuclei at high energy and the early universe.

The phase transitions of gauge theories without quarks are of especial interest, since then the order parameter, and many other aspects of the phase transition, can be characterized precisely [1, 2, 3]. While of course QCD includes quarks, this is not an academic exercise. Recent results from the Lattice on "flavor independence" — for both the pressure [4, 5] and quark susceptibilities [6] — suggest that the results from the pure glue theory may be, in a surprising and unexpected fashion, relevant for QCD. (Whether flavor independence can be generalized when there are many light flavors is not known.)

Albeit indirectly, the Lattice has already told us much about what happens in a pure gauge theory with three colors. By asymptotic freedom, at infinite temperature the pressure is that for an ideal gas of gluons. In the confined phase, the pressure is very small, essentially zero. So the question is, as the pressure turns on at the transition temperature T_c, how rapidly does it approach the ideal gas limit?

J.-P. Blaizot and E. Iancu (eds.), QCD Perspectives on Hot and Dense Matter, 353–384.

The Lattice tells us relatively quickly: by about $2T_c$, it is already 80% of the way to ideality. The "2" in $2T_c$ is meant schematically; it is certainly not, say, $10T_c$. Above $2T_c$, the pressure then approaches ideality slowly, from below.
This suggests that from temperatures of $2T_c$ on up, that the theory is some sort of quasiparticle gas of thermal quarks and gluons; *i.e.*, a Quark-Gluon Plasma. By this I mean that after suitable dressing from bare into quasi-particles, the residual interactions are weak. This is seen from the Lattice: like the ratio of the true to the ideal pressure, the ratio of (gauge-invariant) masses to the temperature also vary slowly above $2T_c$ [7, 8].
It is known that for the free energy, direct perturbative calculations fail at astronomically high temperatures [9]; there is only a perturbative Quark-Gluon Plasma above temperatures of $\sim 10^7$ GeV. Thus from temperatures of 10^7 GeV down to $2T_c$, the theory is what I call a non-perturbative Quark-Gluon Plasma.
One approach to the non-perturbative QGP is to fold in the effects of Debye screening. It is known that the pressure, as obtained from the Lattice, can be fit to an ideal gas of massive gluons all the way down to T_c [10]. The problem is that as introduced, these masses aren't gauge invariant. Further, in the end one is just fitting one function of temperature, the pressure, to another, thermal masses. (Still, it is most intriguing that these fit thermal masses become large near T_c.)
Another approach is provided by the resummation of Hard Thermal Loops (HTL) [11, 12, 13, 14]. For the gauge field, if A_0 is the time-like component of the vector potential, the Debye mass term is just $\sim \mathrm{tr}(A_0^2)$. For gluons, HTL's are the gauge invariant, analytic continuation of this mass term from imaginary to real time. At present, though, HTL resummation has been used mainly to compute the pressure. The crucial test, yet remaining, are the results which it gives for Polyakov loop correlation functions [7, 15].
A method to compute all static correlation functions in the non-perturbative Quark-Gluon Plasma starts with the (perturbative) construction of an effective theory in three dimensions [16]. From the original gauge theory at $T \neq 0$, static magnetic fields produce three dimensional gluons, while static electric fields give A_0, as an adjoint scalar coupled to these gluons. Due to the power-like infrared divergences of gauge theories in three dimensions, perturbative calculations are useless in this effective theory. Since the effective theory is purely bosonic, though, static correlation functions can be efficiently computed by numerical means on the Lattice. While heroic perturbative calculations are required, for three colors this method appears to work down to $2T_c$, where the method itself indicates its failure. To be fair, this method only yields static correlation functions, and not those in real time. With HTL resummation, this continuation is almost automatic.
What about below $2T_c$? If the transition is strongly first order, then presumably the quasiparticle regime extends all of the way down to T_c. Lattice data suggests that for four colors [17], the transition is strongly first order. (Although data on correlation lengths near T_c is absent.) For more than four colors, if the deconfining

transition is like the Potts model, then it becomes more strongly first order as N increases. On the Lattice, "reduced" models have a first order transition at infinite N [18]. In the continuum, for really no good reason, I remain unconvinced [19]. For two colors, however, the deconfining transition is almost certainly of second order [20]. For three colors, the transition is of first order [21], as predicted by Svetitsky and Yaffe [3]. Recent results suggest, however, that the transition is so weakly first order that it is more accurate to speak of a "nearly second order" transition [22].

This is seen most clearly not from the pressure, but from the behavior of electric and magnetic masses. I define the electric, m_{el}, and magnetic masses, m_{mag}, in a gauge invariant way, from the fall-off of the two-point functions for Polyakov loops, and the (trace of the) magnetic field squared, respectively. In the perturbative QGP, $m_{el} > m_{mag}$ [7], while in the non-perturbative QGP, $m_{el} \sim m_{mag}$ [7, 8]. In contrast, in the transition regime m_{mag}/T is approximately constant, but m_{el}/T appears to drop by a factor of ten as $T : 2T_c \rightarrow T_c^+$ [22]. A similar drop in the string tension is seen as $T : 0 \rightarrow T_c^-$ [22].

The broad outlines of the appropriate effective theory in the transition region for a (nearly) second order deconfining transition have been known for some time [1, 2, 3, 23, 24]. In the QGP regime, one deals with A_0. In the transition region, one trades A_0 in for the thermal Wilson line. Traces of powers of the thermal Wilson line give Polyakov loops, which are gauge invariant.

The effective theory of Polyakov loops is just beginning [25, 26, 27, 28, 29, 30, 31]. If true, instead of the A_0 quasiparticles of the QGP, near T_c it is more useful to view the theory as a *condensate* of Polyakov loops. This could produce dramatic signatures in heavy ion collisions, with hadronization at T_c computed semiclassically from the decay of Polyakov loop condensates [26, 27, 28].

In these Lectures I provide some background to understand these questions. After explaining why $SU(N)$ gauge theories have $Z(N)$ degenerate vacua [2], I compute the interface tension between these vacua at high temperature [32, 33, 34, 35, 36, 37, 38, 39]. Viewing the $Z(N)$ vacua as discrete spins, and using results of Giovannangeli and Korthals-Altes [37], I propose a new spin model, distinct from the usual Potts model. (The order of the transition seems to agree with Potts for $N \leq 4$, but is unknown for $N > 4$.) I then develop an effective theory of Polyakov loop spins, considered as continuous variables. Even with the limited amount of relevant Lattice data which exists at present, the form of this effective theory is sharply constrained. Especially intriguing is the possibility that the expectation values of fields which are singlets under the $Z(N)$ symmetry — which I term quarkless baryons — change suddenly about T_c.

The lectures are in part pedagogical, in part base speculation. The latter includes a possible duality in four dimensions, and how $Z(N)$ bubbles, which in perturbation theory can have negative pressure [39], might be ensured of positive pressure non-perturbatively. I also review what is known about renormalization of Wilson loops

and Wilson lines [40, 41, 42, 43].
In these lectures I only discuss the Polyakov Loops Model [25, 26, 27, 28, 29, 30, 31] in passing. I hope to provide a basis for understanding its motivation. At present, a major unsolved problem is the analytic continuation from imaginary to real time. In abelian gauge theories, the analytic continuation of the Debye mass term gives the Random Phase Approximation [44]; in nonabelian theories, it gives Hard Thermal Loops [11]. Absent any results whatsoever, I do have the temerity to coin a phrase for the analytic continuation of the Polyakov Loops Model to real time, as the "*Nonabelian Random Phase Approximation*". Time (dependence) will tell.

1.1. SATURATION

My motivation for studying this subject is its relevance for the collisions of large nuclei at very high energies. Thus at the outset, I wish to add some general comments about nucleus-nucleus collisions, and especially how it might relate to models of saturation.
Consider a completely implausible situation, the collision of two neutron stars (say) at relativistic energies. For a neutron star, the transverse area is essentially infinite on nuclear scales. If the stars completely overlap (zero impact parameter), then at very high energies, a nearly baryon-free region is generated between them (about zero rapidity). The system starts out with energy density, but no pressure. It is then reasonable to think that the system builds up pressure, and thermalizes, before it flies apart.
The crucial question for the collisions of two large nuclei is whether for gold or lead nuclei, with $A \sim 200$, that this finite value of A is close to infinity, or represents some intermediate regime. This will be decided by experiment, at the SPS, RHIC, and the LHC. After one year of running, it appears clear that *something* dramatic has happened between SPS and RHIC energies [45]. Precisely what is still being sorted out.
I wish to comment here on the relevance of "saturation" at these energies [46]. At a very pedestrian level, this can be viewed as a type of finite size effect at $A < \infty$. Consider a collision in the rest frame of one nuclei. The diameter of the other nucleus, with $A \sim 200$, is ~ 15 fm. This distance becomes Lorentz contracted. Thus we can ask, in the rest frame of one nuclei, when does the incident nuclei look like a pancake of negligible width? We want the distance to be really small on typical hadronic scales. If a typical hadronic scale is 1 fm, then, a small scale might be $1/4 \to 1/3$ fm, say. To contract 15 fm down to these sizes then requires a center of mass energy per nucleon pair, $\sqrt{s}/A$, on the order of $\sim (3 \to 4) \times 15 = 45 \to 60$ GeV. While an extremely naive estimate, this does seem to be a reasonable estimate for the $\sqrt{s}/A$ where the distribution of particles in AA collisions changes dramatically, developing a "central plateau" as a function of

pseudo-rapidity. Thus perhaps the appearance of the central plateau is where the effects of saturation first appear.
The details are far more involved, but but our understanding of saturation, which is known formally as the Color Glass [46], gives us some confidence that the system is described in terms of "saturated" gluons, with a characteristic momentum $p_{sat} \sim 1 - 2$ GeV.
The Color Glass changes all assumptions in one fundamental respect. Following Bjorken, the usual assumption is that the system thermalizes on a "typical" hadronic time scale, ~ 1 fm/c. If saturation kicks in, however, all typical scales are then given in terms of $1/p_{sat}$, which is a *much* smaller time scale, $\sim .1 - .2$ fm/c. If this is the relevant scale, then evolution to a thermal state, for a nucleus of size 6 fm, appears much more plausible. Certainly it yields testable predictions, which are testable experimentally.

2. $Z(N)$ symmetries in $SU(N)$

I start by reviewing how, following 't Hooft [2], a global $Z(N)$ symmetry emerges from a local $SU(N)$ gauge theory. The action, including quarks, is

$$\mathcal{L} = \frac{1}{2} \operatorname{tr} G_{\mu\nu}^2 + \overline{q}\, i \, \not{D} q \,, \tag{2.1}$$

where

$$D_\mu = \partial_\mu - igA_\mu \quad , \quad G_{\mu\nu} = \frac{1}{-ig} [D_\mu, D_\nu] \,; \tag{2.2}$$

$A_\mu = A_\mu^a t^a$, with the generators of $SU(N)$ normalized as $\operatorname{tr}(t^a t^b) = \delta^{ab}/2$. The Lagrangian is invariant under $SU(N)$ gauge transformations Ω,

$$D_\mu \to \Omega^\dagger D_\mu \Omega \quad , \quad q \to \Omega^\dagger q \,. \tag{2.3}$$

As an element of $SU(N)$, Ω satisfies

$$\Omega^\dagger \Omega = 1 \,, \; \det \Omega = 1 \,. \tag{2.4}$$

Here Ω, as a local gauge transformation, is a function of space-time.
There is one especially simple gauge transformation — a constant phase times the unit matrix:

$$\Omega_c = e^{i\phi}\, \mathbf{1} \,. \tag{2.5}$$

To be an element of $SU(N)$, the determinant must be one, which requires

$$\phi = \frac{2\pi j}{N} \quad , \quad j = 0, 1 \ldots (N-1) \,. \tag{2.6}$$

Since an integer cannot change continuously from point to point, this defines a global $Z(N)$ symmetry.

2.1. $Z(N)$ AT NONZERO TEMPERATURE

As a particular gauge transformation, $Z(N)$ rotations are always a symmetry of the Lagrangian, either with or without quarks. I now show that with quarks, they are not a symmetry of the theory, because they violate the requisite boundary conditions.

I work in Euclidean space-time at a temperature T, so the imaginary time coordinate τ, is of finite extent, $\tau : 0 \to \beta = 1/T$. The proper boundary conditions in imaginary time are dictated by the quantum statistics which the fields must satisfy. As bosons, gluons must be periodic in τ; as fermions, quarks must be anti-periodic:

$$A_\mu(\vec{x},\beta) = +A_\mu(\vec{x},0) \quad , \quad q(\vec{x},\beta) = -q(\vec{x},0) \; . \tag{2.7}$$

Obviously any gauge transformation which is periodic in τ respects these boundary conditions. 't Hooft noticed, however, that one can consider more general gauge transformations which are only periodic up to Ω_c:

$$\Omega(\vec{x},\beta) = \Omega_c \quad , \quad \Omega(\vec{x},0) = 1 \; . \tag{2.8}$$

Color adjoint fields are invariant under this transformation, while those in the fundamental representation are not:

$$A^\Omega(\vec{x},\beta) = \Omega_c^\dagger A_\mu(\vec{x},\beta)\Omega_c = A_\mu(\vec{x},\beta) = +\,A_\mu(\vec{x},0) \; , \tag{2.9}$$

$$q^\Omega(\vec{x},\beta) = \Omega_c^\dagger\, q(\vec{x},\beta) = e^{-i\phi} q(\vec{x},\beta) \neq -\,q(\vec{x},0) \; . \tag{2.10}$$

Here I have used the fact that Ω_c, as a constant phase times the unit matrix, commutes with any $SU(N)$ matrix. Consequently, pure $SU(N)$ gauge theories have a global $Z(N)$ symmetry which is spoiled by the addition of dynamical quarks.

In the pure glue theory, an order parameter for the $Z(N)$ symmetry is constructed using the thermal Wilson line:

$$\mathbf{L}(\vec{x}) = \mathbf{P} \exp\left(ig \int_0^\beta A_0(\vec{x},\tau) d\tau \right) \; ; \tag{2.11}$$

g is the gauge coupling constant, and A_0 the vector potential in the time direction. The symbol $\mathbf{P}$ denotes path ordering, so that the thermal Wilson line transforms like an adjoint field under local $SU(N)$ gauge transformations:

$$\mathbf{L}(\vec{x}) \to \Omega^\dagger(\vec{x},\beta)\, \mathbf{L}(\vec{x})\; \Omega^\dagger(\vec{x},0) \; . \tag{2.12}$$

The Polyakov loop [1] is the trace of the thermal Wilson line, and is then gauge invariant:

$$\ell = \frac{1}{N} \operatorname{tr} \mathbf{L} . \tag{2.13}$$

Under a global $Z(N)$ transformation, the Polyakov loop ℓ_1 transforms as a field with charge one:

$$\ell \rightarrow e^{i\phi} \ell . \tag{2.14}$$

At very high temperature, the theory is nearly ideal, so $g \approx 0$, and naively one expects that $\langle \ell \rangle \sim 1$. Instead, the allowed vacua exhibit a N-fold degeneracy:

$$\langle \ell \rangle = \exp\left(\frac{2\pi i j}{N}\right) \ell_0 \ , \quad j = 0, 1 \ldots (N-1) \ , \tag{2.15}$$

defining ℓ_0 to be real; $\ell_0 \rightarrow 1$ as $T \rightarrow \infty$. Any value of j is equally good, and signals the spontaneous breakdown of the global $Z(N)$ symmetry.
At zero temperature, confinement implies that ℓ_0 vanishes [2]. The modulus, ℓ_0, is nonzero above T_c:

$$\ell_0 = 0 \ , \quad T < T_c \ ; \quad \ell_0 > 0 \ , \quad T > T_c . \tag{2.16}$$

As is standard, if ℓ_0 turns on continuously at T_c, the transition is of second order; if it jumps at T_c, it is of first order. What is atypical is that the $Z(N)$ symmetry is broken at high, instead of low, temperatures. For a heuristic explanation of this in terms of $Z(N)$ spins, see sec. (3.2).
One can also understand what it means to say that the global $Z(N)$ symmetry is violated by the presence of dynamical quarks. In the pure glue theory, $Z(N)$ rotations take us from one degenerate vacua to another, all of which have the same pressure; see sec. (3.2). Adding dynamical quarks (with real masses), the stable vacuum is that for which $\langle \ell \rangle$ is real, $j = 0$ [23, 24]. As discussed in sec. (3.4), because of quarks the pressure for a $Z(N)$ state with $j \neq 0$ is less than the stable vacuum; thermodynamically, they are unstable. What is exciting to some of us [38], however, is the possibility that these $Z(N)$ rotated states might be *meta*stable. Such "$Z(N)$ bubbles" could have cosmologically interesting consequences [38]. Others contest whether any of this makes any sense [39].
The usual interpretation of the Polyakov loop is as the free energy of an infinitely heavy test quark [47]:

$$\langle \ell \rangle = \exp\left(-F_{test}/T\right) . \tag{2.17}$$

This cannot be quite right: when $N = 2$, the left hand side can be of either sign, while for $N \geq 3$, it is complex. In contrast, free energies are real, so the right hand side is positive.

I suggest a different view. Consider the propagator for a scalar in a background gauge field (the extension to fermions is irrelevant here). This propagator is given by a Feynman sum over paths:

$$\Delta = \int \mathcal{D}x^\mu \exp\left(- \int ds \left(\frac{\dot{x}^2}{2} + m + igA^\mu \dot{x}^\mu \right) \right) , \tag{2.18}$$

with s the length of the path for the worldline of the particle, and $\dot{x} = dx^\mu/ds$. A very heavy quark moves in a straight line; in imaginary time, it sits wherever you put it. As a colored field, however, it also carries a color Aharonov-Bohm phase. This phase is nontrivial, and is precisely the thermal Wilson line. Thus: *the Polyakov loop, ℓ, is the trace of the propagator for a test quark.*

Confinement then means that (the trace of) this propagator vanishes. For two colors, for example, the confining vacuum is $Z(2)$ symmetric: it is composed of domains, of definite size, in which $\ell = +1$ and $\ell = -1$. As the test quark travels through each domain, it picks up one phase or the other. Over a very long path, these phases cancel out, giving zero overall. The same holds for higher N, except that there are then N types of domains. This picture appears analogous to the localization of an electron in a random potential [48].

$Z(N)$ rotations can be expressed in terms of the canonical formalism [2, 35]. In $A_0 = 0$ gauge, usually the partition function is strictly a trace of $\exp(-\mathcal{H}/T)$ ($\mathcal{H}$ is the corresponding Hamiltonian), sandwiched between the same state:

$$Z = \Sigma \, \langle\psi| \exp(-\mathcal{H}/T)|\psi\rangle. \tag{2.19}$$

The sum is over all gauge invariant states in the theory. In a canonical formalism, one inserts a projector to ensure that the states are gauge invariant, satisfying Gauss' Law, although the process is standard. Instead, what enters here is a "twisted" trace:

$$Z(\Omega_c) = \Sigma \, \langle\psi^{\Omega_c}| \exp(-\mathcal{H}/T)|\psi\rangle. \tag{2.20}$$

Here, ψ^{Ω_c} represents the gauge transform of ψ by the gauge transformation Ω_c. This is not automatically equal to ψ, because Gauss' Law only ensures that the state is invariant under local gauge transformations, and does not restrict its behavior under the global gauge transformations. This twisted trace is only possible in a gauge theory.

3. $Z(N)$ Vacua and Bubbles

3.1. TUNNELING BETWEEN DEGENERATE $Z(N)$ VACUA

Typically, discussions of gauge theories at nonzero temperature work up from zero temperature. But the zero temperature theory confines, which is complicated.

Instead, I work down from infinite temperature, in a gas of nearly ideal gluons. I now compute the amplitude to tunnel from one $Z(N)$ vacua to another by semi-classical means [32, 33, 34, 37].

For simplicity, I work with two colors. To compute the interface tension, I put the system in a box:

$$\tau : 0 \to \beta \quad , \quad x, y : 0 \to L_t \quad , \quad z : 0 \to L \, . \tag{3.1}$$

I choose one arbitary direction, say the z-direction, and make that much longer than the transverse spatial directions, x and y, and than the direction in imaginary time, τ. I impose boundary conditions such that the Polyakov loop has one value in $Z(2)$ at one end of the box, and the other value at the other end of the box:

$$\ell(0) = 1 \quad , \quad \ell(L) = -1 \, . \tag{3.2}$$

With these boundary conditions, the system is forced to form an interface between the two ends of the box. The simplest interface is one which is constant in the transverse directions. Thus the natural expectation is that with the above boundary conditions, the action is

$$S_{inter} \;=\; L_t^2 \frac{c}{g^2} \quad , \quad c \;=\; c_0 \;+\; g c_1 \;+ \ldots \tag{3.3}$$

One expects the result to start as $\sim 1/g^2$, as a semiclassical probability for tunneling in weak coupling. This is standard with instantons, *etc.* Higher order corrections to the leading term, c_0, are generated by including quantum effects, and produce the corrections c_1, *etc.*

As an ansatze for the $Z(N)$ interface, I take

$$A_0^{cl}(z) \;=\; \frac{\pi T}{g}\, q(z)\, \sigma_3 \quad , \quad \sigma_3 \;=\; \begin{pmatrix} 1 & 0 \\ 0 & -1 \end{pmatrix} \, . \tag{3.4}$$

With this ansatz, the Polyakov loop is

$$\ell(z) \;=\; \cos\left(\pi q(z)\right) \, . \tag{3.5}$$

Thus the boundary conditions are satisfied by taking

$$q(0) = 0 \quad , \quad q(L) = 1 \, . \tag{3.6}$$

At the classical level, the action of the above configuration is:

$$S^{cl} \;=\; \int_0^\beta d\tau \int d^3x \, \frac{1}{2} \, \mathrm{tr} \left(\left(G_{\mu\nu}^{cl} \right)^2 \right)$$

$$= L_t^2 \, \frac{2\pi^2 T}{g^2} \int dz \left(\frac{dq}{dz} \right)^2 \, . \tag{3.7}$$

Unsurprisingly, the action for the gauge field becomes a kinetic term for the classical field. There is only a kinetic term, since the classical field commutes with itself.

This implies, however, that at the classical level, there is *no* difference between the two vacua, or indeed any state with $q \neq 0$! One can take $q(z) = z/L$; then the action is $\sim L/L^2 \sim 1/L$, and vanishes as $L \to \infty$. In terms of the above,

$$c_0 = 0 \,. \tag{3.8}$$

I now show that quantum corrections generate a nonzero value for c_1. Since c_0 vanishes, this is then the leading term in a semiclassical expansion; the tunneling probability is then not $\sim 1/g^2$, but only $\sim 1/g$.

To show this, it is necessary to compute the quantum corrections about the above semiclassical configuration, taking

$$A_\mu = A^{cl}_\mu + A^q_\mu \,. \tag{3.9}$$

The computation is a bit involved, but an excellent exercise in the use of the background field method [50], which is always good to know.

It is convenient to take background field gauge,

$$D^{cl}_\mu A^q_\mu = \partial_\mu A^q_\mu - ig[A^{cl}_\mu, A^q_\mu] = 0 \,. \tag{3.10}$$

I work in euclidean space-time with $(++++)$ metric. Note that the appropriate covariant derivative is that in the adjoint representation. With this gauge fixing, the Lagriangian density is

$$\mathcal{L} = \frac{1}{2}\,\mathrm{tr}\left(G^{cl}\right)^2 + \frac{1}{\xi}\,\mathrm{tr}\left(D^{cl}\cdot A^q\right)^2 + \overline{\eta}\left(-D^{cl}\cdot D\right)\eta \,, \tag{3.11}$$

suppressing ugly vector indices.

With this form, it is easy integrating out the quantum fields to one loop order, and obtain the quantum action

$$\mathcal{S}^q(A^{cl}) = \frac{1}{2}\,\mathrm{tr}\log\left(\Delta^{-1}_{\mu\nu}\right) - \mathrm{tr}\log\left(\Delta^{-1}_\eta\right) . \tag{3.12}$$

At one loop order, the full effective action is the sum of the classical action, (3.7), and the quantum action, (3.12). The quantum action involves the inverse propagators in a background field. That for the gluon is

$$\Delta^{-1}_{\mu\nu} = -D^2_{cl}\delta_{\mu\nu} + (1-\xi^{-1})D^{cl}_\mu D^{cl}_\nu + 2ig\,[G^{cl}_{\mu\nu},\ ] \,, \tag{3.13}$$

while that for the ghost is (to lowest order in g)

$$\Delta^{-1}_\eta = -D^2_{cl} \,. \tag{3.14}$$

These results are valid for an arbitrary background gauge field.
I now make a crucial assumption, and assume that the field $q(z)$ is constant in space. For the relevant tunneling amplitude, in fact $q(z)$ does depend upon z; what happens, however, is that for the quantum action, this variation only enters to higher order in the coupling constant.
This assumption vastly simplifies the problem. Since A_0 lies in the σ_3 direction, it is a diagonal matrix, and covariant derivatives commute:

$$[D_\mu^{cl}, D_\nu^{cl}] \sim G_{\mu\nu}^{cl} = 0 . \tag{3.15}$$

For example, one can easily show that the quantum action is independent of the gauge fixing condition. The variation of the quantum action with respect to the gauge fixing parameter ξ is

$$\frac{\partial}{\partial \xi^{-1}} \mathcal{S}^q = \frac{1}{2} \text{ tr} \left(-D_\mu^{cl} D_\nu^{cl} \Delta_{\mu\nu}^{cl} \right) , \tag{3.16}$$

with $\Delta_{\mu\nu}^{cl}$ the gluon propagator. Normally, this is difficult to compute. In the present example, however, if covariant derivatives can be assumed to commute with each other, then they can be treated just like ordinary derivates, so that

$$\Delta_{\mu\nu}^{cl} = \frac{\delta^{\mu\nu}}{-D_{cl}^2} + (1-\xi) \frac{D_{cl}^\mu D_{cl}^\nu}{(-D_{cl}^2)^2} . \tag{3.17}$$

Consequently,

$$\frac{\partial}{\partial \xi^{-1}} \mathcal{S}^q = \frac{\xi}{2} \text{ tr}(1) . \tag{3.18}$$

Thus there is gauge dependence in the quantum action, but it is completely independent of the background field, and so can be safely ignored.
Consequently, I take background Feynman gauge, $\xi = 1$, and

$$\mathcal{S}^q = \text{tr} \log \left(-D_{cl}^2 \right) . \tag{3.19}$$

The overall factor is expected for a massless gauge field, with two (spin) degrees of freedom.
To compute the determinant in this background field, I introduce the "ladder" basis,

$$\sigma^+ = \frac{1}{\sqrt{2}} \begin{pmatrix} 0 & 1 \\ 0 & 0 \end{pmatrix} , \quad \sigma^- = \frac{1}{\sqrt{2}} \begin{pmatrix} 0 & 0 \\ 1 & 0 \end{pmatrix} . \tag{3.20}$$

This is useful because of the commutation relations:

$$[\sigma_3, \sigma^\pm] = \pm 2\sigma^\pm , \tag{3.21}$$

so that the covariant derivative becomes

$$D_0^{cl}\sigma^{\mp} = \left(\partial_0 - ig\left(\frac{\pi T}{g}q\right)[\sigma_3, \,]\right)\sigma^{\mp} = i\,(2\pi T)\,(n \pm q)\sigma^{\mp}\,. \quad (3.22)$$

In the last expression, I have gone to momentum space. Remember that given the periodic boundary conditions at nonzero temperature,

$$k_0 = i\,2\pi T\,n\,, \quad (3.23)$$

where for a bosonic field, such as a gluon, the periodic boundary conditions require that n be an integer, $n = 0, \pm 1, \pm 2 \ldots$. In the present case, it is handy to introduce the shifted momentum,

$$k_0^{\pm} = i\,2\pi T\,(n \pm q)\,. \quad (3.24)$$

In the trace, the sign of q doesn't matter, so that in momentum space,

$$\mathcal{S}^q = 2\;\mathrm{tr}\;\log\left((k_0^+)^2 + \vec{k}^2\right)\,. \quad (3.25)$$

In computing integrals at nonzero temperature, the usual approach is to do the sum over the k_0's first by contour integration or the like, and then integrate over the spatial momentum. In the present example, instead it is better to first integrate over the spatial momentum, and then sum over the k_0's, using zeta functions tricks. Only the variation of the action with respect to q,

$$\frac{\partial}{\partial q}\mathcal{S}^q = 8\pi T\,(\beta L_t^2 L)\,T\sum_{n=-\infty}^{+\infty}\int\frac{d^3k}{(2\pi)^3}\,\frac{k_0^+}{(k_0^+)^2 + \vec{k}^2}\,, \quad (3.26)$$

is needed. The integral over $\vec{k}$ can be done using dimensional regularization, viewing k_0^+ like a mass. Using the standard integral,

$$\int d^n k\,\frac{1}{(k^2+m^2)^a} = \frac{\Gamma(a-n/2)}{\Gamma(a)}\,\frac{\pi^{n/2}}{(m^2)^{a-n/2}}\,, \quad (3.27)$$

the result is finite:

$$\frac{\partial}{\partial q}\mathcal{S}^q = 8\pi L_t^2 L\,T\sum_{n=-\infty}^{+\infty} k_0^+\left(\frac{-1}{4\pi}|k_0^+|\right)$$

$$= -8\pi^2 T^3 L_t^2 L\sum_{n=-\infty}^{+\infty}(n+q)|n+q|\,. \quad (3.28)$$

The sum over n, where n runs from minus infinity to plus infinity, is turned into a sum from zero to plus infinity:

$$= -8\pi^2 T^3 L_t^2 L\sum_{n=0}^{+\infty}\left((n+q)^2 - (n+(1-q))^2\right) \quad (3.29)$$

While these sums are very divergent, mathematicians know how to handle them. They are defined by the analytic continuation of the Riemann zeta-function:

$$\zeta(z,q) = \sum_{n=0}^{+\infty} \frac{1}{(n+q)^z} . \tag{3.30}$$

Using

$$\zeta(-2,q) = -\frac{1}{12}\frac{d}{dq}\left(q^2(1-q)^2\right) , \tag{3.31}$$

gives

$$S^q = L_t^2 \frac{4\pi^2 T^3}{3} \int dz\, q^2 (1-q)^2 . \tag{3.32}$$

This expression is only valid for $q : 0 \to 1$; given the derivation, it is a periodic function in q with period one. Also note that in anticipation of later results, I have replaced a factor of the length L by $\int dz$.

Physically, the computation of quantum corrections has lifted the degeneracy in q by generating a potential for q. In the full effective action, it helps to rescale the length in the z direction, introducing

$$z' = \sqrt{\frac{2}{3}}\, gT\, z . \tag{3.33}$$

With this rescaling, the complete effective action at one-loop order is

$$\mathcal{S}^{eff} = \mathcal{S}^{cl} + \mathcal{S}^q = L_t^2 \frac{4\pi^2}{\sqrt{6}} \frac{T^2}{g} S' , \tag{3.34}$$

where

$$S' = \int dz' \left(\left(\frac{dq}{dz'}\right)^2 + q^2 (1-q)^2 \right) . \tag{3.35}$$

The factor of T^2 follows on dimensional grounds, as at high temperature, T is the only natural mass scale in the problem. (The renormalization mass scale doesn't appear until next to leading order [32, 33].)

What is most interesting is that the $1/g^2$, which we had expected, becomes only a $1/g$. This is because our effective action only acquires a potential at one loop order.

This begs the important question, why should this effective action be trusted to one loop order? What about the effects to higher loop order? The point is that in the new effective action, the relevant distance scale is not just $1/T$, but $1/(gT)$: notice the factor of gT in the definition of the rescaled length z'. Thus for small g, any variations in the effective action occur over much larger distance scales than $1/T$. This is why our method of derivation — ignoring the variation of $q(z)$ in the quantum action — works. It does vary in space, but in weak coupling, this variation is very slow, and can be ignored.

Having reduced the effective action to the above form, we merely want the "kink" which interpolates between the two vacua. While the general form of the kink is well known, in fact we only need the action. To compute the action, we note that the "energy" ϵ for this system is conserved:

$$\epsilon = \left(\frac{dq}{dz'}\right)^2 - q^2(1-q)^2 . \tag{3.36}$$

Here we view the spatial coordinate z' as a kind of time; saying the energy is conserved means that it is independent of z'. The energy is then a constant of motion; with these boundary conditions, this energy vanishes. Zero energy implies

$$\frac{dq}{dz'} = q(1-q) . \tag{3.37}$$

Using this,

$$\mathcal{S}' = 2 \int dz'\, q^2(1-q)^2 = 2 \int_0^1 dq\, q(1-q) = \frac{1}{3} . \tag{3.38}$$

Putting everything together,

$$\mathcal{S}^{eff} = L_t^2 \, \frac{4\pi^2}{3\sqrt{6}} \, \frac{T^2}{g} , \tag{3.39}$$

which is the final result for two colors.

This demonstrates that $c_1 \neq 0$; the interface tension vanishes classically, but is generated through quantum effects. The most interesting feature of the analysis is how the tunneling amplitude goes from the expected $1/g^2$ to just $1/g$. There are examples in string theory where tunneling amplitudes are not $1/g^2$, but only $1/g$. These examples appear special to string theory, as the appearance of the coupling constant in this fashion is more or less natural. That is not the case here; the $1/g$ is really novel. The appearance of the inverse distance scale gT is reasonable, as the Debye screening mass in a thermal bath.

3.2. SPINS AND POLYAKOV LOOP SPINS: DUALITY?

Remember the behavior of a usual Ising magnet, in which spins $\sigma_i = \pm$ interact through a coupling constant J_{mag}. The Hamiltonian is:

$$\mathcal{H} = - J_{mag} \, \Sigma_{i,\hat{n}} \, \sigma_i \cdot \sigma_{i+\hat{n}} . \tag{3.40}$$

The sum is over all lattice sites, i, and nearest neighbors to i, $\hat{n}$. The spins align at low temperatures, and disorder at high temperatures. This is just because the partition function is $\mathcal{Z} \sim \exp(-\mathcal{H}/T)$. In magnets, the spin-spin coupling is more or less independent of temperature, so that with $\sim J_{mag}/T$ in the exponential, ordering wins at low temperature, and loses at high temperature.
The interface tension can easily be estimated. The simplest interface is to take all spins on the left hand side spin up, and all on the right hand side, spin down. If a is the lattice spacing, then the interface tension, defined as above, is $\sim J$; by construction, its width is a. This very sharp interface is not the configuration of lowest energy, but the true interface tension is of order $\sim J$, with a width of order, a. Another example, more familiar to field theorists, is given by a scalar field with a double well potential [36].
Now consider an effective lagrangian for Polyakov loops. Over distances $> 1/T$, the four-dimensional theory reduces to an effective theory of spins in three spatial dimensions. For two colors, Polyakov loops are a type of $Z(2)$ spin, with $\ell = \pm$. From the above $\mathcal{S}^{eff}$,

$$J_{Polyakov} \sim \frac{T^2}{g} . \tag{3.41}$$

Now it is easy to understand why Polyakov loop spins order at high, instead of low, temperature. For magnets, the partition function involves $\exp(-J_{mag}/T)$; for Polyakov loop spins, instead we have $\exp(-J_{Polyakov}/T)$; but as $J_{Polyakov} \sim T^2$, in all the temperature dependence in the exponential is not $1/T$, as for ordinary magnets, but T!
This also leads to a natural conjecture of duality: that the temperature for Polyakov loop spins, and an ordinary magnet, are related as

$$T_{Polyakov} \sim \frac{1}{T_{mag}} . \tag{3.42}$$

I have assumed that the variation of the gauge coupling constant with temperature can be neglected.
This argument is extremely heuristic, and carries an important qualification. For ordinary spins, the lattice spacing is of course fixed. (This is true as well for a scalar field with double well potential [36].) From the derivation of the interface tension above, however, the width of the interface is the inverse Debye mass, $\sim 1/(gT)$. Thus the argument fails in the limit of high temperature, since then the size of any single domain is becoming very large, $\sim \sqrt{\log(T)}/T$, as $T \to \infty$. This doesn't contradict the conclusion of ordering at high temperature, since any single domain is, by definition, an ordered state.
Now assume that the transition is of second order, as happens for two colors [20]. Then the correlation length diverges at T_c; with Polyakov loop spins, it decreases as T increases from T_c. This divergence is determined as usual by scaling at a

critical point. Even so, the underlying length scale which fixes the lattice spacing for effective Polyakov loop spins is fixed, set by a mass scale proportional to T_c, *etc.* Near T_c, we have implicitly made the assumption that the interface tension remains proportional to $\sim T^2$. This cannot be true very near T_c, since for a second order transition, the interface tension must vanish at T_c.

So is the argument of any use? Well, if d is the number of space-time dimensions, then simply on geometric grounds, $J_{Polyakov} \sim T^{d-2}$. In three dimensions, then, $J_{Polyakov} \sim T$; depending upon the value of $J_{Polyakov}/T$, the system can still order above T_c, so there is no obvious contradiction. However, it does suggest that the *width* of the critical region is much *narrower* in four, as opposed to three, dimensions. For $SU(2)$ gauge theories, this comparison is of interest in its own right.

For a strongly first order transition, at first sight one might think that one should be able to directly check if $J_{Polyakov} \sim T^{d-2}$. This is complicated by the fact that near T_c, $Z(N)$ states don't tunnel directly from one to another, but from one $Z(N)$ state, to the symmetric vacuum, to another $Z(N)$ state.

3.3. POLYAKOV LOOP SPINS: POTTS VERSUS GKA

The above analysis can be extended to more than two colors. For two colors, there is only one interface tension, between $\ell = +1$ and $\ell = -1$. For N colors, the vacuum is one of the Nth roots of unity, $\ell = \exp(2\pi i j/N)$, $j = 1\ldots(N-1)$. By charge conjugation, under which $\ell \to \ell^*$, the states j and $N-j$ are equivalent. There are then about $\sim N/2$ distinct interface tensions.

At any N, the smallest interface tension is between $j = 0$ and $j = 1$. Defining $S^{eff} = \alpha_1 L_t^2$, then at next to leading order,

$$\alpha_1 = \frac{4(N-1)\pi^2}{3\sqrt{3N}} \frac{T^2}{g(T)} \left(1 - 12.9954\ldots \frac{g^2 N}{(4\pi)^2} + \ldots\right), \tag{3.43}$$

where the running coupling constant $g^2(T)$ is defined using a modified $\overline{MS}$ scheme [37].

I remark that Boorstein and Kutasov [34] argued that due to infrared divergences, σ_1 is not $\sim 1/g$, but one over the magnetic mass scale, $\sigma_1 \sim 1/g^2$. While hardly conclusive, at least at next to leading order, there are no sign of infrared divergences.

The interface tension from $j = 0$ to $j = k$ has been computed by Giovannangeli and Korthals Altes (GKA) [37]. The result is amazingly simple:

$$\alpha_k = \frac{k(N-k)}{N-1}\,\alpha_1 \,. \tag{3.44}$$

Now I construct an effective theory of discrete $Z(N)$ spins. I forget about the factors of temperature which preoccupied me in the previous subsection; all that

I am concerned with is the dependence on the distance between the $Z(N)$ spins. This suggests what I term the GKA model. The spins at each site of the lattice are integers j, $j = 0 \ldots (N-1)$; the Hamiltonian is

$$\mathcal{H}_{GKA} = J_{GKA} \, \Sigma_{j_i} \, k\,(N-k) \quad , \quad k = |j_i - j_{i+\hat{n}}|_{mod\,N} \, . \tag{3.45}$$

Tracing through the factors of N, and holding $g^2 N$ fixed as $N \to \infty$, the coupling constant $J_{GKA} > 0$ is of order one as $N \to \infty$.
The GKA model is in contrast to the Potts model, with Hamiltonian

$$\mathcal{H}_{Potts} = J \, \Sigma_{j_i} \, \delta_{k0} \, . \tag{3.46}$$

For the Potts model with $J > 0$, the energy is lowered if if two spins are equal, while if they differ — no matter by how much — the energy vanishes.
For two and three colors, there is no difference between the GKA model and the Potts model. For example, consider the case of three colors. Then $j = 1$ is equivalent to $j = 2$, so there is only one interface. That is, for the three roots of unity, any root is right next to the other two.
The Potts model is known to be of first order for any number of states greater than, or equal to, three. For the GKA model, in mean field theory the transition is of first order for four colors [37]; after all, the interaction between $j = 0$ and $j = 2$ is $4/3$ that between $j = 0$ and $j = 1$. Thus it would be very surprising if the GKA model wasn't also of first order when $N = 4$. Further, Lattice simulations of $SU(4)$ find a first order deconfining transition [17].
As the number of colors increases, though, the Potts and GKA models become increasingly different. Whatever N is, in the Potts model any spin state interacts equally strongly with any other spin state. In the GKA model, at large N spins only interact significantly with those which are close in spin space. For example, $\sigma_1 \sim N$, while $\sigma_j \sim N^2$ for $j \sim N/2$. It would be interesting to know the order of the phase transition in the GKA model at large N, both in mean field theory and numerically.
This assumes that the interaction between Polyakov loop spins — computed in the limit of very high temperature — remains the same all of the way down to T_c. This certainly is wrong for a second order transition, but the question here is if it is first order. Thus: does the interface tension stay $\sim j(N-j)$ (higher powers of $\sim j^2(N-j)^2$ only make the more less like Potts), or become constant, independent of j?

3.4. $Z(N)$ BUBBLES

From the one-loop effective action, we can define a "potential" for q due to gluons, $\mathcal{V}_{gl}$. I will be schematic, suppressing all inessential details, and taking two colors for now. From the computation of the one-loop effective action, it is clear that the assumption of $L_t \ll L$ was actually a matter of words; one obtains identically the same result in an infinite volume. Thus from $\mathcal{S}^q$, I define

$$\mathcal{V}_{gl}(q) \sim T^4 \left(q^2(1-q)^2 + f_g\right) ; \tag{3.47}$$

$f_g T^4$ is proportional to the free energy of gluons at a temperature T. As noted above, this potential is only valid in the region $0 \leq q \leq 1$; it is periodic, with period one, outside of this region:

$$\mathcal{V}_{gl}(q+1) = \mathcal{V}_{gl}(q) . \tag{3.48}$$

Obviously, $q = 0$ and $q = 1$ are degenerate,

$$\mathcal{V}_{gl}(0) = \mathcal{V}_{gl}(1) . \tag{3.49}$$

This follows from the $Z(2)$ symmetry of the pure glue theory.
The quark contribution is computed similarly:

$$\mathcal{V}_{qk}(q) \sim T^4 \left(2\,q^2 - q^4 + f_{qk}\right) ; \tag{3.50}$$

$f_q k T^4$ is proportional to the quark contribution to the free energy. This expression is only valid for $0 \geq q \geq 1$; else q is defined modulo one. Consequently,

$$\mathcal{V}_{qk}(q+2) = \mathcal{V}_{qk}(q) . \tag{3.51}$$

This must be true for any potential, since $q = 0$ and $q = 2$ both give $\ell = +1$. Moreover, while $q = 1$ is a an extremal point of the potential, it is a local maximum, not a minimum, with

$$\mathcal{V}_{qk}(1) > \mathcal{V}_{qk}(0) . \tag{3.52}$$

This shows how quarks violate the global $Z(2)$ symmetry of the two color theory. To one loop order, the total potential for q is the sum of the gluon and quark contributions. With many (light) quark flavors, the total potential is like the quark contribution, with a maximum at $q = 1$. Dixit and Ogilvie [38] first noticed that if the number of quark flavors isn't too large (or if the quarks are sufficiently heavy), $q = 1$ can be a local minimum; *i.e.*, $q = 1$ is *meta*stable.
The above carries through for an arbitrary number of colors and flavors. If metastable states arise, they are necessarily a $Z(N)$ state, with ℓ a (nontrivial) Nth root of unity. They are termed "$Z(N)$ bubbles" [39, 38].
This all appears to be directly analogous to the usual problem of metastable vacua, but there is one important difference. For an ordinary potential, either in quantum mechanics or in field theory, one never worries about the zero of the potential, as that can be shifted at will. In the present case, however, the zero of the potential *is* physical, and gives the free energy of the stable vacuum. This is because

the "potential" is multiplied by an overall factor of T^4, and thermodynamically, derivatives with respect to the temperature matter.
Thus there is no freedom to change the zero of the potential for q. For some $Z(N)$ bubbles, if the potential at $q \neq 0$ is much higher than $q = 0$, it is well possible that the pressure in the bubble isn't positive, but *negative*! This was noticed first by Belyaev, Kogan, Semenoff, and Weiss [39].
This is a complete disaster. I suggest a possible resolution.
When we deal with q, we are in fact dealing with an *angular* variable. If we write the potential for q in terms of the thermal Wilson line, it is

$$\mathcal{V}_{pert}(q) \sim T^4 \, q^2 \, (1-q)^2 \sim T^4 \, \mathrm{tr}\left((\log \mathbf{L})^2 \left(\mathbf{1} - (\log \mathbf{L})^2 \right) \right) . \tag{3.53}$$

This form is correct in perturbation theory, where at each point in space, $\mathbf{L}(\vec{x})$ is an element of $SU(N)$. As will become clear in the next section, however, if we construct an effective theory for $\mathbf{L}$, it no longer is an element of $SU(N)$. Then this potential, for the purely angular part of $\mathbf{L}$, is ill defined when its modulus vanishes. This ambiguity is easily cured by multiplying by an overall factor of the modulus:

$$\mathcal{V}_{non-pert}(\mathbf{L}) \sim T^4 \left(|\ell|^2 + \ldots \right) \mathrm{tr} \left((\log \mathbf{L})^2 \left(\mathbf{1} - (\log \mathbf{L})^2 \right) \right) . \tag{3.54}$$

This is rank conjecture: it is certainly a *non*-perturbative modification of the potential. There is no reason to exclude terms of higher order in $\sim |\ell|^2$.
This still does not solve the problem of the zero of the potential. I now assume further that the Polyakov Loop Model (PLM) applies [26, 27, 28, 29, 30, 31]. Ignoring the angular variation in $\log \mathbf{L}$, the PLM potential is

$$\mathcal{V}_{PLM}(\ell) \sim T^4 \left(b_2 \, |\ell|^2 + b_4 \, (|\ell|^2)^2 \right) . \tag{3.55}$$

The exact potential depends upon the number of colors and flavors, *etc.*, but this is inessential here. For $q = 0$, the "usual" free energy is given by minimizing $\mathcal{V}_{PLM}(\ell)$ with respect to ℓ; with the above convention, the "mass" squared for ℓ is negative in the deconfined phase, $b_2 < 0$, and positive in the confined phase, $b_2 > 0$.
The complete potential is the sum of $\mathcal{V}_{non-pert}(\mathbf{L})$ and $\mathcal{V}_{PLM}(\ell)$. At fixed ℓ, as before any metastable points are elements of $Z(N)$. The equation which determines ℓ, however, is changed, as any metastable state has an action which acts like a *positive* mass term for ℓ. Thus the expectation value of ℓ in a $Z(N)$ bubble — even deep in the deconfined phase — has $\ell < 1$, not $\ell = 1$. Before a $Z(N)$ bubble develops negative pressure, $\ell = 0$, with zero pressure in the PLM. More likely, $Z(N)$ bubbles become unstable in the ℓ direction before $\ell = 0$.
This could be tested on the Lattice. Compute in a theory in which the splitting between the true vacuum and the metastable state is small; (dynamical) heavy quarks

will do. Then the expectation value of ℓ should be smaller in the metastable state than in the stable vacuum. This holds regardless of the question of renormalization discussed in the next section.

If true, all perturbative calculations of the lifetime of a metastable $Z(N)$ bubble are wrong [38]. At best, they are a upper bound on the true lifetime, which is somewhat useless. On the other hand, it resurrects the possibility that $Z(N)$ bubbles — which spontaneously violate CP symmetry — might have appeared in the early Universe.

4. Renormalization of the Wilson Line

The sections following this deal with mean field theory for the thermal Wilson line. Implicitly, this assumes that it is possible to go from the bare Wilson line, as measured on the Lattice, to the renormalized quantity. How to do this on the Lattice is presently an unsolved problem; in this section I review what is known [40].

In a pure gauge theory, the expectation value of a closed Wilson loop, of length L and area A, is

$$\langle \mathrm{tr}\ \mathcal{P}\ \exp\left(ig\int A_\mu dx^\mu\right)\rangle \;=\; \exp\left(-m_0 L - \sigma A\right)\ . \tag{4.1}$$

The string tension, σ, is nonzero in the confined phase, $T \leq T_c$, and vanishes in the deconfined phase, $T > T_c$.

The concern here is not with the term proportional to the area of the loop, but with the length. This is a type of mass renormalization for an infinitely heavy quark. For example, it is easy to compute this to lowest order in perturbation theory. We are interested in an ultraviolet divergent term, so over short distances, it suffices to assume that the loop is straight. For the sake of discussion, assume that the loop runs in the time direction. (New divergences arise when there are cusps in the loop; these divergences can also be computed perturbatively, by a similar analysis [40].) Then to lowest order, there is a contribution

$$\sim -g^2\,\langle\int_0^\beta d\tau_1 \int_0^\beta d\tau_2\; A_0(\vec{x},\tau_1)A_0(\vec{x},\tau_2)\rangle \sim -\frac{g^2}{T}\int d^3k\,\frac{1}{\vec{k}^2}\ . \tag{4.2}$$

The integral is nominally divergent, but with either dimensional or Pauli-Villars regularization, the divergence vanishes [40]. This cancellation is somewhat trivial at one loop order, arising from having three powers of momentum upstairs, and two powers downstairs.

Thus one would expect divergences to arise at $\sim g^4$, which are found. However, for closed Wilson loops, *all* such divergences can be absorbed into charge renormalization [40]. This is a notable result: in a quantum field theory, generally the renormalization of any composite operator requires the calculation of its mixing with all other operators with the same mass dimension and symmetries. Like the action itself, however, the Wilson loop has a privileged status; it doesn't mix with any other operator.

As noted first by Polyakov [40], this result can be understood on the basis of reparametrization invariance for the Wilson loop. We parametrize the loop as a curve $x^\mu(s)$, where s is the length along the path. Then with $\dot{x} = dx^\mu/ds$, the term

$$\int \sqrt{\dot{x}^2}\, ds \tag{4.3}$$

is invariant under $s \to s'(s)$. Generally, physics shouldn't depend upon how we label path length along the curve. Because it has dimensions of length, however, the coefficient of this term must have dimensions of mass. With dimensional regularization, there is no such mass scale. (The renormalization mass scale only enters to ensure the proper running of the coupling constant.) A term which has no mass dimension is

$$\int \sqrt{\ddot{x}^2}\, ds\ , \tag{4.4}$$

$\ddot{x} = d^2x^\mu/ds^2$. This is not reparametrization invariant, though, and so does arise with dimensional regularization.

On the other hand, assume that the regularization scheme *does* introduce a mass scale. On the Lattice, this is the inverse lattice spacing, $\sim 1/a$. Then a term proportional to the length, L, does appear [40],

$$\sim -g^2\, \frac{L}{a}. \tag{4.5}$$

At nonzero temperature, $L/a = N_t$, the number of lattice steps in the time direction. Clearly, this is the first term in an infinite series in the coupling constant, g.

How to deal with the power divergences generated by the Lattice is at present an unsolved problem. Since in the continuum there are neither logarithmic nor even finite terms to worry about, this appears to be a technical, albeit important, problem to solve.

Why is this important? In the confined phase, this constant is not of any particular consequence, as the Wilson loop is dominated by the string tension. In the high temperature phase, however, the trace of the thermal Wilson line is the order parameter for the phase transition. It would be peculiar if a precise physical definition did not exist. Any composite operator requires a condition to fix its renormalized value; for the thermal Wilson line, the natural prescription is that Polyakov loops approach one as $T \to \infty$.

To compute the leading perturbative correction to the thermal Wilson line, it is necessary to include effects from the Debye mass, $m_D \sim gT$. Replacing the bare propagator for A_0, $1/\vec{k}^2$, by $1/(\vec{k}^2 + m_D^2)$,

$$\sim -\frac{g^2}{T} \int d^3k \, \frac{1}{\vec{k}^2 + m_D^2} \, . \tag{4.6}$$

This divergent integral can be computed using either dimensional or Pauli-Villars regularization. Or, one can just subtract the integral with $m_D = 0$:

$$-\frac{g^2}{T} \int d^3k \left(\frac{1}{\vec{k}^2 + m_D^2} - \frac{1}{\vec{k}^2} \right) \sim -\frac{g^2}{T}(-m_D) \sim +g^3 \, . \tag{4.7}$$

This was first demonstrated by Gava and Jengo [41]. That is, while the leading term is negative in the bare theory, it is positive after regularization. This change in sign is unremarkable, as the sign of a renormalized operator is not preserved under regularization.

This appears to indicate that the renormalized thermal Wilson line is not a unimodular matrix. One concern is that any quantity $\sim g^3$ really arises from a two-loop graph, $\sim g^4$, times an infrared singular piece $\sim 1/m_D \sim 1/g$. Thus it is necessary to ensure that the above is the only infrared singular term at this order.

A different calculation was performed by Korthals Altes [33]. He computed the one loop corrections to the thermal Wilson line in a background A_0 field. The method is identical to that used in sec. II to compute the interface tension. Classically, the thermal Wilson line is a special unitary matrix. Korthals Altes finds that the one-loop corrections to the thermal Wilson line are not only infinite (!), but generate a matrix which is neither unitary nor special. On the other hand, all Polyakov loops are finite.

Thus even in the continuum, the renormalization of the thermal Wilson line, and Polyakov loops, remains an unsolved problem.

A way of measuring renormalized Polyakov loops on the Lattice has been proposed by Zantow *et al.* [43]. They compute only two-point functions of the Polyakov loop. At short distances, the static potential can be computed perturbatively, which allows one to extract the renormalized Polyakov loop.

5. Deconfining Transition for Two, Three, and Four Colors

5.1. POLYAKOV LOOPS AND QUARKLESS BARYONS

So far I have been concerned with the (pure glue) theory at very high temperatures. Now I turn to the question of its behavior near the critical temperature. I review Lattice results on the order of the phase transition for two [20], three [21], and four [17] colors, and then ask what constraints it places on the mean field theory for Polyakov loops.

Up to this point, I have only considered the trace of the thermal Wilson line in the fundamental represenation, which is the the Polyakov loop ℓ. By a local gauge

transformation, at each point in space one can diagonalize the thermal Wilson line. These eigenvalues are gauge invariant, so since $\mathbf{L}(\vec{x})$ is an $SU(N)$ matrix, at each point there are $N-1$ independent degrees of freedom. Another $N-1$ degrees of freedom are given by the trace of powers of $\mathbf{L}$, $\mathrm{tr}\,\mathbf{L}^j$, $j = 1\ldots(N-1)$.
Under a global $Z(N)$ transformation, the "usual" Polyakov loop transforms as a field with charge-one, eq. (2.14); thus I relabel it ℓ_1. Polyakov loops with higher $Z(N)$ charge are easy to construct. I introduce the traceless part of $\mathbf{L}$:

$$\widetilde{\mathbf{L}} = \mathbf{L} - \ell_1 \mathbf{1} \, . \tag{5.1}$$

Then I define the charge-two Polyakov loop to be

$$\ell_2 = \frac{1}{N}\,\mathrm{tr}\,\widetilde{\mathbf{L}}^2 = \frac{1}{N}\,\mathrm{tr}\,\mathbf{L}^2 - \frac{1}{N^2}\,(\mathrm{tr}\,\mathbf{L})^2 \, , \tag{5.2}$$

where

$$\ell_2 \to e^{2i\phi}\ell_2 \, , \tag{5.3}$$

with ϕ as in eq. (2.6). There are two operators with charge-two, ℓ_2 and ℓ_1^2.
For example, consider two colors, and the parametrization of the thermal Wilson line in the strict perturbative regime, (3.4). The charge-one loop is $\ell_1 = \cos(\pi q)$, eq. (3.5), while the charge-two loop is $\ell_2 = -\sin^2(\pi q)$. At high temperature, where $q = 0, 1$, $\langle \ell_1 \rangle \to \pm 1$, while $\langle \ell_2 \rangle \to 0$.
I note that Polyakov loops of charge-two and beyond are related to the trace of the thermal Wilson line in higher $SU(N)$ representations. For two colors, in perturbation theory the trace of the Wilson line in the adjoint representation is $\mathrm{tr}(\mathbf{L}_{adj}) = 1 + 2\ell_2$. So far, though, I haven't found this particularly useful.
Continuing on,

$$\ell_3 = \frac{1}{N}\,\mathrm{tr}\,\widetilde{\mathbf{L}}^3 \tag{5.4}$$

has charge three under the global $Z(N)$ symmetry. Other charge-three operators are ℓ_1^3 and $\ell_1\ell_2$. The construction of operators with higher $Z(N)$ charge proceeds similarly. For example, operators with charge four, independent of the singlet part, are given by $\mathrm{tr}\,\widetilde{\mathbf{L}}^4$ and $(\mathrm{tr}\,\widetilde{\mathbf{L}}^2)^2$, *etc.*
I stress that both the expectation values, and correlations functions of, Polyakov loops of arbitrary charge are well worth measuring on the Lattice. When the $Z(N)$ symmetry is spontaneously broken at $T > T_c$, Polyakov loops ℓ_j with charge $j = 1\ldots(N-1)$ all acquire nonzero expectation values. As Polyakov loops of charge-two and beyond are constructed from the traceless part of the thermal Wilson line, they aren't that interesting at high temperature; as $T \to \infty$, their expectation values are proportional to nonzero powers of g^2, times powers of T to make up the mass dimension. Near T_c, however, there is nothing general which can be said about their expectation values.

I will assume that for a second order deconfining transition, the only critical field is the charge-one loop [3]. Even so, Polyakov loops of charge-two and beyond will certainly affect non-universal behavior. One notable example is the Polyakov Loops Model [25, 26, 27, 28, 29, 30, 31], which conjectures a relationship between the expectation value of Polyakov loops and the pressure. The original model assumed that only the charge-one loop mattered, but I no longer see why the expectation values of higher-charge loops are not important as well.
There are certain Polyakov loops which have a privileged status: these are those with charge-N. As they are neutral under $Z(N)$, their expectation values are nonzero at any temperature. I term such operators *quarkless baryons*.
In QCD, a baryon is N quarks tied together through an antisymmetric tensor in color space. One can also consider a more general object, a baryon "junction" [49]. This is an antisymmetric color tensor, with N Wilson lines coming out of it. Putting quarks at the end of each line gives the usual QCD baryon, since in the confined phase, Wilson lines are short, on the order of $\sim 1/\sqrt{\sigma}$, where σ is the string tension. While directly related to QCD baryons, without quarks, baryon junctions are not gauge invariant: only junction anti-junction pairs are.
In contrast, all quarkless baryons are gauge invariant. The simplest quarkless baryon is ℓ_1^N. In mean field theory, this is zero in the confined phase, and nonzero above. In the full quantum theory, $\langle \ell_1^N \rangle \neq 0$ at all T. While there is not good Lattice data on this expectation value, I assume that it is small below T_c, and large above, but this is just a guess. Since junction anti-junction pairs involve N Wilson lines, they are directly related to the operators for quarkless baryons, ℓ_1^N, *etc.*.

5.2. EFFECTIVE THEORIES FOR POLYAKOV LOOPS

I next turn to the construction of an effective theory for the thermal Wilson line. Remember how this proceeds with an Ising model on a lattice. While the value of the spin on each site is ± 1, after an effective spin is computed by averaging over a domain of fixed size, the result effective spin is a continuous variable, $\phi(\vec{x})$. The effective theory is just the usual ϕ^4 theory. By the renormalization group, it is in the same universality class as the original Ising model.
The analogous proceedure can be carried through for the thermal Wilson line. The effective thermal Wilson line, constructed by a gauge invariant [25] average over a domain of some size, is not an $SU(N)$ matrix, but has more degrees of freedom. I then consider *all* Polyakov loops, from charge-one up to charge-N. Of course there is no reason to stop there, but presumably the number of effective fields which really matters is limited.
In an effective Lagrangian, the first thing to ask about are the mass terms:

$$\mathcal{L}^{eff} = m_1^2 |\ell_1|^2 + m_2^2 |\ell_2|^2 + \dots \tag{5.5}$$

The simplest assumption is that for $T \geq T_c$, condensation is driven by a negative mass term for the charge-one loop:

$$m_1^2 < 0 \quad , \quad T > T_c \quad , \quad m_1^2 > 0 \quad , \quad T < T_c \, , \tag{5.6}$$

and that the masses for all higher loops are positive at all temperatures,

$$m_2^2 > 0 \quad , \quad m_3^2 > 0 \ldots \tag{5.7}$$

If so, then the charge-one loop controls the critical behavior [3].
There is good reason why one expects that condensation is driven by that of the charge-one loop, and *not* by loops with higher charge. If the mass for the charge-one loop is negative, the favored vacuum is given by maximizing $|\text{tr}\mathbf{L}|^2$. After a global gauge rotation, we can always choose the expectation value of $\mathbf{L}$ to be a diagonal matrix. If $\mathbf{L}$ were a $U(N)$, instead of an $SU(N)$ matrix, then $|\text{tr}\mathbf{L}|^2$ is maximized when $\mathbf{L}$ is a constant phase times the unit matrix. This remains true when $\mathbf{L}$ is a $SU(N)$ matrix; for it to be a unit matrix, however, it must be an element of the center of the gauge group,

$$\langle \mathbf{L} \rangle \;=\; \ell_0 \, \exp{(i\phi)} \; \mathbf{1} \, . \tag{5.8}$$

with ϕ a $Z(N)$ phase, $\phi = 2\pi j/N, j = 1 \ldots (N-1)$.
For this particular expectation value, the vacuum does *not* spontaneously break the (global) $SU(N)$ symmetry above T_c. This accords with naive expectation: the high temperature vacuum is not in a Higgs phase. The possibility of having an expectation value which doesn't break $SU(N)$ is special to a field in the adjoint, as opposed to the fundamental, representation.
On the other hand, assume that $m_1^2 > 0$, and $m_2^2 < 0$ in the deconfined phase, so that symmetry breaking is driven by condensation of the charge-two, instead of the charge-one, loop. Then the vacuum is given by maximizing $|\text{tr}\mathbf{L}^2|^2$; this means that the expectation value of $\mathbf{L}^2$ is an element of the center. But if so, besides the $SU(N)$ invariant vacuum, there are also vacua which are only invariant under $SU(N-1)$. At present, there is no evidence to suggest that the high temperature vacuum is one where $SU(N)$ spontaneously breaks to $SU(N-1)$.
The above description can be extended beyond mean field theory, at least if the deconfining transition is of second order. A transition driven by the charge-one loop is one where $m_1^2 \to 0$ at T_c; one driven by the charge-two loop is where $m_2^2 \to 0$ at T_c, *etc.* To be precise, for a transition driven by a charge-k loop, both its mass, and that of the charge-$(N-k)$ loop, vanish at T_c.
The masses for Polyakov loops of all charges can be directly measured on the Lattice. Even for the charge-one loop, data near T_c is, at present, limited [8, 22]. There is also some data for higher charge loops for three colors in 2+1 dimensions [51].
Given this (crucial!) assumption about the masses, I next turn to the order of the phase transition for a small number of colors.

5.3. TWO COLORS: SECOND ORDER, AND QUARKLESS BARYONS

For two colors, all Polyakov loops are real. For the charge-one loop, I take the potential

$$\mathcal{V}_1 = \frac{m_1^2}{2}\,\ell_1^2 + \frac{\lambda_1}{4}\,\ell_1^4 \,, \tag{5.9}$$

with a positive quartic coupling, $\lambda_1 > 0$. Of course higher powers in ℓ_1 are also possible. Near a second order phase transition, however, the most relevant operators, with the fewest powers of ℓ_1, dominate.

Invariant terms involving the charge-two loop include

$$\mathcal{V}_2 = h\,\ell_2 + \frac{1}{2}\,m_2^2\,\ell_2^2 + \ldots \tag{5.10}$$

plus terms $\sim \ell_2^3$, $\sim \ell_2^4$, *etc.* All powers of ℓ_2 are allowed because it is a singlet. The potential which mixes the charge-one and charge-two loops starts out as

$$\mathcal{V}_{mix} = \xi\,\ell_1^2\,\ell_2 + \ldots \,. \tag{5.11}$$

plus many other terms; this has the lowest mass dimension.

There is extensive Lattice data on the nature of the deconfining phase transition [20]. Especially from the work of Engels *et al*, it appears that the transition is of second order. To wit, the critical exponents are within $\sim 1\%$ of the values expected for the Ising model [3].

There is a surprise, however. As first stressed by Damgaard [20], the expectation value of the Polyakov loop in the adjoint representation is an approximate order parameter. In perturbation theory, the adjoint Polyakov loop is $1 + 2\ell_2$, so from the Lattice data, the expectation value of ℓ_2 presumably jumps at T_c.

From the terms above, the expectation value of the charge-two loop is

$$\langle \ell_2 \rangle = -\,\frac{h + \xi^2\,\ell_0^2}{m_2^2}\,. \tag{5.12}$$

To explain the jump in $\langle \ell_2 \rangle$ about T_c, there are then two possibilities. If the charge-two loop is heavy, then the coupling constant of the charge-two loop to the charge-one loop, ξ, must be large. This means that changes in the density of the charge-two loop is driven by condensation of the charge-one loop.

The other possibility is that h and ξ are not especially large, but that the charge-two loop becomes light near T_c. The latter doesn't violate universality, as long as the charge-two loop isn't massless at T_c.

There is no lattice data on $\langle \ell_1^2 \rangle$. I presume that as suggested by mean field theory, ℓ_1^2 quarkless baryons are rare below T_c, and common above. This is seperate from the changes in ℓ_2.

5.4. THREE COLORS

5.4.1. *A "Nearly" Second Order Transition*

In an asymptotically free gauge theory, it is natural to form the ratio of the true pressure to that of an ideal gas. In principle, positivity of entropy does not require this ratio to be less than one. In practice, Lattice data with improved actions finds that this ratio is less than one at all temperatures [5].

Numerical simulations find that for three colors, the deconfining transition is of first order [21], in agreement with general arguments [3]. For a first order transition, the pressure is continuous at T_c, but the energy density jumps. Thus consider the ratio of the jump in the energy density to that of an ideal gas.

This ratio is not bounded by one. To illustrate this, consider a bag model. Above T_c, the pressure is that of an ideal gas, minus a bag constant, b:

$$p_{bag} = c_0 T^4 - b . \tag{5.13}$$

The pressure is assumed to vanish below T_c, with T_c fixed by $p_{bag} = 0$. This bag model does not describe the Lattice data near T_c, but is a useful construct. In the bag model, the ratio of energies is:

$$\frac{\delta e}{e_{ideal}}|_{Bag} = \frac{4}{3} . \tag{5.14}$$

In contrast, Lattice data appears to give a result which is much smaller:

$$\frac{\delta e}{e_{ideal}}|_{Lattice} \sim \frac{1}{3} . \tag{5.15}$$

The mass of the charge-one loop has also been measured on the Lattice [22]. It goes from $m_1/T \sim 2.5$ at $\sim 2T_c$, down to $m_1/T \sim .25$ at $\sim T_c^+$. This decrease in the screening mass, apparently by a factor of ten, strongly suggests the in fact the transition is even weaker than the above comparison with the bag model suggests. Instead of weakly first order, I prefer to call the deconfining transition *nearly second order*.

(It is necessary to measure correlation functions of Polyakov loops to see this decrease. Masses measured from other operators, such as plaquettes, do not decrease dramatically about T_c [8]. This implies that the mixing between Polyakov loops and and plaquette operators are small. This small mixing is found in related problems [7].)

An effective theory cannot explain why the deconfining transition is weakly first order; it merely requires that certain coupling constants are small.

5.4.2. *Polyakov Loops with Charge One and Minus One*
For three colors, I consider Polyakov loops with charge one, two and three.
There is no data on the expectation values for the quarkless baryons, ℓ_1^3, $\ell_1\ell_2$, and ℓ_3. I presume that as indicated by mean field theory, $\langle \ell_1^3 \rangle$ is small below T_c, and large above. It would be interesting to knowhow the density of the other quarkless baryons, $\ell_1\ell_2$ and ℓ_3, change about T_c. Regardless of the renormalization issues discussed in sec. (4), changes in these expectation values are presumably physical.
Thus I concentrate on the interaction between the charge-one and the charge-two loops. Remember that for three (or more) colors, the ℓ_j's are all complex valued fields. The potential for charge-one loops is dictated by the global $Z(3)$ symmetry:

$$\mathcal{V}_1 = m_1^2 \, |\ell_1|^2 + \kappa_1 \left(\ell_1^3 + (\ell_1^*)^3 \right) + \lambda_1 \left(|\ell_1|^2 \right)^2 . \tag{5.16}$$

The notable feature is the appearance of a cubic term, which necessarily ensures a first order transition [3].
The charge-two loop has charge minus one under $Z(3)$, so its potential is the same, albeit with different masses and coupling constants:

$$\mathcal{V}_2 = m_2^2 \, |\ell_2|^2 + \kappa_2 \left(\ell_2^3 + (\ell_2^*)^3 \right) + \lambda_2 \left(|\ell_2|^2 \right)^2 . \tag{5.17}$$

There are many terms by which the charge-one and charge-two fields can mix. The most important is that with the smallest mass dimension:

$$\mathcal{V}_{mix} = \xi \left(\ell_1 \ell_2 + \ell_1^* \ell_2^* \right) . \tag{5.18}$$

In terms of the original thermal Wilson line, this term is $\sim (\mathrm{tr}\, \mathbf{L})(\mathrm{tr}\, \mathbf{L}^2)$, *etc.*
If the charge-two field is heavy near T_c, we can integrate it out. While it may be a mess to do so analytically, any resulting potential, involving only ℓ_1, must still respect the overall $Z(3)$ symmetry. This produces a potential identical in form to $\mathcal{V}_1$, but with different values for the mass and coupling constants. A weakly first order requires that the effective cubic coupling in the resulting effective theory is small, $\widetilde{\kappa}_1 \ll 1$.
If the charge-two loop becomes light near T_c, and if it mixes strongly with the charge-one loop through a large coupling constant ξ, then its effects cannot be neglected.
There is another possibility. The mass and quartic terms in the potentials are invariant not just under $Z(3)$, but under a global $U(1)$ symmetry. Assume that the charge-two field is always heavy. Then all terms in both potentials are invariant under a global $U(1)$, with the exception of the cubic terms, with couplings κ_1 and κ_2, *and* the mixing term between ℓ_1 and ℓ_2, with coupling ξ. Thus perhaps *all* terms invariant under $Z(3)$, but not $U(1)$, are small. That is, the heavy charge-two loop mixes weakly with the charge-one loop. This doesn't explain why all $Z(3)$ couplings are small, but hints at a more general principle.

It will be interesting to see what detailed numerical studies on the Lattice tell us.

5.5. FOUR COLORS: FIRST ORDER FROM CHARGE-TWO LOOPS

For four colors, I consider just the charge-one and charge-two loops. Under $Z(4)$, $\ell_1 \to i\ell_1$, so the potential for the charge-one field alone is

$$\mathcal{V}_1 = \frac{m_1^2}{2} |\ell_1|^2 + \lambda_1 (|\ell_1|^2)^2 + \kappa_1 \left(\ell_1^4 + (\ell_1^*)^4\right) . \tag{5.19}$$

The term $\sim \lambda_1$ is $O(2)$ invariant, while that $\sim \kappa_1$ is only invariant under $Z(4)$. Under $Z(4)$, $\ell_2 \to -\ell_2$, so the potential for the charge-two field by itself is just like that for the charge-one field:

$$\mathcal{V}_2 = + m_2^2 |\ell_2|^2 + \lambda_2 \left(|\ell_2|^2\right)^2 + \kappa_2 \left(\ell_2^4 + (\ell_2^*)^4\right) . \tag{5.20}$$

The allowed terms which mix the two fields are:

$$\mathcal{V}_{mix} = \zeta_1 \left(\ell_2^* \ell_1^2 + \ell_2 (\ell_1^*)^2\right) + \zeta_2 \left(\ell_2 \ell_1^2 + \ell_2^* (\ell_1^*)^2\right) . \tag{5.21}$$

Unlike two colors, a term linear in ℓ_2 is not allowed by the $Z(4)$ symmetry. For $\zeta_2 = 0$, and assuming that the charge-two field reamins heavy about T_c, ℓ_2 can be integrated out to give:

$$\sim - \frac{\zeta_1^2}{m_2^2} (|\ell_1|^2)^2 . \tag{5.22}$$

One can convince oneself that this term is necessarily negative. That is, the quartic coupling for the charge-one field is shifted downward:

$$\widetilde{\lambda}_1 \equiv \lambda_1 - \frac{\zeta_1^2}{m_2^2} . \tag{5.23}$$

The same holds if both ζ_1 and ζ_2 are nonzero: integrating out the charge-two field generates corrections to the quartic coupling constants of the charge-one field which are uniformly negative, shifting them downwards.

The transition for four colors appears to be of first order [17]. One explanation for this is that the original coupling constants λ_1 and λ_2 are positive, but after integrating out the charge-two field, they become negative, and thus drive the transition first order.

The analysis for higher numbers of colors is then immediate. The leading term which couples a charge-j loop to the charge-one loop is

$$\mathcal{V}_{mix} = \xi \left(\ell_j^* \ell_1^j + \ell_j (\ell_1^*)^j\right) . \tag{5.24}$$

Assuming that the charge-j loop is heavy about T_c, we can integrate it out, which produces a term in the potential for the charge-one loop $\sim (|\ell_1|^2)^j$. For charges greater than two, this is less relevant (has smaller mass dimension) than the quartic terms expected to dominate.
Thus if the deconfining transition is of first order for more than four colors, in the present language it is uniquely due to how the coupling between charge-two and charge-one loops affects the effective quartic coupling constant for the charge-one loop. Polyakov loops with charge greater than two do not affect the order of the transition.

6. A Parting Comment

While it is true that, in equilibrium, all thermodynamic quantities follow from the pressure, experience with the perturbative calculation of processes near equilibrium — such as transport coefficients, real photon production, *etc.* — teaches us they often depend on the details of equilibrium correlation functions. Thus regardless of theoretical prejudice, such as the Polyakov Loops Model, it is important to measure as many gauge invariant correlation functions as possible.
There is an astounding amount of superb data which is pouring out of RHIC [45]. Many features, including the change in the spectrum of "hard" particles, details of Hanbury-Brown-Twiss interferometry, chemical composition, *etc.*, appear to defy explanation by any conventional mechanisms [52]. We may very well need a detailed understanding of the theory, near T_c and above, in order to sort out these amazing results.

References

1. A. M. Polyakov, Phys. Lett. B 72 (1978) 477.
2. G. 't Hooft, Nucl. Phys. B 138 (1978) 1; *ibid.* 153 (1979) 141.
3. B. Svetitsky and L. G. Yaffe, Nucl. Phys. B 210 (1982) 423.
4. F. Karsch, E. Laermann and A. Peikert, Phys. Lett. B 478 (2000) 447; Nucl. Phys. B 605 (2001) 579.
5. F. Karsch, lectures in this volume.
6. R. V. Gavai and S. Gupta, Phys. Rev. D 64 (2001) 074506.
7. K. Kajantie, M. Laine, J. Peisa, A. Rajantie, K. Rummukainen and M. E. Shaposhnikov, Phys. Rev. Lett. 79 (1997) 3130; M. Laine and O. Philipsen, Nucl. Phys. B 523 (1998) 267; Phys. Lett. B 459 (1999) 259; A. Hart and O. Philipsen, Nucl. Phys. B 572 (2000) 243; A. Hart, M. Laine and O. Philipsen, Nucl. Phys. B 586 (2000) 443.
8. R. V. Gavai and S. Gupta, Phys. Rev. Lett. 85 (2000) 2068.
9. P. Arnold and C. Zhai, Phys. Rev. D 50 (1994) 7603; *ibid.* 51 (1995) 1906; E. Braaten, Phys. Rev. Lett. 74 (1995) 2164; E. Braaten and A. Nieto, Phys. Rev. D 53 (1996) 3421; B. Kastening and C. Zhai, *ibid.* 52 (1995) 7232; R. R. Parwani and C. Coriano, Phys. Rev. Lett. 73 (1994) 2398; Nucl. Phys. B 434 (1995) 56.
10. F. Karsch, Z. Phys. C 38 (1988) 147; D. H. Rischke, M. I. Gorenstein, H. Stöcker, and W. Greiner, Phys. Lett. B 237 (1990) 153; A. Peshier, B. Kampfer, O.P. Pavlenko, and G. Soff, Phys. Rev. D 54 (1996) 2399; P. Levai and U. Heinz, Phys. Rev. C 57 (1998) 1879.

11. R. D. Pisarski, Phys. Rev. Lett. 63 (1989) 1129; E. Braaten and R. D. Pisarski,Nucl. Phys. B 337 (1990) 569; J. C. Taylor and S. M. Wong, *ibid.* 346 (1990) 115; E. Braaten and R. D. Pisarski, Phys. Rev. D 45 (1992) 1827. J. P. Blaizot and E. Iancu, Phys. Rept. 359 (2002) 355, and references therein.
12. A. Rebhan, lectures in this volume.
13. J.-P. Blaizot, E. Iancu, and A. Rebhan, Phys. Rev. Lett. 83 (1999) 2906; Phys. Lett. B 470 (1999) 181; Phys. Rev. D 63 (2001) 065003; arXiv:hep-ph/0104033; arXiv:hep-ph/0110369.
14. J. O. Andersen, E. Braaten, and M. Strickland, Phys. Rev. Lett. 83 (1999) 2139; Phys. Rev. D 61 (2000) 014017; *ibid.*, D61 (2000) 074016; *ibid.*, D62 (2000) 045004; *ibid.*, D63 (2001) 105008.
15. P. Arnold and L. G. Yaffe, Phys. Rev. D 52 (1995) 7208.
16. M. Laine, Nucl. Phys. B 451 (1995) 484; K. Kajantie, M. Laine, K. Rummukainen and M. E. Shaposhnikov, Nucl. Phys. B 458 (1996) 90; Nucl. Phys. B 502 (1997) 357; M. Laine and A. Rajantie, Nucl. Phys. B 513 (1998) 471; K. Kajantie, M. Laine, K. Rummukainen and Y. Schroder, Phys. Rev. Lett. 86 (2001) 10; K. Kajantie, M. Laine and Y. Schroder, Phys. Rev. D 65 (2002) 045008. K. Kajantie, M. Laine, K. Rummukainen and Y. Schroder, arXiv:hep-lat/0110122.
17. G. G. Batrouni and B. Svetitsky, Phys. Rev. Lett. 52 (1984) 2205; A. Gocksch and M. Okawa, *ibid.* 52 (1984) 1751; F. Green and F. Karsch, Phys. Rev. D 29 (1984) 2986; J. F. Wheater and M. Gross, Phys. Lett. B 144 (1984) 409; Nucl. Phys. B 240 (1984) 253; S. Ohta and M. Wingate, Nucl. Phys. Proc. Suppl. B 73 (1999) 435;*ibid.* 83 (2000) 381; Phys. Rev. D 63 (2001) 094502; R. V. Gavai, hep-lat/0110054.
18. A. Gocksch and F. Neri, Phys. Rev. Lett. 50 (1983) 1099; M. Billo, M. Caselle, A. D'Adda, and S. Panzeri, Int. J. Mod. Phys. A12 (1997) 1783.
19. R. D. Pisarski, Phys. Rev. D 29 (1984) 1222; R. D. Pisarski and M. H. G. Tytgat, in "Hirschegg '97: QCD Phase Transitions", edited by H. Feldmeier *et al.* (GSI Publishing, Darmstadt, 1997); arXiv:hep-ph/9702340.
20. P. H. Damgaard, Phys. Lett. B 194 (1987) 107; J. Engels, J. Fingberg, K. Redlich, H. Satz, and M. Weber, Z. Phys. C 42 (1989) 341; J. Kiskis, Phys. Rev. D 41 (1990) 3204; J. Fingberg, D. E. Miller, K. Redlich, J. Seixas, and M. Weber, Phys. Lett. B 248 (1990) 347; J. Christensen and P. H. Damgaard, Nucl. Phys. B 348 (1991) 226; P. H. Damgaard and M. Hasenbusch, Phys. Lett. B 331 (1994) 400; J. Kiskis and P. Vranas, Phys. Rev. D 49 (1994) 528; J. Engels, F. Karsch, K. Redlich,Nucl. Phys. B 435 (1995) 295; J. Engels, S. Mashkevich, T. Scheideler, and G.Zinovev, Phys. Lett. B 365 (1996) 219; J. Engels and T. Scheideler, Phys. Lett. B 394 (1997) 147; Nucl. Phys. B 539 (1999) 557.
21. P. Bacilieri *et al.*, Phys. Rev. Lett. 61 (1988) 1545; F.R. Brown *et al.*, *ibid.* 61 (1988) 2058; A. Ukawa, Nucl. Phys. Proc. Suppl. B 17 (1990) 118; *ibid.* 53 (1997) 106; E. Laermann, *ibid.* 63 (1998) 114; F. Karsch, *ibid.* 83 (2000) 14.
22. O. Kaczmarek, F. Karsch, E. Laermann and M. Lutgemeier, Phys. Rev. D 62 (2000) 034021.
23. T. Banks and A. Ukawa, Nucl. Phys. B 225 (1983) 145.
24. A. Patel, Nucl. Phys. B 243 (1984) 411; Phys. Lett. B 139 (1984) 394; F. Takagi, Phys. Rev. D 34 (1986) 1646 C. DeTar, Phys. Rev. D 37 (1988) 2328; C. Rosenzweig and A. M. Srivastava, Phys. Rev. D 42 (1990) 4228; A. Momen and C. Rosenzweig, Phys. Rev. D 56 (1997) 1437; M. Engelhardt, K. Langfeld, H. Reinhardt and O. Tennert, Phys. Rev. D 61 (2000) 054504; M. Engelhardt and H. Reinhardt, Nucl. Phys. B 585 (2000) 591; J. Condella and C. DeTar, Phys. Rev. D 61 (2000) 074023.
25. R. D. Pisarski, Phys. Rev. D 62 (2000) 111501.
26. A. Dumitru and R. D. Pisarski, Phys. Lett. B 504 (2001) 282.
27. A. Dumitru and R. D. Pisarski, arXiv:hep-ph/0106176.
28. A. Dumitru, O. Scavenius, and A. D. Jackson, Phys. Rev. Lett. 87 (2001) 182302.

29. J. Wirstam, Phys. Rev. D 65 (2002) 014020.
30. P. N. Meisinger, T. R. Miller, and M. C. Ogilvie, hep-ph/0108009; P. N. Meisinger and M. C. Ogilvie, hep-ph/0108026.
31. A. Dumitru and R. D. Pisarski, Nucl. Phys. Proc. Suppl. B 106 (2002) 483; R. D. Pisarski, hep-ph/0112037.
32. T. Bhattacharya, A. Gocksch, C. Korthals Altes and R. D. Pisarski, Phys. Rev. Lett. 66 (1991) 998; Nucl. Phys. B 383 (1992) 497.
33. C. P. Korthals Altes, Nucl. Phys. B 420 (1994) 637.
34. J. Polchinski, Phys. Rev. Lett. 68 (1992) 1267; J. Boorstein and D. Kutasov, Phys. Rev. D 51 (1995) 7111; C. Korthals Altes, A. Michels, M. Stephanov and M. Teper, Phys. Rev. D 55 (1997) 1047; C. Korthals-Altes, A. Kovner and M. Stephanov,Phys. Lett. B 469 (1999) 205; C. Korthals-Altes and A. Kovner,Phys. Rev. D 62 (2000) 096008.
35. A. Gocksch and R. D. Pisarski, Nucl. Phys. B 402 (1993) 657.
36. C. P. Korthals Altes, in "Festschrift for M. Veltman", eds R. Akhoury et al. (World Scientific Press, Singapore, 1992); C. P. Korthals Altes, hep-th/9402028.
37. P. Giovannangeli and C. P. Korthals Altes, Nucl. Phys. B 608 (2001) 203.
38. V. Dixit and M. C. Ogilvie, Phys. Lett. B 269 (1991) 353; J. Ignatius, K. Kajantie and K. Rummukainen,Phys. Rev. Lett. 68 (1992) 737; C. P. Korthals Altes, K. Y. Lee and R. D. Pisarski,Phys. Rev. Lett. 73 (1994) 1754; C. P. Korthals Altes and N. J. Watson, Phys. Rev. Lett. 75 (1995) 2799; S. Bronoff, G.Dvali, K. Farakos, and C. P. Korthals Altes, "Proceedings Strong and Electrowoweak Matter (SEWM 97)", pg 192-212, Ed. Z. Fodor, (World Scientific Press, Singapore, 1997); C. P. Korthals Altes, R. D. Pisarski and A. Sinkovics,Phys. Rev. D 61 (2000) 056007.
39. V. M. Belyaev, I. I. Kogan, G. W. Semenoff and N. Weiss,Phys. Lett. B 277 (1992) 331; W. Chen, M. I. Dobroliubov and G. W. Semenoff,Phys. Rev. D 46 (1992) 1223; I. I. Kogan, Phys. Rev. D 49 (1994) 6799; A. V. Smilga, Annals Phys. 234 (1994) 1; A. V. Smilga, Surveys High Energ. Phys. 10 (1997) 233; J. E. Kiskis, Nucl. Phys. Proc. Suppl. 53 (1997) 465.
40. A. M. Polyakov, Nucl. Phys. B 164 (1980) 171; I. Y. Arefeva, Phys. Lett. B 93 (1980) 347; V. S. Dotsenko and S. N. Vergeles, Nucl. Phys. B 169 (1980) 527; L. Gervais and A. Neveu, Nucl. Phys. B 163 (1980) 189.
41. E. Gava and R. Jengo, Phys. Lett. B 105 (1981) 285.
42. K. Enqvist, K. Kajantie, L. Karkkainen and K. Rummukainen, Phys. Lett. B 249 (1990) 107.
43. F. Zantow, O. Kaczmarek, F. Karsch and P. Petreczky, arXiv:hep-lat/0110103; arXiv:hep-lat/0110106.
44. A. M. Tsvelik, "Quantum Field Theory in Condensed Matter Physics" (Cambridge Univ. Press, Cambridge, 1996).
45. J. Nagle and T. Ulrich, lectures in this volume.
46. Y. Dokshitzer, E. Iancu, D. Kharzeev, L. McLerran, and R. Venugopalan, lectures in this volume.
47. L. D. McLerran and B. Svetitsky, Phys. Rev. D 24 (1981) 450.
48. P. W. Anderson, Phys. Rev. 109 (1958) 1498.
49. G. C. Rossi and G. Veneziano, Nucl. Phys. B 123 (1977) 507; Nucl. Phys. B **123**, 507 (1977). E. Witten, Nucl. Phys. B 160 (1979) 57; L. Montanet, G. C. Rossi and G. Veneziano, Phys. Rep. 63 (1980) 149; D. Kharzeev, Phys. Lett. B 378 (1996) 238.
50. L. F. Abbott, Nucl. Phys. B 185 (1981) 189, and references therein.
51. P. Bialas, A. Morel, B. Petersson, K. Petrov and T. Reisz, Nucl. Phys. B 581 (2000) 477; P. Bialas, A. Morel, B. Petersson and K. Petrov, Nucl. Phys. B 603 (2001) 369; Nucl. Phys. Proc. Suppl. B 106 (2002) 882; arXiv:hep-lat/0112008.
52. M. Gyulassy, nucl-th/0106072.

LATTICE QCD AT FINITE TEMPERATURE

FRITHJOF KARSCH (karsch@physik.uni-bielefeld.de)
Fakultät für Physik, Universität Bielefeld,
D-33615 Bielefeld, Germany

Abstract. After a brief introduction into basic aspects of the formulation of lattice regularized QCD at finite temperature and density we discuss our current understanding of the QCD phase diagram at finite temperature, present results on the QCD equation of state, the transition temperature and the heavy quark free energy. Furthermore, we discuss in-medium properties of hadrons and present first results from a calculation of thermal dilepton rates resulting from quark anti-quark annihilation in the QCD plasma phase.

1. Introduction

Almost immediately after the ground-breaking demonstration that the numerical analysis of lattice regularized quantum field theories [1] can also provide quantitative results on fundamental non-perturbative properties of QCD [2] it has been realized that this approach will also allow to study the QCD phase transition [3, 4] and the equation of state of the quark-gluon plasma [5]. During the last 20 years we have learned a lot from lattice calculations about the phase structure of QCD at finite temperature. However, it is only now that numerical calculations start to reach a level of accuracy that allows us to seriously consider quantitative studies of QCD with a realistic light quark mass spectrum. An important ingredient in the preparation of such calculations has been the development of new regularization schemes in the fermion sector of the QCD Lagrangian, which reduce discretization errors and also improve the flavour symmetry of the lattice actions. The currently performed investigations of QCD thermodynamics provide first results with such improved actions and prepare the ground for calculations with a realistic light quark mass spectrum.
One of the central goals in analyzing the properties of QCD under extreme conditions is to reach a quantitative description of the behaviour of matter at high temperature and density. This does provide important input for a quantitative description of experimental signatures for the occurrence of a phase transition in heavy ion collisions and should also help to better understand the phase transitions

J.-P. Blaizot and E. Iancu (eds.), QCD Perspectives on Hot and Dense Matter, 385–417.

that occurred during the early times of the evolution of the universe. Eventually it also may allow to answer the question whether a quark-gluon plasma can exist in the interior of dense neutron stars or did exist in early stages of supernovae explosions. To answer such questions requires a quantitative understanding of the QCD equation of state, the determination of critical parameters such as the critical temperature and the critical energy density and an analysis of possible thermal modifications of basic hadron properties (masses, decay widths).
In the next section we give a short introduction into the lattice formulation of QCD thermodynamics. In Section 3 we discuss basic lattice results on the structure of the QCD phase diagram at finite temperature. Sections 4 and 5 are devoted to a discussion of the quark mass and flavour dependence of the QCD equation of state and the phase transition temperature. In Section 6 we analyze the temperature dependence of the heavy quark free energy and discuss consequences for the formation of heavy quark bound states. Thermal modifications of the light quark spectrum, on the other hand, are discussed in Section 7. Here we also present results from a calculation of dilepton rates in the high temperature plasma phase. Finally we give our conclusions in Section 8.

2. The Lattice Formulation of QCD Thermodynamics

Starting point for the discussion of the equilibrium thermodynamics of QCD is the Euclidean path integral formulation of the partition function. This is given as an integral over the fundamental quark ($\bar{\psi}$, ψ) and gluon (A_ν) fields. In addition to its dependence on the bare couplings of QCD, *i.e.* the gauge coupling g^2 and the quark masses m_f for $f = 1,..,n_f$ different quark flavours, the partition function also depends explicitly on the thermodynamic parameters, volume (V) and temperature (T)[1],

$$Z(V,T) = \int \mathcal{D}A_\nu \mathcal{D}\bar{\psi}\mathcal{D}\psi \, \mathrm{e}^{-S_E(V,T)} \quad . \tag{2.1}$$

Here A_ν and $\bar{\psi}$, ψ obey periodic and anti-periodic boundary conditions in Euclidean time, respectively. The Euclidean action $S_E \equiv S_G + S_F$ contains a purely gluonic contribution (S_G) expressed in terms of the field strength tensor, $F_{\mu\nu} = \partial_\mu A_\nu - \partial_\nu A_\mu - ig[A_\mu, A_\nu]$, and a fermionic part ($S_F$), which couples the gauge and fermions field through the standard minimal substitution,

$$S_E(V,T) \equiv S_G(V,T) \; + \; S_F(V,T) \tag{2.2}$$

$$S_G(V,T) = \int_0^{1/T} \mathrm{d}x_4 \int_V \mathrm{d}^3\mathbf{x} \, \frac{1}{2}\mathrm{Tr} \, F_{\mu\nu}F_{\mu\nu} \tag{2.3}$$

[1] We will discuss here only the thermodynamics for vanishing baryon number density or equivalently vanishing baryon chemical potential.

$$S_F(V,T) = \int_0^{1/T} \mathrm{d}x_4 \int_V \mathrm{d}^3\mathbf{x} \sum_{f=1}^{n_f} \bar{\psi}_f \left(\gamma_\mu[\partial_\mu - igA_\mu] + m_f\right) \psi_f . \quad (2.4)$$

The path integral appearing in Eq. 2.1 is regularized by introducing a four dimensional space-time lattice of size $N_\sigma^3 \times N_\tau$ with a lattice spacing a. Volume and temperature are then related to the number of points in space and time directions, respectively,

$$V = (N_\sigma \, a)^3 \quad , \quad T^{-1} = N_\tau \, a \quad . \quad (2.5)$$

The pure gauge sector: The most important step towards a lattice formulation of QCD that also could be handled in computer simulations was to provide a gauge invariant, discretized version of the field strength tensor in terms of compact *link variables* $U_{x,\mu}$ which are elements of the $SU(3)$ group colour [1]. The link variables $U_{x,\mu}$ are associated with the link between two neighbouring sites of the lattice and describe the parallel transport of the field A_μ from site x to $x + \hat{\mu}a$,

$$U_{x,\mu} = \mathrm{P}\exp\left(ig \int_x^{x+\hat{\mu}a} \mathrm{d}x^\mu A_\mu(x) \right) \quad , \quad (2.6)$$

where $\hat{\mu}$ is a unit vector in the μ-direction of the 4-dimensional lattice and P denotes the path ordering. A product of link variables around an elementary plaquette (Wilson loop of length 4) may be used to define an approximation to the continuum gauge action,

$$W^{(1,1)}_{n,\mu\nu} = 1 - \frac{1}{3}\mathrm{Re} \, \square_{\,n,\mu\nu} \equiv \mathrm{Re}\,\mathrm{Tr}\, U_{n,\mu} U_{n+\hat{\mu},\nu} U^\dagger_{n+\hat{\nu},\mu} U^\dagger_{n,\nu}$$
$$= \frac{g^2 a^4}{2} F^\alpha_{\mu\nu} F^\alpha_{\mu\nu} + \mathcal{O}(a^6) \quad . \quad (2.7)$$

A discretized version of the Euclidean gauge action, which reproduces the continuum version up to cut-off errors of order a^2, thus is given by the *Wilson action* [1],

$$\beta S_G = \beta \sum_{\substack{n \\ 1\le\mu<\nu\le4}} W^{(1,1)}_{n,\mu\nu} \quad \Longrightarrow \quad \int \mathrm{d}^4x \, \frac{1}{2} \mathrm{Tr} F_{\mu\nu} F_{\mu\nu} \, + \, \mathcal{O}(a^2) \quad , \quad (2.8)$$

where we have introduced the lattice gauge coupling $\beta = 6/g^2$.

One may use more involved discretization schemes, which also eliminate the leading $\mathcal{O}(a^2)$ errors. In general this can be achieved by adding in Eq. 2.8 larger Wilson loops. The simplest extension of the Wilson one-plaquette action thus is to include an additional contribution from a planar six-link Wilson loop,

$$W^{(1,2)}_{n,\mu\nu} = 1 - \frac{1}{6}\,\mathrm{Re}\left(\square\!\square + \begin{array}{c}\square\\[-4pt]\square\end{array} \right)_{n,\mu\nu} . \quad (2.9)$$

A suitable combination of these six-link loops and the four-link plaquette term eliminates the leading $\mathcal{O}(a^2)$ corrections and yields a formulation that reproduces the continuum action up to $\mathcal{O}(a^4)$ corrections,

$$\beta S_G = \beta \sum_{\substack{n \\ 0\le\mu<\nu\le 3}} c_{1,1} W^{(1,1)}_{n,\mu\nu} + c_{1,2} W^{(1,2)}_{n,\mu\nu} \quad , \tag{2.10}$$

with $c_{1,1} = 5/3$ and $c_{1,2} = -1/6$. This choice of coefficients defines the so-called tree-level improved (1×2)-action. It insures that the leading cut-off effects are removed on the classical level, *i.e.* at $\mathcal{O}(g^0)$. The action may be further improved perturbatively by either eliminating the leading lattice cut-off effects also at $\mathcal{O}(g^2)$, *i.e.* $c_{i,j} \Rightarrow c^{(0)}_{i,j} + g^2 c^{(1)}_{i,j}$, or by introducing non-perturbative modifications of the coefficients $c_{i,j}$ (*tadpole improvement*). Another well-studied gluon action with non-perturbative corrections is the RG-improved action introduced by Y. Iwasaki [6]. The *RG-action* also has the structure of Eq. 2.10 but with coefficients $c^{RG}_{1,1} = 3.648$ and $c^{RG}_{1,2} = -0.662$. This choice of parameters is motivated by an analysis of scaling properties of the pure gauge action at zero temperature and finite values of the cut-off. In the continuum limit it, however, still leads to $\mathcal{O}(a^2)$ discretization errors.

The fermion sector: Although the discretization of the fermion sector of QCD does at first look much simpler, it turns out that additional problems arise. In the lattice formulation it is by no means easy to reproduce the correct number of fermionic degrees of freedom without violating essential symmetries (chiral symmetry) of the continuum action. The naive discretization of the fermionic part of the action, which is obtained by introducing a simple finite difference scheme to discretize the derivative appearing in the fermion Lagrangian, *i.e.* $\partial_\mu \psi_f(x) = (\psi_{n+\hat\mu} - \psi_{n-\hat\mu})/2a$, does not reproduce the particle content one started with in the continuum. The massless lattice fermion propagator has poles not only at zero momentum but also at all other corners of the Brillouin zone and thus generates 16 rather than a single fermion species in the continuum limit ($a \to 0$). One therefore faces a severe species doubling problem. The way out has been to either introduce a higher derivative term, $a\partial^2_\mu \psi_f(x)$, which is proportional to the lattice spacing a and thus vanishes in the continuum limit (Wilson fermions [1]), or to distribute the components of the fermion Dirac spinors over several lattice sites (staggered fermions) [7]. While the Wilson formulation removes all the unwanted doublers in the continuum limit by giving them a mass of $\mathcal{O}(a^{-1})$ on the lattice, the staggered fermion formulation aims at a reduction of the number of fermion doublers. It does not eliminate the problem completely. One still obtains four degenerate fermion species in the continuum limit. On the other hand, the staggered fermion formulation has the advantage that it preserves a continuous subgroup of the original global chiral symmetry, while this symmetry is explicitly broken in the Wilson formulation. The former thus has the advantage that it still provides a

true order parameter for critical behaviour at finite temperature even for non-zero values of the lattice spacing. In the massless limit the chiral condensate still is an order parameter.
In the following we will present results from calculations based on the Wilson as well as staggered fermion formulations. Progress has also been made in formulating lattice QCD with chiral fermion actions which do avoid the species doubling and at the same time preserve the chiral symmetry of the QCD Lagrangian [8]. At present, however, very little has been done to study QCD thermodynamics on the lattice with these actions [9]. As an example for fermion actions we will discuss in a bit more detail the staggered fermion formulation [7]. This action has been used extensively in studies of the QCD phase transition. It can be written as

$$S_F^{KS} = \sum_{nm} \bar{\chi}_n Q_{nm}^{KS} \chi_m \quad , \tag{2.11}$$

where the staggered fermion matrix Q^{KS} is given by

$$\begin{aligned} Q_{nm}^{KS}(m_q) &= \frac{1}{2} \sum_{\mu=1}^{3} (-1)^{x_1+\ldots x_\mu} (\delta_{n+\hat{\mu},m} U_{n,\mu} - \delta_{n,m+\hat{\mu}} U_{m,\mu}^\dagger) \\ &\quad + \frac{1}{2} (\delta_{n+\hat{4},m} U_{n,4} - \delta_{n,m+\hat{4}} U_{m,4}^\dagger) + \delta_{nm} m_q \quad . \end{aligned} \tag{2.12}$$

As the fermion action is quadratic in the Grassmann valued quark fields $\bar{\chi}$ and χ we can integrate them out in the partition function and finally arrive at a representation of $Z(V,T)$ on a 4-dimensional lattice of size $N_\sigma^3 \times N_\tau$,

$$Z(N_\sigma, N_\tau, \beta, m_q) = \int \prod_{n,\nu} \mathrm{d}U_{n,\nu} (\det Q^{KS}(m_q))^{n_f/4} \mathrm{e}^{-\beta S_G} \quad . \tag{2.13}$$

We have made explicit the fact that the staggered fermion action does lead to four degenerate fermion flavours in the continuum limit, *i.e.* taking the continuum limit with the action given in Eqs. 2.11 and 2.12 corresponds to $n_f = 4$ in Eq. 2.13. As the number of fermion species does appear only as an appropriate power of the fermion determinant, which is true also in the continuum limit, one also may choose $n_f \neq 4$ in Eq. 2.13. This is the approach used to perform simulations with different number of flavours in the staggered fermion formulation.
Within the framework of staggered fermions one again is free to choose other discretization schemes, e.g. higher order difference schemes, which may lead to a smoother approach towards the continuum limit. Unlike in the pure gauge sector it, however, is difficult to systematically eliminate $\mathcal{O}(a^2)$ errors beyond the tree level. A particular form of action, which improves the rotational symmetry of the fermion propagator and reduces cut-off effects in bulk thermodynamic observables, is the *p4-action* [10]. In addition to the standard one-link term that appears in Eq. 2.12 it also includes a set of bended three-link terms,

$$
\begin{aligned}
S_F(m_q) &= c_1^F S_{1-link,fat}(\omega) + c_3^F S_{3-link} + m_q \sum_x \bar{\chi}_x^f \chi_x^f \\
&\equiv \sum_x \bar{\chi}_x^f \sum_\mu \eta_\mu(x) \left(\frac{3}{8} \left[\text{[1-link path } x \to y\text{]} + \omega \sum_{\nu\neq\mu} \text{[staples } x \to y\text{]} \right] \right. \\
&\quad \left. + \frac{1}{96} \sum_{\nu\neq\mu} \left[\text{[3-link path]} + \text{[3-link path]} + \text{[3-link path]} + \text{[3-link path]} \right] \right) \chi_y^f \\
&\quad + m_q \sum_x \bar{\chi}_x^f \chi_x^f \quad .
\end{aligned}
\qquad (2.14)
$$

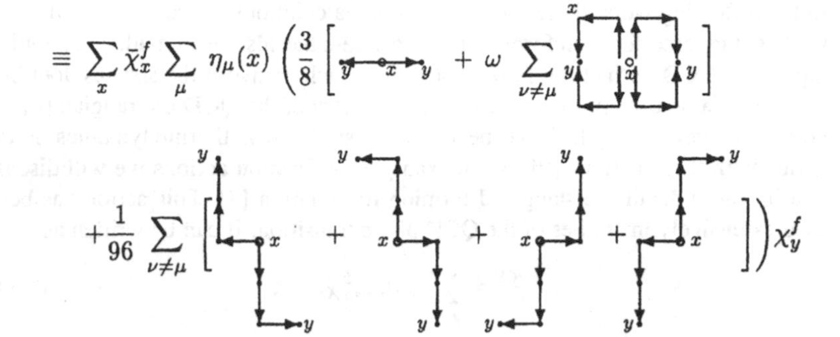

The action contains an additional free parameter ω which can be optimized to improve the flavour symmetry of the action.

Continuum limit: The lattice discretized QCD actions discussed above reproduce the continuum action in the limit $a \to 0$. The actual value of the lattice spacing a is controlled through the bare couplings appearing in the action. In order to reproduce continuum physics at constant temperature without errors emerging from systematic cut-off effects, we will have to take the limit $(a \to 0,\ N_\tau \to \infty)$ with $T = 1/N_\tau a$ fixed. Eventually we thus have to analyze our observables on different size lattices and extrapolate results to $N_\tau \to \infty$ at fixed temperature. Unless we perform calculations at a well defined temperature, e.g. the critical temperature, we will have to determine the temperature scale from an independent (zero-temperature) calculation of an observable for which we know its physical value (in MeV). This requires a calculation at the same value of the cut-off, i.e. for the same values of bare couplings. Of course, we know such a quantity only for the physical case realized in nature, *i.e.* QCD with two light up and down quark flavours and a heavier strange quark. Nonetheless, we have good reason to believe that certain observables are quite insensitive to changes in the quark masses, e.g. quenched hadron masses[2] ($\tilde{m}_H$) or the string tension ($\tilde{\sigma}$) are believed to be suitable observables to set a physical scale. Even in the limit of infinite quark masses, *i.e.* in calculations within the pure $SU(3)$ gauge theory (quenched QCD), they differ from experimentally known values only on the 10% level. We thus may

[2] A physical observable O is calculated on the lattice as dimensionless quantity, which we denote here by $\tilde{O}$. Quite often, however, we will also adopt the customary lattice notation, which explicitly specifies the cut-off dependence in the continuum limit, e.g. $\tilde{m}_H \equiv m_H a$ or $\tilde{\sigma} \equiv \sigma a^2$.

use calculations of these quantities to define a temperature scale,

$$T/\sqrt{\sigma} = 1/\sqrt{\tilde{\sigma}}N_\tau \quad \text{or} \quad T/m_H = 1/\tilde{m}_H N_\tau \quad . \tag{2.15}$$

3. The QCD Phase Diagram at Finite Temperature

At vanishing baryon number density the properties of the QCD phase transition depend on the number of quark flavours and their masses. While it is a detailed quantitative question at which temperature the transition to the high temperature plasma phase occurs, we do expect that qualitative aspects of the transition, e.g. its order and details of the critical behaviour, are controlled by global symmetries of the QCD Lagrangian. Such symmetries only exist in the limits of either infinite or vanishing quark masses. For any non-zero, finite value of quark masses the global symmetries are explicitly broken. In fact, the explicit symmetry breaking induced by the finite quark masses is very much similar to that induced by an external ferromagnetic field in spin models. We thus expect that a continuous phase transition, which may exist in the zero or infinite quark mass limit, will turn into a non-singular crossover behaviour for any finite value of the quark mass. First order transitions, on the other hand, may persist for some time before they end in a continuous transition. However, whether in QCD with the physical quark mass spectrum a true phase transition exists or whether in this case the transition is just a (rapid) crossover, again becomes a quantitative question which we have to answer through direct numerical calculations.

Our current understanding of the qualitative aspects of QCD phase diagram is based on universality arguments for the symmetry breaking patterns in the heavy [11] as well as the light quark mass regime [12, 13]. In the limit of infinitely heavy quarks, the pure $SU(3)$ gauge theory, the large distance behaviour of the heavy quark free energy, $F_{\bar{q}q}$, provides a unique characterization of confinement below T_c and deconfinement above T_c. On a lattice of size $N_\sigma^3 \times N_\tau$ the heavy quark free energy[3] can be calculated from the expectation value of the Polyakov loop correlation function

$$\exp\left(-\frac{F_{\bar{q}q}(r,T)}{T}\right) = \langle \mathrm{Tr} L_{\vec{x}} \mathrm{Tr} L^\dagger_{\vec{y}} \rangle \quad , \quad rT = |\vec{x} - \vec{y}|/N_\tau \tag{3.1}$$

where the Polyakov loops $L_{\vec{x}}$ and $L^\dagger_{\vec{y}}$ represent static quark and anti-quark sources located at the spatial points $\vec{x}$ and $\vec{y}$, respectively,

$$L_{\vec{x}} = \prod_{x_4=1}^{N_\tau} U_{n,4} \quad , \quad n \equiv (\vec{x}, x_4) \quad . \tag{3.2}$$

[3] In the $T \to 0$ limit this is just the heavy quark potential; at non-zero temperature $F_{\bar{q}q}$ does, however, also include a contribution resulting from the overall change of entropy that arises from the presence of external quark and anti-quark sources.

For large separations ($r \to \infty$) the correlation function approaches $|\langle L\rangle|^2$, where $\langle L\rangle = N_\sigma^{-3}\langle\sum_{\vec{x}} \mathrm{Tr} L_{\vec{x}}\rangle$ denotes the Polyakov loop expectation value, which therefore characterizes the behaviour of the heavy quark free energy at large distances and is an order parameter for deconfinement in the $SU(3)$ gauge theory,

$$\langle L\rangle \begin{cases} = 0 \Leftrightarrow \text{confined phase}, & T < T_c \\ > 0 \Leftrightarrow \text{deconfined phase}, & T > T_c \end{cases} . \tag{3.3}$$

The effective theory for the order parameter is a 3-dimensional spin model with global $Z(3)$ symmetry. Universality arguments then suggest that the phase transition is first order in the infinite quark mass limit [11].

In the limit of vanishing quark masses the classical QCD Lagrangian is invariant under chiral symmetry transformations; for n_f massless quark flavours the symmetry is

$$U_A(1) \times SU_L(n_f) \times SU_R(n_f).$$

The $SU_{L/R}(n_f)$ flavour part of this symmetry is spontaneously broken in the vacuum. This gives rise to $(n_f^2 - 1)$ massless Goldstone particles, the pions. The axial $U_A(1)$ symmetry, however, only is a symmetry of the classical Lagrangian. It is explicitly broken due to quantum corrections in the QCD partition function and the true symmetry therefore only is a discrete $Z(n_f)$ symmetry. The basic observable which reflects the chiral properties of QCD is the chiral condensate,

$$\langle\bar{\chi}\chi\rangle = \frac{1}{N_\sigma^3 N_\tau}\frac{\partial}{\partial m_q}\ln Z \quad . \tag{3.4}$$

In the limit of vanishing quark masses the chiral condensate stays non zero as long as chiral symmetry is spontaneously broken. The chiral condensate thus is an obvious order parameter in the chiral limit,

$$\langle\bar{\chi}\chi\rangle \begin{cases} > 0 \Leftrightarrow \text{symmetry broken phase}, & T < T_c \\ = 0 \Leftrightarrow \text{symmetric phase}, & T > T_c \end{cases} . \tag{3.5}$$

For light quarks the global chiral symmetry is expected to control the critical behaviour of the QCD phase transition. In particular, the order of the transition is expected to depend on the number of light or massless flavours. The basic aspects of the n_f-dependence of the phase diagram have been derived by Pisarski and Wilczek [12] from an effective, 3-dimensional Lagrangian for the order parameter[4],

$$\begin{aligned} \mathcal{L}_{eff} = & -\frac{1}{2}\mathrm{Tr}(\partial_\mu\Phi^\dagger\partial^\mu\Phi) - \frac{1}{2}m^2\mathrm{Tr}(\Phi^\dagger\Phi) + \frac{\pi^2}{3}g_1\left(\mathrm{Tr}(\Phi^\dagger\Phi)\right)^2 \\ & + \frac{\pi^2}{3}g_2\mathrm{Tr}\left((\Phi^\dagger\Phi)^2\right) + c\left(\det\Phi + \det\Phi^\dagger\right) \quad , \end{aligned} \tag{3.6}$$

[4] It should be noted that this ansatz assumes that chiral symmetry is broken at low temperatures. Instanton model calculations suggest that the vacuum, in fact, is chirally symmetric already for $n_f \geq 5$ [14].

with $\Phi \equiv (\Phi_{ij}), i, j = 1, ..., n_f$. $\mathcal{L}_{eff}$ has the same global symmetry as the QCD Lagrangian. A renormalization group analysis of this Lagrangian suggests that the transition is first order for $n_f \geq 3$ and second order for $n_f = 2$. The latter, however, is expected to hold only if the axial $U_A(1)$ symmetry breaking, related to the det Φ terms in Eq. 3.6, does not become too weak at T_c so that the occurrence of a fluctuation induced first order transition would also become possible.
This basic pattern indeed is supported by lattice calculations. So far no indication for a discontinuous transition has been observed for $n_f = 2$. The transition is found to be first order for $n_f \geq 3$. Moreover, the transition temperature is decreasing with increasing n_f and there are indications that chiral symmetry is already restored in the vacuum above a critical number of flavours [15].
The anticipated phase diagram of 3-flavour QCD at vanishing baryon number density is shown in Fig. 1.

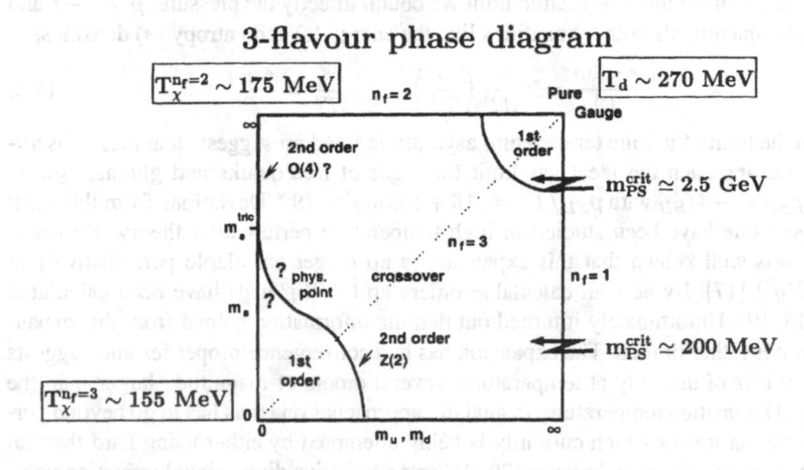

Figure 1. The QCD phase diagram of 3-flavour QCD with degenerate (u,d)-quark masses and a strange quark mass m_s.

An interesting aspect of the phase diagram is the presence of a second order transition line in the light quark mass regime which forms the boundary of the region of first order phase transitions. On this line the transition is controlled by an effective 3-dimensional theory with global $Z(2)$ symmetry [13]. As this is not a symmetry of the QCD Lagrangian neither the chiral condensate nor the Polyakov loop will be the order parameter. The critical behaviour (large fluctuations) caused by this transition may, however, equally important for the critical or crossover behaviour of QCD with a realistic quark mass spectrum as the nearby critical point in the chiral limit. It therefore is important to determine in detail the chiral critical line in the QCD phase diagram and the location of the physical point relative to it. A first step along this line has been the verification that the chiral point in three flavour QCD indeed belongs to the universality class of the 3-d Ising model [16]. The determination of the pseudo-scalar meson mass at this critical point indicates that the first order transition regime is rather small and only persists for $m_{PS} \lesssim 200$ MeV.

4. The QCD Equation of State

The most fundamental quantity in equilibrium thermodynamics is, of course, the partition function itself, or the free energy density,

$$f = -\frac{T}{V} \ln Z(T,V) \quad . \tag{4.1}$$

All basic bulk thermodynamic observables can be derived from the free energy density. In the thermodynamic limit we obtain directly the pressure, $p = -f$ and subsequently also other quantities like the energy (ϵ) and entropy (s) densities,

$$\frac{\epsilon - 3p}{T^4} = T \frac{\mathrm{d}}{\mathrm{d}T}\left(\frac{p}{T^4}\right) \quad , \quad \frac{s}{T^3} = \frac{\epsilon + p}{T^4} \quad . \tag{4.2}$$

In the limit of infinite temperature asymptotic freedom suggests that these observables approach the ideal gas limit for a gas of free quarks and gluons, $\epsilon_{SB} = 3p_{SB} = -3f_{SB}$ with $p_{SB}/T^4 = (16 + 10.5n_f)\pi^2/90$. Deviations from this ideal gas value have been studied in high temperature perturbation theory. However, it was well-known that this expansion is no longer calculable perturbatively at $\mathcal{O}(g^6)$ [17]. By now all calculable orders up to $\mathcal{O}(g^5 \ln g)$ have been calculated [18, 19]. Unfortunately it turned out that the information gained from this expansion is rather limited. The expansion has bad convergence properties and suggests that it is of use only at temperatures several orders of magnitude larger than the QCD transition temperature. In analytic approaches one thus has to go beyond perturbation theory which currently is being attempted by either using hard thermal loop resummation techniques [20, 21] or perturbative dimensional reduction combined with numerical simulations of the resulting effective 3-dimensional theory [23].

Lattice calculations allow to calculate the basic thermodynamic quantities non-perturbatively. However, in order to make use of the basic thermodynamic relations, Eqs. 4.1 and 4.2, in numerical calculations one has to go through an additional intermediate step. The free energy density itself is not directly accessible in Monte Carlo calculations; e.g. only expectation values can be calculated easily. One thus proceeds by calculating differences of the free energy density at two different temperatures. These are obtained by taking a suitable derivative of $\ln Z$ followed by an integration, e.g.

$$\left.\frac{f}{T^4}\right|_{T_o}^{T} = -\frac{1}{V}\int_{T_o}^{T} \mathrm{d}x \, \frac{\partial x^{-3} \ln Z(x,V)}{\partial x} \quad . \tag{4.3}$$

This ansatz readily translates to the lattice. Taking derivatives with respect to the gauge coupling, $\beta = 6/g^2$, rather than the temperature as was done in Eq. 4.3, we obtain expectation values of the Euclidean action which can be integrated again to give the free energy density,

$$\frac{f}{T^4}\bigg|_{\beta_o}^{\beta} = N_\tau^4 \int_{\beta_o}^{\beta} \mathrm{d}\beta' (\langle \tilde{S} \rangle - \langle \tilde{S} \rangle_{T=0}) \quad . \tag{4.4}$$

Here

$$\langle \tilde{S} \rangle \equiv -\frac{1}{N_\sigma^3 N_\tau} \frac{\partial \ln Z}{\partial \beta} = \langle S_E + \frac{\partial S_E}{\partial \beta} \rangle \quad , \tag{4.5}$$

is calculated on a lattice of size $N_\sigma^3 \times N_\tau$ and $\langle ... \rangle_{T=0}$ denotes expectation values calculated on zero temperature lattices, which usually are approximated by symmetric lattices with $N_\tau \equiv N_\sigma$. The lower integration limit is chosen at low temperatures so that f/T_o^4 is small and may be ignored[5].

A little bit more involved is the calculation of the energy density as we have to take derivatives with respect to the temperature, $T = 1/N_\tau a$. On lattices with fixed temporal extent N_τ we rewrite this in terms of a derivative with respect to the lattice spacing a which in turn is controlled through the bare couplings of the QCD Lagrangian, $a \equiv a(\beta, m_q)$. We thus find for the case of degenerate quark flavours of mass m_q

$$\begin{aligned} \frac{(\epsilon - 3p)}{T^4} = N_\tau^4 \Bigg[&\left(\frac{\mathrm{d}\beta(a)}{\mathrm{d}\ln a}\right)\left(\langle \tilde{S} \rangle - \langle \tilde{S} \rangle_{T=0}\right) \\ &- \left(\frac{\mathrm{d}m_q(a)}{\mathrm{d}\ln a}\right)\left(\langle \bar{\chi}\chi \rangle - \langle \bar{\chi}\chi \rangle_{T=0}\right)\Bigg] \quad . \end{aligned} \tag{4.6}$$

An evaluation of the energy density e.g. requires the knowledge of the two β-functions appearing in Eq. 4.6. These may be determined by calculating two physical observables in lattice units for given values of β and m_q; for instance, the string tension, $\tilde{\sigma}$ and a ratio of hadron masses, $m_{PS}/m_V \equiv m_\pi/m_\rho$. These quantities have to be calculated at zero temperature. Using Eq. 2.15 one then can determine also a temperature scale in physical units.

The numerical calculation of thermodynamic quantities is done on finite lattices with spatial extent N_σ and temporal extent N_τ. In order to perform calculations close to the thermodynamic limit we want to use a large spatial extent of the lattice. In general it has been found that lattices with $N_\sigma \gtrsim 4N_\tau$ provide a good

[5] In the gluonic sector the relevant degrees of freedom at low temperature are glueballs. Even the lightest ones calculated on the lattice have large masses, $m_G \simeq 1.5$ GeV. The free energy density thus is exponentially suppressed already close to T_c. In QCD with light quarks the dominant contribution to the free energy density comes from pions. As long as we are dealing with massive quarks also this contribution gets suppressed exponentially. However, in the massless limit clearly some care has to be taken with the normalization of the free energy density.

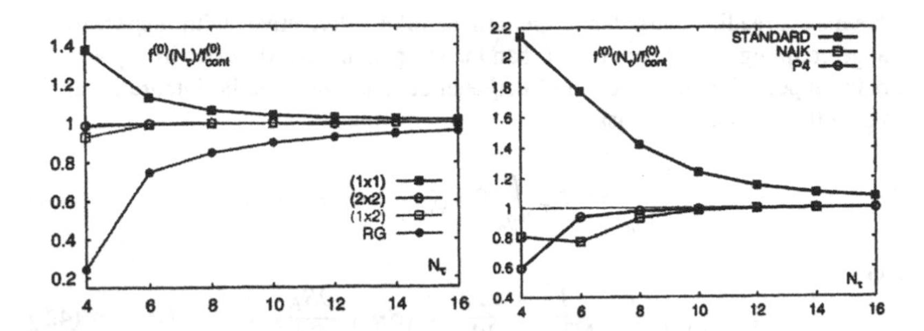

Figure 2. Cut-off dependence of the ideal gas pressure or free energy density ($p \equiv -f$) for the $SU(3)$ gauge theory (left) and several staggered fermion actions (right). Cut-off effects for the Wilson fermion action are compatible with those of the standard staggered fermion action defined by Eqs.2.11 and 2.12. The p4-action is defined by Eq. 2.14 and the Naik action has a similar structure with the bended 3-link terms replaced by straight 3-link terms [24].

approximation to the infinite volume limit. In addition, we want to get close to the continuum limit in order to eliminate discretization errors and at the same time keep the temperature constant; we thus have to combine the limit of vanishing lattice spacing ($a \rightarrow 0$) with the limit of infinite lattice size ($N_\tau \rightarrow \infty$). In order to perform this limit in a controlled way we have to analyze in how far lattice calculations of bulk thermodynamic observables are influenced by the introduction of a finite lattice cut-off ($a > 0$). These cut-off effects are largest in the high (infinite) temperature limit which can be analyzed analytically in weak coupling lattice perturbation theory. In Fig. 2 we show the results of such a calculation for lattices with different temporal extent. This clearly shows the importance of improved actions for the reduction of cut-off effects on finite lattices.

Using the tree level improved gauge action in combination with the improved staggered fermion action in numerical simulations at finite temperature it is possible to perform calculations with small systematic cut-off errors already on lattices with small temporal extent, e.g. $N_\tau = 4$ or 6. In actual calculations performed with various actions in the pure gauge sector one finds that for temperatures $T \lesssim 5T_c$ the cut-off dependence of thermodynamic observables shows the pattern predicted by the weak coupling perturbative calculation. At these temperatures the absolute magnitude of the cut-off effects, however, is smaller by about a factor of two. Although such a detailed systematic study of the cut-off dependence does not exist for QCD with light quarks it is reasonable to expect that at least at high temperature the situation will be similar to that in the pure gauge sector.

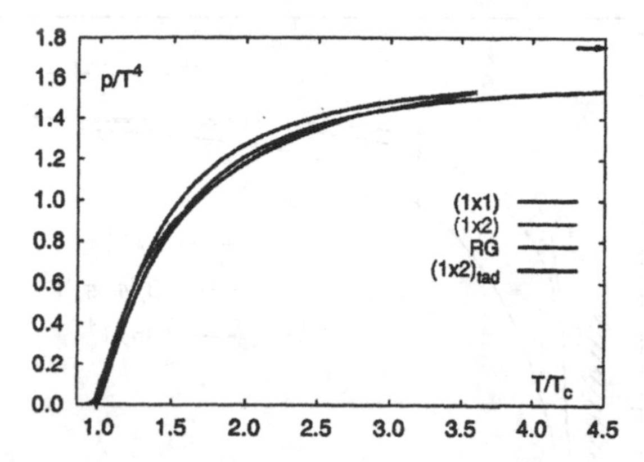

Figure 3. Pressure of the SU(3) gauge theory calculated on lattices with different temporal extent and extrapolated to the continuum limit. Shown are results from calculations with the standard Wilson (1×1)-action [25] and several improved actions [27, 28]. The broad band shows the approximately self-consistent HTL calculation of [22].

4.1. THERMODYNAMICS OF THE SU(3) GAUGE THEORY

Before entering a discussion of bulk thermodynamics in two and three flavour QCD it is worthwhile to discuss some results on the equation of state in the heavy quark mass limit of QCD – the SU(3) gauge theory. In this case the temperature dependence of the pressure and energy density has been studied in great detail, calculations with the standard action [25] and various improved actions [26, 27, 28] have been performed, the cut-off dependence has explicitly been analyzed through calculations on lattices with varying temporal extent N_τ and results have been extrapolated to the continuum limit. In Fig. 3 we show some results for the pressure obtained from such detailed analyzes with different actions [25, 27, 28]. This figure shows the basic features of the temperature dependence of bulk thermodynamic quantities in QCD, which also carry over to the case of QCD with light quarks. The pressure stays small for almost all temperatures below T_c; this is expected, as the only degrees of freedom in the low temperature phase are glueballs which are rather heavy and thus lead to an exponential suppression of pressure and energy density at low temperature. Above T_c the pressure rises rapidly and reaches about 70% of the asymptotic ideal gas value at $T = 2\ T_c$. For even larger temperatures the approach to this limiting value proceeds rather slowly. In fact, even at $T \simeq 4\ T_c$ deviations from the ideal gas value are larger than 10%. This is still too much to be described in terms of weakly interacting massless gluons as it is done in ordinary high temperature perturbation theory [18, 19]. Even at these high temperatures non-perturbative effects have to be taken into ac-

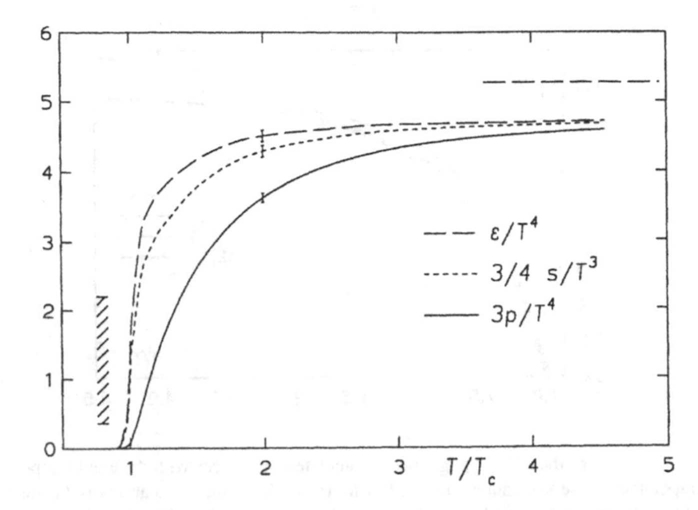

Figure 4. Energy density, entropy density and pressure of the SU(3) gauge theory calculated on lattices with different temporal extent and extrapolated to the continuum limit. The dashed band indicates the size of the latent heat gap in energy and entropy density

count which may be described in terms of interactions among quasi-particles. The broad band in Fig. 3 shows the result of a self-consistent HTL resummation [22], which leads to good agreement with the lattice calculations for $T \gtrsim 3T_c$. Other approaches [23, 29] reach a similarly good agreement in the high temperature regime.

Compared to the pressure the energy density rises much more rapidly in the vicinity of T_c. In fact, as the transition is first order in the $SU(3)$ gauge theory the energy density is discontinuous at T_c with a latent heat of about $1.5T_c^4$ [30]. In Fig. 4 we show results for the energy density, entropy density and the pressure obtained from calculations with the Wilson action which have been extrapolated to the continuum limit [25]. The delayed rise of the pressure compared to that of the energy density has consequences for the velocity of sound in the QCD plasma. In the vicinity of T_c it is substantially smaller than in the high temperature ideal gas limit.

4.2. FLAVOUR DEPENDENCE OF THE QCD EQUATION OF STATE

The calculation of the QCD pressure based on Eq. 4.4 is independent of the quark mass. The analysis of the pressure in QCD with light quarks thus proceeds along the same line as in the pure gauge sector. Unlike in the pure gauge case it, however, will be difficult to perform calculations on lattices with large temporal extent. In fact, at present all calculations of the equation of state are restricted to lattices with

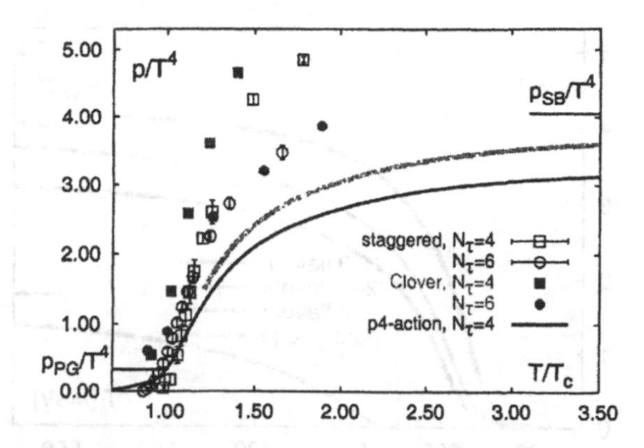

Figure 5. The pressure in two flavour QCD calculated with unimproved gauge and staggered fermion actions (open symbols) [32], RG-improved gauge and clover improved Wilson action (full symbols) [33] and the p4-action (improved gauge and improved staggered fermions (full line) [31]. The broad line estimates the results in the continuum limit as described in the text. The horizontal lines to the right and left show the Stefan-Boltzmann values for an ideal pion gas and a free quark-gluon gas, respectively.

$N_\tau = 4$ and 6 [31, 32, 33]. The use of an improved fermion action thus seems to be even more important in this case. Of course, an additional problem arises from insufficient chiral properties of staggered and Wilson fermion actions. This will mainly be of importance in the low temperature phase and in the vicinity of the transition temperature. The continuum extrapolation thus will be more involved in the case of QCD with light quarks than in the pure gauge theory and we will have to perform calculations closer to the continuum limit. Nonetheless, in particular for small number of flavours, we may expect that the flavour symmetry breaking only has a small effect on the overall magnitude of bulk thermodynamic observables. After all, for $n_f = 2$, the pressure of an ideal massless pion gas contributes less than 10% of that of an ideal quark-gluon gas in the high temperature limit. For our discussion of bulk thermodynamic observables the main source for lattice artifacts thus still seems to arise from the short distance cut-off effects, which we have to control. Additional confidence in the numerical results can be gained by comparing simulations performed with different fermion actions.

The importance of an improved lattice action, which leads to small cut-off errors at least in the high temperature ideal gas limit is apparent from Fig. 5, where we compare the results of a calculation of the pressure in 2-flavour QCD performed with unimproved gauge and staggered fermion actions [32], the RG-improved gauge action combined with improved Wilson fermions (clover action) [33] and an improved staggered fermion action (p4-action) [31]. At temperatures above

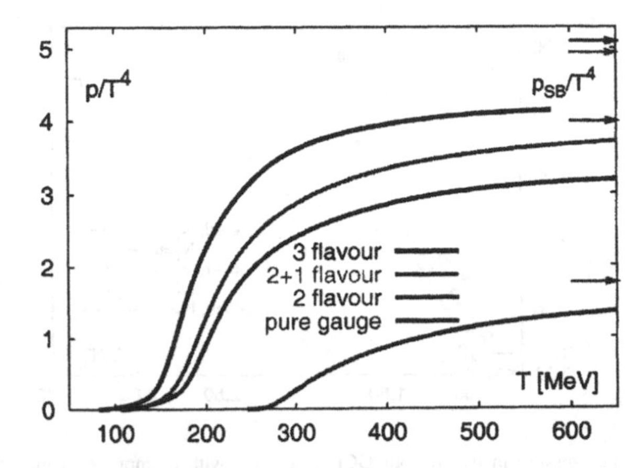

Figure 6. The pressure in QCD with different number of degrees of freedom as a function of temperature. The curve labeled (2+1)-flavour corresponds to a calculation with two light and a four times heavier strange quark mass (Karsch et al., 2000).

$T \simeq 2\ T_c$ these actions qualitatively reproduce the cut-off effects calculated analytically in the infinite temperature limit. In particular, it is evident that also the clover improved Wilson action leads to an overshooting of the continuum ideal gas limit. This is expected as the clover term in the Wilson action does eliminate $\mathcal{O}(ag^2)$ cut-off effects but does not improve the high temperature ideal gas limit, which is $\mathcal{O}(g^0)$. The clover improved Wilson action thus leads to the same large $\mathcal{O}(a^2)$ cut-off effects as the unimproved Wilson action. The influence of cut-off effects in bulk thermodynamic observables thus is similar in calculations with light quarks and in the $SU(3)$ gauge theory. This observation may also help to estimate the cut-off effects still present in current calculations with light quarks. In particular, we know from the analysis performed in the pure gauge sector that in the interesting temperature regime of a few times T_c the cut-off dependence seems to be about a factor two smaller than calculated analytically in the infinite temperature limit; we may expect that this carries over to the case of QCD with light quarks. This is the basis for the estimated continuum extrapolation of the $n_f = 2$ results shown as a broad line in Fig. 5.

In Fig. 6 we show results for the pressure obtained in calculations with different number of flavours. This figure clearly shows that the transition region shifts to smaller temperatures as the number of degrees of freedom is increased. Such a conclusion, of course, requires the determination of a temperature scale that is common to all *QCD-like* theories which have a particle content different from that realized in nature. We have determined this temperature scale by assuming

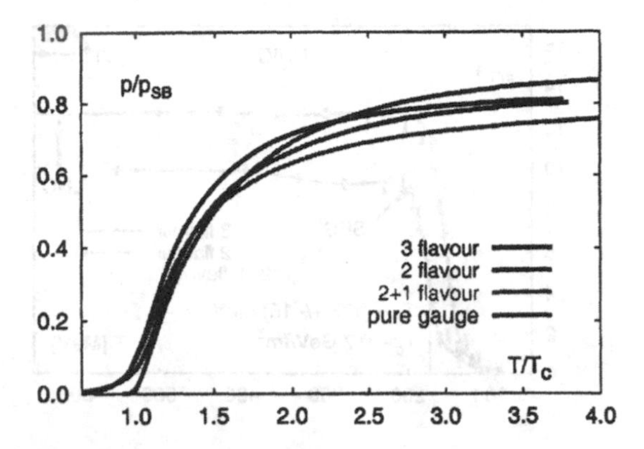

Figure 7. The pressure in units of the ideal gas pressure for the $SU(3)$ gauge theory and QCD with various number of flavours. The latter calculations have been performed on lattices with temporal extent $N_\tau = 4$ using the p4-action. Results are not yet extrapolated to the continuum limit.

that the string tension is flavour and quark mass independent. This assumption is supported by the observation that already in the heavy quark mass limit the string tension calculated in units of quenched hadron masses, e.g. $m_\rho/\sqrt{\sigma} = 1.81\ (4)$ [34], is in good agreement with values required in QCD phenomenology, $\sqrt{\sigma} \simeq 425\,\mathrm{MeV}$.

At high temperature the magnitude of p/T^4 clearly reflects the change in the number of light degrees of freedom present in the ideal gas limit. When we rescale the pressure by the corresponding ideal gas values it becomes, however, apparent that the overall pattern of the temperature dependence of p/T^4 is quite similar in all cases. This is shown in Fig. 7. In particular, when one takes into account that a proper continuum extrapolation in QCD with light quarks is still missing this agreement achieved with improved staggered fermions is quite remarkable.

We also note that the pressure at low temperature is enhanced in QCD with light quarks compared to the pure gauge case. This is an indication for the contribution of hadronic states, which are significantly lighter than the heavy glueballs of the $SU(3)$ gauge theory. Their contribution is even more clearly visible in the behaviour of the energy density. In Fig. 8 we show the energy density of two and three flavour QCD obtained with improved staggered[6] fermions. These calculations yield as an estimate for the energy density at T_c,

$$\epsilon_c \simeq (6 \pm 2) T_c^4 \quad . \tag{4.7}$$

[6] The figure for staggered fermions is based on data from Ref. [31]. Here a contribution to ϵ/T^4 which is proportional to the bare quark mass and vanishes in the chiral limit is not taken into account.

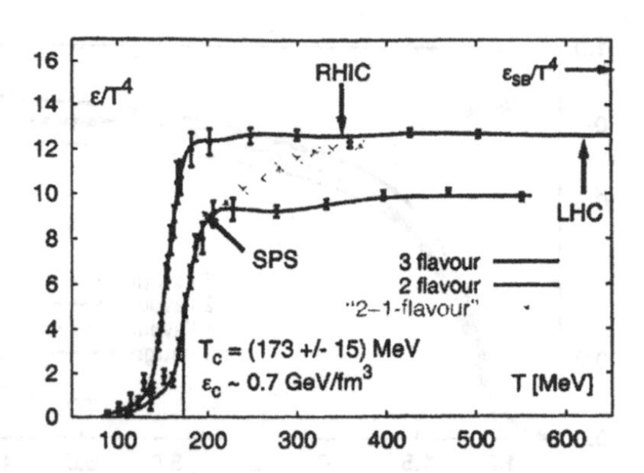

Figure 8. The energy density in QCD for two and three flavour QCD. The calculations have been performed on a lattice with temporal extent $N_\tau = 4$ and a pseudo-scalar to vector meson mass ratio of $m_{PS}/m_V = 0.7$. The crosses give a guess for the temperature dependence of QCD with a realistic strange mass value (see text). Also indicated are energy densities expected to be reached at the SPS at CERN, the RHIC at Brookhaven as well as the future LHC at CERN.

Similar results for the critical energy density can be deduced from calculations with Wilson [33] fermions and unimproved staggered fermions[32]. The estimate for ϵ_c/T_c^4 is an order of magnitude larger than the critical value on the hadronic side of the transition in the pure gauge theory (see Fig. 4). It is, however, interesting to note that this difference gets to a large extent compensated by the shift in T_c to smaller values. When we convert the result for ϵ_c in units of [MeV/fm^3] the transitions in the infinite quark mass limit and in QCD with light quarks seem to take place at compatible values of the energy density, $\epsilon_c \simeq (0.3 - 1.3)\mathrm{GeV/fm}^3$. At present the largest uncertainty on this number arises from uncertainties on the value of T_c (see next section). However, also the magnitude of ϵ_c/T_c^4 still has to be determined more accurately. Here two competing effects will be relevant. On the one hand we expect ϵ_c/T_c^4 to increase with decreasing quark masses, *i.e.* closer to the chiral limit. On the other hand, it is likely that finite volume effects are similar to those in the pure gauge sector, which suggests that ϵ_c/T_c^4 will still decrease closer to the thermodynamic limit, *i.e.* for $N_\sigma \to \infty$.
In the case of 2-flavour QCD simulations have been performed also with improved Wilson fermions [33]. Here the quark mass dependence of bulk thermodynamic observables has been analyzed in a wide range of masses, which is controlled by the ratio of the pseudo-scalar to vector meson mass m_{PS}/m_V. The results show no significant quark mass dependence up to $m_{PS}/m_V \simeq 0.9$, which corresponds to pseudo-scalar meson masses of about 1.5 GeV. As this mass is somewhat larger than the Φ-meson mass the corresponding quark mass is compatible with that

of the strange quark. The approximate quark mass independence of the equation of state observed in the high temperature phase thus is consistent with our expectation that quark mass effects should become significant only when the quark masses get larger than the temperature. This observation led to the estimate for the temperature dependence of the energy density in QCD with realistic (u, d, s)-quark masses which is indicated with crosses in Fig. 8. While the strange quarks will not contribute to the thermodynamics close to T_c they rapidly will contribute like an additional massless degree of freedom above T_c. The pressure and energy density will then be close to that of massless 3-flavour QCD.

5. The Critical Temperature of the QCD Transition

As discussed in Section 3 the transition to the high temperature phase is continuous and non-singular for a large range of quark masses. Nonetheless, for all quark masses the transition proceeds rather rapidly in a small temperature interval. A definite transition point thus can be identified, for instance through the location of maxima of the susceptibilities of the Polyakov loop or the chiral condensate,

$$\chi_L = N_\sigma^3\Big(\langle L^2\rangle - \langle L\rangle^2\Big) \quad , \quad \chi_m = \frac{\partial}{\partial m_q}\langle\bar\psi\psi\rangle \quad . \qquad (5.1)$$

For a given value of the quark mass this defines a pseudo-critical coupling, $\beta_{pc}(m_q)$, on a lattice with temporal extent N_τ. An additional calculation of an experimentally or phenomenologically known observable at zero temperature, e.g. a hadron mass or the string tension, is still needed to determine the transition temperature from Eq. 2.15. In the pure gauge theory the transition temperature, again has been analyzed in great detail and the influence of cut-off effects has been examined through calculations on different size lattices and with different actions. From this one finds for the critical temperature of the first order phase transition,

$$\underline{\mathrm{SU(3)\ gauge\ theory:}} \quad T_c/\sqrt{\sigma} = 0.637 \pm 0.005$$
$$T_c = (271 \pm 2)\ \mathrm{MeV} \qquad (5.2)$$

Already the early calculations for the transition temperature in QCD with dynamical quark degrees of freedom [35, 36] indicated that the inclusion of light quarks leads to a significant decrease of the transition temperature. However, these early calculations, which have been performed with standard Wilson [35] and staggered [36] fermion actions, also led to significant discrepancies in the results for T_c as well as the order of the transition. These differences strongly diminished in the newer calculations which are based on improved Wilson fermions (clover action) [36, 37, 38], domain wall fermions as well as improved staggered fermions (p4-action) [39]. A compilation of these newer results is shown in Fig. 9 for various values of the quark masses. In order to compare calculations performed with different actions the results are presented in terms of a *physical observable*, the meson mass ratio m_{PS}/m_V. In Fig. 9a we show T_c/m_V obtained for 2-flavour QCD while Fig. 9b gives a comparison of results obtained with improved staggered fermions [39] for 2 and 3-flavour QCD. Also shown there is a result for the case of (2+1)-flavour QCD, *i.e.* for two light and one heavier quark flavour degree of freedom. Unfortunately the quark masses in this latter case are still too large to be compared directly with the situation realized in nature. We note however, that the results obtained so far suggest that the transition temperature in (2+1)-flavour QCD is close to that of 2-flavour QCD. The 3-flavour theory, on the other hand, leads to consistently smaller values of the critical temperature, $T_c(n_f = 2) - T_c(n_f = 3) \simeq 20$ MeV. Extrapolations of the transition temperatures to the chiral limit gave

$$\underline{2 - \text{flavour QCD}:} \quad T_c = \begin{cases} (171 \pm 4)\ \text{MeV}, & \text{clover-improved Wilson fermions [37]} \\ (173 \pm 8)\ \text{MeV}, & \text{improved staggered fermions [39]} \end{cases}$$

$$\underline{3 - \text{flavour QCD}:} \quad T_c = (154 \pm 8)\ \text{MeV}, \quad \text{improved staggered fermions (Karsch et al., 2001)}$$

Here m_ρ has been used to set the scale for T_c. Although the agreement between results obtained with Wilson and staggered fermions is striking, one should bear in mind that all these results have been obtained on lattice with temporal extent $N_\tau = 4$, *i.e.* at rather large lattice spacing, $a \simeq 0.3$ fm. Moreover, there are uncertainties involved in the ansatz used to extrapolate to the chiral limit. We thus estimate that the systematic error on the value of T_c/m_ρ still is of similar magnitude as the purely statistical error quoted above.

We note from Fig. 9 that T_c/m_V drops with increasing ratio m_{PS}/m_V, *i.e.* with increasing quark mass. This may not be too surprising as m_V, of course, does not take on the physical ρ-meson mass value as long as m_{PS}/m_V did not reach its physical value (vertical line in Fig. 9a). In fact, m_V will increase with increasing quark mass. The ratio T_c/m_V thus will approach zero for $m_{PS}/m_V = 1$ as T_c will stay finite and take on the value calculated in the pure $SU(3)$ gauge theory. Fig. 9 thus does not yet allow to quantify how T_c depends on the quark mass. A simple percolation picture for the QCD transition would suggest that $T_c(m_q)$ or better $T_c(m_{PS})$ will increase with increasing m_q; with increasing m_q also the hadron masses increase and it becomes more difficult to excite the low lying hadronic states. It thus becomes more difficult to create a sufficiently high particle/energy density in the hadronic phase that can trigger a phase (percolation) transition. Such a picture also follows from chiral model calculations [41, 42].

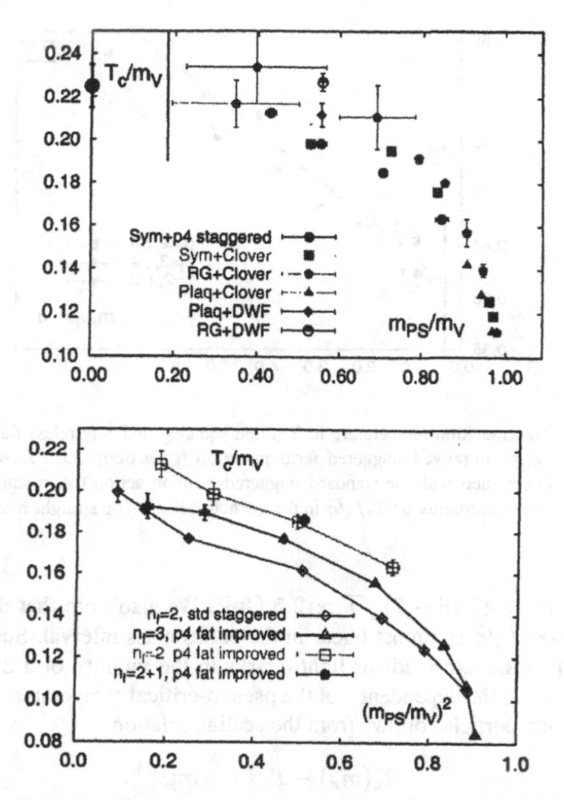

Figure 9. Transition temperatures in units of m_V. The upper figure shows a collection of results obtained for 2-flavour QCD with various fermion actions (for references see [40]) while in the lower figure we compare results obtained in 2 and 3-flavour QCD with the p4-action. All results are from simulations on lattices with temporal extent $N_\tau = 4$. The large dot drawn for $m_{PS}/m_V = 0$ indicates the result of chiral extrapolations based on calculations with improved Wilson [37] as well as improved staggered [39] fermions. The vertical line in the upper figure shows the location of the physical limit, $m_{PS} \equiv m_\pi = 140$ MeV.

As argued previously we should express T_c in units of an observable, which itself is not dependent on m_q; the string tension (or also a quenched hadron mass) seems to be suitable for this purpose. In fact, this is what tacitly has been assumed when one converts the critical temperature of the SU(3) gauge theory, $T_c/\sqrt{\sigma} \simeq 0.63$, into physical units as it has been done in Eq. 5.2.

To quantify the quark mass dependence of the transition temperature one may express T_c in units of $\sqrt{\sigma}$. This ratio is shown in Fig. 10 as a function of $m_{PS}/\sqrt{\sigma}$. As can be seen the transition temperature starts deviating from the quenched

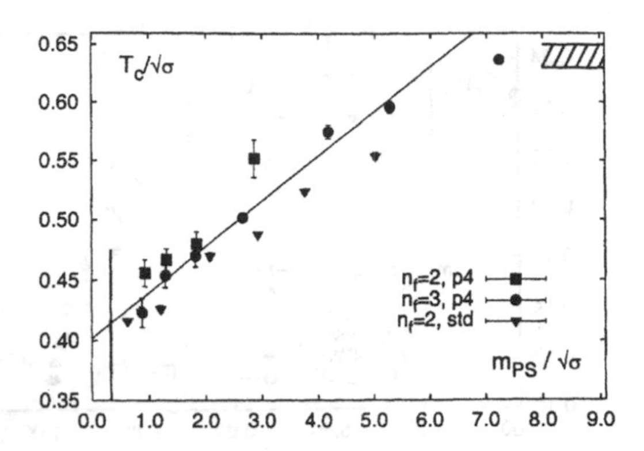

Figure 10. The transition temperature in 2 (filled squares) and 3 (circles) flavour QCD versus $m_{PS}/\sqrt{\sigma}$ using an improved staggered fermion action (p4-action). Also shown are results for 2-flavour QCD obtained with the standard staggered fermion action (open squares). The dashed band indicates the uncertainty on $T_c/\sqrt{\sigma}$ in the quenched limit. The straight line is the fit given in Eq. 5.4.

values for $m_{PS} \lesssim (6-7)\sqrt{\sigma} \simeq 2.5$ GeV. We also note that the dependence of T_c on $m_{PS}/\sqrt{\sigma}$ is almost linear in the entire mass interval. Such a behaviour might, in fact, be expected for light quarks in the vicinity of a 2nd order chiral transition where the dependence of the pseudo-critical temperature on the mass of the Goldstone-particle follows from the scaling relation

$$T_c(m_\pi) - T_c(0) \sim m_\pi^{2/\beta\delta} \,. \tag{5.3}$$

For 2-flavour QCD the critical indices β and δ are expected to belong to the universality class of 3-d, $O(4)$ symmetric spin models and one thus indeed would expect to find $2/\beta\delta \simeq 1.1$. However, this clearly cannot be the origin for the quasi linear behaviour which is observed for rather large hadron masses and seems to be independent of n_f. Moreover, unlike in chiral models [41, 42] the dependence of T_c on m_{PS} turns out to be rather weak. The line shown in Fig. 10 is a fit to the 3-flavour data, which gave

$$\left(\frac{T_c}{\sqrt{\sigma}}\right)_{m_{PS}/\sqrt{\sigma}} = \left(\frac{T_c}{\sqrt{\sigma}}\right)_0 + 0.04(1)\left(\frac{m_{PS}}{\sqrt{\sigma}}\right) \quad . \tag{5.4}$$

It seems that the transition temperature does not react strongly to changes of the lightest hadron masses. This favours the interpretation that the contributions of heavy resonance masses are equally important for the occurrence of the transition. In fact, this also can explain why in the heavy quark mass limit the transition still sets in at quite low temperatures even when all hadron masses, including the

pseudo-scalars, attain masses of the order of 1 GeV or more. Such an interpretation also is consistent with the weak quark mass dependence of the critical energy density which we found from the analysis of the QCD equation of state in the previous section.
For the quark masses currently used in lattice calculations a resonance gas model combined with a percolation criterion thus provides an appropriate description of the thermodynamics close to T_c. It remains to be seen whether the role of the light meson sector becomes more dominant when we get closer to the chiral limit.

6. The Heavy Quark Free Energy

The heavy quark free energy defined in Eq. 3.1 plays a central role in our understanding of the QCD phase transition and the thermal properties of the high temperature plasma phase. As discussed in Section 2 the correlation function for static quark anti-quark sources defines the excess free energy due the presence of these sources in a thermal medium [3],

$$\frac{F_{\bar{q}q}(r,T)}{T} = -\ln\left(\langle \mathrm{Tr} L_{\vec{x}} \mathrm{Tr} L^{\dagger}_{\vec{y}}\rangle\right) + c(T) \quad , \quad rT = |\vec{x}-\vec{y}|/N_\tau \quad , \tag{6.1}$$

with a so far unspecified normalization constant $c(T)$ related to the self-energy of the quark anti-quark sources. The heavy quark free energy probes the confining properties of the thermal medium. From the large distance behaviour of $F_{\bar{q}q}$ one can extract a screening mass. Its T-dependence at has extensively been used in the analysis of heavy quark bound state formation in the plasma phase and the discussion of J/ψ-suppression [43]. Equally important for this discussion, however, is to understand the short distance properties of $F_{\bar{q}q}$ which is an statistical average of contributions arising from $(\bar{q}q)$-pairs in singlet and octet states (colour averaged free energy) while the potential relevant for heavy quark bound state formation refers to the singlet potential only. For this reason it is important to understand in detail the structure of $F_{\bar{q}q}$ and firmly establish its connection to the heavy quark potential.
The long distance behaviour of $F_{\bar{q}q}$ is given by $-\ln|\langle L\rangle|^2$ (see Eqs. 3.2 and 3.3). In the pure SU(3) gauge theory this provides an order parameter for the deconfinement phase transition as indicated in Eq. 3.3. In the presence of dynamical quarks, *i.e.* for QCD with arbitrary but finite quark masses, the heavy quark free energy stays finite at all distances and all temperatures,

$$F_\infty(T) = \lim_{r\to\infty} F_{\bar{q}q}(r,T) \quad . \tag{6.2}$$

We would like to interpret F_∞ as the change in free energy due to the presence of two well separated quarks, each of which is screened by a cloud of quarks and gluons. In order to do so we have to clarify the normalization condition for

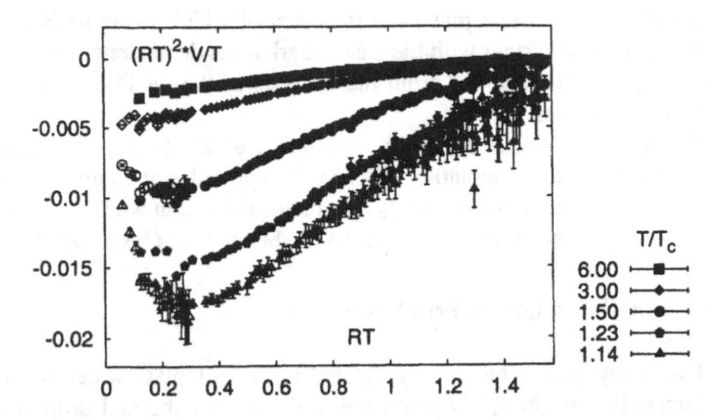

Figure 11. The heavy quark free energy at various temperatures in the deconfined phase of the $SU(3)$ gauge theory. The asymptotic value at infinite quark anti-quark separation has been subtracted. Calculations have been performed on lattices of size $32^3 \times 8$ (filled symbols) and $32^3 \times 16$ (open symbols) [44].

$F_{\bar{q}q}(r,T)$ which unambiguously fixes the T-dependent constant in Eq. 6.1. This can be achieved through an analysis of the short distance behaviour of $F_{\bar{q}q}(r,T)$. At short distances the quark anti-quark pair interacts through the exchange of gluons, which may be calculated perturbatively. More importantly we expect that this interaction is essentially T-independent for separations r which are smaller than the average separation between partons in the thermal medium. We thus would like to fix the normalization constant $c(T)$ through a matching of the short distance behaviour of the heavy quark free energy with that of the zero temperature heavy quark potential.

We may split $F_{\bar{q}q}$ in contributions arising from quark anti-quark pairs in singlet (F_1) and octet (F_8) states [3],

$$e^{-F_{\bar{q}q}(r,T)/T} = \frac{1}{9}\, e^{-F_1(r,T)/T} + \frac{8}{9}\, e^{-F_8(r,T)/T} \quad . \tag{6.3}$$

In zero temperature perturbation theory one finds that the singlet free energy is attractive whereas the octet free energy is repulsive. To leading order (1-gluon exchange) they are given by

$$F_1(r,T) = -\frac{g^2}{3\pi}\frac{1}{r} \quad , \quad F_8(r,T) = \frac{g^2}{24\pi}\frac{1}{r} \quad . \tag{6.4}$$

The relative strength of F_1 and F_8 is such that the 1-gluon exchange contribution cancels in the colour averaged heavy quark free energy $F_{\bar{q}q}(r,T)$. The leading contribution then arises from the exchange of two gluons. This leads to the well known perturbative result $F_{\bar{q}q}(r,T)/T \sim (F_1(r,T)/T)^2 \sim 1/(rT)^2$, which can also directly be obtained from a weak coupling expansion of the Polyakov loop correlation function. From Eq. 6.3 it is, however, apparent, that the cancellation of singlet and octet contributions only occurs at large distances and/or temperatures ($rT >> 1$) where it is justified to expand the exponentials appearing in Eq. 6.3. At short distances the repulsive octet term is exponentially suppressed and the contribution from the attractive singlet channel will dominate the heavy quark free energy,

$$\begin{aligned}\frac{F_{\bar{q}q}(r,T)}{T} &= \frac{F_1(r,T)}{T} - \ln 9 \\ &= -\frac{g^2(T)}{3\pi}\frac{1}{rT} - \ln 9 \qquad \text{for} \quad rT << 1 \ . \end{aligned} \tag{6.5}$$

We thus expect that at short distances the colour averaged free energy coincides with the colour singlet free energy and shows a Coulomb-like $1/r$ singularity. Such a behaviour indeed has been found in recent numerical calculations which aimed at an analysis of the short distance structure of $F_{\bar{q}q}$. Some results of this analysis which has been performed in the $SU(3)$ gauge theory [44] are shown in Fig. 11. In order to eliminate the anticipated power-like behaviour at large distances we show $(rT)^2\ V(r,T)/T$, where $V(r,T) \equiv F_{\bar{q}q}(r,T) - F_\infty(T)$. The rise at short distances, which is particularly pronounced at temperatures close to T_c, indicates that $F_{\bar{q}q}(r,T)$ drops less rapid than $1/(rT)^2$; in fact, it is consistent with a Coulomb-like behaviour. Furthermore, one can establish that the singlet and colour averaged free energies differ by the statistical factor $T\ln 9$ at short distances [44].

At large distances the heavy quark free energy is exponentially screened. As can be seen the change from the Coulomb-like behaviour at short distances to the exponential screening at large distances is well localized. For $T_c \leq T \lesssim 2T_c$ it occurs already for $rT \simeq 0.2$ or $r \simeq 0.15\ (T_c/T)$ fm and shifts slightly to smaller rT with increasing temperature. A consequence of this efficient screening at short distances is that heavy quark bound states get destroyed close to T_c in the plasma phase ([43, 45, 46]).

As we have established that the short distance part of the colour averaged free energy is dominated by the singlet contribution we may proceed and try to normalize the finite-T heavy quark free energy by matching it to the zero temperature heavy quark potential. This is shown in Fig. 12 for QCD with two degenerate, light quark flavours [39]. We notice that the asymptotic, large distance value of the colour averaged heavy quark free energy rapidly decreases in the vicinity of T_c. In the lower part of Fig. 12 we show the change in free energy needed to separate a quark anti-quark pair from a distance typical for the J/ψ radius, *i.e.* $\bar{r} = 0.5/\sqrt{\sigma} \simeq 0.2$ fm, to infinity. At T_c the change in free energy is only about 200 MeV, which is of the same magnitude as typical parton momenta in

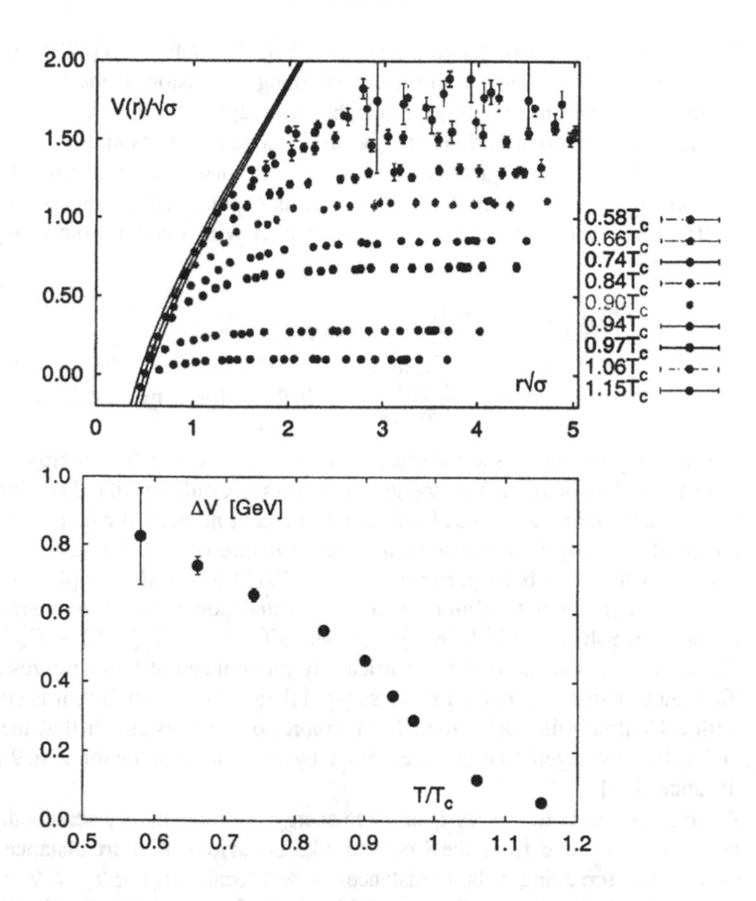

Figure 12. The heavy quark free energy at various temperatures in the high temperature phase of three flavour QCD (upper figure). The free energy has been normalized at the shortest distance available ($r = 1/T$) to the zero temperature Cornell potential, $V(r)/\sqrt{\sigma} = -\alpha/r\sqrt{\sigma} + r\sqrt{\sigma}$ with $\alpha = 0.25 \pm 0.05$ (solid band). The lower figure shows the temperature dependence of the change in free energy when separating the quark anti-quark pair from a distance $\bar{r} = 0.5/\sqrt{\sigma}$ to infinity.

the thermal medium. It thus is likely that a thermal break-up of $\bar{c}c$-bound states is possible at this temperature and provides the explanation for the experimentally observed suppression of J/ψ bound states [47].

7. Thermal modifications of hadron properties

7.1. SPECTRAL FUNCTIONS

A central issue in the discussion of the QCD phase transition and possible signatures for its occurrence in heavy ion collisions is to understand the thermal modifications of hadron properties. In the previous section we have discussed modifications of the heavy quark free energy which indicate drastic changes of the heavy quark potential in the QCD plasma phase. As a consequence heavy quark bound states cannot form above certain critical temperatures. Similarly it is expected that the QCD plasma cannot support the formation of light quark bound states. In the pseudo-scalar sector the disappearance of the light pions clearly is related to the vanishing of the chiral condensate at T_c. For $T > T_c$ the pions are no longer (nearly massless) Goldstone bosons. In the plasma phase one thus may expect to find only massive quasi-particle excitations in the pseudo-scalar quantum number channel. However, also below T_c it is expected that the gradual disappearance of the spontaneous breaking of chiral flavour symmetry as well as the weakening of the explicit breaking of the axial $U_A(1)$ symmetry will lead to thermal modifications of hadron properties, e.g. their masses and widths. While the breaking of the $SU_L(n_f) \times SU_R(n_f)$ flavour symmetry leads, for instance, to the splitting of scalar (f_0) and pseudo-scalar (pion) particle masses, the $U_A(1)$ symmetry breaking is visible in the splitting of the pion and the scalar a_0 meson. The effect of the breaking/restoration of chiral symmetries does not only lead to a temperature dependence of the splitting of physical particle states. The symmetry transformations can also be performed directly on the hadronic currents[7], $H(\tau, \vec{x}) = \bar{q}(\tau, \vec{x})\Gamma q(\tau, \vec{x})$, and the hadron correlation functions constructed from them,

$$G_H(\tau, \vec{x}) = \langle H(\tau, \vec{x}) H^{\dagger}(0) \rangle \quad , \tag{7.1}$$

or equivalently the correlation functions at fixed momentum,

$$G_H(\tau, \vec{p}) = \int d^3x \, \exp{(i\, \vec{p}\, \vec{x})}\, G_H(\tau, \vec{x}) \quad . \tag{7.2}$$

One thus may analyze directly the temperature dependence of hadronic correlation functions. Such a comparison is given in Fig. 13 for scalar (a_0) and pseudo-scalar correlation functions calculated in the pure $SU(3)$ gauge theory (quenched QCD) at a temperature below ($T = 0.6T_c$) and above ($T = 3T_c$) the deconfinement phase transition. The effective restoration of the chiral $U_A(1)$ symmetry is clearly visible in this figure; scalar and pseudo-scalar correlators coincide above T_c within statistical accuracy.

[7] Here we denote with Γ a combination of Dirac and flavour matrices appropriate for the given quantum number channel H.

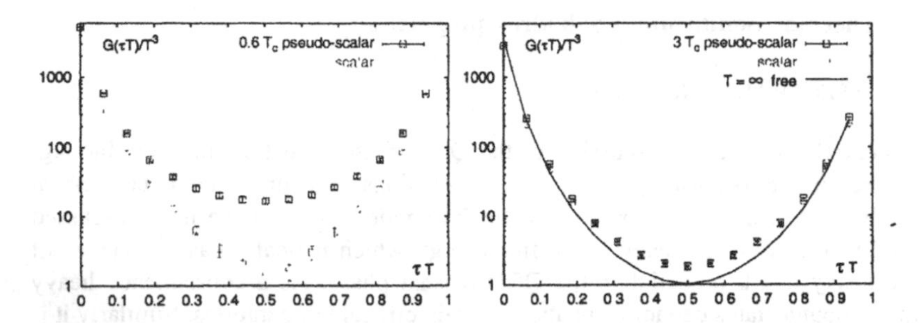

Figure 13. Zero momentum projected temporal correlation functions, $G_H(\tau T) \equiv G_H(\tau T, \vec{p} = 0)$, for scalar ($a_0$) and pseudo-scalar mesons at $T = 0.6T_c$ and $3T_c$. The correlation functions have been calculated in the quenched approximation of QCD using clover improved Wilson fermions on $32^3 \times 16$ and $64^3 \times 16$ lattices, respectively.

Similar results hold for vector and pseudo-vector correlation functions. The crucial question, of course, is to what extent the obvious temperature dependence of hadronic correlation functions also results in modifications of hadron masses. The answer to this can be obtained through a study of the spectral representation of the Euclidean time correlation functions,

$$G_H(\tau, \vec{p}) = \int_0^\infty d\omega \, \sigma_H(\omega, \vec{p}) \, \frac{\cosh(\omega(\tau - 1/2T))}{\sinh(\omega/2T)} \quad . \tag{7.3}$$

Here we have introduced the spectral function σ_H. On the other hand $G_H(\tau, \vec{p})$ can be represented as the Fourier transform of the momentum space correlation function,

$$G_H(\tau, \vec{p}) = T \sum_n \mathrm{e}^{-i\omega_n \tau} \, \tilde{G}_H(\omega_n, \vec{p}) \quad , \tag{7.4}$$

with Matsubara frequencies $\omega_n \equiv 2\pi n T$. From Eqs. 7.3 and 7.4 one finds

$$\tilde{G}_H(\omega_n, \vec{p}) = \int_{-\infty}^{\infty} \mathrm{d}\omega \, \frac{\sigma_H(\omega, \vec{p}, T)}{\omega - i\omega_n} \quad , \tag{7.5}$$

which shows that the spectral function appearing in the Euclidean correlation function indeed coincides with the spectral function introduced in Minkowski space through the retarded correlation function [48],
$\sigma_H(\omega, \vec{p}, T) = \pi^{-1} \mathrm{Im}\, \tilde{G}_H^R(\omega, \vec{p})$, with $\tilde{G}_H^R(\omega, \vec{p}) = \tilde{G}_H(-i\omega, \vec{p})$.
In the infinite temperature limit the normalized scalar and pseudo-scalar correlation functions shown in Fig. 13 approach the correlation function for a freely propagating quark anti-quark pair. Introducing dimensionless variables, $\tilde{\omega} = \omega/T$ and $\tilde{\tau} = \tau T$ the free correlator is given by

$$\frac{G_H^{free}(\tau, \vec{p})}{T^3} \equiv \lim_{T \to \infty} \frac{G_H(\tau, \vec{p})}{T^3}$$

$$= \int_0^\infty d\tilde{\omega}\ \sigma_H^{free}(\tilde{\omega},\vec{p})\ \frac{\cosh(\tilde{\omega}(\tilde{\tau}-1/2))}{\sinh(\tilde{\omega}/2)} \quad , \tag{7.6}$$

where the free spectral function can be constructed from the quark spectral functions projected to the pair-momentum $\vec{p}$ [49, 50]. It takes on a simple form for vanishing quark masses and $\vec{p} = 0$; $\sigma_H^{free}(\tilde{\omega},\vec{0}) = 0.375 a_H \pi^{-2}\tilde{\omega}^2 \tanh(\tilde{\omega}/4)$ with $a_H = 1$ or 2 in the pseudo-scalar and vector channels, respectively. In the free (infinite temperature) limit, the spectral function only receives contributions from the two particle continuum. In general, however, σ_H will also receive contributions from T-dependent bound states, which show up as δ-function singularities or resonance peaks in addition to a T-dependent continuum which, moreover, may only contribute beyond a certain T-dependent threshold.
In order to control these different T-dependent features of the spectral function and, in particular, in order to separate modifications of the pole and resonance structure from thermal modifications of the continuum, it will not be sufficient to analyze only the correlation functions. A direct determination of σ_H, on the other hand, requires the inversion of the integral equation, Eq. 7.3. This can, in general, quite efficiently be achieved with statistical techniques like the maximum entropy method (MEM) [51, 52]. An additional problem arises in lattice calculations where the correlation functions can only be determined at a discrete set of lattice grid points and, moreover, temperature is controlled through the finite lattice extent in temporal direction. At finite temperature the temporal correlation function are thus only determined for Euclidean times $\tau T = k/N_\tau$, $k = 0, 1,.., N_\tau - 1$. Nonetheless, recent attempts [53, 54] to reconstruct hadronic spectral functions using MEM have shown that this becomes possible already with a moderate number of grid points in the temporal direction of the lattice. The reconstruction of the spectral functions corresponding to the correlators shown in Fig. 13 is shown in Fig. 14. This clearly reflects the drastic, qualitative change which occurs in the pseudo-scalar channel when one crosses T_c. The large peak which corresponds to the Goldstone below T_c pole[8] drastically reduces in size above T_c. In the high temperature phase the spectral functions reconstructed from the scalar and pseudo-scalar correlation functions coincide. This, of course, simply reflects the almost perfect agreement of the correlation functions themselves (Fig. 13). The peak, still visible in the spectral functions shifts wit hincreasing temperature. This suggests that the pion pole indeed disappears above T_c.

7.2. DILEPTON RATES

Of particular interest is the influence of a rising temperature on the spectral properties in the vector meson channel. This is directly related to the thermal dilepton

[8] Below T_c calculations have been performed with non-zero quark masses. For this reason, the pseudo-scalar is not massless.

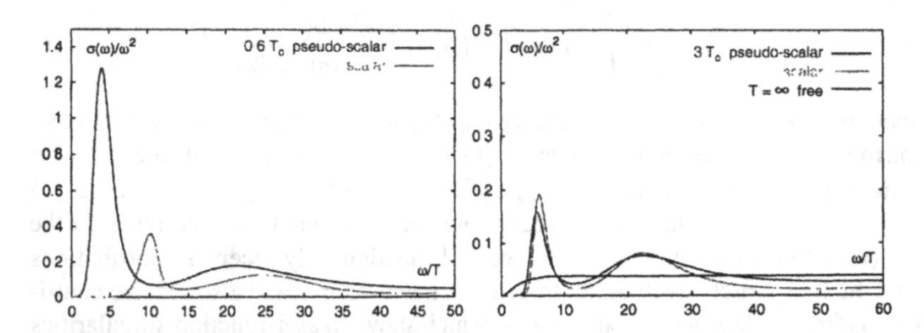

Figure 14. Scalar and pseudo-scalar spectral functions at $T = 0.6T_c$ and $3T_c$ reconstructed from the correlation functions shown in Fig. 13.

production which can be analyzed in heavy ion experiments. In fact, an enhanced production of low mass dileptons has been observed [55]. For two massless quark flavours the thermal rate resulting from quark anti-quark annihilation in the QCD plasma phase is related to the vector spectral function σ_V through

$$\frac{\mathrm{d}W}{\mathrm{d}\omega\mathrm{d}^3p} = \frac{5\alpha^2}{27\pi^2}\frac{1}{\omega^2(\mathrm{e}^{\omega/T}-1)}\sigma_V(\omega,\vec{p},T) \quad . \tag{7.7}$$

While in the (pseudo-)scalar channel the correlation functions above T_c still differ substantially from that of the free case the discrepancy is much smaller in the vector channel. Already for $T \simeq 1.5T_c$ the vector correlator $G_V(\tau T)$ differs by less than 15% from the limiting form expected at infinite temperature, *i.e.* the correlation function of freely propagating quark anti-quark pairs. This is shown in Fig. 15a. As a consequence also the spectral functions reconstructed from the correlators (Fig. 15b) and in turn the dilepton rates (Fig. 15c) stay close to the leading order perturbative results [56].

8. Conclusions

We have given a brief introduction into the lattice formulation of QCD thermodynamics and presented a few of the basic results on the equation of state, the critical parameters for the transition to the QCD plasma phase and the fate of hadrons in this new phase of matter. We now have reached a first quantitative understanding of the quark mass dependence of QCD thermodynamics which justifies a first attempt to extrapolate the results to the physical quark spectrum. At present calculations with light quarks still correspond to a world in which the pion would have a mass of about (300-500) MeV. This mass still is too large to become sensitive to details of the physics of chiral symmetry breaking. Nonetheless, lattice calculations performed with different lattice fermion formulations start to produce

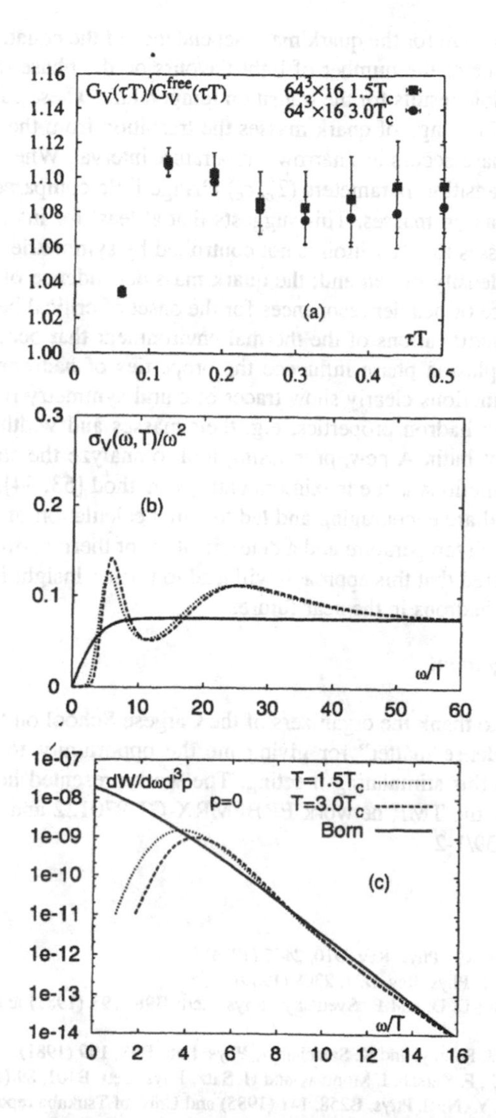

Figure 15. The ratio of the vector correlation function $G_V(\tau T)$ and the corresponding free correlation function for massless quark anti-quark pairs, $G_V^{free}(\tau T)$ versus Euclidean time in units of the temperature (a), reconstructed vector spectral functions σ_V in units of ω^2 at zero momentum (b) and the resulting zero momentum differential dilepton rate (c) at $T/T_c = 1.5$ (dots) and 3 (dashes). Also shown with solid lines is the free spectral function in (b) and the resulting Born rate for thermal dilepton production in (c).

a consistent picture for the quark mass dependence of the equation of state as well as the influence of the number of light flavours on the phase transition and they yield compatible results for the transition temperature. These calculations suggest that over a wide range of quark masses the transition from the hadronic phase to the plasma phase occurs in a narrow temperature interval. When varying the quark masses the transition parameters (T_c, ϵ_c) change little compared to the change in the lightest hadron masses. This suggests that at least for intermediate values of the quark masses the transition is not controlled by symmetries of QCD. It rather seems to be density driven and; the quark mass dependence of T_c and ϵ_c reflects the importance of heavier resonances for the onset of critical behaviour.
The strong modifications of the thermal environment that occur during the transition to the plasma phase influence the properties of hadrons. Thermal hadron correlation functions clearly show traces of chiral symmetry restoration and suggest that basic hadron properties, e.g. their masses and widths, are modified in a thermal heat bath. A new, promising tool to analyze the structure of thermal correlation functions is the maximum entropy method [53, 54]. First applications of this method are encouraging and led to a first calculation of the vector spectral function at high temperature and a determination of thermal dilepton rates [56]. It is to be expected that this approach will lead to further insight into the in-medium properties of hadrons in the near future.

Acknowledgement

I would like to thank the organizers of the Cargese School on "QCD perspectives on hot and dense matter" for giving me the opportunity to participate in and contribute to this stimulating meeting. The work presented here has partly been supported by the TMR network ERBFMRX-CT-970122 and by the DFG under grant FOR 339/1-2.

References

1. Wilson, K. G., Phys. Rev. D10, 2445 (1974).
2. Creutz, M. Phys. Rev. D21, 2308 (1980).
3. McLerran, L. D. and B. Svetitsky, Phys. Lett. B98, 195 (1981) and Phys. Rev. D24, 450 (1981).
4. Kuti, J., J. Polonyi and K. Szlachanyi, Phys. Lett. B98, 199 (1981).
5. Engels, J., F. Karsch, I. Montvay and H. Satz, Phys. Lett. B101, 89 (1981).
6. Iwasaki, Y. Nucl. Phys. B258, 141 (1985) and Univ. of Tsukuba report UTHEP-118 (1983).
7. Kogut, J. and L. Susskind, Phys. Rev. D11, 395 (1975) .
8. Kaplan, D. B. Phys. Lett. B288, 342 (1992) .
9. Chen, P. et al., Phys. Rev. D64 (2001) 014503.
10. Heller, U. M., F. Karsch and B. Sturm, Phys. Rev. D60 (1999) 114502 .
11. Svetitsky, B. and L. G. Yaffe, L. G. Nucl. Phys. B210 [FS6], 423 (1982) .
12. Pisarski, R. D. and F. Wilczek, Phys. Rev. D29, 338 (1984) .
13. Gavin, S., A. Gocksch, and R. D. Pisarski, Phys. Rev. D49, 3079 (1994) .

14. Shuryak, E. and T. Schaefer, Phys. Rev. Lett. 75, 1707 (1995) .
15. Iwasaki, Y., K. Kanaya, S. Sakai and T. Yoshie, Nucl. Phys. (Proc. Suppl.) B42, 261 (1995) .
16. Karsch, F., E. Laermann and C. Schmidt, Phys. Lett. B520, 41 (2001).
17. Linde, A. D. Phys. Lett. B96, 289 (1980) .
18. Arnold, P. and C.-X. Zhai, Phys. Rev. D50, 7603 (1994).
19. Kastening, B. and C. Zhai, Phys. Rev. D52, 7232 (1995) .
20. Blaizot, J.-P., E. Iancu and A. Rebhan, Phys. Rev. Lett. 83, 2906 (1999) .
21. for a review see J.-P. Blaizot, E. Iancu, Phys. Rept. 359, 355 (2002); for further details see also the contributions of J.-P. Blaizot and A. Rebhan to this volume.
22. Blaizot, J.-P., E. Iancu and A. Rebhan, Phys. Lett. B470, 181 (1999) .
23. Kajantie K., M. Laine, K. Rummukainen, Y. Schröder, Phys. Rev. Lett. 86, 10 (2001) .
24. Naik, S., Nucl. Phys. B316 (1989) 238 .
25. Boyd, G. et al., Phys. Rev. Let. 75, 4169 (1995) and Nucl. Phys. B469, 419 (1996).
26. Papa, A. Nucl. Phys. B478, 335 (1996) .
27. Okamoto, M. et al. (CP-PACS). Phys. Rev. D60, 094510 (1999) .
28. Beinlich, B. et al., Eur. Phys. J. C6, 133 (1999) .
29. Peshier, A., B. Kämpfer, O. P. Pavlenko and G. Soff, Phys. Rev. D54,2399 (1996). Lévai, P. and U. Heinz, Phys. Rev. C57,1879 (1998) .
30. Beinlich, B., F. Karsch and A. Peikert, Phys. Lett. B390, 268 (1997) .
31. Karsch, F., E. Laermann and A. Peikert, Phys. Lett. B478, 447 (2000) .
32. Bernard, C. et al., Phys. Rev. D55, 6861 (1997) .
33. Ali Khan, A. et al., Phys. Rev. D64 (2001) 074510.
34. Wittig, H. Int. J. Mod. Phys. A12, 4477 (1997) .
35. Bitar, K. M. et al., Phys. Rev. D43, 2396 (1991) .
36. Bernard, C. et al., Phys. Rev. D56, 5584 (1997) and references therein .
37. Ali Khan, A. et al.(CP-PACS), Phys. Rev. D63, 034502 (2001) .
38. Edwards, R. G. and U. H. Heller, Phys. Lett. B462, 132 (1999 .
39. Karsch, F., E. Laermann and A. Peikert, Nucl. Phys. B605, 579 (2001).
40. Karsch, F., *Lattice results on QCD thermodynamics*, hep-ph/0103314.
41. Meyer-Ortmanns, H. and B.-J. Schaefer, Phys. Rev. D53, 6586 (1996).
42. Berges, J., D. U. Jungnickel and C. Wetterich, Phys. Rev. D59, 034010 (1999).
43. Matsui, T. and H. Satz, Phys. Lett. B178, 416 (1986) .
44. Zantow, F., O. Kaczmarek, F. Karsch and P. Petreczky, hep-lat/0110103 and hep-lat/0110106.
45. Karsch, F., T. Mehr and H. Satz, Z. Phys. C 37 (1988) 617 .
46. Digal, S., P. Petreczky and H. Satz, Phys. Lett. 514 (2001) 57 and Phys. Rev. D 64 (2001) 094015 .
47. Abreu, M.C., et al. (NA50 Collaboration), Phys. Lett. B450 (1999) 4R56 .
48. Le Bellac, M., Thermal Field Theory, Cambridge University Press 1996.
49. Florkowski, W. and B. L. Friman, Z. Phys. A 347, 271 (1994).
50. Karsch, F., M.G. Mustafa and M.H. Thoma, Nucl. Phys. B497, 249 (2001).
51. Bryan, R.K., Eur. Biophys. J. 18, 165 (1990). .
52. Asakawa, M., T. Hatsuda and Y. Nakahara, Prog. Part. Nucl. Phys. 46, 459 (2001).
53. Nakahara, Y., M. Asakawa and T. Hatsuda, Phys. Rev. D60, 091503 (1999) .
54. Wetzorke, I., and F. Karsch, Proceedings of the International Workshop on Strong and Electroweak Matter 2000 (SEWM 2000) (Edt. C.P. Korthals-Altes, World Scientific 2001), p.193, hep-lat/0008008 .
55. Agakichiev, G., et al. (CERES Collaboration), Phys. Lett. B422 (1998) 405 .
56. Karsch, F., E. Laermann, P. Petreczky, S. Stickan and I. Wetzorke, A Lattice Calculation of Thermal Dilepton Rates, hep-lat/0110208.

VARIATIONAL APPROACH TO THE DYNAMICS OF QUANTUM FIELDS

T. MATSUI
Institute of Physics, University of Tokyo, Komaba
3-8-1 Komaba, Meguro-ku, Tokyo 153-8902, Japan

1. Introduction

In the summer of 1995, Dominique Vautherin came to Kyoto and stayed three months at the Yukawa Institute of Kyoto University where he delivered a series of very informal, pedagogical lectures on the application of variational methods to quantum field theories. [1] This initiated our long lasting enjoyable collaboration on the subject and Yasuhiko Tsue joined the force later. Throughout this collaboration Vautherin was always enthusiastic to explore new physics problems and always came up with new innovative ideas, much of which I suspect have origins in his expertise in nuclear many-body theory. We are very sorry that this fruitful collaboration with Vautherin came to end so soon by his untimely death and that we are no longer able to be inspired by his deep physics insights, charmed by his elegance, and most of all cheered up by his very warm presence. I like to dedicate these lectures to the memory of Dominique Vautherin from whom I learned most of the material presented below.

First, I like to say a few words about physics motivations of our works. In recent years, several authors have constructed non-trivial time-dependent solutions of classical field equations of effective meson fields. [2, 3] Such solutions are relevant in considering the fate of defects which might be produced in dynamical order-disorder phase transitions in ultrarelativistic nucleus-nucleus collisions[4] or in the evolution of very early universe.[5] We like to study *the effect of quantum fluctuations* which has been ignored in these classical analyses. We may consider the classical solution as a collective mode of underlying microscopic degrees of freedom. It is well-known in nuclear many-body theory that the internal microscopic motions of nucleons are influenced by the motion of the nuclear mean field such as rotation or vibration. One may expect similar effects would arise in the field theory.

J.-P. Blaizot and E. Iancu (eds.), QCD Perspectives on Hot and Dense Matter, 419–431.

Our method is based on the Schrödinger wave functional representation of the quantum field theories in which the time-evolution of the wave function is explicitly considered. Another advantage of using the Schrödinger picture is that one can introduce the variational method to construct approximate but non-perturbative solutions of the problem as in similar problems in quantum mechanics. The extension of the variational principle for time-dependent wave function has also been developed. [6, 7] The method can also be extended to deal with statistical ensembles described by Gaussian form of density matrices. [8]

2. A Simple Exercise in Quantum Mechanics

We start with a very elementary example in quantum mechanics which all students learn in an introductory course of quantum mechanics: the harmonic oscillator problem. Although our interests lies in the time-dependent solution of quantum field theories, it would be instructive to first study the time-dependent solution of this simple exactly soluble problem in order to gain some physical insights into our more difficult problems since quantum field theories are nothing but an assemble of infinite number of coupled harmonic oscillators.
Let us consider a simple one-dimensional harmonic oscillator whose Hamiltonian is given by

$$H_0 = \frac{1}{2}\left(p^2 + \omega^2 x^2\right), \tag{2.1}$$

where we set the mass of the particle $m = 1$ for simplicity. We look for solutions of the time-dependent Schrödinger equation:

$$i\frac{\partial}{\partial t}\Psi(t) = H\Psi(t)$$

with the quantization condition: $[x, p] = i$.
There are two ways to solve this problem. The easier one is the *algebraic* method in which one introduces "creation" and "annihilation" operators:

$$a^\dagger = \frac{1}{\sqrt{2\omega}}(-ip + \omega x), \qquad a = \frac{1}{\sqrt{2\omega}}(ip + \omega x)$$

which satisfy a usual commutation relation: $\left[a, a^\dagger\right] = 1$ and diagonalize the Hamiltonian: $H = \omega\left(a^\dagger a + \frac{1}{2}\right)$. The ground state of the Hamiltonian is given by the condition: $a|0\rangle = 0$. This method is usually transcribed to the quantization of fields and one obtains particle creation and annihilation operators as basic building blocks in describing physical processes. In this approach, one usually does not refer to the wave function of the system explicitly but instead concentrates on amplitudes of particular process for given initial and final states specified by the particle number and other quantum numbers.

Alternatively, one can solve the problem by the *analytic* method in which one expresses the Schrödinger equation in coordinate representation with the differential operator $p = -i\partial/\partial x$ and solve the resultant second order differential equation. The eigenfunction of this differential equation are Hermite's polynomials with the Gaussian ground state wave function:

$$\Psi_0(x) = \langle x|0\rangle = \left(\frac{\omega}{\pi}\right)^{1/4} e^{-\frac{1}{2}\omega x^2} \tag{2.2}$$

It is straightforward to write down the ground state wave functional of the quantized free field $\varphi(x)$. For one component scalar field theory, it becomes just a product of the Gaussian wave functions of normal modes each specified by the momentum with the oscillator frequency $\omega_k = \sqrt{m^2 + k^2}$:

$$\Psi_0[\varphi(x)] = \mathcal{N} \exp\left[-\frac{1}{2}\sum_k \omega_k \varphi_k^* \varphi_k\right].$$

with a proper normalization condition.

Time-dependent variational wave function:

Now let us consider time-dependent solutions of the harmonic oscillator. We first modify the ground state wave function by adding extra complex phase factor $e^{i(p_0 - i\omega x_0)x}$. One then obtains the Gaussian wave function with its the center shifted

$$\Psi(x, t; x_0, p_0) = \exp\left[-\frac{1}{2}\omega(x - x_0(t))^2 + ip_0(t)x\right] \tag{2.3}$$

Here $p_0(t), x_0(t)$ are time-dependent parameters which are to be determined by imposing that the above function is a solution of the time-dependent Schrödinger equation. Here we use the variational method to derive the equations of motion of $p_0(t), x_0(t)$.

The Schrödinger equation can be obtained by imposing a stationary condition $\delta S = 0$ for the action:

$$S[\Psi] = \int dt \langle\Psi(t)|H - i\frac{\partial}{\partial t}|\Psi(t)\rangle$$

with respect to variation of the wave function $\langle\Psi(t)|$. For the variational wave function in the form of 2.3, the integrand can be computed easily with $\langle\Psi|p^2|\Psi\rangle = p_0^2$, $\langle\Psi|x^2|\Psi\rangle = x_0^2$, $\langle\Psi|i\frac{\partial}{\partial t}|\Psi\rangle = -p_0 x_0$:

$$\langle\Psi|H - i\frac{\partial}{\partial t}|\Psi\rangle = p_0^2 + \omega^2 x_0^2 + \dot{p}_0 x_0. \tag{2.4}$$

Taking stationary conditions of the action with respect to two time-dependent parameters, $\delta S/\delta p_0 = \delta S/\delta x_0 = 0$, one finds

$$\dot{p}_0 = -\omega^2 x_0, \qquad \dot{x}_0 = p_0 \tag{2.5}$$

which are just the *classical* equation of motion of the harmonic oscillator. The center-shifted Gaussian wave function may be obtained, beside the phase factor e^{-ip_0x}, by operating a translation operator on the ground state:

$$|0; x_0, p_0\rangle \sim \exp[-ix_0p]|0\rangle = \exp\left[-i\sqrt{\frac{\omega}{2}}x_0(a^\dagger - a)\right]|0\rangle.$$

It thus describes a phase-coherent mixture of infinite number of excited states $|n\rangle$. The corresponding states in quantum field theory are called *coherent states*. One thus sees that classical field equations arises if one describes the wave functional of the quantized fields in terms of coherent states.

Keeping Gaussian form of variational wave function one can go one step further by introducing additional phase factor quadratic in x. The wave function now takes the form of

$$\Psi_{\text{sq.}}(x,t) = \left(\frac{\mu}{\pi}\right)^{1/4} \exp\left[-\frac{1}{2}(\mu + i\sigma)(x - x_0)^2 - ip_0x\right] \tag{2.6}$$

where we have introduced two new real parameters, μ, σ, characterizing the width of the wave function, which we shall consider as time-dependent variational parameters. The time-dependent parameter $\mu(t)$ is related to the quantum fluctuation of the position of the particle around its mean $x_0 = \langle x\rangle$ by $\langle(x - x_0)^2\rangle = 1/\mu(t)$. The another parameter σ is related to the rate of change of μ as we shall see below. This modification of the wave function thus describes the breathing motion (squeezing and stretching) of the wave function centered at $x_0(t)$. In the quantum field theory, these generalized coherent states are called *squeezed states*.

With the new variational wave function the action integrand becomes

$$\langle\Psi(t)|H - i\frac{\partial}{\partial t}|\Psi(t)\rangle_{\text{sq.}} = \frac{1}{2}\left(p_0^2 + \omega^2 x_0^2\right) + \frac{1}{4}\left(\mu + \frac{\sigma^2}{\mu} + \frac{\omega^2}{\mu}\right) + \dot{p}_0x_0 - \frac{\dot{\sigma}}{4\mu}$$

Taking the stationary conditions with respect to the variations of $\mu(t), \sigma(t)$, in addition to classical parameters, $x_0(t), p_0(t)$, one finds

$$\dot{\mu} = 2\sigma\mu, \qquad \dot{\sigma} = \sigma^2 + \omega^2 - \mu^2, \tag{2.7}$$

while the equations of motion of $x_0(t), p_0(t)$ are unchanged. We see that the imaginary part σ of the Gaussian width parameter plays a role similar to the velocity of the motion of the center of the Gaussian x_0. For the harmonic oscillator Hamiltonian, coherent states and squeezed states are exact solutions of the Schrödinger equations. The classical motion and the quantum fluctuation decouple in this exactly soluble problem. This is not the case, however, when the potential is not quadratic in x.

Anharmonicity:

To illustrate the effect of non-harmonic part of the potential, we add a term quartic in x in our Hamiltonian:

$$H_I = \frac{\lambda}{4!}x^4$$

In the scalar field theory, this corresponds to adding a φ^4 self-interaction term. This anharmonic term generates new terms in the integrand of the action:

$$\langle\Psi(t)|H_I|\Psi(t)\rangle_{\text{sq.}} = \frac{\lambda}{4!}\left(\frac{3}{4\mu^2} + 3\frac{x_0^2}{\mu} + x_0^4\right)$$

which cause coupling between the classical motion of the mean of particle position and the quantum fluctuation around it:

$$\dot{x}_0 = p_0, \qquad \dot{p}_0 = -\omega^2 x_0 - \frac{\lambda}{6}x_0^3 - \frac{\lambda}{4\mu}x_0,$$

$$\dot{\mu} = 2\sigma\mu, \qquad \dot{\sigma} = \sigma^2 + \omega^2 - \mu^2 + \frac{\lambda}{4\mu} + \frac{\lambda}{2}x_0^2$$

Let us see how the time independent solution (the ground state) is modified by the anharmonic term. The conditions $\dot{x}_0 = \dot{\mu}_0 = 0$ demand that $p_0 = \sigma = 0$, while the remaining two conditions $\dot{p}_0 = \dot{\sigma}_0 = 0$ determine the values of x_0 and μ. There are two types of solutions depending on the sign of ω^2. When $\omega^2 > 0$ we have a "normal" solution centered at the origin $x_0 = 0$ but the width is modified slightly as determined by:

$$\mu^2 = \omega^2 + \frac{\lambda}{4\mu}$$

This equation may be called the *gap equation* for the reason to be discussed in the next section. If $\omega^2 < 0$, the potential has double minima at $x_{\text{min.}} = \pm\sqrt{6/\lambda}\omega$ and the $x_0 = 0$ point becomes local maximum of the potential. In this case, there are two "symmetry breaking" solutions centered at the two solutions of

$$\omega^2 + \frac{\lambda}{4\mu} + \frac{\lambda}{6}x_0^2 = 0$$

where the value of the width parameter μ is determined self-consistently with the modified gap equation:

$$\mu^2 = -2\omega^2 - \frac{\lambda}{2\mu}.$$

We note that the position of the center of our Gaussian variational wave function shift slightly toward the origin from the position of the minimum of the potential.

3. Scalar Field Theory: φ^4 model

Having been warmed up by a much simpler problem in quantum mechanics, it is now our task to transcribe the result to a problem in quantum field theories. We consider first a prototype scalar field theory with φ^4 self-interaction.
The Hamiltonian density of the theory is given by

$$\mathcal{H}(\mathbf{x}) = \frac{1}{2}\pi^2(\mathbf{x}) + \frac{1}{2}\left(\nabla\varphi(\mathbf{x})\right)^2 + \frac{m_0^2}{2}\varphi^2(\mathbf{x}) + \frac{\lambda}{24}\varphi^4(\mathbf{x}), \tag{3.1}$$

where $\pi(\mathbf{x})$ is an operator conjugate to the field $\varphi(x)$ and is expressed by a functional derivative $-i\delta/\delta\varphi(\mathbf{x})$. [1] We write a Gaussian time-dependent variational wave functional for this Hamiltonian formally as

$$\Psi\left[\varphi(\mathbf{x})\right] = \mathcal{N}\exp\left(i\langle\bar{\pi}|\varphi - \bar{\varphi}\rangle - \langle\varphi - \bar{\varphi}|\frac{1}{4G} + i\Sigma|\varphi - \bar{\varphi}\rangle\right), \tag{3.2}$$

where G, Σ, $\bar{\varphi}$, $\bar{\pi}$ define respectively the real and imaginary part of the kernel of the Gaussian width and its average position and momentum. We have used the short hand notation $\langle\bar{\pi}|\varphi\rangle = \int\bar{\pi}(\mathbf{x},t)\varphi(\mathbf{x})dx$. Although it looks a little horrible, it is just a straightforward generalization of (2.6). The correspondences between the previous quantum mechanical example and the present case are summarized in the Table 1.

TABLE 1. Correspondence between quantum mechanics and scalar field theory

quantum mechanics	scalar field theory
x	$\varphi(\mathbf{x})$
ω^2	$m_0^2 - \Delta$ (or $m_0^2 + \mathbf{k}^2$)
$\frac{\lambda}{4}x^4$	$\frac{\lambda}{4}\varphi^4(\mathbf{x})$
$x_0(t), p_0(t)$	$\bar{\varphi}(\mathbf{x},t), \bar{\pi}(\mathbf{x},t)$
$\mu(t), \sigma(t)$	$\frac{1}{2}G^{-1}(\mathbf{x},\mathbf{y},t), 2\Sigma(\mathbf{x},\mathbf{y},t)$
$\mu\xi$	$G^{-1/2}\xi G^{-1/2}$

[1] A care must be taken to give precise mathematical meanings for these expressions[9], but we do not go into such problems here.

The equations of motion are found to be

$$\begin{aligned}
\dot{\bar{\varphi}} &= -\bar{\pi}, \\
\dot{\bar{\pi}} &= \left(-\Delta + m_0^2 + \tfrac{\lambda}{6}\bar{\varphi}^2(\mathbf{x}) + \tfrac{\lambda}{2}G(\mathbf{x},\mathbf{x})\right)\bar{\varphi}, \\
\dot{G} &= 2(G\Sigma + \Sigma G), \\
\dot{\Sigma} &= \tfrac{1}{8}G^{-2} - 2\Sigma^2 - \tfrac{1}{2}\left(-\Delta + m_0^2 + \tfrac{\lambda}{2}\bar{\varphi}^2 + \tfrac{\lambda}{2}G(\mathbf{x},\mathbf{x})\right).
\end{aligned} \tag{3.3}$$

where it is understood that $\bar{\varphi}$ and $\bar{\pi}$ denote vectors with components $\bar{\varphi}(\mathbf{x})$ and $\bar{\pi}(\mathbf{x})$, G and Σ denote matrices with matrix elements $G(\mathbf{x},\mathbf{y})$ and $\Sigma(\mathbf{x},\mathbf{y})$, and the matrix product $G\Sigma$ is given by $\int d\mathbf{z} G(\mathbf{x},\mathbf{z})\Sigma(\mathbf{z},\mathbf{y})$.
For the vacuum, we should have time-independent solution so that $\bar{\pi} = \Sigma = 0$, $\bar{\varphi}(\mathbf{x}) = \varphi_0$, and

$$G(\mathbf{x},\mathbf{y},t) = G_0(\mathbf{x}-\mathbf{y}) = \int d\mathbf{k}\frac{e^{i(\mathbf{x}-\mathbf{y})\cdot\mathbf{k}}}{2\sqrt{\mu^2+k^2}}. \tag{3.4}$$

where the effective mass μ is determined self-consistently by the non-linear integral equation

$$\mu^2 = m_0^2 + \frac{\lambda}{2}G_0(0) + \frac{\lambda}{2}\varphi_0^2\,. \tag{3.5}$$

which is usually called the gap equation because it determines the mass gap self-consistently. In the symmetric phase the expectation value of the field φ_0 vanishes while in the symmetry broken phase it must be such that

$$m_0^2 + \frac{\lambda}{2}G_0(0) + \frac{\lambda}{6}\varphi_0^2 = 0. \tag{3.6}$$

This last equation implies that $\mu^2 = \lambda\varphi_0^2/3$. The equations of motion (3.3) determine the time-evolution of the variational wave functional (3.2).
We comment on two well-known problems which are absent in quantum mechanics but are characteristic in quantum field theory: divergences and covariance. In the above expression the momentum integral in $G_0(0)$ is quadratically divergent. As well-known, this divergence originates from couplings of infinite degrees of freedom (all momentum modes) in the interaction of local fields. In the renormalizable field theories, these divergences (or cut-off dependence) may be absorbed into an appropriate redefinition of finite number of physical parameters, such as the mass and the coupling constant. This renormalization procedure works at least order by order in the power series expansion in terms of the coupling constant. In our non-perturbative calculation scheme with the variational method, we may absorb the cut-off dependence into the mass and the coupling constant.[10, 11] We

encounter, however, well-known triviality problem: renormalized coupling constant λ_R becomes zero as one sends the cut-off Λ to infinity keeping the original coupling λ positive for the stability of the ground state. Despite this well-known pathology, the model still makes sense as an effective theory with the finite cut-off. The another problem is the lack of manifest covariance in our formulation: the time coordinate t have been treated differently from spatial coordinates throughout calculations. This is an old problem which was originally solved in QED by Tomonaga, Schwinger, Feynman and Dyson who developed the covariant perturbation theory based on the interaction representation. Vautherin invented a new ingenious trick to derive manifest covariant form of the equations of motion which I shall now describe. [12, 14]

Mean field equations in the Hartree-Bogoliubov form:

A key ingredient of his method is the reduced density matrix defined by

$$\mathcal{M}(\mathbf{x},\mathbf{y};t) = \begin{pmatrix} i\langle\hat{\varphi}(\mathbf{x})\hat{\pi}(\mathbf{y})\rangle - 1/2 & \langle\hat{\varphi}(\mathbf{x})\hat{\varphi}(\mathbf{y})\rangle \\ \langle\hat{\pi}(\mathbf{x})\hat{\pi}(\mathbf{y})\rangle & -i\langle\hat{\pi}(\mathbf{x})\hat{\varphi}(\mathbf{y})\rangle - 1/2 \end{pmatrix}, \tag{3.7}$$

where $\hat{\varphi} = \varphi - \bar{\varphi}$, $\hat{\pi} = \pi - \bar{\pi}$, and expectation values are calculated with the Gaussian functional $\Psi(t)$ and is given by

$$\mathcal{M} = \begin{pmatrix} -2iG\Sigma & G \\ \frac{1}{4}G^{-1} + 4\Sigma G\Sigma & 2i\Sigma G \end{pmatrix}. \tag{3.8}$$

Using (3.3), one can show that the equation of motion of the reduced density matrix can be cast into the Liouville-von Neumann form

$$i\dot{\mathcal{M}} = [\mathcal{H}, \mathcal{M}], \tag{3.9}$$

where the generalized Hamiltonian $\mathcal{H}$ is given by

$$\mathcal{H} = \begin{pmatrix} 0 & 1 \\ \Gamma & 0 \end{pmatrix}. \tag{3.10}$$

with

$$\Gamma = -\Delta + m_0^2 + \frac{\lambda}{2}\bar{\varphi}^2 + \frac{\lambda}{2}G(\mathbf{x},\mathbf{x}). \tag{3.11}$$

This form of equations is known in many-body theory as the time-dependent Hartree-Bogoliubov equations.

The reduced density matrix satisfies $\mathcal{M}^2 = \frac{1}{4}\mathbf{I}$ so that it has two eigenvalues of $\pm 1/2$. We write the n-th eigenvector of $\mathcal{M}$ with eigenvalue $1/2$ as $(u_n(\mathbf{x}), v_n(\mathbf{x}))$:

$$\mathcal{M}\begin{pmatrix} u_n \\ v_n \end{pmatrix} = \frac{1}{2}\begin{pmatrix} u_n \\ v_n \end{pmatrix}, \tag{3.12}$$

then one can show that vectors $(u_n^*(\mathbf{x}), -v_n^*(\mathbf{x}))$ give eigenvectors for eigenvalue $-1/2$. The u and v components of eigenvectors are called *mode functions.* With these eigenvectors the reduced density matrix has a spectral decomposition as

$$\mathcal{M} = \frac{1}{2}\sum_{n>0}\left[\begin{pmatrix} u_n \\ v_n \end{pmatrix}(v_n^*,\, u_n^*) + \begin{pmatrix} u_n^* \\ -v_n^* \end{pmatrix}(-v_n,\, u_n)\right]. \tag{3.13}$$

The Liouville-von Neumann equation (3.9) can be rewritten in terms of the mode functions as

$$i\partial_t \begin{pmatrix} u_n \\ v_n \end{pmatrix} = \begin{pmatrix} 0 & 1 \\ \Gamma & 0 \end{pmatrix}\begin{pmatrix} u_n \\ v_n \end{pmatrix}. \tag{3.14}$$

Eliminating v_n and inserting (3.11) we obtain a modified Klein-Gordon type equation for the mode functions

$$\left(\Box + m_0^2 + \frac{\lambda}{2}\bar{\varphi}^2 + \frac{\lambda}{2}G(\mathbf{x},\mathbf{x})\right)u_n = 0, \tag{3.15}$$

where the spectral representation (3.13) implies

$$G(\mathbf{x},\mathbf{x}) = \langle \mathbf{x}|G(t)|\mathbf{x}\rangle = \frac{1}{2}\sum_n |u_n(\mathbf{x},t)|^2.$$

To write the equation of motion fully covariant way, we introduce a Feynman propagator in terms of the mode functions

$$\begin{aligned}\langle x|S|y\rangle \;=\; & \theta(x_0 - y_0)\textstyle\sum_{n>0} u_n^*(\mathbf{x},x_0)u_n(\mathbf{y},y_0) \\ & + \theta(y_0 - x_0)\textstyle\sum_{n<0} u_n^*(\mathbf{x},x_0)u_n(\mathbf{y},y_0).\end{aligned} \tag{3.16}$$

so that

$$\langle \mathbf{x}|G(x_0)|\mathbf{x}\rangle = \langle x|S|x\rangle \tag{3.17}$$

Then we finally arrive at a fully covariant, self-consistent equations of motion for the mode functions[12, 14] which reproduces the same equations obtained earlier by the functional integral method.[15]

This way of writing the equations of motion also paves a way to generalize the calculation at finite temperatures.

4. Statistical Ensembles

Foregoing discussions are limited to evolution of a single coherent (Gaussian) state. In realistic physical situations, we are more interested in the evolution of the statistical ensemble which is described by the density matrix:

$$\hat{\rho}(t) = \sum_n |\Psi_n(t)\rangle p_n(t)\langle\Psi_n(t)| \tag{4.1}$$

where $p_n(t)$ is a probability distribution specifying the ensemble hence It should satisfy $\sum_n p_n(t) = 1$. The expectation value of an observable O is given by $\langle O \rangle = \mathrm{Tr}\hat{\rho}O$ and the statistical entropy S is given by

$$S = -\mathrm{Tr}\hat{\rho}\ln\hat{\rho} = -\sum_n p_n \ln p_n \tag{4.2}$$

Hence if p_n is time-independent then the entropy is conserved. In equilibrium at temperature $T = 1/\beta$, the density matrix is given by the canonical ensemble: $p_n^{eq.}(t) = e^{-\beta E_n}/Z$ where $Z = \sum_n e^{-\beta E_n}$ which maximizes the entropy $S(t)$ under the condition of fixed expectation value of the energy $E = \langle H \rangle$. The density matrix with time-independent p_n obeys the Liouville-von Neumann equation: $i\partial_t\hat{\rho} = [H, \hat{\rho}]$.

The variational method has been extended for the time-dependent density matrix by Eboli, Jackiw and Pi. [8] Without going into detail, we illustrate the essence of their method in terms of simple quantum mechanical example with harmonic oscillator Hamiltonian (2.1). We first introduce the coordinate representation of the density matrix by

$$\rho(x, y; t) = \langle x|\hat{\rho}(t)|y\rangle = \sum_n \Psi_n(x,t)p_n(t)\Psi_n^*(y,t) \tag{4.3}$$

and observe that with single Gaussian variational wave function the density matrix is just a product of two Gaussian: $\rho(x, y; t) \sim e^{-\omega(x^2+y^2)/2}$. One can show that in the other extreme limit of thermal equilibrium, the density matrix $\rho(x, y; t)$ can also be expressed by a mixed Gaussian form. This is so because the equilibrium density matrix $\hat{\rho}_{eq.}(\beta) = e^{-\beta H}$ obeys the imaginary time Schrödinger equation (the Bloch equation): $-\partial_\tau\hat{\rho}_{eq.}(\tau) = H\hat{\rho}_{eq.}$. Using $\rho_{eq.}(x, y) = \langle x|\hat{U}(-i\tau)|y\rangle$ and well-known path-integral expression of the matrix elements of the unitary evolution operator $\hat{U}(t) = e^{-itH}$, one finds

$$\rho_{eq.}(x, y) = \left(\frac{\omega}{2\pi\sinh\omega\beta}\right)^{1/2} \exp\left[-\frac{\omega}{2\sinh\omega\beta}\left\{(x^2+y^2)\cosh\omega\beta - 2xy\right\}\right] \tag{4.4}$$

which is again Gaussian with an extra term containing a product of two coordinates xy. For more general mixed states we introduce a generalized Gaussian density matrix[8]

$$\rho(x, y; t) = \mathcal{N}\exp\left[ip_0(\hat{x}-\hat{y}) - \frac{\mu}{2}\left(\hat{x}^2+\hat{y}^2-2\xi\hat{x}\hat{y}\right) + i\frac{\sigma}{2}(\hat{x}^2-\hat{y}^2)\right] \tag{4.5}$$

where we have used the time-dependent shifted coordinates: $\hat{x} = x - x_0$ and $\hat{y} = y - x_0$. Equations of motion of the time-dependent parameters of the generalized Gaussian density matrix (4.5) can be derived from the Liouville equation of the density matrix and we obtain a set of equation similar to (2.7) with a small modification:

$$\dot{\mu} = 2\sigma\mu, \quad \dot{\sigma} = \sigma^2 + \omega^2 - (1-\xi^2)\mu^2. \tag{4.6}$$

The parameter ξ is called the mixing parameter which measures the degree of mixture of different pure states in the ensemble; it remains constant for an adiabatic evolution of the system. In equilibrium, $p_0 = x_0 = \sigma = 0$ and other two parameters are given by the specific functions of temperature:

$$\mu_{eq.} = \omega \coth \omega\beta, \quad \xi_{eq.} = \cosh^{-1} \omega\beta \tag{4.7}$$

as indicated by the formula (4.4). Extension of the Gaussian density matrix in quantum mechanics to that in quantum field theories is straightforward as indicated in the last row of the Table 1.

The reduced density matrices we have introduced in the previous section for a pure Gaussian state can be extended for a mixed state immediately by the replacements:

$$\langle\hat{\varphi}(\mathbf{x})\hat{\pi}(\mathbf{y})\rangle = \mathrm{Tr}(\hat{\rho}(t)\hat{\varphi}(\mathbf{x})\hat{\pi}(\mathbf{y})), \langle\hat{\varphi}(\mathbf{x})\hat{\varphi}(\mathbf{y})\rangle = \mathrm{Tr}(\hat{\rho}(t)\hat{\varphi}(\mathbf{x})\hat{\varphi}(\mathbf{y})), \mathrm{etc.} \tag{4.8}$$

and one can derive equations of motion of the reduced density matrix similar to the pure state case. In the case of equilibrium distribution, this amounts to introduce a factor containing the occupation number in the sum over the mode functions.

5. Rotating chiral condensate in the $O(N)$ sigma model

We briefly mention about an application of the above method to the $O(N)$ sigma model which is composed of N-components coupled scalar fields φ_n with continuous $O(N)$ symmetry. We expect that this global symmetry of the model is broken spontaneously at low temperatures, characterized by non vanishing expectation value of one component of fields, say $\bar{\varphi}_0 = \mathrm{Tr}\hat{\rho}\hat{\varphi} \neq 0$, and the system exhibits an order-disorder phase transition to a state with $\bar{\varphi}_0 = 0$. This is what is expected in QCD where the chiral symmetry is broken in vacuum and is expected to be restored at finite temperature. This chiral phase transition has been studied by an effective theory with pion and sigma fields with $O(4)$ global symmetry.

We have applied our method to describe a special kind of time-dependent condensate which rotates in a subspace of the internal symmetry space:

$$\begin{pmatrix} \bar{\varphi}_1(x) \\ \bar{\varphi}_2(x) \end{pmatrix} = \begin{pmatrix} \cos(q\cdot x) & \sin(q\cdot x) \\ -\sin(q\cdot x) & \cos(q\cdot x) \end{pmatrix} \begin{pmatrix} \varphi_0 \\ 0 \end{pmatrix} = \exp[i(q\cdot x)\tau_2] \begin{pmatrix} \varphi_0 \\ 0 \end{pmatrix} \tag{5.1}$$

where $\tau_2 = \begin{pmatrix} 0 & -i \\ i & 0 \end{pmatrix}$ and $q_\mu = (\omega, \mathbf{q})$ is a four vector specifying the direction of rotation: for a spatially uniform rotation in time it is a pure time-like vector; while a static condensate with oscillation in space, it is a pure space-like vector.

The matrix $U(x) = \exp[i(q \cdot x)\tau_2]$ can be considered as a "gauge transformation" to the local rest frame of the rotating condensate. In the rotating frame, the four derivative which appear in the equation of motion of the mode functions is "gauge transformed" to $U\partial_\mu U^\dagger = \partial_\mu - iq_\mu\tau_2$ and the effect of the rotation is seen in this frame as an appearance of the apparent "centrifugal force". Indeed, our result of phase diagram for such dynamical condensates shows that the amplitude of the chiral condensate with uniform time-like rotation increases due to the centrifugal force. This effect has been observed in the classical solutions of Anselm and Ryskin [2]; our quantum generalization of their solution shows that this effect is amplified by the coupling of quantum fluctuations to rotations, which the static condensate with spatial oscillations are more suppressed by the quantum fluctuations.[13] A phase diagram of rotating condensate was obtained in [14] and the damping of the rotating condensate due to the symmetry breaking perturbation was computed by the method of the response function in [16].

6. Outlook

Vautherin started to work on the variational approach to quantum field theories many years ago with Arthur Kerman. They developed many important ideas in their unpublished works and tried to solve QCD non-perturbatively with their method with a hope to gain new insights in the quark confinement problem.[10] The Gaussian Ansatz for the variational wave functional however has difficulty of breaking the local gauge invariance and the projection to color singlet state destroys a nice feature of the Gaussian wave functional.[17] Vautherin continued to work on the problem with his students and brought a new insight into the problem again introducing a technique developed in nuclear many-body theory in his last paper.[18] His efforts in this direction may be carried over to study the dynamical evolution of the quark-gluon plasma in ultrarelativistic nucleus-nucleus collisions. Vautherin was also interested in the recent experimental breakthrough of creating weakly interacting Bose-Einstein condensates in well-controlled laboratory environments. Our method can be also applied to such problem to investigate the effect of quantum fluctuations which are usually ignored in theoretical descriptions.[19]

I am much indebted to Yasuhiko Tsue as well as to Dominique Vautherin for our works quoted above. I thank them for sharing the joy of physics in our collaboration.

References

1. D. Vautherin, Lectures given at Yukawa Institute, Kyoto, 1995, (unpublished).
2. A. A. Anselm and M.G.Ryskin, Phys. Lett. **B266**, 482 (1991) .
3. J. -P. Blaizot and A. Krzywicki, Phys. Rev. **D 46**, 246 (1992).
4. K. Rajagopal and F. Wilczek, Nucl. Phys. **B404**, 577 (1993) .

5. A. Guth and S. Y. Pi, Phys. Rev. **D32** ,1899 (1985) .
6. R. Jackiw and A. Kerman,Phys. Lett. **A71**,158 (1979) .
7. R. Balian and M. Vènéroni, Phys. Rev. Lett. **47** (1981) 1357; 1765(E).
8. O. Eboli, R. Jackiw and S. Y. Pi, Phys. Rev. **D37**, 3557 (1988) .
9. M. Lüscher, Nucl. Phys. **B254**, 52 (1985).
10. A. Kerman and D. Vautherin, Ann. Phys. **192**, 408 (1989) .
11. S. -Y. Pi and M. Samiullah, Phys. Rev. **D 36**, 3128 (1987); F. Cooper and E. Mottola, Phys. Rev. **D 36**, 3114 (1987).
12. D. Vautherin and T. Matsui, Phys. Rev. **D 55**, 4492 (1997).
13. D. Vautherin and T. Matsui, Phys. Lett. **B437**,173 (1998) .
14. Y. Tsue, D. Vautherin and T. Matsui, Prog. Theor. Phys. **102**, 313 (1999).
15. J. Cornwall, R. Jackiw and E. Tomboulis, Phys. Rev. **D 10**, 2428 (1974).
16. Y. Tsue, D. Vautherin and T. Matsui, Phys. Rev. **D 61**, 076006 (2000).
17. I. I. Kogan and A. Kovner, Phys. Rev. **D 52**, 3719 (1995).
18. C. Heinemann, E. Iancu, C. Martin and D. Vautherin, Phys. Rev. **D 61**, 116008 (2000).
19. Y. Tsue, D. Vautherin and T. Matsui, in preparation.

LECTURES ON CHIRAL DISORDER IN QCD

MACIEJ A. NOWAK (nowak@th.if.uj.edu.pl)
M. Smoluchowski Institute of Physics,
Jagellonian University, Cracow, Poland
Gesellschaft fuer Schwerionenforschung, Darmstadt, Germany

Abstract. I explain the concept that light quarks diffuse in the QCD vacuum following the spontaneous breakdown of chiral symmetry. I exploit the striking analogy with disordered electrons in metals, identifying, among others, the universal regime described by random matrix theory, diffusive regime described by chiral perturbation theory and the crossover between these two domains.

Introduction

In these lectures I review spectral aspects of the mechanism of the spontaneous breakdown of the chiral symmetry in Quantum Chromodynamics. Most probably, the spontaneous breakdown of the chiral symmetry is a collective phenomenon caused by the microscopic disorder, in striking resemblance to the diffusive phenomena appearing in disordered metals. Despite the microscopic, detailed nature of this disorder is still unknown, the constraints arising from realizations of chiral symmetry in QCD are so strong, that allow us to predict several non-trivial consequences of this phenomenon. Among most profound, are the predictions on the spectral properties of the Dirac operator. As such, these are amenable to test using the lattice calculations.

The lectures are intended to be elementary and self-contained. The outline of the lectures is as follows:

• In Part 1, I introduce the basic facts on symmetries and anomalies of the QCD, leading to fundamental low energy constraints.

• In Part 2, I explore the analogy between the metal viewed as a complex quantum system sharing universal properties with so-called chaotic systems from one side, and Euclidean QCD with light quarks diffusing in disordered medium built from the lumps of the gauge field, on the other side. In particular, I identify the relevant scales and determine the "diffusion constant" of the QCD vacuum. Then I comment on the close relation of this picture to several known descriptions of chiral symmetry breaking.

J.-P. Blaizot and E. Iancu (eds.), QCD Perspectives on Hot and Dense Matter, 433–460.

• In Part 3, I demonstrate how the hierarchy of spectral scales explains the origin of the appearance of random matrix ensembles in QCD. Then I show how random matrix regime regime breaks down at a certain spectral scale (analog of Thouless energy), leading to various versions of chiral perturbation theory. I also point at some explicit confirmations of this picture coming from the recent lattice calculations. Next, I suggest how this picture can be modified/generalized in the presence of external parameters of the QCD. I mention the possibility of (multicritical) scaling. I conclude with an (incomplete) list of works on disorder and QCD, that could be used as a guide for further reading and also as a good starting point for the original research.

1. Part 1

Let me refresh here some low energy theorems of the QCD, which will be needed for the second part of these lectures. Then I switch to the Euclidean regime, and I recall the Banks-Casher relation.

QCD

Quantum Chromodynamics is a Yang Mills theory based on the local gauge group $SU(3)_{\rm color}$. This means that all six quarks of different flavors interact with gluons, as well gluons interact with themselves, with the same universal coupling g. Looking at the typical masses of the quarks, we see that they cover several orders of magnitude – u,d,s,c,b,t, $\sim$ 5,7,150,1400,4400, 175000 MeV, – respectively. The origin of this hierarchy goes beyond the strong interaction sector and is unknown. The fact, that three masses are less or at most comparable to the scale of the strong interaction, and three others are well beyond the scale, suggests a simplification, in which we put three heavy masses equal to infinity, and three light equal to zero. The heavy sector decouples from the light, and the light one reveals a series of essential symmetries of the theory. We denote this idealization of the QCD as QCD$_\chi$. First, let us note that such classical theory lacks any dimension-full parameter, therefore it is scale invariant. Second, if we introduce the notion of left and right-handed quarks,

$$q_R = \frac{1}{2}(1+\gamma_5)q \qquad q_L = \frac{1}{2}(1-\gamma_5)q \tag{1.1}$$

we see that in the presence of massless quarks the left and the right handed quarks interact independently with the gluons, i.e. we have two decoupled chiral copies of the initial theory:

$$S_{QCD} = \int dt d^3x \sum_f \bar{q}_R^{(f)} i\not{D}(A) q_R^{(f)} + (R \leftrightarrow L) + S_{\rm glue}(A) \tag{1.2}$$

Constructing the standard set of the Noether currents corresponding to the symmetries yields:

$$\begin{aligned}
V_a^\mu &= \bar{q}\gamma^\mu t_a q \\
V_0^\mu &= \bar{q}\gamma^\mu q \\
A_a^\mu &= \bar{q}\gamma^\mu \gamma_5 t_a q \\
A_0^\mu &= \bar{q}\gamma^\mu \gamma_5 q \\
J^\mu_{\text{scale}} &= x_\nu \Theta^{\mu\nu}
\end{aligned} \tag{1.3}$$

where $a = 1, 2, 3, ...8$ for $SU(3)_{\text{flavor}}$, t_a are corresponding generators (here Gell-Mann matrices, modulo normalization), $\bar{q} = (\bar{u}, \bar{d}, \bar{s})$, $\Theta^{\mu\nu}$ denotes symmetric traceless energy momentum tensor, and we used the linear transformation for the original left and right currents ($V = R + L, A = R - L$).
For three light flavors, we are left with a nonet of the conserved vector currents, a nonet of the conserved axial currents, and conserved currents corresponding to the scale invariance.

$$SU(3)_A \times SU(3)_V \times U(1)_A \times U(1)_V \times [\text{SCALE} \quad \text{INV.}] \tag{1.4}$$

This classical picture is however strongly distorted at the quantum level. First, QCD_χ has anomalies - i.e. certain classical symmetries are violated at the quantum level. The scale-invariance is broken (scale anomaly),

$$\partial_\mu J^\mu_{\text{scale}} = \Theta^\mu_\mu = \frac{2\beta(g)}{g} \cdot \frac{1}{4} \left[G^{\mu\nu i} G^i_{\mu\nu} \right]_{\text{renorml.}} \tag{1.5}$$

and the appearance of beta function introduces the scale of the strong interactions, $\Lambda_{\text{QCD}} \sim 200$ MeV.
Second, the singlet component of the axial current is also anomalous

$$\partial_\mu A_0^\mu = \frac{N_f}{2} \frac{g^2}{8\pi^2} \tilde{G}^{\mu\nu i} G^i_{\mu\nu} \tag{1.6}$$

A similar anomaly appears in QED, with $N_f/2 \to 1$, $g \to e$, and non-Abelian $G^{\mu\nu i}$ replaced by Abelian electromagnetic $F^{\mu\nu}$ tensor. These are the all anomalies of the QCD, if no external (e.g. electromagnetic) currents are added. The appearance of the Gell-Mann function β in the scale anomaly reminds about another feature of QCD: the interactions between quarks and gluons get stronger at smaller energy scales, invalidating the perturbative calculation in low energy domain, contrary to the precisely opposite behavior in QED. Quarks and gluons interact strongly forming colorless states, and the unraveled nature of long-wavelength limit of these interactions is usually coded under the name "confinement".
The closer look at the experimental spectrum of elementary excitations (hadrons) of the QCD vacuum shows that the picture is more complicated, and another, on

top of confinement, nonperturbative phenomenon has to take place in QCD. If we look at the remaining symmetries after quantizing the QCD_χ, we see that they are

$$SU(3)_A \times SU(3)_V \times U(1)_V \tag{1.7}$$

where the last one corresponds to the baryon number conservation ($3V^\mu = B^\mu$). The interaction preserves baryon number and is invariant under $SU(3)_V$ and $SU(3)_A$ symmetries. Since the masses of up and down quarks are 20 times smaller than the scale of the QCD, the lightest particles in real QCD should show the traces of the exact symmetry (1.7). This is at odds with the experiment, which shows that the light vector-like particles differ from axial-like particles, despite similar flavor content - e.g. $\rho(770)$ is much lighter then axial $a_1(1200)$. This asymmetry holds also for baryons of opposite parity, (nucleon $n(940)$ versus $N(1535)S_{11}$) and manifests dramatically at the level of the scalars - the lowest pseudoscalar (pion) $\pi(140)$ seems to not have a narrow chiral partner at all - when comparing pion to $f_0(400-1200)$ with full width $600-1000$ MeV, an interpretation of f_0 as a *particle* is controversial... This suggests that the vacuum state of the QCD is not respecting all the symmetries of the interaction of the QCD. This phenomenon is called a spontaneous breakdown of the symmetry. Then, chirally invariant interactions of the QCD_χ, acting on chirally non-invariant vacuum, can indeed produce such an asymmetry in the hadronic spectrum.
The important hint comes from the observation by Vafa and Witten [1], that theories with vector-like couplings (e.g. QCD) cannot break spontaneously vector symmetries. We are therefore left with the alternative:

$$Q_A^a|0> \equiv |PS^a> \neq 0 \tag{1.8}$$

where Q_A^a are axial charges corresponding to the currents A_a^μ. Vacuum state respects therefore only vector symmetries. Since the Hamiltonian of the QCD_χ still commutes with all the 16 generators Q_V^a, Q_A^a we see that the vacuum state ($H|0> = E_0|0>$) is degenerated with the octet of the states $Q_A^a|0>$, ($a = 1, 2, ...8$). Indeed

$$H|PS^a> = HQ_A^a|0> = Q_A^a H|0> \; = \; E_0 Q_A^a|0> = E_0|PS^a> \tag{1.9}$$

This is a basic message of the Goldstone theorem [2]. To each broken generator of the axial current corresponds a massless, spinless excitation corresponding to quantum number of the generator. In the case of three flavors, the Goldstone theorem predicts the appearance of an octet of massless, pseudoscalar mesons. They correspond to massless pions (isotriplet), kaons (two isodoublets) and an eta (isosinglet).
A priori, from the point of view of chiral symmetry, QCD allows (at least) two phases: asymmetric (Nambu-Goldstone phase), described above, and the symmetric one (Wigner-Weyl phase), where vacuum is respecting all symmetries,

$Q_V^a|0>=Q_A^a|0>=0$. Hence, a phase transition may happen between these two phases. Such phase transition can be characterized by the appearance of an order parameter. The lowest-dimensional order parameter is the expectation value of $\bar{q}_R q_L + \bar{q}_L q_R$. It carries zero baryon number, is a scalar (vacuum respects space reflection), and is diagonal in flavor. Note that e.g. $<\bar{u}d>$ is not invariant under rotations generated by isospin matrices, therefore is not invariant under the vector, unbroken group of the vacuum. There are infinitely many other operators, which are order parameters, e.g. $\bar{q}_L q_R \bar{q}_R q_L$, but they carry higher canonical dimension. The lowest dimensional order parameter $<\bar{q}q>$ is called "quark condensate". It is believed that under the action of external parameters, like e.g. temperature and/or density, the crossover to other phases of QCD is possible (see lectures by Rob Pisarski and Krishna Rajagopal), including the phase with restored (approximate) chiral symmetry. It is important to mention, that despite some experimental signals point at the restoration of the chiral symmetry (cf.[3]), there is, in my opinion, no "smoking gun" evidence for this phenomenon and most of our understanding of chiral phase transition comes from the lattice studies (see lectures by Frithjof Karsch). We should also remember, that in real QCD, the masses of u,d,s quarks are light, but non zero, therefore on top of the phenomenon of spontaneous breakdown of the symmetry we have also an explicit, albeit small, explicit breakdown of the chiral symmetry due to the explicit presence of mass terms $m\bar{q}q$ in the Hamiltonian. Therefore the chiral restoration is not really a phase transition, but rather a crossover process.

CONDENSATE, PION, GMOR

We still consider a QCD_χ. Let us come back to the Goldstone theorem, and show that indeed the presence of the condensate forces the presence of massless excitations in the spectrum. To demonstrate this, we use the Ward identity, relating the correlator of the axial SU(3) current A_a^μ and pseudoscalar SU(3) current $P^b = i\bar{q}t^b\gamma_5 q$ to scalar densities,

$$\partial^\mu < 0|TA_\mu^i(x)P^j(0)|0> = -i\delta(x) < 0|\bar{q}[t^i, t^j]_+ q|0> \quad (1.10)$$

<u>Exercise</u>. *Justify (1.10), making use of equal-time commutation relations (cf. the lecture of Jean-Paul Blaizot).*

Since Lorentz invariance implies that Fourier transformation of the correlator has to take a form

$$\int dx e^{ipx} < 0|TA_\mu^i(x)P^j(0)|0> = p_\mu \Pi^{ij}(p^2) \quad (1.11)$$

the Ward identity says that $p^2\Pi^{ij}(p^2) = -\delta^{ij} < \bar{q}q >$. (Taking a derivative corresponds to multiplying the matrix element by ip_μ. Note that the vacuum expectation value of the second term in $t^i t^j = \delta^{ij} + d^{ijk}t^k$ has to vanish due to the

invariance under vector transformations). Since the condensate (expectation value of r.h.s. of (1.10)) is non-zero in the broken phase, the spectral function $\Pi(p^2)$ has to contain a massless particle corresponding to the pole at $p^2 = 0$.
Now we can make use of a powerful condition, (formulated by 't Hooft), relating fundamental theories with theories in which particles are bound states of the fundamental constituents. This condition, known as anomaly matching condition, states that a composite particle has to reproduce exactly the anomaly present in the fundamental theory.
To see how this condition works let us consider the axial $SU(3)_A$ current corresponding to π^0, i.e. $A^3_\mu = \bar{q}\gamma_\mu t^3 \gamma_5 q$, with $q = (u, d, s)$ and $t^3 = \mathrm{diag}(1, -1, 0)$. This current is anomaly free in QCD, but if we allow quarks couple to photons, the matrix element

$$\begin{aligned} < 0|\partial_\mu A^\mu_3|\gamma\gamma> &= < 0|\partial_\mu \bar{u}\gamma^\mu\gamma_5 u|\gamma\gamma> - < 0|\partial_\mu \bar{d}\gamma^\mu\gamma_5 d|\gamma\gamma> \\ &= \frac{\alpha}{2\pi} N_c(e_u^2 - e_d^2) < 0|F\tilde{F}|\gamma\gamma> \end{aligned} \tag{1.12}$$

It is easy to understand the r.h.s. of (1.12). This is simply the difference of two *electromagnetic* anomalous U(1) currents, corresponding to the charges with fractions $e_u = 2/3$ and $e_d = -1/3$, respectively. This equation represents the contribution from the celebrated triangle anomaly.
From the above identity we infer immediately the form of the matrix element

$$< 0|A^\mu_3|\gamma\gamma> = -i\frac{p^\mu}{p^2}\frac{\alpha}{2\pi} N_c(e_u^2 - e_d^2) < 0|F\tilde{F}|\gamma\gamma> + ... \tag{1.13}$$

where p is the momentum of both photons and the dots stand for less singular terms. The pole here is simply the remnant of the massless quark circulating in a triangle graph. If we now *assume* confinement, anomaly matching condition says that in the hadronic (composite) world, there must exist a massless, colorless object which couples to the axial current and photons and which reproduces exactly the pole in the anomaly above. Since massless baryons are excluded (cf. Vafa-Witten theorem), this object has to be a meson. Mesons fulfill the anomaly condition in an easy way. We do not need to circulate them inside the triangle loops like quarks, it is enough if there is a relation between the axial current and the pion. This well-known relation reads

$$< 0|A^a_\mu|\pi^b> = i\delta^{ab} F_\pi p_\mu \tag{1.14}$$

where F_π is pion decay constant. So anomaly matching means, that we read out the r.h.s. of (1.13) as a product of (1.14), massless pion propagator $-i/p^2$ and the amplitude for $\pi^0 \to \gamma\gamma$ decay (rest of the r.h.s. of (1.13). The similar matching of the trace anomaly between the quark and hadronic worlds was discussed in Dima Kharzeev lectures). Let us come back to Ward identity (1.10). In the $p^2 \to 0$ limit the correlator is fully determined by the condensate,

$$\int dx e^{ipx} < 0|TA^i_\mu(x)P^j(0)|0> = -\frac{p_\mu}{p^2} < \bar{q}q > \delta^{ij} \tag{1.15}$$

Saturating the correlator with pion states, and using (1.14), we read out comparing left and right h.s, the value of the another matrix element

$$< 0|P^j|\pi^k> = \delta^{jk}\frac{< \bar{q}q >}{F_\pi} \tag{1.16}$$

With the help of the above relations we can now prove the Gell-Mann Oakes Renner relation [4]. Introducing the small quark masses, we see that Dirac equation implies

$$\partial_\mu(\bar{u}(x)\gamma^\mu\gamma_5 d(x)) = (m_u + m_d)\bar{u}(x)i\gamma_5 d(x) \tag{1.17}$$

Calculating the vacuum to pion matrix element of the above identity , and using formulae (1.14) and (1.16), we get

$$M_\pi^2 F_\pi^2 = (m_u + m_d) < \bar{q}q > \tag{1.18}$$

This relation shows how the gap in the Goldstone mode appears due to the presence of the nonzero quark mass.

EUCLIDEAN WORLD

In this subsection we present the formulae that allow the transcription from Minkowski to Euclidean space. The advantages of working in Euclidean space are two-fold: first, several mathematical operations are well defined, second, the formulation is comparable to the lattice simulations. The following set of rules defines the transition (l.h.s. denotes Minkowski, r.h.s. denote Euclidean)

- Space time: $ix_0 \equiv x_4$, $x_i \equiv x_i$
- Vector potentials: $A_0 \equiv iA_4$, $A_i \equiv A_i$
- Gamma matrices: $\gamma_0 \equiv \gamma_4$, $\gamma_i \equiv i\gamma_i$, $\gamma_5 \equiv \gamma_5$
- Fermi fields: $i\bar{q} \equiv \bar{q}$, $q \equiv q$
- Action: $iS \equiv -S$

Then, Euclidean Dirac matrices obey $\gamma_\mu\gamma_\nu + \gamma_\nu\gamma_\mu = 2\delta_{\mu\nu}$, all four matrices are hermitian, as well as the Dirac operator $i\partial_\mu\gamma_\mu$. Finally, the Euclidean action for QCD reads:

$$S = \int d^4x \left[\frac{1}{4}G^a_{\mu\nu}G^a_{\mu\nu} - \bar{q}(i\gamma_\mu D_\mu(A) + im)q\right] \tag{1.19}$$

and partition function $Z = \exp(-S)$.

BANKS CASHER RELATION

We will now demonstrate that the chiral condensate is related to the Dirac operator spectrum in Euclidean space-time. Consider a Dirac propagator in a Euclidan box $V = L^4$ in the presence of the gluonic background field A_μ. From the action above, we read that propagator is $S_F = (\not{D}(A) + m)^{-1}$. We can write then:

$$V < 0|\bar{q}(0)q(0); A|0 > = -\text{TR } S_F(x,x) = -\sum_k \frac{1}{i\lambda_k(A) + m}$$
$$\longrightarrow \lim_{V\to\infty} i \int d\lambda \rho(\lambda) \frac{1}{\lambda - im} \qquad (1.20)$$

where TR denotes trace over coordinates, color, spin and Dirac indices, and spectral density is defined as $\rho(\lambda) = \sum_k \delta(\lambda - \lambda_k)$ where $i\not{D}q_k = \lambda_k q_k$. Now, we average the above equation over the gluonic configuration weighted with the full QCD measure (including standard gauge fixing etc.)

$$< (...) >_{\text{QCD}} \equiv \int [dA](...)\det(\not{D}(A) + m)e^{-S_{\text{glue}}} \qquad (1.21)$$

As a result, we get

$$V < \bar{q}q >= i \int d\lambda \frac{\bar{\rho}(\lambda)}{\lambda - im} \qquad (1.22)$$

where $\bar{\rho}(\lambda) =< \rho >_{\text{QCD}}$ is the average spectral density, i.e. the density $\rho(\lambda)$ averaged over the full QCD measure. As a final step, we take a chiral limit and use the relation

$$\lim_{m\to 0} \frac{1}{\lambda \pm im} = \text{PV}\frac{1}{\lambda} \mp i\pi\delta(\lambda) \qquad (1.23)$$

As a result, we relate the chiral condensate to the average spectral density around zero eigenvalues

$$V < \bar{q}q >= -\pi\bar{\rho}(0) \qquad (1.24)$$

Note that the contribution from principle value (PV) part drops, since due to the chiral property $[\not{D}, \gamma_5]_+$ the eigenvalues come in pairs (corresponding to eigenfunctions q_k and $\gamma_5 q_k$), so the average spectral density is an even function $\bar{\rho}(\lambda) = \bar{\rho}(-\lambda)$. Note that the same property guarantees the positivity of the QCD measure, despite at the first look $\not{D} + m$ seems to create complex measure.
The relation (1.24) was first suggested by Banks and Casher [5]. (For the careful discussion of the UV part of the spectrum, see [6]). It is very important that the chiral limit $m \to 0$ is taken after the thermodynamical limit $V = L^4 \to \infty$, for otherwise the average spectral density would be zero. The result states that in a

vector like theory with chirally symmetric spectra, the quark condensate is related to mean spectral density at zero virtuality (i.e. at $\lambda = 0$).
We can now ask the crucial question, what kind of mechanism can cause spectral density to be non-vanishing at zero virtuality? Note that the levels of free particle closed in the box scale like $2\pi n/L$, with n integer, so the mean level spacing goes like L^{-1}, and the average spectral density (proportional to the inverse of the mean level spacing) for free particle scales like $L = V^{1/4}$, therefore will never be able to balance in the thermodynamical limit the l.h.s. of (1.24) provided the condensate is non zero. The only solution is that spontaneous symmetry breakdown requires enormous accumulation of the eigenvalues in the vicinity of zero, with the condensation rate scaling like V, so $\bar{\rho}(0) \sim V$. This at the first look obvious fact was first emphasized and exploited by Leutwyler and Smilga [6] and forms the cornerstone of the spectral analysis of the Dirac operator.
We will devote the next lecture to unravel the most plausible microscopic mechanism responsible for such a spectral behavior.

2. Part 2

In this part, we outline the basic concepts of diffusion and translate them into the Euclidean QCD language, identifying in this way the hierarchy of the spectral scales of the Dirac operator.

PRIMER ON THE DIFFUSION

The Banks-Casher relation is reminiscent of the Einstein relation describing the conductivity σ of degenerate gas of electrons,

$$\sigma = 2e^2 D\bar{\rho}(E_F) \tag{2.1}$$

where D is a diffusion constant, $\bar{\rho}(E_F)$ is the density of states (per spin direction) at the Fermi surface. Is it possible that the spontaneous breakdown of the chiral symmetry has also the diffusive nature, with the Fermi surface replaced by the zero virtuality band? We will show explicitly in this chapter, that this indeed is the case.
In order to prove this conjecture, we have to remind some basic facts on the diffusion [7, 8]. The diffusion is a process, in which a typical distance covered by the diffusing particle in a time t varies as

$$\bar{r}^2(t) = Dt \tag{2.2}$$

where D is the diffusion constant characterizing the medium. If we consider the diffusive motion in a cube L^d of the linear size L, it is natural to define the time scale $t_c = L^2/D$, characterizing the time during which the particle can probe the whole system. The energy scale corresponding to this time, known as a Thouless energy [9] (in units where $h = 1$) is $E_c = 1/t_c = D/L^2$. For times shorter than t_c, the diffusing particle can probe only part of the volume. For even shorter times (shorter then the time of mean free path between the "dirt" causing the diffusion, t_e) the diffusion concept is meaningless. On the other side, for very large times ($t_H \sim 1/\Delta$), the diffusive particle will always leave the volume (see Edmond Iancu's lectures). When such times start corresponding to the inverse of mean quantum spacing of the quantum mechanical levels, the classical concept of the diffusion also becomes meaningless. The above hierarchy of scales could be pictured by a cartoon, where we introduced also the names of four different regimes using the terminology borrowed from the mesoscopic physics [7, 8]: Quantum ($t > t_H$), Ergodic (universal) ($t_H < t < t_c$), Diffusive ($t_c < t < t_e$) and Ballistic ($t < t_e$).

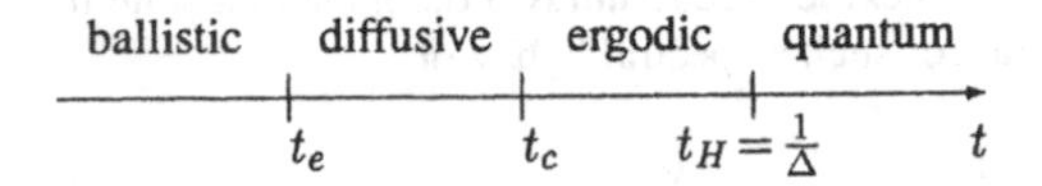

Figure 1. Schematic ordering of the diffusion time-scales.

In metallic grains, in the ergodic window the spectral properties of the metals are universal, and described by the random matrix theory (universal conductance fluctuations), hence the second name of this regime.
Let us try to make this description more quantitative. Let us introduce the retarded Green's function $G^R(x, y; E) =< x|(E-H+i\epsilon)^{-1}|y >$. The time-Fourier transformation $G^R(x, y; t)$ describes the amplitude, that the diffusing particle propagates from x to y in a time t, under the influence of the dynamics governed by some Hamiltonian H. The Hamiltonian is the microscopic source of the disorder. A classic example is the Anderson Hamiltonian [10]

$$H_A = -\frac{1}{2m}(\nabla - ieA)^2 + V(\vec{x}) \tag{2.3}$$

where $< V(\vec{x}) >= 0$ and $< V(\vec{x})V(\vec{y} >= \delta(\vec{x} - \vec{y})/(2\pi\bar{\rho}(E_F)t_e)$. Note that the disorder is *static*, i.e. time independent. We can now define a crucial concept of the return probability. This is simply the square of the amplitude of returning to the same point x in a time t, averaged over the disorder. For a particle at a Fermi surface, it reads

$$P(t) = \lim_{x \to y} \frac{V}{2\pi\bar{\rho}} \int d\lambda e^{i\lambda t} \left\langle G^R(x, y; E_F + \lambda/2)[G^R(x, y; E_F + \lambda/2)]^* \right\rangle_{H_A} \tag{2.4}$$

The prefactor V comes from the translational invariance of the return probability, and the denominator $2\pi\bar{\rho}(E_F)$ guarantees the normalization of the probability to 1.

For a *static* random potential of the type above one can perform the averaging and then the integration, with the following result:

$$P(x,y,t) = \sum_{\vec{q}} e^{Dq^2t} e^{i\vec{q}(\vec{x}-\vec{y})} \overrightarrow{\lim_{V\to\infty}} \frac{V}{(4\pi Dt)^{d/2}} e^{-|\vec{x}-\vec{y}|^2/4Dt} \tag{2.5}$$

Hence

$$P(t) = \sum_{\vec{q}} e^{-Dq^2t} \tag{2.6}$$

or, equivalently, in Fourier space

$$P(\lambda) = \sum_{\vec{q}} \frac{1}{-i\lambda + Dq^2} \tag{2.7}$$

In the above, while performing the averaging over the microscopic disorder, we integrated out fast degrees of freedom, getting the effective description in terms soft modes q_i. Details of fast degrees of freedom are now hidden in effective parameters like the diffusion constant D, and the obvious name "diffuson" for a soft modes q_i is natural in the light of the form $P(x,y,t)$, being the Green's function of a diffusion operator $(\partial_t + D\nabla^2)$.
Formally, for very large times ($\lambda \to 0$), the return probability develops a pole (diffuson pole at $q^2 = 0$). However, for such times, quantum effects become relevant, introducing the natural cutoff γ at the energy scale of the average quantum spacing Δ. This cutoff suppresses infinitely long diffusive orbits, and regulates the pole

$$P(t) = e^{-\gamma t} \sum_{\vec{q}} e^{-Dq^2t} \tag{2.8}$$

or, after Fourier transformation

$$P(t) = \sum_{\vec{q}} \frac{1}{-i\lambda + \gamma + Dq^2} \tag{2.9}$$

In the next section, we will demonstrate, that all the above listed concepts are directly applicable to QCD.

EUCLIDEAN QCD IS DIFFUSIVE IN D=4

Let us look at the Dirac operator in the background of some Euclidean gluonic configuration A

$$(i\not{D}(A) + im)q_k = \lambda_k(A)q_k \tag{2.10}$$

as a "Hamiltonian" corresponding to the "eigenenergy" λ. The imaginary constant shift im does not spoil the analogy. For static Hamiltonians in $d = 1, 2, 3$ we know from the quantum mechanics, that the evolution in time t is governed by $q_k(t) = \exp(iE_kt)q_k(0)$. Time t is dual to the energy E. We will parallel this construction here, introducing dual "time" τ to the virtuality ("energy") λ. Since the "Hamiltonian" $i\not{D}(A) + im$ is by definition independent on this "time" (*static* in τ), we can write down the eigenmode evolution in this "time"

$$q_k(\tau) = e^{i(\lambda_k + im)\tau} q_k(0) \tag{2.11}$$

corresponding to $4 + 1$ dimensional Schroedinger-like equation with static (τ independent) potential

$$i\partial_\tau q(\tau) = (i\not{D}(A) + im)q(\tau). \tag{2.12}$$

Note that this construction does not modify any properties of the Dirac operator, and is basically equivalent to introducing the Schwinger proper time τ.
We will prove now, following [14], that the dynamics of the Euclidean $d = 4$ Dirac operator in Schwinger time τ parallels the dynamics of the usual diffusion $d = 1, 2, 3$ in the real time t. This means, that we will identify, in the spectrum of the Dirac operator, four distinct regimes corresponding to quantum, ergodic, diffusive and ballistic regimes of the mesoscopic physics.
From now on, we follow, step by step, the construction outlined in the previous chapter. Instead of Green's function for a diffusing electron we take a Green's function for a quark in some unknown gluonic background, which is constant in Schwinger time τ

$$G^R(x, y; E) \longrightarrow S_F(x, y; \lambda) \tag{2.13}$$

We can now define by analogy to (2.4) the return probability in time τ for a quark at the *zero virtuality surface*

$$P(\tau) = \frac{V}{2\pi\bar{\rho}(0)} \int d\lambda e^{i\lambda\tau} \left\langle \text{Tr } S_F S_F^\dagger \right\rangle_{\text{QCD}} \tag{2.14}$$

Here $S_F \equiv S_F(x, y; m - i\lambda/2)$, trace appear due to the γ_μ matrix structure of the Dirac operator and the averaging is done over the full QCD measure (1.21), representing the analog of Anderson Hamiltonian representing the disorder of the system. The parameter m appears here due to the uniform shift im of the original spectrum of the massless Dirac operator.
Till now the analogy was exact. But now we have to perform the averaging over the a priori unknown measure of the QCD. The first alternative is to choose some model of the disorder. Actually the analytical instanton model of Diakonov and Petrov [11] and the numerical simulation by Shuryak [12] are the first realizations of this scenario. We will return to this point later.

Surprisingly, we can calculate $P(\tau)$ *without assuming any model of the disorder*, but making use of almost exact low energy theorems of the QCD, introduced in Part 1. First, let us note, that due to the chiral properties of the Dirac operator

$$< \mathrm{Tr} S_F S_F^\dagger >= - < \mathrm{Tr} S_F(x,y;z)\gamma_5 S_F(y,x;z^*\gamma_5 > \tag{2.15}$$

where $z = m - \lambda/2$. We may introduce also the isospin sources. One then immediately recognizes the similarity to the pion correlation function structure

$$\begin{aligned} C_\pi^{ab}(x,y;m) &= < S_F(x,y;m) i\gamma_5 \tau^a S_F(y,x;m) i\gamma_5 \tau^b >_{\rm QCD} \\ &= \frac{1}{V}\sum_{Q_\mu} e^{iQ(x-y)} \left[\frac{<\bar{q}q>}{F_\pi}\right]^2 \cdot \frac{1}{Q^2+M_\pi^2} \end{aligned} \tag{2.16}$$

where we used the pion dominance formulae (1.16).
Second, we use another low-energy theorem (1.18), to replace the M_π on the r.h.s. of the above equation by the current quark mass m. Third, we analytically continue the $C_\pi^{ab}(x,y;m)$, by replacing the mass m by $z = m - i\lambda/2$. We recognize then that the integrand in the return probability involves the analytically continued pion correlation function, so averaging is equivalent to the analytical continuation of the r.h.s. of C_π

$$\begin{aligned} P(x,y,\tau) &= \frac{1}{2\pi\bar{\rho}}\int d\lambda e^{i\lambda\tau} C_\pi(x,y;z(\lambda)) \\ &\approx \sum_{Q_\mu}\int d\lambda e^{i\lambda\tau} e^{iQ(x-y)} \frac{1}{\frac{F_\pi^2}{|<\bar{q}q>|}Q^2 + 2m + i\lambda} \end{aligned} \tag{2.17}$$

Finally, after integrating over λ by the residue method, we see already the diffusive structure of $P(x,y,\tau)$, and the return probability (limit $x \to y$) reads

$$P(\tau) = e^{-2m\tau} \sum_{Q_\mu} e^{-DQ^2\tau} \tag{2.18}$$

or, after Fourier transforming

$$P(\lambda) = \sum_{Q_\mu} \frac{1}{-i\lambda + 2m + DQ^2} \tag{2.19}$$

We constructed in this way the precise analog of the diffusive return probability, with diffusion constant $D = F_\pi^2/|<\bar{q}q>| \sim 0.22$ fm and the slow "diffuson modes" $Q_\mu = 2\pi n_\mu/L$, with $\mu = 1,2,3,4$ and n_μ integers.
In fact, using the low energy theorems and the above construction, we *implicitly* integrated out the fast (gluonic) degrees of freedom, getting in this way the diffusion constant. The remaining dynamics of slow modes (diffusons) is the dynamics of the pions - longest wave excitations of the QCD. The Goldstone nature of the

pion manifest itself as a pole in the chiral limit and $\lambda \to 0$ limit. We see easily that the very heavy quarks do not diffuse at all (due to the exponential damping). Note also that for light but massive quarks we can read GOR as an expression of the coherence length of the QCD vacuum. Indeed, by definition, coherence length is

$$L_{coh} \equiv \sqrt{D/\gamma} = 1/M_{\pi} \tag{2.20}$$

where we used that for QCD $\gamma = 2m$ and the expression for the diffusion constant calculated above. Coherence length of the QCD vacuum is related to the pion mass, so in the massless limit, pions are indeed much more vacuum modes than the pairs of bounded constituent quark and antiquark.
Now we can identify the different regimes in the spectrum of the Dirac operator. Since the Schwinger time τ is dual to the eigenvalue of the Dirac operator, we immediately identify the analog of Thouless energy, here the Thouless virtuality [14, 15]

$$\Lambda_c = D/L^2 = F_{\pi}^2/(\Sigma\sqrt{V}) \tag{2.21}$$

where $| < \bar{q}q > | \equiv \Sigma$. Therefore the eigenvalues smaller than Λ_c are expected to belong to the ergodic (universal) regime, whereas larger than Λ_c to the diffusive. Since the Banks Casher relation gives us an estimate of the mean level spacing, we know also the borderline of the quantum and ergodic regime. Finally, for very short times (large eigenvalues) the concept of the diffusion becomes meaningless. It is not very difficult to argue, that the borderline between the diffusive and ballistic regime is given by twice the mass of the constituent quark, basically the mass of the ρ meson. Indeed, for times shorter than the required to travel one mean-free path between lumps of the gauge field the concept of dressing (via multiple scatterings) of the current mass is void.

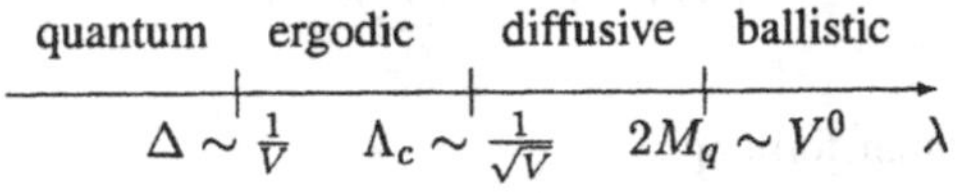

Figure 2. Disorder regimes in the eigenvalue spectrum of the Dirac operator. Note the thermodynamical ordering.

The above hierarchy of scales could be summarized again by the cartoon, where we also remind about the thermodynamical ordering of the scales. On the basis of the analogy with the condensed matter systems, we may suspect, that similarly to the universal conductance fluctuation in the ergodic regime, we may see universal spectral fluctuations in the ergodic window of the QCD, hopefully described by some sort of random matrix theory. In the next chapters, we will see that this indeed is the case.

ERGODIC (UNIVERSAL) REGIME OF THE QCD

Let us consider in more detail the ergodic regime. We expand the sum over the diffuson (pion) modes valid in diffusive regime, i.e. we sum over the quadruples of integers $Q_\mu = (n_1, n_2, n_3, n_4)2\pi/L$

$$P(\tau) = e^{-2m\tau}(1 + 8e^{-4\pi^2 D/L^2\tau} + ...) \quad (2.22)$$

where the first term comes from all $n_i = 0$, second from all combinations of n of the type $(1, 0, 0, 0)$ etc. Since $D/L^2 = \tau_c^{-1}$, the return probability reads

$$P(\tau) = e^{-2m\tau}(1 + 8e^{-4\pi^2\tau/\tau_c} + ...) \quad (2.23)$$

For the times larger than t_c all terms except of the first one vanish exponentially, so we obtain, that in the ergodic regime the return probability is simply

$$P(\tau) = e^{-2m\tau} \quad (2.24)$$

approaching constant for very large τ. Only the softest modes (zero modes $n_1 = n_2 = n_3 = n_4 = 0$) determine the return probability in the ergodic regime. This points that the properties of the ergodic regime are universal – they are independent on the details of space-time interactions, since the Goldstone bosons interaction involves derivative terms, which vanish in long range limit.

This result was known since long ago, although rephrased in a different way. In the usual chiral perturbation theory, pion momenta are of order $1/L$, the mass of the pion scales therefore like $1/L^2$, and since $m \sim M_\pi^2$, the combination $m^2V \sim O(1)$ is fixed. The systematic expansion based on this counting is the clue of the Weinberg [13] chiral perturbation expansion and was practically realized by Gasser and Leutwyler [16]. What is less known, that Gasser and Leutwyler looked also at finite volume, and what happens, when one keeps only zero modes of the pion propagator. Since the propagator in the final volume reads

$$G = \frac{1}{V}\sum e^{iQ_n x}\frac{1}{Q_n^2 + M_\pi^2}$$
$$= \frac{1}{VM_\pi^2} + \frac{1}{V}\sum{}' e^{iQ_n x}\frac{1}{Q_n^2 + M_\pi^2} \quad (2.25)$$

the counting based on the concept of having mV fixed [17](equivalent to $M_\pi^2 V \sim O(1)$) kills all the terms in the primed sum (non-zero modes) and leaves the zero mode. This is very different to the counting based on before-mentioned $m^2V = 1$ principle (chiral perturbation theory).

Gasser and Leutwyler managed to re-sum the contribution of all the zero modes based on $mV \sim O(1)$ counting, and obtained the exact partition function

$$Z(m) = \int dU e^{V\Sigma \mathrm{Tr}\, m(U+U^\dagger)} \quad (2.26)$$

with $U = \exp(i\pi\tau)$ and $\Sigma = | < \bar{q}q > |$. This formula is universal: it depends only on the ways how chiral symmetry is to be broken spontaneously (choice of the measure dU of the Goldstone modes) and how chiral symmetry is being broken explicitly, by the $(N_f, \bar{N}_f) + (\bar{N}_f, N_f)$ representation in the exponent. In the ergodic regime any theory/model sharing the same global symmetries as QCD belongs to the same universality class, i.e. any theory/model leading to the return probability $P(\tau) = e^{-2m\tau}$ will exhibit the same universal spectral properties in the ergodic window. Since the information about the dimensionality is being lost in the ergodic regime, this could be even a zero-dimensional version of the field theory. But field-theory in zero-dimensions is a matrix model. So in the ergodic window QCD is equivalent to a certain matrix model. Before we will unravel the details of this model, let us remind once more that this equivalence between QCD and Random Matrix Model (RMM) happened only for eigenvalues *below* Λ_c. Since the edges of the ergodic window scale with volume as $\Delta \sim 1/V$ (Banks-Casher) and as $\Lambda_c \sim 1/\sqrt{V}$ (Thouless virtuality in four dimensions) the ergodic, universal window shrinks to a point in the infinite volume limit.

WHERE DOES THE "COLOR DIRT" COME FROM?

At this moment we may start to worry, what plays the role of the diffusive dirt in the case of the QCD? QCD is a fundamental theory, so the "dirt" has to be an immanent feature of the QCD itself. Whereas we do not have an exact answer what is the color dirt in QCD, let us observe, that most of the localized, Euclidean (i.e. static from the point of view of Schwinger time) gluonic configurations do the job. Typical and perhaps the most natural are instantons [18]. First, they are the classical, localized stable Euclidean solutions of the QCD. Second, Dirac equations in the presence of instanton background possesses the chiral zero modes, therefore the instanton vacuum immediately provides a microscopic "hopping" mechanism from one instanton field to another and guarantees flipping the chirality at each "scattering" in "time" τ. Quantitatively, each instanton provides a seed of non-conservation of the chiral charge, by integrated form of the anomaly:

$$\delta Q_A = 2N_f Q \tag{2.27}$$

where Q is the topological charge of the instanton. The instanton model involves basically two parameters, the average size of the instanton ($\rho \sim 1/3$ fm) and the concentration $n \sim fm^{-4}$ (the typical density of instantons). Therefore every dimension-full quantity in the instanton model depends parametrically on the combination of the two scales. In particular, the smallness of the diffusion constant calculated in the previous chapter, finds a natural explanation as a diluteness of the instanton medium. Note that the finite value of the condensation requires that thermodynamical limit $V \to \infty$, $N \to \infty$, with N/V fixed, where N is the number of instantons and antiinstantons.

Each instanton vacuum configuration is a snapshot in a time τ. Each snapshot violates the Lorentz invariance ("static", particular distribution of instantons and antiinstantons in four volume) and gauge invariance (each instanton freezes a direction in color). But the averaging over collective coordinates of the instantons (here over the centers of instantons and over their color measure) restores the gauge and Lorentz invariance. We would like to stress, that instantons are sufficient, but not necessary configurations to realize the diffusive scenario. Several other models may also provide chiral disorder, e.g. family of stochastic vacuum models [19]. Each of these models comes with a certain correlation length scale, which corresponds quantitatively to the diffusion constant.
In other ways, the spontaneous breakdown of the chiral symmetry in QCD is a very robust phenomenon, comparing to the confinement. Usually, the models of color dirt either ignore confinement, or introduce it by hand, or as instantons, seem to be (at least naively) not related to confinement at all. Mysteriously, lattice evidence suggests strongly that both phenomena (confinement and chiral symmetry breakdown) are correlated and vanish at the same temperature. Since the confining configurations are the topic of vivid speculations, the fundamental understanding of the "color dirt" and chiral disorder is still missing.

3. Part 3

I analyze in more detail two regimes of the QCD, the ergodic one and the diffusive one. Then I confront few sample predictions with the "experimental data" obtained from lattice simulations.

RANDOM MATRICES - FIELD THEORY IN 0 DIMENSION

As stated before, in the ergodic regime any model obeying the global symmetries of the QCD belongs to same universality class as QCD. Since this is the regime where only the constant pionic modes matter, we can ignore space-time dependence and stick to the field theory in zero dimensions, i.e. the theory where fields are numbers and do not have any space-time dependence. The exact form of such theory in the QCD can be inferred in numerous ways. Here we follow the historical route [11, 12, 20]. Imagine that we have two lumps of the "dirt", e.g. instanton and anti-instanton, separated at the very large distance. Then, the Dirac equation in the field of the instanton has an exact fermionic right-handed zero mode ϕ_R, and the Dirac equation for the anti-instanton has a similar, but left-handed zero mode ϕ_L. When we decrease the distance between instantons, the degenerated pair of zero modes is replaced by the pair of eigenvalues $(T, -T)$, where the overlap $T = \int d^4x \phi_R^{\dagger} i\partial\!\!\!/ \phi_L$ depends on the distance and the mutual orientation of the instantons. Let us now add more and more instantons into this medium, and still work in a dilute gas approximation. The infinite fermion determinant, when calculated in the basis of left and right handed quark zero modes, is now approximated by the matrix of overlaps between the I-th instanton and J-th anti-instanton T_{IJ}. The off-diagonal block structure comes from the chirality flipping mechanism. The diagonal blocks are zero in the chiral limit. In this way, instanton picture trades the a priori unknown QCD measure (1.21) into the approximate measure

$$< (...) >= \int \prod d[\Omega_i](\ldots)e^{-S_{\rm glue}(\Omega_i)} \prod_{N_f} \det \begin{vmatrix} 0 & T(\Omega_i) \\ T^\dagger(\Omega_i) & 0 \end{vmatrix} \tag{3.1}$$

where Ω_i is the set of collective coordinates of the instantons (positions and color orientations), and $S_{\rm glue}$ is the gluonic part of the QCD action saturated with the initial instanton vacuum ansatz. Let us truncate now all the space time dependence in the above action. We are left with the model of the type

$$Z = \int dT e^{-N\sigma {\rm Tr} TT^\dagger} \det \begin{vmatrix} 0 & T \\ T^\dagger & 0 \end{vmatrix}^{N_f} \tag{3.2}$$

Here T is an N by N matrix built out of complex numbers (one could generalize the matrices to the rectangular ones as well). The gluonic measure is replaced by the polynomial measure with some potential $v(T^\dagger T)$, here being the simplest - the harmonic potential. The factor N in front of the potential guarantees, that each integration over T is appropriately weighted. The model has one scale. i.e. the "width" σ of the Gaussian matrix measure. The partition function Z defines the chiral Gaussian random matrix model [21]. The name chiral comes from the off-diagonal block structure, which mimics the original chiral structure of the Dirac operator, $[\not{D}(A), \gamma_5]_+ = 0$.
We will show now, that such defined model is exactly solvable for any finite N. The main problem is the integration over the measure dT. We will use here the trick, similar to the change of the variables from Cartesian coordinates to the spherical ones. The simplest way to change usual coordinate variables is to look at the infinitesimal interval

$$ds^2 = dx^2 + dy^2 + dz^2 = dr^2 + r^2 d\theta^2 + r^2 \sin^2\theta d\phi^2 \tag{3.3}$$

so the metric tensor is $g_{lk} = {\rm diag}(1, r^2, r^2\sin^2\theta)$. Hence the Jacobian (the square root of the determinant of the metric tensor) reads $J = \sqrt{\det g_{lk}} \equiv \sqrt{g} = r^2 \sin\theta$. We observe now that the integrand of the partition function (3.2) depends only on the "radial" combination $X = T^\dagger T$, and we follow the trivial example above. Every hermitian matrix can be diagonalized by a unitary transformation. Since X is hermitian, introducing $X = URU^\dagger$, where the unitary matrix $U = \exp(iH)$, H hermitian, and $R_{ij} = \delta_{ij} r_i$ is positive diagonal, we calculate

$$\begin{aligned} {\rm tr}(dX^2) &= {\rm tr}(U^\dagger(dR + i[R, dH])U)^2 = {\rm tr}(dR + i[R, dH])^2 \\ &= \sum_k dr_k^2 + \sum_{i\neq j}(r_i - r_j)^2 |dH_{ij}|^2 \end{aligned} \tag{3.4}$$

Hence the metric tensor reads

$$g_{lk} = \text{diag}(\underbrace{1, ...1}_{N}, \underbrace{..., (r_i - r_j)^2, ...}_{N(N-1)}) \tag{3.5}$$

so $\det g_{lk} = \prod_{i \neq j}(r_i - r_j)^2$ and Jacobian $J = \sqrt{g} = \prod_{i<j}(r_i - r_j)^2 \equiv \Delta(R)^2$. Since the integrand and the Jacobian do not depend on the angles parameterizing the unitary matrices, we are left with N integrations over the diagonal eigenvalues ("radial" variables) r_i

$$Z = \int \prod_i dr_i \Delta(R)^2 \exp(-N\sigma \sum r_i) \prod r_i^{N_f} \tag{3.6}$$

Let us note, that the expression $\Delta(R)$ can be rewritten as a determinant (Vandermonde determinant).

$$\Delta(R) = \prod_{i<j}(r_i - r_j) = \begin{vmatrix} 1 & r_1 & r_1^2 & \dots & r_1^{N-1} \\ 1 & r_2 & r_2^2 & \dots & r_2^{N-1} \\ \vdots & \vdots & \vdots & \ddots & \vdots \\ 1 & r_N & r_N^2 & \dots & r_N^{N-1} \end{vmatrix} \tag{3.7}$$

[Exercise. *Prove* (3.7).] Since we can add to each row of the determinant an arbitrary combination of the other rows (without changing the value of the determinant), we can replace the original Vandermondian by the determinant build out of polynomials.

$$\Delta(R) = \begin{vmatrix} P_0(r_1) & P_1(r_1) & P_2(r_1) & \dots & P_{N-1}(r_1) \\ P_0(r_2) & P_1(r_2) & P_2(r_2) & \dots & P_{N-1}(r_2) \\ \vdots & \vdots & \vdots & \ddots & \vdots \\ P_0(r_N) & P_1(r_N) & P_2(r_N) & \dots & P_{N-1}(r_N) \end{vmatrix} \tag{3.8}$$

These polynomials are a priori arbitrary, modulo the fact that the coefficient at the highest power is always 1. We choose now the polynomials in such a way, that they are orthogonal with respect to the measure. In our case we require, that

$$\int_o^\infty dr e^{-N\sigma r} r^{N_f} P_i(r) P_k(r) = \delta_{ik} \tag{3.9}$$

It is obvious, that in the case of $N_f = 0$ such polynomials are Laguerre polynomials (modulo trivial rescaling of the coefficients, so the highest power is multiplied by 1) and in case of $N_f \neq 0$ are associated Laguerre polynomials $L_k^{(N_f)}$. Due to the orthogonality of the polynomials all integrations in the partition function factorize, and we can write down arbitrary k-correlation function between the k

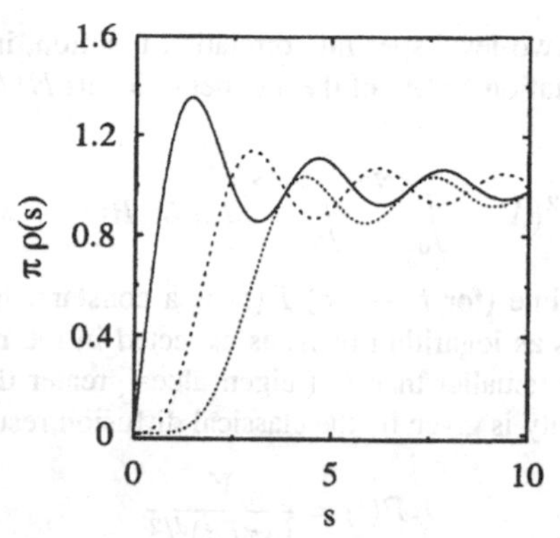

Figure 3. Microscopic spectral density for $N_f = 0, 1, 2$ (solid, dashed dotted lines, respectively).

CROSSOVER TO THE DIFFUSIVE REGIME - CHIRAL PERTURBATION

The universal (ergodic) regime ends at the scale corresponding to the Thouless virtuality. For the scales larger than Thouless virtuality neglecting non-zero modes is no longer justified – all modes have to be taken into account. Let us look at the scaling properties of the eigenvalues. The first eigenvalue which falls off from the ergodic window (i.e. by definition, Thouless virtuality) has to scale [14, 15] (in units of Δ) as

$$\frac{\Lambda_c}{\Delta} = \frac{D}{\Delta L^2} = \frac{F_\pi^2}{\Delta | < \bar{q}q > | L^2} \sim \frac{F_\pi^2 V}{L^2} = F_\pi^2 L^2 \tag{3.13}$$

where we used the Banks-Casher relation and the definitions of the Thouless virtuality and the diffusion constant.
It is remarkable, that the analogy with the diffusion can lead to a quantitative description of the spectral correlations in the diffusive regime of the QCD as well. Let us define two-point correlation density of states

$$R(s) = \frac{1}{< \rho(\Lambda) >^2} \langle \rho(\Lambda - \epsilon/2) \rho(\Lambda + \epsilon/2) \rangle - 1 \tag{3.14}$$

where $s = \epsilon/\Delta$ and, as usual, $\Delta \sim 1/\rho V$. Then we define the spectral formfactor

$$K(t) = \frac{1}{2\pi} \int d\epsilon R(\epsilon) e^{-i\epsilon t} \tag{3.15}$$

For times shorter than the Heisenberg time $t_H \sim \Delta^{-1}$, the standard semi-classical argument developed for mesoscopic systems by [25] allows to relate spectral formfactor to the return probability $P(t)$

$$K(t) = \frac{1}{4\pi^2} t \Delta^2 P(t) \tag{3.16}$$

By definition, the two-level spectral correlation function, integrated from 0 to Λ measures the fluctuation $\Sigma^2(\Lambda)$ of the number of levels $N(\Lambda)$ in a strip of a width Λ

$$\Sigma^2(\Lambda) = \int_0^{<N>} \int_0^{<N>} d\lambda_1 d\lambda_2 R(\lambda_1 - \lambda_2) \tag{3.17}$$

In the ergodic regime (for $t \to \infty$) $P(t)$ is a constant, hence $K(t) \sim t$. The variance Σ^2 grows as logarithm of Λ, as expected in the random matrix theory. However, for times smaller than t_d (eigenvalues greater then Thouless energy), the return probability is given by the classical diffusion result

$$P(t) = \frac{V}{(4\pi Dt)^{d/2}} \tag{3.18}$$

Then $K(t) \sim t^{1-d/2}$, which corresponds, after Fourier transforming, to

$$R(s) \sim -\text{Re } \Delta^2 \sum_Q \frac{1}{(s\Delta + iDQ^2)^2} \tag{3.19}$$

For mesoscopic systems (e.g. for disordered electrons in $d = 3$ metallic grains), this is the seminal result obtained by Altshuler and Shklovskii [26], demonstrating, that the two-point correlation function comes diagrammatically from the two-diffuson exchange. The variance $\Sigma^2(E)$ reads then

$$\Sigma^2(E) \sim \left(\frac{E}{E_c}\right)^{d/2} \sim \left(\frac{L}{L(E)}\right)^d \tag{3.20}$$

The second equation comes from definition of the Thouless scale and from $E \sim t^{-1} = D/L^2(E)$. The *power behavior* reflects the fact, that in the diffusive regime diffusion of a particle with energy E is non-homogeneous, and takes place independently in the number of sub-blocks $(L/L(E))^d$ of the original block $V = L^d$.

The above mentioned crossover from universal regime (logarithmic behavior of the variance) to diffusive regime (power behavior $(E/E_c)^{3/2}$) was confirmed by [27] in numerical simulation of the metallic regime for Anderson model with 20^3 sites.

On the basis of the diffusive scenario presented in these notes, we expect similar behavior for QCD, but with $d = 4$ and $t \to \tau$. In our case, the square denominator in (3.19) follows from the exchange of two pions in the double ring diagram corresponding to density-density correlation function, reflecting on disconnected quark susceptibility. Note that we have obtained this result on the basic of spectral analysis, and not by using a standard diagrammatics of the chiral perturbation. The diffusive regime of Altschuler-Shklovskii in mesoscopic systems is equivalent to chiral perturbation theory. The effective models of diffusons (sigma models) correspond to sigma models of the pions, and supersymmetric formulation of the diffuson models by Efetov [28] resembles the family of so-called (partially) quenched chiral perturbation theories [29]. Finite volume QCD plays a role of a 4-dimensional *quantum dot.*

For QCD, the diffusive regime should show the power behavior $\Sigma^2(\lambda) \sim (\lambda/\Lambda_c)^2$, since the diffusion takes place in $d = 4$ ([14]). Recent lattice simulation [30] of the $SU(3)$ QCD has confirmed this scenario, demonstrating clearly the expected crossover from ergodic regime (logarithmic behavior) to diffusive regime (power-like behavior) predicted for the QCD. It is illuminating to compare the character of the crossover in metallic grains (e.g. Fig. 3 in [27]) to crossover regime of the disconnected quark susceptibility (e.g. Fig. 8 in [30]).

Let me mention another intriguing analogy. In the case of Hamiltonians invariant under the time reflection, one does not differentiate between the direction, in which the loop contributing to return probability is traversed. Hence the naive contribution to the return probability is doubled, which corresponds to additional diffuson-like contribution in the formula for return probability. This quantum interference of the identical orbits traversed in opposite directions is called in condensed matter *coherent backscattering (weak localization)*, and the collective diffuson like-excitation corresponding to this effect is called a *cooperon*, since the charges add during this interference. This phenomenon has also an analogy in QCD [14, 31]. In case of a real color group (e.g. $SU(2)$), we do not distinguish between the quark and antiquark, and on top of the usual pions (diffusons) we can form the collective states corresponding to *qq* pairs, so called baryonic pions - hence QCD-cooperons. The above mentioned lattice analysis [30] clearly identified the baryonic pions in quenched lattice simulation, in agreement with theoretical predictions based on the diffusive scenario of the QCD.

In the light of the above results, it is tempting to speculate, that the ideas borrowed from mesoscopic systems may have much broader domain of applications in the Euclidean QCD. Some preliminary analysis of the chiral disorder influenced by *external sources* seems to show, that this is indeed the case [31]. The effects on disorder of quark chemical potential resemble a complex electric Aharonov-Bohm effect, breaking particle-antiparticle symmetry, accumulating two flux lines and leading to rupture of "baryonic" quark-antiquark pair at $\mu = M_\pi/2$. This is observed in quenched lattice simulations and in random matrix models [32]. The phenomenon of persistence currents in disordered media finds an analogy to diffusion of light quarks in the presence of several Abelian Aharonov-Bohm fluxes. Some magnetic properties of chiral condensate can be explained by replacing the diffuson/cooperon trajectories by the four-dimensional Landau orbits. Low energy theorems obtained in [33] can be interpreted as *negative magnetoresistance* of the QCD vacuum (quark condensate grows with the magnetic field). Low temperatures correspond to the replacement of initial cube L^4 by asymmetric box βL^3, and the diffuson modes reproduce the lowest temperature corrections for the chiral

observables.
An extremely interesting case corresponds to the situation, when the critical temperature is reached. In the case of a second order phase transition, pion wavefunction, susceptibility and the condensate undergo the following scaling

$$Z_\pi = \Sigma^{\nu\eta/\beta}, \qquad \chi_\pi = \Sigma^{1-\delta}, \qquad \Sigma = m^{1/\delta} \tag{3.21}$$

where I used the standard notation for the critical exponents. These effects have an obvious quantitative effects on the diffuson (pion), modifying the return probability. In particular, the probability of return at the critical point tends to universal behavior $P(\tau) = (\tau_H/\tau)^{1/\delta}$. This has to be contrasted with the vacuum result $P(\tau) \rightarrow const.$ in the ergodic regime and vacuum result $P(\tau) \rightarrow (\tau_c/\tau)^2$ in the diffusive regime. Finally, let me point that the concept of the return probability at the Schwinger time allows to check scenarios beyond the rigorous theory of phase transitions. For metal-insulator type transition, $P(\tau) \sim \tau^{1-\eta/4}$. The multifractal exponent η estimated on the basis of chiral disorder is $\eta/4 = 0.057$ [31]. The smallness of η makes the return probability to look "diffusive" with $d = 2$. Finally, the asymmetry of the box may cause the appearance of two diffusion processes: "temporal" one and the "spatial" one. Since the spatial diffusion constant D_s is small in the high temperature phase (typically, $D_s/D_t \sim 0.1$ at $T \sim 180$ MeV), the asymmetry in conduction properties may cause a "percolation" from $d = 4$ to $d = 1$, with one-dimensional diffusive behavior $P(\tau) \sim 1/\sqrt{\tau}$. These speculations are interesting from the point of the view of lattice spectral analysis in the vicinity of the critical temperature, since the return probability in the Schwinger time can be explicitly expressed in terms of the eigenfunctions and eigenvalues of the Dirac operator.

THREEFOLD WAYS AND LATTICE SPECTRA

Let me finish these lectures reminding the classification of the global symmetries of the Dirac operator, hence the classification of the chiral universality classes.
For complex quarks, the Dirac operator has no symmetries. For real quarks (two colors), Dirac has an anti-unitary symmetry, $[\gamma_2\gamma_4\sigma_2 K, i\not{D}_F] = 0$, where K denotes complex conjugation. For adjoint quarks Dirac operator has another anti-unitary symmetry, $[\gamma_2\gamma_4 K, i\not{D}_A] = 0$, where subscripts F, A denote the fundamental and adjoint representations, respectively. This completes the list, forming the threefold way, how global symmetries can be realized for the Dirac Hamiltonian [34]:
I. For complex quarks with N_f flavors , the pattern of spontaneous breakdown of the chiral symmetry corresponds to $SU(N_f) \times SU(N_f) \rightarrow SU(N_f)$, and the random matrix realization is obtained by filling the matrix with complex numbers (chiral (denoted by χ) Gaussian unitary ensemble (χGue)). This is the case of the QCD, analyzed in these lectures.

II. For real quarks with N_f flavors, the pattern of spontaneous breakdown of the chiral symmetry corresponds to $SU(2N_f) \rightarrow Sp(2N_f)$, and the random matrix realization is obtained by filling the matrix with real numbers (chiral Gaussian orthogonal ensemble (χGoe)). This is the case of QCD with two colors.
III. For adjoint quarks with even N_f, the pattern of spontaneous breakdown of the chiral symmetry corresponds to $SU(N_f) \rightarrow O(2N_f)$, and the random matrix realization is obtained by filling the matrix with quaternions (chiral Gaussian symplectic ensemble ($\chi Gspe$)). This is the case of QCD with adjoint fermions, for any number of colors.
For each of these ensembles, one can obtain exact spectral formulae. Historically, the first one was (3.12), obtained in [23]. Technically, the complex case is the easiest one.
This triad is distorted on the lattice, due to the known problems of incompatibility of having chiral fermions on the discrete lattice with local action [2].
For example, the additional symmetries of Kogut-Susskind Hamiltonian cause, that for two colors the lattice universality class is $\chi Gspe$, and not, as expected from II, χGoe. Historically, the microscopic spectrum for this ensemble was the first one explicitly confirmed by lattice, in [35]. The formula (3.12) was confirmed successfully on the lattice ($N_f = 0$ case) only recently [36].
Nowadays, there exist an impressive list of theoretical predictions for several spectral (also higher-point) correlations and plethora of lattice evidence for various combinations of gauge groups, fermion representations, topological sectors, with quenched and unquenched and "double scaled" determinants etc. We refer the curious reader to original literature. In all cases, the agreement with RMM is excellent.
There is also a growing interest in moving toward the diffusive regime. First, the scaling (3.13) was confirmed by lattice study [37]. In the previous subsection we mentioned the measurements of the crossover between the ergodic and diffusive regime. They are also first measurements on the chiral parameters, motivated by the spectral properties of the Dirac operator.

FURTHER READING

An excellent review on chiral dynamics is [38], where the reader will find deeper justification of several statements made in Part 1. The original arguments on diffusion in QCD , based on rather compact paper [14] and the sequels [31], will become more obvious after reading condensed matter literature on this subject [7, 8]. The consequences of Banks-Casher relation are discussed in [6]. Basic informations on the phenomenology of the instanton vacuum could be found in [18]. Mehta' s book [39] is a classics in random matrix theory, for the reviews on the chiral ensembles we refer to [40, 41]. Readers interested in broader aspects of

[2] The original classification can be however reproduced using nonlocal actions.

RMM we refer to [42, 43]. I did not discuss here at all the schematic use of random models in QCD phenomenology, concentrating here on exact, quantitative predictions in the ergodic window. Sample applications could be found in a review [44] and references therein. Finally, a guidebook to advanced details (supersymmetric technique, replicas etc.) of the partially quenched perturbation techniques could be found in recent reviews [41, 45] and references therein.

4. Conclusions

In these lectures, I tried to demonstrate in *a physical way* how much the concepts of the spontaneous breakdown of the chiral symmetry share with the concepts of the disorder in mesoscopic systems. This should comply with the introductory character of these lectures. I perhaps omitted several important works to quote, and I apologize for this. The main idea of these lectures was to introduce a non-expert to the fascinating field of chiral disorder in the QCD. I therefore sacrificed several advanced technical details and mathematical aspects, which are often necessary to prove the analogies presented here in a strict way. Luckily, there exist reviews, some quoted above, which exhaustively can guide the reader through the technicalities of random matrices and various versions of chiral perturbation theories adapted for finite volume systems. They also provide the more complete bibliography.

I pointed, that the ergodic (universal) domain, is to large extent understood and documented in an impressive way. Universality of the ergodic regime helped lattice QCD to check to what extend chiral properties (including the global features e.g. zero modes) of the quarks are reflected by simulation. They also allowed to quantify the discrepancies, in the light of the exact analytical predictions of the RMM. I tried to emphasize, how challenging is to look at the spectra of the diffusive regime. The lattice analogs of the diffuson theories open a way to systematic extraction of the effective parameters of the chiral Lagrangian, following step by step the Weinberg expansion. (The first example is a pion decay constant). It is also possible to study the dependence of these constants of the chiral Lagrangian on the external parameters and it is important to see what happens to them when the critical regime is approached. It is also plausible, that the spectral analysis in the diffusive regime may serve as an important tool for studying the nonperturbative nature of the gluon fields and their quantitative rearrangements in the vicinity of the phase transition, hopefully shading some more light on the nature of the lumps of the "color dirt".

A more general message, which the reader should infer from these lectures and from several other lectures at this school is, how broadly and how fruitfully the theory of strong interactions can borrow from the concepts and ideas of the condensed matter theory.

Acknowledgements

The philosophy of these lectures (if one is permitted to use such an expression) is greatly influenced by my present collaborators on the chiral disorder: Romuald Janik, Jurek Jurkiewicz, Gabor Papp and Ismail Zahed. Some early studies of chiral disorder and on random matrices in instanton models, which triggered my interest in this subject, were done in collaboration with Reinhard Alkofer, Jac Verbaarschot and Ismail Zahed. It is a pleasure to notice, that these and other early attempts [46] have come logically full circle, with so powerful applications and predictions.
I would like to thank Jean-Paul Blaizot, Edmond Iancu, Andrei Leonidov and Larry Mc Lerran for an invitation to deliver these lectures and for creating an imaginative and inspiring atmosphere of the 2001 Cargése Summer School.
I would like to thank also Wolfgang Noerenberg, Hans Feldmeier, Bengt Friman, Joern Knoll, Volker Koch and Matthias Lutz for discussions and for providing the excellent opportunity to complete these lectures during my visit at GSI.

References

1. C. Vafa and E. Witten, Nucl. Phys. **B234** (1984) 173.
2. J. Goldstone, Nuovo Cimento **19** (1961) 154.
3. G.E. Brown and M. Rho, e-print hep-ph/0103102.
4. M. Gell-Mann, R.J. Oakes and B. Renner, Phys. Rev. **175** (1968) 2195.
5. T. Banks and A. Casher, Nucl. Phys. **B169** (1980) 103; see also E. Marinari, G. Parisi and C. Rebbi, Phys. Rev. Lett. **47** (1981) 1795.
6. H. Leutwyler and A. Smilga, Phys. Rev. **D46** (1992) 5607.
7. Y. Imry, *Introduction to mezoscopic physics*, Oxford Univ. Press (1997).
8. G. Montambaux, in Proceedings *"Quantum Fluctuations"*, Les Houches, Session LXIII, eds. E. Giacobino, S. Reynaud and J. Zinn-Justin, Elsevier Science (1997).
9. D.J. Thouless, Phys. Rep. **13** (1974) 93.
10. P. Anderson, Phys. Rev. **109** (1958) 1492.
11. D. Diakonov and V. Petrov, Nucl. Phys. **B272** (1986) 457.
12. E. Shuryak, Phys. Lett. **B193** (1987) 319.
13. S. Weinberg, Physica **96A** (1979) 327.
14. R. Janik, M.A. Nowak, G. Papp and I. Zahed, Phys. Rev. Lett. **81** (1998) 264.
15. J. Osborn and J.J.M. Verbaarschot, Nucl. Phys. **B525** (1998) 738.
16. J. Gasser and H. Leutwyler, Ann. Phys. **158** (1984) 142; Phys. Lett. **B184** (1987) 83; Phys. Lett. **B188** (1987) 477; Nucl. Phys. **B307** (1988) 763.
17. see also T. Jolicoeur and A. Morel, Nucl. Phys. **B262** (1985) 627.
18. see e.g. D.Diakonov, in Proceedings *"Selected Topics in QCD"*, ed. A. Di Giacomo and D. Diakonov, IOS Pr., Amsterdam (1996); T. Schafer and E. Shuryak, Rev. Mod. Phys. **70** (1998) 323; M.A. Nowak, M. Rho and I. Zahed, *Chiral Nuclear Dynamics*, World Scientific, Singapore (1996).
19. see e.g. A. Di Giacomo, H. Dosch, V.I. Shevchenko and Yu. A. Simonov, e-print hep-ph/0007223.

20. M.A. Nowak, J.J.M. Verbaarschot and I. Zahed, Nucl. Phys. **B324** (1989) 1; Phys. Lett. B217 (1989) 157; Phys. Lett. **B228** (1989) 251; R. Alkofer, M.A. Nowak, J.J.M. Verbaarschot and I. Zahed, Phys. Lett. **B233** (1989) 205.
21. E. Shuryak and J.J.M. Verbaarschot, Nucl. Phys. **A560** (1993) 306.
22. P. Damgaard, J. Osborn, D. Toublan and J.J.M. Verbaarschot, Nucl. Phys. **B547** (1999) 305.
23. J.J.M. Verbaarschot and I. Zahed, Phys. Rev. Lett. **70** (1993) 3852.
24. G. Akemann, P.H. Damgaard, U. Magnea and S. Nishigaki, Nucl. Phys. **B487** (1997) 721.
25. N. Argaman, Y. Imry and U. Smilansky, Phys. Rev. **B47** (1993) 4440.
26. B.L. Altschuler and B. Shklovskii, Sov. Phys. JETP **64** (1986) 127.
27. D. Braun and G. Montambaux, Phys. Rev. **52** (1995) 13903.
28. see e.g. K.B. Efetov, *Supersymeetry in disorder and chaos*, Cambridge Univ. Press, Cambridge (1997).
29. see e.g. M. Golterman, Acta Phys. Pol. **B25** (1994) 1731.
30. M. Goekeler et al., e-print hep-lat/0105011.
31. R.A. Janik, M.A. Nowak, G. Papp and I. Zahed, Phys. Lett. **B440** (1998) 123; Phys. Lett. **B442** (1998) 300; Nucl. Phys. Proc. Suppl. **83** (2000) 977.
32. M. Stephanov, Phys. Rev. Lett. **76** (1996) 4472.
33. I.A. Shushpanov and A. Smilga, Phys. Lett. **B402** (1997) 351.
34. J.J.M. Verbaarschot, Phys. Rev. Lett. **72** (1994) 2531.
35. M.E. Berbenni-Bitsch et al., Phys. Rev. Lett. **80** (1998) 1146.
36. P. Damgaard, U. M. Heller and A. Krasnitz, Phys. Lett. **B445** (1999) 366.
37. M.E. Berbenni-Bitsch et al., Phys. Lett. **B438** (1998) 14.
38. H. Leutwyler, e-print hep-ph/0008124.
39. M.L. Mehta, *Random Matrices*, Acad. Press, NY (1991).
40. R.A. Janik, M.A. Nowak, G. Papp and I. Zahed, Acta Phys. Pol. **B29** (1998) 3957.
41. J.J.M. Verbaarschot and T. Wettig, Ann. Rev. Nucl. Part. Sci. **50** (2000) 343.
42. T. Guhr, A. Müller-Groeling and H.A. Weidenmüller, Phys. Rep. **299**, (1998) 189.
43. P. Di Francesco, P. Ginsparg and J. Zinn-Justin, Phys. Rep. **254** (1995) 1.
44. R.A. Janik, M.A. Nowak, G. Papp and I. Zahed, Acta Phys. Pol. **B28** (1997) 2949.
45. P.H. Damgaard, eprint hep-lat/0110192 and references therein.
46. see [11, 12, 20], Y.A. Simonov, Phys. Rev. **D43** (1991) 3534.

HEAVY ION PHYSICS AT THE LHC

J. SCHUKRAFT (juergen.schukraft@cern.ch)
CERN, CH-1211, Geneva 23, Switzerland

Abstract. The field of ultra-relativistic heavy ion physics, which started some 15 years ago at the Brookhaven AGS and the CERN SPS with fixed target experiments, is entering today a new era with the recent start-up of the Relativistic Heavy Ion Collider RHIC and preparations well under way for a new large heavy ion experiment at the Large Hadron Collider LHC. At this crossroads, the article will give a summary of the experimental program and our current view of heavy ion physics at the LHC, concentrating in particular on physics topics that are different or unique compared to current facilities.

1. Introduction

The aim of high-energy heavy-ion physics is the study of strongly interacting matter at extreme energy densities. Statistical QCD predicts that, at sufficiently high density, there will be a transition from hadronic matter to a plasma of deconfined quarks and gluons — a transition which in the early universe took place in the inverse direction some 10^{-5} s after the Big Bang and which might play a role still today in the core of collapsing neutron stars. The study of the phase diagram of nuclear matter (see Fig. 1), utilising methods and concepts from both nuclear and high-energy physics, constitutes a new and interdisciplinary approach in investigating matter and its interactions. In high-energy physics, interactions are derived from first principles (gauge theories), and the matter concerned consists mostly of single particles (hadrons/quarks). In contrast, on nuclear physics scales the strong interaction is shielded and can, therefore, to date only be described in effective theories, whereas matter consists of extended systems with collective features. Combining the *elementary-interaction* aspect of high-energy physics with the *macroscopic-matter* aspect of nuclear physics, the subject of heavy-ion collisions is the study of bulk matter consisting of strongly interacting particles (hadrons/partons), i.e. QCD thermodynamics.
The study of the QGP is of interest to explore and test QCD on its natural scale (Λ_{QCD}) and addresses the fundamental questions of confinement and chiral-symmetry breaking, which are connected to the existence and properties of the quark-gluon plasma. Moreover, it is of general relevance in understanding the dynamical

J.-P. Blaizot and E. Iancu (eds.), QCD Perspectives on Hot and Dense Matter, 461–473.

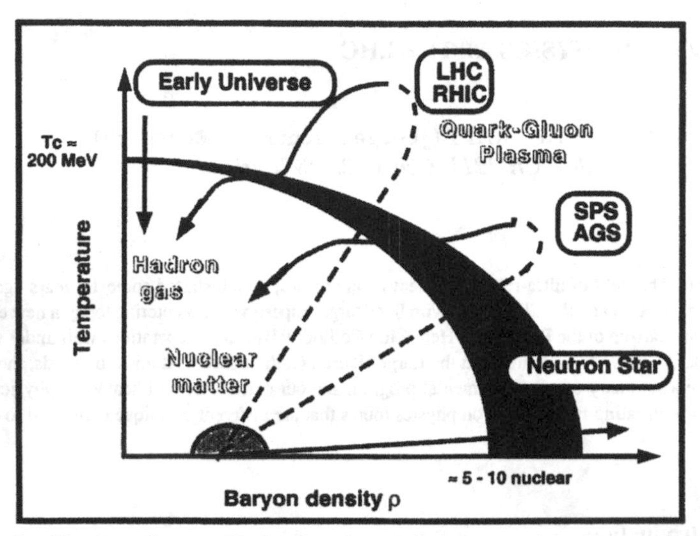

Figure 1. The phase diagram of hadronic matter and the hadron gas - quark-gluon plasma phase transition.

nature of phase transitions involving elementary quantum fields, as the QCD phase transition is the only one accessible to laboratory experiments.

2. Current status and results

The study of ultra-relativistic heavy-ion collisions is a rather new, but rapidly evolving field. After the pioneering experiments at the BEVALAC and in DUBNA with relativistic heavy ions ($E/m \approx 1$), the first experiments started in 1986 with light ions almost simultaneously in Brookhaven (AGS) and at CERN (SPS). Really heavy ions (A $\approx$ 200) have been available in the AGS since the end of 1992 and at the SPS since the end of 1994.

The year 2000 was an extremely important and fruitful one for ultra-relativistic heavy ion physics. It started of with an appraisal of the CERN SPS Pb beam results [1] which concluded that 'compelling evidence has been found for a new state of matter' featuring many of the characteristics expected for a Quark–Gluon Plasma (QGP). Later that year the RHIC collider and its four experiments at BNL began operation with Au–Au collisions at up to 130 GeV/nucleon. Already this first short run was a success beyond even the most optimistic expectations, with physics results appearing in print and essentially flooding the relevant conferences within months. It is therefore timely to ask what, if anything, the LHC heavy ion program can contribute beyond existing facilities to the study of strongly interacting matter under extreme conditions.

2.1. INITIAL CONDITIONS AND GLOBAL FEATURES

The predictions of lattice QCD are rather firm in that a transition to the QGP should exist in the vicinity of a critical temperature T_c of $\approx$ 150 - 200 MeV (whether the transition is of first order, second order, or only 'rapid' is still a matter of debate). However, whether the QGP is actually created in heavy-ion collisions at current energies is a different question and will depend on the dynamics of the reactions and in particular on the initial conditions of the system shortly after the collision. In order to reach the QGP, or even only to use macroscopic concepts (such as 'phase transition') and the language and variables of thermodynamics (such as 'temperature' or 'density'), the system has to be *extended* — i.e. its dimensions ought to be much larger than the typical scale of strong interactions — it has to be in (or near) *equilibrium* — i.e. its lifetime has to be larger than the relevant relaxation times — and the *energy density* ϵ has to exceed the critical threshold for QGP formation. This threshold is predicted by lattice QCD to be of the order of 1 – 3 GeV/fm^3, equivalent to a temperature T_c of 150 – 200 MeV or a baryon density ρ_c of 5 to 10 times normal nuclear matter density (see Fig. 1). Present results from the ongoing fixed-target program indicate that the initial conditions realized in these reactions could indeed be favourable for QGP formation. In head-on central collisions, hundreds of particles are produced per unit of rapidity, the system expands to a size of the order of 1000 fm^3 (as measured by particle interferometry), and initial energy densities are estimated to exceed 2 GeV/fm^3. However, the expansion is also extremely fast, with an estimated total lifetime of only a few fm/c from the first instance of the collision until the final freeze-out of hadrons.
While these results show that we are certainly *close* to the requirements listed above for QGP formation, they are by no means *sufficient*. In particular the energy density estimates are inversely proportional to the assumed 'formation time', i.e. the time needed to reach thermal equilibrium, and might well be smaller (or bigger?) by a factor of the order of two. Also, the lifetime of the system seems marginal, and even if a QGP is formed it might simply not live long enough for its signals to clearly stand out from the background created in later, hadronic phases of the evolution. The existence of a QGP phase can only be settled experimentally by searching for direct and specific signals.

2.2. SPECIFIC SIGNALS

Significant progress has been achieved over the last few years in addressing the main questions at the heart of the quest for the QGP: Are there experimental indications for *equilibrated hadronic matter, chiral symmetry restoration*, and *deconfinement*? Results from both AGS and SPS have produced puzzling results which strongly hint at a picture of high energy nuclear reactions almost too good to be true: i) a premordial phase of deconfined partons – the QGP ? – responsible for quarkonium suppression (deconfinement), followed by ii) a transition regime with gradual onset of chiral symmetry breaking, leading to changes in the properties of light hadrons (as inferred from dilepton spectra) , concluded by iii) a gas of hadronic matter governed by the simple laws of thermodynamics.

2.3. FUTURE FIXED TARGET PROGRAM

Given the recent exciting developments, the future directions are perfectly clear. The SPS fixed target program has entered an extremely productive phase and it is now being brought to its full potential. With the exception of the Hyperon and the low mass lepton pair measurements, statistics is, in general, not a problem. A run at the lowest possible SPS energy, around 40 GeV/nucleon and still being analyzed, has increased the baryon density, possibly close to its maximum value. Signals related to chiral symmetry restoration, in particular the low mass lepton pairs, will in general be rather sensitive to baryon density. The low energy run can also make contact with the AGS regime and will allow the CERN experiments, which are quite distinct in their capabilities from the AGS detectors, to compare with and complement the program at lower energies. The SPS program has been extended beyond the year 2001 to address in particular charm production and the intermediate mass lepton pair excess with the new NA60 experiment.

3. Heavy ion physics of the 21st century

With the colliders RHIC ($\sqrt{s}$ = 200 GeV/n) and LHC ($\sqrt{s}$ = 5.5 TeV/n) coming into operation in 2000 and 2006, respectively, the available energy in the centre-of-mass will have increased by almost five orders of magnitude within 20 years. This unprecedented pace was made possible only by (re)using accelerators, and to some extent even detectors, built over a much longer time scale for use in high-energy physics. The following sections will summarise the physics and experiments to come in these latest (and possible last) heavy-ion machines.

3.1. INITIAL CONDITIONS AND GLOBAL EVENT FEATURES

In order to get some qualitative feeling about the changes in global event features to be expected when going up in energy by over two orders of magnitude from SPS via RHIC to LHC, some of the relevant parameters and predictions are listed in Table 1 and described below:
In elementary proton-proton reactions, both the charged particle multiplicity dN_{ch}/dy [2] and the average transverse momentum $< p_t >$ [3] grow only slowly (logarithmically) with $\sqrt{s}$. This would lead only to a modest growth in dN_{ch}/dy (SPS:RHIC:LHC = 0.75:1:2.1), $< p_t >$ (0.92:1:1.35) and energy density (roughly the product $dN_{ch}/dy \cdot p_t$). However, also the scaling when going from pp to central

TABLE II.

Table1: Qualitative extrapolation for particle production from pp to AA and estimated global features in AA. For details see text.

	SPS	RHIC	LHC	Pb–Pb	SPS	RHIC	LHC
$\sqrt{s}/A$ (GeV)	17	200	5500	dN_{ch}/dy	400	800	2-8x10^3
pp dN_{ch}/dy	1.8	2.4	5	τ_0^{QGP} (fm/c)	1	0.2	0.1
pp p_t (GeV)	0.36	0.39	0.53	ϵ (GeV/fm^3)	3	35	500
AA dN/dy	$\propto N_{part}$	?	$\propto N_{col}$	τ_{QGP} (fm/c)	$\lesssim$2	2-4	$\gtrsim$10
partons in p	4	10	30	τ_{fo} (fm/c)	10	20-30	30-40
shadowing	1	0.8	0.5	V_{loc}(fm^3)	few 10^3	few 10^4	few 10^5
dN_p/dy [9]		1400	5100	dE_t/dy	(GeV)	2000	17000
dN_p/dy [10]		80-120	1.5-5x10^3	dE_t/dy	(GeV)	200-300	5-15x10^3

AA reactions changes drastically over this energy range. While at low cms energy particle production is, to a very good approximation, proportional to the number of participants N_{part}, i.e. scales like A in central AA collisions, hard processes, which dominate at high energies, scale proportional to the number of collisions N_{col} ($\propto A^{4/3}$). Compared to SPS, this leads to an additional enhancement factor for dN/dy in central AA reactions of about three (assuming equal contributions of soft and hard processes) or six (assuming hard dominance) for RHIC and LHC, respectively. In addition, as structure functions rise strongly at small x, the number of partons available for collisions increases a function of energy. As an illustration, the number of partons inside the proton above 2 GeV is listed in the table as row 'partons p' [4]. While some of this increase is reduced by shadowing in heavy nuclei, the net effect for particle production, roughly proportional to the square of parton number times shadowing reduction, is still very sizeable: A heavy nucleus contains about twice (four times) as many effective partons at RHIC (LHC) than at the SPS. The fact that nuclei get 'denser' at high energy is a great bonus which was not anticipated only a few years ago!

The right hand side of Table 1 summarises a selection of different measurements and estimates for global event features in Pb-Pb collisions (for a similar table, including some original references, see [5]). While dN_{ch}/dy has been measured at the SPS and can be extrapolated with confidence from the current RHIC results [6] to full RHIC energy, the predictions for LHC vary wildly between some 2000 and up to 8000 charged particles per unit rapidity. The main reasons for this spread are connected e.g. with uncertainties in gluon shadowing, final state parton saturation [7] and jet-quenching [8], all of which are expected to be rather different at RHIC and LHC. Therefore we might not be able to narrow down significantly this large range before LHC start-up even if more detailed results will become available from RHIC. Both the quantum mechanically determined 'formation time' of harder partons ($\propto 1/p_t$) and the increased interaction rate at high parton density decrease the thermalization time τ_0^{QGP} of the dense medium, leading to extreme energy densities ($\varepsilon(\tau_0) \propto 1/\tau_0$) and temperatures at very early times during the collision at the LHC. As the system has to expand and cool to freeze-out conditions which are probably similar at all energies, QGP lifetime τ_{QGP}, freeze-out time τ_{fo} and final (local) freeze-out volume V_{loc} likewise increase substantially. In particular the strong increase in QGP lifetime, both in absolute terms, in units of relaxation times (which in general decrease in denser systems), as well as compared to the hadronic phase, might be particularly relevant as it should enhance the relative contribution of signals generated in and sensitive to the QGP phase.

Finally, the last two rows show two representative pQCD calculations of initial parton density dN_p/dy, dominantly gluons, and transverse energy production dE_t/dy [9] [10]. While they differ in detail in their treatment of shadowing, parton saturation effects and p_t cut-off, they qualitatively agree that densities should increase by about an order of magnitude from RHIC to LHC, much stronger than the logarithmic rise seen in pp reactions.

In summary, by increasing the energy from soft to hard dominated regimes we gain independently on several fronts: More partons are available, their inelastic interactions get both more effective (already in pp) and in addition more frequent (in nuclei), and the thermalized high temperature phase is established more rapidly and lasts longer.

3.2. QUALITATIVE CHANGES AT LHC

Besides *better* initial conditions, a number of qualitative changes are expected at high energy leading to a very *different* system and observables which are either *new* or can be studied with *greater precision.*

Parton saturation: At LHC, and to a smaller extend at RHIC, the density of low$-x$ partons both in the incoming nuclei and in the final state after scattering are so high that the phase space for low momentum partons comes close to saturation and gluon merging becomes important to an extend that the reactions are sometimes referred to as a collision between two 'gluon walls' [7]. In the initial state, structure functions are modified by shadowing, and also in the final state the gluon density is reduced, limiting in a self-consistent way the infrared divergence of pQCD cross sections. Due to the large occupation numbers, the scattering process might be described by classical field theory, a new regime that becomes accessible for the first time with heavy ions at LHC. This 'gluon saturation' is expected to set in around 2 GeV at LHC [9] where pQCD might be applicable and theoretical concepts can be tested.

Baryon density: The incoming nuclei are slowed down during the collisions and the net baryon number is distributed over typically 3-5 units in rapidity. This leads to a high baryon density at SPS and below, whereas it is not sufficient to

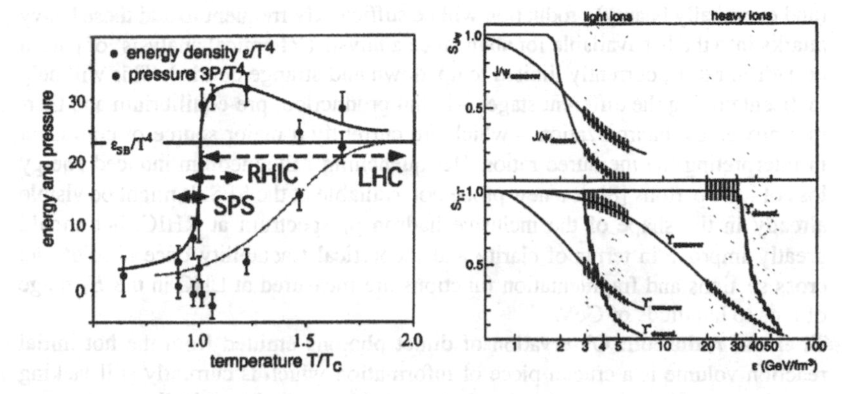

Figure 2 Energy density and pressure as calculated on the lattice as a function of temperature T in units of the critical temperature T_C. The ideal gas Stefan-Boltzman limit is labeled ε_{SB}. The temperature range accessible at SPS, RHIC and LHC are indicated with arrows

Figure 3. Energy density dependency of charmonium and bottonium survival probabilities as expected from melting in a deconfined medium and from the comover model. The precision of the measurement in ALICE after one month of running is shown as 1σ error bands.

populate the central region at LHC, which will have a vanishing baryochemical potential μ_b comparable to the early universe. Therefore very different regions of the phase diagram will be explored at low energy ('compression') and high energy ('heating'). At RHIC, first indications are that a baryon excess persists, which is however small and probably not very relevant ($\mu_b/T \ll 1$).

Energy density: Energy density and pressure, as calculated on the lattice, are shown in Figure 2, together with the estimated temperature range accessible at different machines. The SPS is certainly very close to, and likely already reaches beyond the hadronic phase; RHIC is ideally suited to explore the phase boundary in great detail; but only LHC will reach deep into the QGP and could approach the Stefan-Boltzman limit of an 'ideal gas' of QCD quanta. The region between T_c and the SB limit is of particular interest in order to study the interactions responsible for the deviation from the ideal parton gas case.

Hard processes: Hard scattering with large momentum transfer becomes significant at RHIC (measurable up to about 20 GeV) and copious at LHC (up to several hundred GeV)[11]. These hard probes, in particular high p_t jets and heavy quark production, are formed at very early times and therefore will test the surrounding dense medium. The study of the classical deconfinement signal, i.e. J/Ψ suppression, can be extended at LHC to the Υ family. A full 'spectral analysis' of heavy quarkonia states [12] will be much more difficult at RHIC, as the cross sections are lower by at least an order of magnitude [10] and the initial temperature is likely to be below the melting point of the tightly bound Υ state (Figure 3). Open charm (and eventually beauty) production will be sufficiently frequent to add these heavy quarks into the list available for abundance analysis ('chemical analysis' of parton or hadron ratios, currently limited to up, down and strange quarks). This will help in disentangling the different stages – initial production, pre-equilibrium and thermal processes, hadronization – which are currently a major source of confusion in interpreting the measured ratios. 'Jet-quenching', the medium induced energy loss of hard partons [8], is a new probe not available at the SPS. It might be visible already in the shape of the inclusive hadron p_t spectrum at RHIC, but should greatly improve in terms of clarity and theoretical tractability once absolute jet cross sections and fragmentation functions are measured at LHC in the E_t range of tens to hundreds of GeV.

Thermal radiation: Observation of direct photons emitted from the hot initial reaction volume is a crucial piece of information which is currently still lacking or at least ambiguous. As the signal strength improves dramatically with energy density and plasma life-time, it might be observable at the colliders, in particular if the background from high p_t π^0's is reduced by thermalization and jet-quenching.

Event-by-Event physics: A unique feature of heavy ion collisions is the possibility to measure a number of observables on an even-by-event basis. Non-statistical fluctuations in these observables are of interest [13] because they are, in general, associated with critical phenomena in the vicinity of a phase transition, they can be related to thermal quantities (eg the heat capacity), or might be indicative of 'anomalous' events as suggested by cosmic ray observations (Centauro's etc.). Currently, with clear guidance and precise predictions from theory missing, this subject is essentially data driven. However, as the accuracy will in general increase inversely with the number of observed particles ($\propto 1/\sqrt{N}$), EbE physics will become a precision instrument at high particle multiplicity and any observation of non-trivial fluctuating would be a clean 'smoking gun'. Even in the absence of anomalous fluctuations, some EbE measurements will be extremely useful to study the correlation with other inclusive observables, like the azimuthal dependence of jet-quenching with respect to the event plane which can be determined via the elliptic flow.

4. Heavy ions in the LHC

4.1. THE LHC MACHINE

The LHC will start to collide protons at $\sqrt{s} = 14TeV$ in early 2006 and will provide the first heavy ion collisions (Pb-Pb) towards the end of its first year of operation at a total cms energy of 1148 TeV ($\sqrt{s} = 5.5$ TeV per nucleon for Pb-Pb). The sharing between proton and ion runs is expected to be similar to the one used in the past at the SPS, i.e. a long proton run (10^7 s effective time) followed by a few weeks (10^6 s) of heavy ion or proton-nucleus collisions. The ion luminosity is limited for heavy systems by the short beam life-time and energy deposition in the supra-conducting machine magnets (quench protection), and for light ions by the injector chain. It reaches values between $3 \cdot 10^{31} cm^{-2}s^{-1}$ for light ions (O^{16}) and $10^{27} cm^{-2}s^{-1}$ for Pb ions. Proton- or deuteron-nucleus collisions are feasible and foreseen, however only at equal magnetic rigidity per beam as the bending field is identical in both rings ('two-in-one' magnet design of the LHC). Further information on heavy ion operation of the LHC can be found in ref. [14].

4.2. HEAVY ION PHYSICS WITH ATLAS AND CMS

While dedicated and optimized for the high-energy frontier in particle physics, both ATLAS and CMS have expressed to various degrees an interest in making use of the LHC heavy ion beam. While discussions within ATLAS are still at a very preliminary stage, CMS has a long declared interest and should be able in particular to investigate quarkonium production and very hard processes (jets, Z production) at or above a scale of 100 GeV [15].

4.3. THE DEDICATED HEAVY ION EXPERIMENT ALICE

ALICE is a general-purpose heavy-ion detector designed to study the physics of strongly interacting matter and the quark-gluon plasma in nucleus-nucleus collisions at the LHC. It currently includes more than 900 physicists - both from nuclear and high energy physics - from about 70 institutions in 25 countries.
The detector is designed to cope with the highest particle multiplicities anticipated for Pb-Pb reactions (dN/dy $\approx$ 8000) and it will be operational at the start-up of the LHC. In addition to heavy systems, the ALICE Collaboration will study collisions of lower-mass ions, which are a means of varying the energy density, and protons (both pp and p-nucleus), which provide reference data for the nucleus-nucleus collisions.
ALICE (Fig. 4) consists of a central part, which measures hadrons, electrons and photons, and a forward spectrometer to measure muons. The central part, which covers polar angles from 45° to 135° over the full azimuth, is embedded in the large L3 solenoidal magnet. It consists of an inner tracking system (ITS) of high-resolution silicon tracking detectors, a cylindrical TPC, three particle identification arrays of time-of-flight (TOF), Ring Imaging Cerenkov (HMPID) and Transition Radiation (TRD) counters and a single-arm electromagnetic calorimeter (PHOS). The forward muon arm (2°-9°) consists of a complex arrangement of absorbers, a large dipole magnet, and fourteen stations of tracking and triggering chambers. Several smaller detectors (ZDC, PMD, FMD, CASTOR, T0) are located at very forward angles.
Magnet: The optimal choice for ALICE is a large solenoid with a weak field (<0.5 T) allowing full tracking and particle identification inside the magnet. The available space has to be sufficiently large to accommodate the PHOS, which must

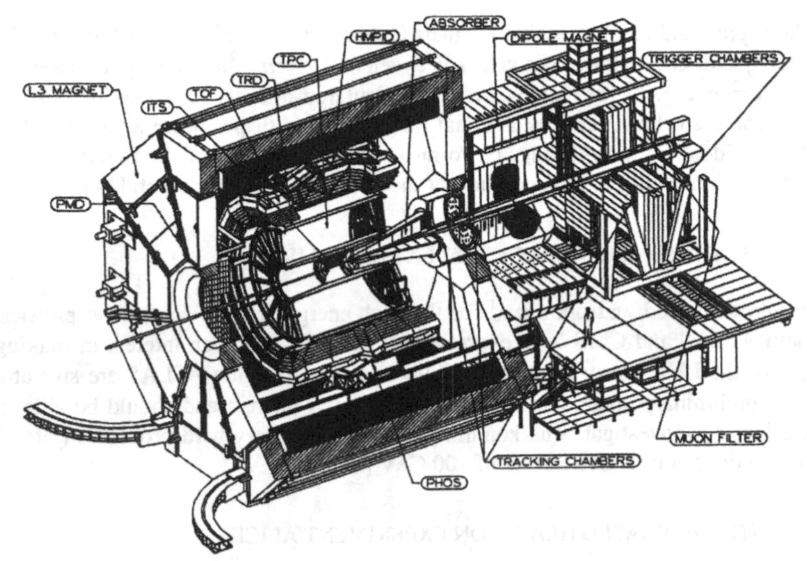

Figure 4. Artists view of the ALCE detector.

be placed at a distance of $\approx$ 5 m from the vertex, because of the large particle density. The magnet of the L3 experiment fulfils all these requirements.

Inner Tracking System: The basic functions of the inner tracker - secondary vertex reconstruction of hyperon and charm decays, particle identification and tracking of low-momentum particles, and improvement of the momentum resolution - are achieved with six barrels of high-resolution detectors. Because of the high particle density, the innermost four layers need to be truly two-dimensional devices, i.e. silicon pixel and silicon drift detectors. The outer layers, at r $\approx$ 50 cm, will be equipped with double-sided silicon micro-strip detectors. Four of the layers will have analog readout for independent particle identification via dE/dx in the non-relativistic region, which will give the inner tracking system a stand-alone capability as a low p_t particle spectrometer.

Time Projection Chamber: The need for efficient and robust tracking of up to 12000 charged particles within the acceptance has led to the choice of a TPC as the main tracking system. The inner radius of the TPC (r $\approx$ 90 cm) is given by the maximum acceptable hit density, the outer radius of 250 cm by the length required for a dE/dx resolution of $<10\%$. The design of the readout chambers and electronics, as well as the choice of the operating gas, is optimised for good double-track resolution.

Particle Identification System: Particle identification over a large part of the phase space and for many different particles is an important design feature of ALICE. A large Time-of-Flight array (> 100m^2) at a radius of about 3.5 m is made with novel multigap resistive plate chambers (MRPC) with an intrinsic time resolution of < 100 ps. The proximity focusing RICH detector HMPID, of smaller acceptance and at larger radii, is optimised for the detection of high p_t particles and will extend the accessible momentum range for inclusive particle spectra into the semi-hard region. The six layer Transition Radiation Detector (TRD) will identify electrons with momenta above 1 GeV/c to study quarkonia suppression and heavy quark production (charm, beauty) in the central acceptance.

Photon Spectrometer: The electromagnetic calorimeter will be located below the interaction region at 4.6 m from the vertex, and covers 8 m^2 with 17k channels of scintillating $PbWO_4$ crystals to measure direct photons and high p_t neutral mesons. These very dense crystals are needed to cope with the large particle density, and to have sufficient light-output to allow readout with silicon photodiodes.

Forward Muon Arm: The forward muon arm is designed in order to cover the complete spectrum of heavy quark resonances, i.e., J/Ψ, Ψ', Y, Y', Y" (Figure 3). It will measure the decay of these resonances, both in proton-proton and in heavy-ion collisions. The angular acceptance of the muon spectrometer is from 2° to 9° ($\eta = 2.5 - 4$). Its mass resolution will be around 100 MeV at 10 GeV, sufficient to separate all resonance states. It consists of a composite absorber, made with layers of both high- and low-Z materials, starting 90 cm from the vertex, a large dipole magnet with a 3 Tm field integral placed outside the L3 magnet, and 10 planes of thin, high-granularity tracking stations (Cathode Strip and Cathode Pad Chambers). A second absorber at the end of the spectrometer and four more detector planes (RPC's) are used for muon identification and triggering. The spectrometer is shielded throughout its length by a dense absorber tube, of about 60 cm outer diameter, which surrounds the beam pipe.

Forward Detectors: ALICE uses a number of smaller detector systems (ZDC, PMD, FMD, CASTOR, T0) located at forward angles to define and trigger on global event characteristics. Four small and very dense calorimeters (Zero Degree Calorimeters, ZDC) are located about 100 m inside the machine tunnels on both sides of the interaction to define the impact parameter of the collision. A similar calorimeter (CASTOR), located closer to the interaction region on side of the experiment, will measure electromagnetic and hadronic transverse energy at large rapidity. The Forward Multiplicity Detector (FMD) measures charge particle production over a large fraction of phase space ($|\eta| < 4$). The Photon Multiplicity Detector (PMD) will search for non statistical fluctuations in the ratio of photons to charged particles, measure collective flow and transverse energy of neutral particles, and in addition determine the reaction plane. The T0 detector will measure the event time with great precision.

Trigger, Data acquisition and Offline: The ALICE trigger is foreseen to work on five levels, starting with a fast minimum bias interaction pre-trigger (issued after $< 1\mu s$) to strobe some of the front-end electronics and ending with a high level online computing farm of several hundred PC's (L3 trigger) which is intended

for further event selection (for example on high p_t jets or high mass lepton pairs) as well as online pre-tracking and event compression. Several detectors provide input to the different trigger levels to select e.g. for centrality, high p_t electrons, muons, or photons.
The relatively short heavy-ion running period and the very large event sizes (up to 80 Mbyte even after zero suppression) determine the main features of the DAQ. In order to collect a sufficient number of events for physics analysis, the DAQ system has to be designed with a very large bandwidth up to 1.25 GByte/s on mass-storage. The DAQ architecture is based on a network of high-speed links linking all the data sources and the data destinations through a switch. This architecture provides the required flexibility and scalability to run in very different modes.
A new Off-line framework (ALIRoot) has been developed since 1998 based on all new C++ code and the OO paradigm. The ROOT framework was adopted as a base for this development, integrating currently the GEANT 3 and later also the GEANT 4 simulation package. At the moment a complete OO simulation of ALICE exists and the OO reconstruction code is being developed in this framework.

5. Summary

RHIC and LHC, building on the success of the heavy ion fixed target programmes, will provide a unique opportunity for exploring the physics of QCD matter in a qualitatively very different region of extremely high energy density. RHIC and its four major experiments (STAR, PHENIX, BRAHMS and PHOBOS) are making the first step, starting in 2000 with Au+Au collisions at an energy an order of magnitude above what is currently available at the SPS.
Some 5 years from now, a new regime of very high energy density but low baryon density will be accessible with heavy ion collisions at the LHC. At some 30 times RHIC design energy, the step in energy from RHIC to LHC will be enormous, in fact larger than the one going from SPS to RHIC. LHC will reach, and even extend, the energy range probed by cosmic ray nucleus–nucleus collisions. Extrapolating from present results, all parameters relevant to the formation of the Quark–Gluon Plasma (QGP) will be more favourable: the energy density, the size and lifetime of the system, as well as relaxation times should all improve by a large factor, typically by an order of magnitude compared to SPS and even RICH. It should then be possible to obtain energy densities far above the deconfinement threshold, and to probe the QGP in its asymptotically free 'ideal gas' form. Unlike at lower energies, the central rapidity region will have a nearly vanishing baryon number density, similar to the state of the early universe. Reactions will be dominated in the early pre-equilibrium stage by a very dense system of semi-hard partons ('mini-jets'), which would lead to rapid thermalization and extremely high initial temperatures. Hard probes (heavy quark production, jets, even weak

bosons) will be copiously produced well into the 100 GeV mass and momentum range providing a new and, with pQCD, well calibrated tool to study the QGP. The LHC will therefore complement, extend and eventually succeed the ongoing heavy ion program by providing a very different, and in many instances significantly better, environment for the study of strongly interacting matter.
With the past at AGS/SPS very productive and the present at RHIC extremely exciting, the field of ultra-relativistic heavy ion physics may confidently look forward to a promising future at LHC.

References

1. U. Heinz and M. Jacob, preprint nucl-th/0002042.
2. F. Abe et al., Nucl. Phys A661 (1991) 690.
3. C. Albajar et al., Nucl. Phys B335 (1990) 261.
4. K.J. Eskola, private communication.
5. K. J. Eskola, International Europhysics Conference on High-energy Physics : EPS-HEP '99, Tampere, Finland, 15 - 21 Jul 1999 - 1999, preprint hep-ph/9911350.
6. B. B. Back *et al.* [PHOBOS Collaboration], Phys. Rev. Lett. **85** (2000) 3100.
7. A.H. Mueller and J. Qiu, Nucl. Phys. B268 (1986) 427; L.D. McLerran and R. Venugopalan, Phys. Rev. D49 (1994) 3352. A.H. Mueller, Nucl. Phys. B572 (2000) 227; W. Pöschl and B. Müller, Phys. Rev. D60 (1999) 114505.
8. R. Baier, Y. L. Dokshitzer, A. H. Mueller, S. Peigne and D. Schiff, Nucl. Phys. B483 (1997) 291; M. Gyulassy, P. Levai and I. Vitev, Phys. Rev. Lett. 85 (2000) 5535; U.A. Wiedemann, Nucl. Phys. B588 (2000) 303.
9. K.J. Eskola, K. Kajantie, P.V. Ruuskanen, and K. Tuominen, Nucl. Phys. B570 (2000) 379.
10. V. Emel'yanov, A. Khodinov, S.R. Klein, and R. Vogt, Phys. Rev. C 61 (2000) 044904.
11. X.N. Wang, Phys. Rep. 280 (1997) 287.
12. F. Karsch and H. Satz, Z. Phys. C **51** (1991) 209.
13. H. Heiselberg, subm to Phys. Rep., preprint nucl-th/0003046; M. Bleicher, S. Jeon and V. Koch, Phys. Rev. C **62** (2000) 061902; M. Asakawa, U. Heinz and B. Muller, Phys. Rev. Lett. **85** (2000) 2072.
14. D. Brandt CERN-LHC-Project-Report-450 ; CERN , 20 Dec 2000 http://documents.cern.ch/archive/electronic/cern/preprints/lhc/lhc-project-report-450.pdf.
15. P. P. Yepes [CMS Collaboration], nucl-ex/0104026.

COLOR SUPERCONDUCTIVITY

KRISHNA RAJAGOPAL (krishna@ctp.mit.edu)
Center for Theoretical Physics, MIT
Cambridge, MA, USA 02139

Abstract. I review the physics of cold, dense quark matter. I describe the phenomena of color superconductivity, color-flavor locking and crystalline color superconductivity, focussing on the physical properties of these condensed phases of QCD rather than on calculational methods. I close with a look at consequences for the physics of compact stars.

1. Introduction

In the decades that have passed since its inception, QCD has become firmly established as the theory underlying all of strong-interaction physics—a pillar of the standard model. Perturbative QCD has been verified in deep inelastic scattering, and the spectrum and structural properties of the hadrons are gradually being calculated by the nonperturbative lattice formulation of QCD.

Even so, there remain tantalizing questions. As well as predicting the behavior of small numbers of particles, QCD should also be able to tell us about the thermodynamics of matter in the realm of extraordinarily high temperatures ($\gtrsim 100$ MeV) and densities at which it comes to dominate the physics. These regions are of more than academic interest: neutron stars are believed to consist of matter squeezed beyond nuclear density by gravitational forces, and the whole universe was hotter than 100 MeV for the first crucial microseconds of its history. However, only in the last few years have these regions begun to be probed experimentally in heavy ion collisions and astrophysical observations of neutron stars, and our theoretical understanding of them remains elementary.

High densities have proven particularly difficult to study, in part because lattice gauge theory has been blocked by the complexity of the fermion determinant. We are still trying to establish the symmetries of the ground state, and find effective theories for its lowest excitations. These questions are of direct physical relevance: an understanding of the symmetry properties of dense matter can be expected to inform our understanding of neutron star astrophysics and conceivably also heavy

J.-P. Blaizot and E. Iancu (eds.), QCD Perspectives on Hot and Dense Matter, 475–522.

ion collisions which somehow achieve high baryon densities without reaching high temperatures.

In these lectures, we explore the progress that has been made in the last few years in understanding the possible phases of QCD at low temperatures and high densities, and go on to discuss the possible consequences for in compact stars phenomenology. Other reviews, with different emphases, will also prove useful to the reader [1].

2. The Fermi surface and Cooper instability

One of the most striking features of QCD is asymptotic freedom [2], the fact that the force between quarks becomes arbitrarily weak as the characteristic momentum scale of their interaction grows larger. This immediately suggests that at sufficiently high densities and low temperatures, QCD will become tractable [3]. High density brings in a large energy scale, the chemical potential μ, and one might hope that the relevant coupling to describe the dynamics is $g(\mu)$. Matter will then consist of a Fermi sea of essentially free quarks, whose behavior is dominated by the freest of them all: the high-momentum quarks that live at the Fermi surface. This naive expectation does not stand up to critical scrutiny. As we shall discuss at length below, perturbation theory around the naive ground state (free quark Fermi spheres) encounters infrared divergences. Furthermore, the naive perturbative ground state is unstable. Fortunately, related difficulties have been met and overcome previously, in the theory of superconductivity. There we learn that arbitrarily weak attractive interactions can change the ground state qualitatively. In the true ground state there is an effective mass for photons — the Meissner effect — and energy gaps for charged excitations. These phenomena remove potential infrared divergences, and render the perturbation theory around the true ground state regular (nondegenerate).

This can be seen intuitively in the following way. Consider a system of free particles. The Helmholtz free energy is $F = E - \mu N$, where E is the total energy of the system, μ is the chemical potential, and N is the number of particles. The Fermi surface is defined by a Fermi energy $E_F = \mu$, at which the free energy is minimized, so adding or subtracting a single particle costs zero free energy. Now, suppose a weak attractive interaction is introduced. Bardeen, Cooper, and Schrieffer (BCS) [4] showed that this favors a complete rearrangement of the states near the Fermi surface, because it costs no free energy to make a pair of particles (or holes), and the attractive interaction makes it favorable to do so. Many such pairs will therefore be created, in all the modes near the Fermi surface, and these pairs, being bosonic, will form a condensate. The ground state will be a superposition of states with all numbers of pairs, breaking the fermion number symmetry. An arbitrarily weak interaction has lead to spontaneous symmetry breaking.

Since pairs of quarks cannot be color singlets, the resulting condensate will break the local color symmetry $SU(3)_{\text{color}}$. This is the definition of color superconductivity [5, 6, 7, 8, 9]. Note that the quark pairs play the same role here as the Higgs particle does in the standard model: the color-superconducting phase can be thought of as the Higgsed (as opposed to confined) phase of QCD.

It is important to remember that the breaking of a gauge symmetry cannot be characterized by a gauge-invariant local order parameter which vanishes on one side of a phase boundary [10]. The superconducting phase can be characterized rigorously only by its global symmetries. In electromagnetism there is a non-local order parameter, the mass of the magnetic photons, that corresponds physically to the Meissner effect and distinguishes the free phase from the superconducting one. In nonabelian theories like QCD there is no free phase: even without pairing the gluons are not states in the spectrum. No order parameter distinguishes the Higgsed phase from a confined phase or a plasma, so we have to look at the global symmetries. This is just what happens in the standard electroweak model, whose Higgs phase can also be understood as a confined phase [11]. The absence of massless gauge bosons and of long-range forces is the essence of the Meissner-Anderson-Higgs effect, and it is also the essence of confinement.

For detailed weak-coupling calculations it is more convenient to work in a fixed gauge, where *after* gauge-fixing the gauge potentials make only small fluctuations around zero. Of course at the end of any calculation we must restore the gauge symmetry, by averaging over the gauge fixing parameters, and only gauge-invariant quantities will survive. However, in the intermediate steps, within a fixed gauge, one can capture important correlations that characterize the ground state by specifying the existence of nonzero condensates relative to the ambient gauge. This is the procedure used in the standard model, where intermediate steps in the calculations involve a nonzero vacuum expectation value for a Higgs doublet field $\langle\phi^a\rangle = v\delta_1^a$, which is not gauge invariant. In superconductivity, the essence of the physics is the correlation in the fermionic wave function which describes the Cooper pairs, and the resulting modification of the dispersion relations which describe the excitation spectrum. In particular, the gap in the spectrum of fermionic excitations at the Fermi surface is a gauge invariant quantity. Describing this physics within a fixed gauge as a condensate which "breaks" the gauge symmetry is a convenient fiction. By forging a connection with superconductivity and condensate formation, it brings the universality class of confinement down to earth, and makes it accessible to weak coupling methods. These condensates need not break any true (i.e. global) symmetries. If a global symmetry *is* broken, some combination of the condensates themselves is a gauge invariant physical observable, and not just a convenient fiction.

Compared to ordinary superconductivity, color superconductivity, though it appears superficially to be more complex mathematically, is in a profound sense simpler and more directly related to fundamentals. Ordinary superconductivity takes place in solids and the accurate effective interactions are determined by band structure and other complicated effects. Furthermore, ordinary superconductivity in a metal involves electron pairing, and the fundamental interaction between electrons (the screened Coulomb interaction) is repulsive. The effective attraction near the Fermi surface that leads to superconductivity arises in classic superconductors only as a subtle consequence of retarded phonon interactions, and in the cuprate superconductors through some mechanism yet unknown. In color superconductivity, by contrast, the attractive interaction can arise already from the primary, strong, interactions. This has two consequences. First, the accurate form of these interactions can be calculated from first principles, using asymptotic freedom. This makes calculations at high enough density robust. Second, at accessible densities, where the strong interactions are much stronger than the electromagnetic interactions, we expect the color superconductors themselves to be robust in the sense that the ratio of their gaps and critical temperatures to the Fermi energy should be quite large.

In QCD with three colors and three flavors, we find an improved ground state at high density, based on color superconductivity, around which weak-coupling perturbation theory is valid. In particular, all the colored degrees of freedom acquire gaps. Thus, the improved ground state differs qualitatively from the naive one.

The resulting predictions regarding the low-energy spectrum and dynamics are striking. Color symmetry and chiral symmetry are spontaneously broken. The spectrum of elementary excitations is quite different from that found in naive perturbation theory. Nominally massless quarks and gluons become massive, new massless collective modes appear, and various quantum numbers get modified. All the elementary excitations carry integral electric charges [12]. Altogether, one finds an uncanny resemblance between the properties one computes at asymptotic densities, directly from the microscopic Lagrangian, and the properties one expects to hold at low density, based on the known phenomenology of hadrons. In particular, the traditional "mysteries" of confinement and chiral symmetry breaking are fully embodied in a controlled, fully microscopic, weak-coupling (but nonperturbative!) calculation, that accurately describes a directly physical, intrinsically interesting regime [12, 13].

3. The gap equation

We shall return to the main line of these lectures, namely a description of the physical properties of color superconducting quark matter, in Section 4. In this section, we give an incomplete introduction to the methods used to calculate the gap, which is the fundamental energy scale characterizing a (color) superconductor. The gap Δ is the free energy of a fermionic excitation about the ground state. Also, the critical temperature T_c below which quark matter is a color superconductor and above which the condensate melts, yielding a quark-gluon plasma, is given by 0.57Δ, as in any BCS superconductor.

It would be ideal if the calculation of the gap were within the scope of lattice gauge theory as is, for example, the calculation of the critical temperature for the QCD phase transition at zero baryon density. Unfortunately, lattice methods relying on importance sampling have to this point been rendered exponentially impractical at nonzero baryon density by the complex action at nonzero μ. Various lattice methods *can* be applied for $\mu \neq 0$ as long as T/μ is large enough [14]; so far, though, none have proved applicable at temperatures which are low enough that color superconductivity occurs. Lattice simulations are possible in two-color QCD and in QCD at large isospin density. Finally, new algorithms have recently allowed the simulation of theories which are simpler than QCD at nonzero chemical potential but which have just as severe a fermion sign problem [15]. This bodes well for the future.

To date, in the absence of suitable lattice methods, quantitative analyses of color superconductivity have followed two distinct strategies. The first approach is utilitarian and semi-phenomenological, emphasizing the use of simplified models. The overarching theme here is to define models which incorporate the salient physical effects, yet are tractable using known mathematical techniques of quantum many-body theory. Free parameters within a model of choice are chosen to give reasonable vacuum physics. Examples include analyses in which the interaction between quarks is replaced simply by four-fermion interactions with the quantum numbers of the instanton interaction [8, 9, 16] or of one-gluon exchange [12, 17], random matrix models [18], and more sophisticated analyses done using the instanton liquid model [19, 20]. Renormalization group methods have also been used to explore the space of all possible effective four-fermion interactions.[21, 22] These methods yield results which are in qualitative agreement: the gaps range between several tens of MeV up to of order 100 MeV; the associated critical temperatures (above which the diquark condensates vanish) can therefore be as large as about $T_c \sim 50$ MeV. This agreement between different models reflects the fact that what matters most is simply the strength of the attraction between quarks in the color $\bar{\mathbf{3}}$ channel, and by fixing the parameters of the model interaction to fit, say, the magnitude of the vacuum chiral condensate, one ends up with attractions of similar strengths in different models.

The second, more ambitious approach is fully microscopic. Such an approach is feasible, for high-density QCD, due to asymptotic freedom. Several important results have been obtained from the microscopic approach, perhaps most notably the asymptotic form of the gap [23]. Very significant challenges remain, however. It is not really known, for example, how to calculate corrections to the leading term in a systematic way.

These approaches have complementary virtues – simplicity versus rigor, openness to phenomenological input versus quantitative predictive power at asymptotically high density. Fortunately, they broadly agree in their main conclusions as to the patterns of symmetry breaking and the magnitude of the gap at accessible densities.

In a field-theoretic approach to fermion pairing, the relevant quantity is the fermion self energy, i.e. the one-particle irreducible (1PI) Green function of two quark fields. Its poles give the gaps, namely the gauge-invariant masses of the quasi-quarks, the lowest energy fermionic excitations around the quark Fermi surface. To see if quark condensation occurs in some channel, one writes down a self-consistency equation, the gap equation, for a self energy with that structure, and solves it to find the actual self energy (the gap). If it is zero, there is no condensation in that channel. If not, there can be condensation, but it may just be a local minimum of the free energy. There may be other solutions to the gap equation, and the one with the lowest free energy is the true ground state.

There are several possible choices for the interaction to be used in the gap equation. At asymptotically high densities QCD is weakly coupled, so one gluon exchange is appropriate. Such calculations [23, 24, 25, 26, 27, 28, 29, 30, 31, 32, 33, 34, 35, 36] are extremely important, since they demonstrate from first principles that color superconductivity occurs in QCD. However, the density regime of physical interest for neutron stars is several to ten times nuclear density ($\mu \sim 400 - 500$ MeV) and weak coupling calculations are unlikely to be trustworthy in that regime. In fact, current weak-coupling calculations cannot be extrapolated below about 10^8 MeV because of gauge dependence arising from the neglect of vertex corrections [34].

The alternative is to use some phenomenological interaction that can be argued to capture the essential physics of QCD in the regime of interest. The interaction can be normalized to reproduce known low-density physics such as the vacuum chiral condensate, and then extrapolated to the desired chemical potential. In two-flavor theories, the instanton vertex is a natural choice [8, 9, 16, 19], since it is a four-fermion interaction. With more flavors, the one gluon exchange vertex without a gluon propagator [7, 37, 12] is more convenient. It has been found that these both give the same results, to within a factor of 2 or less. This is well within the inherent uncertainties of such phenomenological approaches. One caveat to bear in mind is that the single-gluon-exchange interaction is symmetric under $U(1)_A$, and so it sees no distinction between condensates of the form $\langle qCq \rangle$ and $\langle qC\gamma_5 q \rangle$ (C is the Dirac charge conjugation matrix). However, once instantons are included the Lorentz scalar $\langle qC\gamma_5 q \rangle$ is favored [8, 9], so in single-gluon-exchange calculations the parity-violating condensate is usually ignored.

The mean-field approximation to the Schwinger-Dyson equations is shown diagramatically in Fig. 1, relating the full propagator to the self-energy. In the mean-field approximation, only daisy-type diagrams are included in the resummation, vertex corrections are excluded. Algebraically, the equation takes the

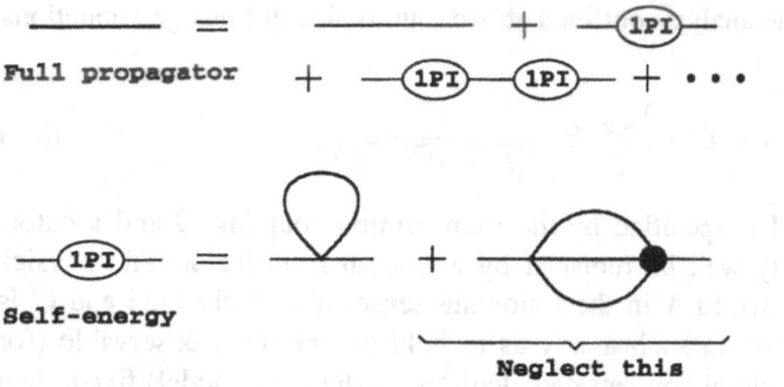

Full propagator

1PI 1PI 1PI

1PI

Self-energy

Neglect this

Figure 1. Mean-field Schwinger-Dyson (gap) equations

form

$$\Sigma(k) = -\frac{1}{(2\pi)^4} \int d^4q\, S(q) D(k-q), \tag{3.1}$$

where $\Sigma(k)$ is the self-energy, S is the full fermion propagator, and $D(k-q)$ is the vertex, which in NJL models will be momentum-independent, but in a weak-coupling QCD calculation will include the gluon propagator and couplings. Since we want to study quark-quark condensation, we have to write propagators in a form that allows for this possibility, just as to study chiral symmetry breaking it is necessary to use 4-component Dirac spinors rather than 2-component Weyl spinors, even if there is no mass term in the action. We therefore use Nambu-Gorkov 8-component spinors, $\Psi = (\psi, \bar\psi^T)$, so the self-energy Σ can include a quark-quark pairing term Δ. The fermion inverse propagator S^{-1} then takes the form

$$S^{-1}(q) = S^{-1}_{\text{free}}(q) + \Sigma = \begin{pmatrix} \not{q} + \mu\gamma_0 & \gamma_0 \Delta^\dagger \gamma_0 \\ \Delta & (\not{q} - \mu\gamma_0)^T \end{pmatrix}. \tag{3.2}$$

Equations (3.1) and (3.2) can be combined to give a self-consistency condition for Δ, the gap equation. If the interaction is a point-like four-fermion NJL interaction then the gap parameter Δ will be a color-flavor-spin matrix, independent of momentum. If the gluon propagator is included, Δ will be momentum-dependent, complicating the analysis considerably.

In NJL models, the simplicity of the model has allowed renormalization group analyses [21, 22] that include a large class of four-fermion interactions, and follow their running couplings as modes are integrated out. This confirms that in QCD with two and three massless quarks the most attractive channels for condensation are those corresponding to the two-flavor superconducting (2SC) and color-flavor locked (CFL) phases studied below.

Following through the analysis outlined above, one typically finds gap equations of the form

$$1 = G \int_0^\Lambda k^2 dk \, \frac{1}{\sqrt{(k-\mu)^2 + \Delta^2}}, \tag{3.3}$$

Here, the NJL model is specified by the four-fermion coupling G and a cutoff Λ, which may equally well be replaced by a smooth form-factor. The physics is reasonably insensitive to Λ in the following sense: if Λ is changed and G is changed simultaneously in such a way as to hold one physical observable (for example the vacuum chiral condensate calculated in the same model) fixed, then other physical observables (for example the gap Δ) change little. In the limit of small gap, the integral can be evaluated, giving

$$\Delta \sim \Lambda \exp\Big(-\frac{\text{const}}{G\mu^2}\Big). \tag{3.4}$$

This shows the non-analytic dependence of the gap on the coupling G. Condensation is a nonperturbative effect that cannot be seen to any order in perturbation theory. The reason it can be seen in the diagrammatic Schwinger-Dyson approach is that there is an additional ingredient: an ansatz for the form of the self energy. This corresponds to guessing the form of the ground state wavefunction in a many-body variational approach. All solutions to gap equations therefore represent possible stable ground states, but to find the favored ground state their free energies must be compared, and even then one can never be sure that the true ground state has been found, since there is always the possibility that there is another vacuum that solves its own gap equation and has an even lower free energy. When G is chosen to give a reasonable value of the vacuum chiral condensate, the gap Δ in quark matter at densities of interest (say $\mu \sim 400 - 500\tilde{\text{M}}\text{eV}$) turns out to be of order 100 MeV [8, 9].

In weak-coupling QCD calculations, where the full single-gluon-exchange vertex complete with gluon propagator is used, the solution to the gap equation takes the form [5, 6, 23, 25]

$$\Delta \sim \mu \frac{1}{g^5} \exp\Big(-\frac{3\pi^2}{\sqrt{2}}\frac{1}{g}\Big), \tag{3.5}$$

or, making the weak-coupling expansion in the QCD gauge coupling g more explicit,

$$\ln\Big(\frac{\Delta}{\mu}\Big) = -\frac{3\pi^2}{\sqrt{2}}\frac{1}{g} - 5\ln g + \text{const} + \mathcal{O}(g). \tag{3.6}$$

This gap equation has two interesting features. Firstly, it does not correspond to what you would naively expect from the NJL model of single gluon exchange, in which the gluon propagator is discarded and $G \propto g^2$, yielding $\Delta \sim \exp(-1/g^2)$. The reason [6, 23] is that at high density the gluon propagator has an infrared divergence at very small angle scattering, since magnetic gluons are only Landau damped, not screened. This divergence is regulated by the gap itself, weakening its dependence on the coupling.

Secondly, in (3.5) we have left unspecified the energy scale at which the coupling g is to be evaluated. Natural guesses range between μ and Δ. If we use $g(\mu)$ (this is pessimistic; using a lower energy scale results in significantly larger gaps) and assume $g(\mu)$ runs according to the one-loop formula $1/g^2 \sim \ln\mu$ then the exponential factor in (3.5) gives very weak suppression, and is in fact overwhelmed by the initial factor μ, so that the gap rises without limit at asymptotically high density, although Δ/μ shrinks to zero so that weak-coupling methods are still self-consistent. This means that color superconductivity will inevitably dominate the physics at high enough densities.

Although the value of Δ is under control asymptotically, it seems fair to say that applying asymptotic results at $\mu = 400 - 500$ MeV is at least as uncertain a proposition as applying estimates made using phenomenologically normalized models with point-like interactions. Nevertheless, if we take the estimates for the prefactor b provided by Schäfer and Wilczek's numerical results [28] and apply them at $\mu \sim 400$ MeV, they predict gaps of order 100 MeV. The consequent critical temperatures are related to the zero temperature gap Δ by the standard weak-coupling BCS result $T_c = 0.57\Delta$,[25, 29] and are therefore of order 50 MeV. Some known corrections (like the fact that g must be evaluated at some scale lower than μ) push this estimate up while others (like the corrections described in [29, 36]) push it down. And, regardless, the asymptotic calculation is of quantitative value only for $\mu \gg 10^8$ MeV [34]. It is nevertheless satisfying that two very different approaches, one using zero density phenomenology to normalize models, the other using weak-coupling methods valid at asymptotically high density, yield predictions for the gaps and critical temperatures at accessible densities which are in qualitative agreement. Neither can be trusted quantitatively for quark number chemical potentials $\mu \sim 400 - 500$ MeV, as appropriate for the quark matter which may occur in compact stars. Still, both methods agree that the gaps at the Fermi surface are of order tens to 100 MeV, with critical temperatures about half as large.

The ratio of the critical temperature to the Fermi energy in a color superconductor is $T_c/E_F = T_c/\mu \sim 0.1$, which is about three orders of magnitude larger than in a traditional metallic BCS superconductor. This reflects the fact that color superconductivity is induced by an attraction due to the primary, strong, interaction in the theory, rather than having to rely on much weaker secondary interactions, as in phonon mediated superconductivity in metals. Quark matter is a high-T_c superconductor by any reasonable definition. It is unfortunate that its T_c is nevertheless low enough that it is unlikely the phenomenon can be realized in heavy ion collisions.

4. Color-Flavor Locking

In this Section we shall analyze the high-density, zero-temperature behavior of a slight idealization of QCD, in which the masses of the u, d and s quarks are set to zero, and those of the c, b and t quarks to infinity. This idealization gives rise to an especially clear and beautiful form of the theory. Also, as we shall discuss below, the analysis applies with only minor modifications to real-world QCD at high enough density. We concentrate on the physical properties of dense quark matter in the idealized three-flavor world. For a more quantitative discussion of the calculation of the magnitude of the gap than that provided in the previous section, the reader should consult the literature cited in that section. The attractive interaction between quarks, even if arbitrarily weak, renders the Fermi surface unstable to the formation of a condensate of quark Cooper pairs. We expect those quark quasiparticles which interact with the condensate to acquire an energy gap, and we expect a Meissner effect to occur for all gauge bosons except those which see the condensate as neutral.

4.1. FORM OF THE CONDENSATE

The most attractive color channel for quark pairing is the $\bar{\mathbf{3}}$. This is true at weak coupling, where single gluon exchange dominates, at intermediate coupling where instanton interaction becomes important, and at strong coupling, where the color $\bar{\mathbf{3}}$ combines two flux strings into one, lowering the energy of the pair. The relevant gap equation has been studied for pointlike 4-fermion interactions with the index structure of single gluon exchange [12, 20, 13] as well as a weakly coupled gluon propagator [31, 32, 33]. They agree that the true ground state contains nonzero condensates approximately of the form [12]

$$\langle\psi^{a\alpha}_{iL}(\vec{p})\psi^{b\beta}_{jL}(-\vec{p})\epsilon_{ab}\rangle = -\langle\psi(\vec{p})^{a\alpha}_{iR}\psi^{b\beta}_{jR}(-\vec{p})\epsilon_{ab}\rangle = \Delta(p^2)\epsilon^{\alpha\beta A}\epsilon_{ijA} \,. \quad (4.1)$$

We have explicitly displayed color (α,β), flavor (i,j), and spinor (a,b) indices. The A-index is summed and therefore links color and flavor. We have used a two-component spinor notation; note that properly the right-helicity fields should involve dotted spinors. The important information conveyed by the spinors is that the condensation does not violate rotational invariance. The relative minus sign between left-helicity and right-helicity condensates signifies that the ground state is a scalar, rather than a pseudoscalar, so that parity is unbroken.
In reality, condensation in the color $\bar{\mathbf{3}}$ channel (4.1) induces a small but nonzero condensate in the color $\mathbf{6}$ channel even if the interaction is repulsive in this channel [12], because this additional condensation breaks no further symmetries [17]. This

means that the right hand side of (4.1) is slightly more complicated and should, in fact, be written in terms of two gap parameters κ_1 and κ_2,

$$\begin{aligned}\langle \psi^{a\alpha}_{iL}(\vec{p})\psi^{b\beta}_{jL}(-\vec{p})\varepsilon_{ab}\rangle &= -\langle \psi(\vec{p})^{a\alpha}_{iR}\psi^{b\beta}_{jR}(-\vec{p})\epsilon_{ab}\rangle \\ &= \kappa_1(p^2)\delta^\alpha_i\delta^\beta_j + \kappa_2(p^2)\delta^\alpha_j\delta^\beta_i\end{aligned} \tag{4.2}$$

The pure color $\bar{\mathbf{3}}$ condensate displayed in (4.1) has $\kappa_2 = -\kappa_1$. Using (4.1) is a good approximation because the induced color $\mathbf{6}$ condensate is much smaller than the dominant color $\bar{\mathbf{3}}$ condensate mandated by the attraction in this channel.[12, 31, 32]

We now explain the term "color-flavor locking". The Kronecker delta functions in (4.2) link color and flavor indices. This means that the condensates transform nontrivially under separate color and flavor transformations, but remain invariant if we simultaneously rotate both color and flavor. Thus these symmetries are locked together. (Color-flavor locking is analogous to the B-phase of superfluid helium 3, where orbital and spin rotations, which are separate symmetries in a nonrelativistic system, are locked together.)

4.2. SYMMETRY BREAKING

The color-flavor locked phase (4.1) features two condensates, one involving left-handed quarks alone and one involving right-handed quarks alone. Each is invariant only under equal and opposite color and flavor rotations, but since color is a vector symmetry, the combined effect is to break the color and flavor symmetries down to a global vector "diagonal" symmetry, that makes equal transformations in all three sectors – color, left-handed flavor, and right-handed flavor. Baryon number symmetry is broken down to a discrete Z_2 symmetry under which all quark fields are multiplied by -1. The symmetry breaking pattern is [12]

$$[SU(3)_{\rm color}] \times \underbrace{SU(3)_L \times SU(3)_R \times U(1)_B}_{\supset [U(1)_Q]} \longrightarrow \underbrace{SU(3)_{C+L+R}}_{\supset [U(1)_{\tilde{Q}}]} \times \mathbb{Z}_2 \tag{4.3}$$

where gauge symmetries are in square brackets, $SU(3)_L$ and $SU(3)_R$ are the global chiral flavor symmetries, and $U(1)_B$ is baryon number. (See also Ref. [38] in which a similar pattern was considered at zero density.)

We see that chiral symmetry has been broken, but not by the usual low-density $\langle\bar{\psi}\psi\rangle$ condensate which pairs left-handed quarks with right-handed antiquarks. Even though the condensates in (4.1,4.2) only pair left-handed quarks with left-handed quarks, and right-handed quarks with right-handed quarks, and do not *appear* to lock $SU(3)_L$ to $SU(3)_R$, they manage to do so by locking both to $SU(3)_{\rm color}$. Color-flavor locking, therefore, provides a mechanism by which chiral symmetry can be broken.

Even without doing any detailed calculations, we can see that the symmetry breaking pattern (4.3) will have profound effects on the physics.

1. The color gauge group is completely broken. All eight gluons become massive. This ensures that there are no infrared divergences associated with gluon propagators. Since the quark modes are also gapped (see below), we conclude that weak coupling perturbation theory *around the correct, condensed ground state* is free of the difficulties that appeared around the naive ground state.
2. All the quark modes are gapped. This removes the other potential source of infrared divergences, from integration over low-energy excitations around the Fermi surface. The nine quasiquarks (three colors times three flavors) fall into an $\mathbf{8} \oplus \mathbf{1}$ of the unbroken global $SU(3)$, so there are two gap parameters. The singlet has a larger gap than the octet.
3. Electromagnetism is no longer a separate symmetry, but corresponds to gauging one of the flavor generators. A rotated electromagnetism ("$\tilde{Q}$"), a combination of the original photon and one of the gluons, survives unbroken.
4. Two global symmetries are broken, the chiral symmetry and baryon number, so there are two gauge-invariant order parameters that distinguish the CFL phase from the QGP. There are corresponding Goldstone bosons which are long-wavelength symmetry rotations of the order parameter, whose energy cost is arbitrarily low. The chiral Goldstone bosons form a pseudoscalar octet, like the zero-density $SU(3)_{\text{flavor}}$ pion octet. The breaking of the baryon number symmetry leads to a singlet scalar Goldstone boson which makes the CFL phase a superfluid, and remains massless even when quark masses are introduced.
5. Quark-hadron continuity. It is striking that the symmetries of the 3-flavor CFL phase are the same as those one might expect for 3-flavor hypernuclear matter [13]. In hypernuclear matter one would expect the hyperons to pair in an $SU(3)_{\text{flavor}}$ singlet ($\langle\Lambda\Lambda\rangle, \langle\Sigma\Sigma\rangle, \langle N\Xi\rangle$), breaking baryon number but leaving flavor and electromagnetism unbroken. Chiral symmetry would be broken by the chiral condensate. This means that one might be able to follow the spectrum continuously from hypernuclear matter to the CFL phase of quark matter. The pions would evolve into the pseudoscalar octet mentioned above. The vector mesons would evolve into the massive gauge bosons. This will be discussed in more detail below, but we note already now that nuclear matter (as opposed to the hypernuclear matter in a world with three degenerate quarks) *cannot* be continuously connected to the CFL phase.

4.3. GLOBAL SYMMETRIES AND ORDER PARAMETERS

Color-flavor locking, unlike the Higgs mechanism in the electroweak sector of the standard model, breaks global symmetries as well as gauge symmetries. Physically, this implies that there are sharp differences between the color-flavor locked phase and the quark-gluon plasma phase (in which all symmetries of the QCD lagrangian are unbroken), so that any passage between them must be marked by one or more phase transitions. It is a simple matter to construct the corresponding gauge invariant order parameters, which have a strict meaning valid at any coupling, from our primary, gauge variant condensate at weak coupling. For instance, to form a gauge invariant order parameter capturing chiral symmetry breaking we may take the product of the left-handed version of (4.1) with the right-handed version and saturate the color indices, to obtain

$$\langle \psi^{\alpha}_{Li}\psi^{\beta}_{Lj}\bar{\psi}^{k}_{R\alpha}\bar{\psi}^{l}_{R\beta}\rangle \sim \langle \psi^{\alpha}_{Li}\psi^{\beta}_{Lj}\rangle\langle \bar{\psi}^{k}_{R\alpha}\bar{\psi}^{l}_{R\beta}\rangle \sim \Delta^2 \epsilon_{ijm}\epsilon^{klm} . \tag{4.4}$$

Likewise we can take a product of three copies of the condensate and saturate the color indices, to obtain a gauge invariant order parameter for the baryon-number violating superfluid order parameter. These secondary order parameters will survive gauge unfixing unscathed. Unlike the primary condensate, from which they were derived, they are more than convenient fictions.

The spontaneous violation of baryon number does not mean that compact stars can change the baryon number of the universe, any more than electron pairing in ordinary superconductors can change the lepton number of the universe. Actually, ordinary nuclear matter is a superfluid in which nucleon-nucleon pairing violates baryon number symmetry. The essential point in all these cases is that in a finite sized sample there is no true violation of the conservation laws, since a Gaussian surface can be constructed enclosing the sample. The correct interpretation of the formal "violation" of these symmetries is that there can be large fluctuations and easy transport of the corresponding quantum numbers within the sample. These are precisely the phenomena of superconductivity and superfluidity.

If we turn on a common mass for all the quarks, the chiral $SU(3)_L \times SU(3)_R$ flavor symmetry of the Lagrangian will be reduced to the diagonal $SU(3)_{L+R}$. If we turn on unequal masses, the symmetry will be even less. In any case, however, the $U(1)$ of baryon number a good microscopic symmetry, and the corresponding six-quark order parameter remains a strict signature of the color-flavor locked phase, distinguishing it from the quark-gluon plasma phase.

As it stands the order parameter (4.4) is not quite the usual one, but roughly speaking its square. It leaves invariant an additional Z_2, under which only the left-handed quark fields change sign. Actually this Z_2 is not a legitimate symmetry of the full theory, but suffers from an anomaly. So we might expect that the usual chiral order parameter is induced by the anomalous interactions that violate the axial baryon number symmetry of the classical Lagrangian. To put this another way, because axial baryon number is not a symmetry of QCD, once chiral symmetry is broken by color-flavor locking there is no symmetry argument precluding the existence of an ordinary chiral condensate. Indeed, instanton effects do induce a nonzero $\langle \bar{\psi}_R \psi_L \rangle$ because the instanton-induced interaction is a six-fermion operator which can be written as a product of $\bar{\psi}_R \psi_L$ and the operator in (4.4) which already has a nonzero expectation value [12], but this turns out to be a small effect [20, 31].

At weak coupling, we can be more specific about these matters. The most important interactions near the Fermi surface, quantitatively, arise from gluon exchange. These are responsible for the primary condensation. The instanton interaction is much less important asymptotically because the gauge fields which make up the instantons themselves are screened, the effects of instantons are intrinsically smaller for more energetic quarks, and because the instanton-induced interaction involves six fermion fields, and hence (one can show) becomes irrelevant upon renormalization toward the Fermi surface. The instanton interaction is qualitatively important, however, because it represents the leading contribution to axial baryon number violation. It is only such $U(1)_A$ violating interactions that remove the degeneracy among states with different relative phases between the left- and right-handed condensates in (4.1). In the absence of intrinsic $U(1)_A$ breaking, the spontaneous violation of this symmetry in the color-flavor locked phase would be accompanied by the existence of a pseudoscalar $SU(3)_{\text{color}+L+R}$ singlet Nambu-Goldstone bosons. Since the intrinsic violation of this symmetry is parametrically small, the corresponding boson will not be strictly massless, but only very light. Quantum fluctuations in this light η'-field, among other things, will keep the conventional chiral symmetry breaking order parameter small compared to (4.4) at high density.

4.4. ELEMENTARY EXCITATIONS

The physics of the excitations in the CFL phase has been the focus of much recent work.[12, 13, 17, 39, 20, 40, 41, 31, 32, 33, 42, 43, 44, 45, 46, 47, 48, 49, 50, 51, 52, 53, 54, 55, 56, 57, 58] There are three sorts of elementary excitations. They are the modes produced directly by the fundamental quark and gluon fields, and the collective modes associated with spontaneous symmetry breaking. These modes can be classified under the unbroken $SU(3) \times Z_2$ symmetry, and the unbroken rotation and parity symmetries.

The quark fields of course produce spin 1/2 fermions. Some of these are true long-lived quasiparticles, since there are no lighter states of half-integer spin that they might decay into. With the conventions we have been using, as expressed in (4.1), the quark fields are triplets and antitriplets under color and flavor, respectively. Thus they decompose into an octet and a singlet under the diagonal $SU(3)_{\text{color}+L+R}$. There is an energy gap for production of pairs above the ground state. More precisely, there are two gaps: a smaller one for the octet, and a larger one for the singlet [12]. The dispersion relations describing these fermionic quasiparticle excitations in the CFL phase have been described in some detail [17, 41, 55].

The gluon fields produce an $SU(3)_{\text{color}+L+R}$ octet of spin 1 bosons. As previously mentioned, they acquire a common mass by the Meissner-Anderson-Higgs mechanism. The quantitative expressions for the masses of these vector mesons which have been computed at weak coupling [42, 59, 49, 55] and in an instanton-liquid model [60].

The fermionic excitations have a gap; the vector mesons have mass; but, the Nambu-Goldstone bosons are massless. These bosons form a pseudoscalar octet associated with chiral symmetry breaking, and a scalar singlet associated with baryon number superfluidity. The octet, but not the singlet, will be lifted from zero mass if the quarks are massive. Finally there is the parametrically light, but never strictly massless, pseudoscalar singlet associated with $U(1)_A$ breaking.

The Nambu-Goldstone bosons arising from chiral symmetry breaking in the CFL phase are Fermi surface excitations in which the orientation of the left-handed and right-handed diquark condensates oscillate out of phase in flavor space. The effective field theory describing these oscillations has been constructed [40, 42, 47]. Up to two derivatives, it is given by

$$\mathcal{L}_{\text{eff}} = \frac{f_\pi^2}{4}\text{Tr}\left(\partial_0\Sigma\partial_0\Sigma^\dagger + v_\pi^2\partial_i\Sigma\partial^i\Sigma^\dagger\right) - c\left(\det M\,\text{Tr}(M^{-1}\Sigma) + \text{h.c.}\right). \quad (4.5)$$

The Nambu-Goldstone boson field matrix Σ is a color singlet and transforms under $SU(3)_L \times SU(3)_R$ as $\Sigma \to U_L\Sigma U_R^\dagger$ as usual. $M = \text{diag}(m_u, m_d, m_s)$ is the quark mass matrix. The construction of Σ from rotations of the CFL condensates can be found in Refs. [40, 42]: one first finds the 17 putative Nambu-Goldstone bosons expected when the full symmetry group is broken to $SU(3)_{C+L+R}$; one then identifies 8 of these which become the longitudinal parts of massive vector bosons; the remaining 9 are the octet described by (4.5), and the superfluid mode. In addition, although the singlet η' is not a true Goldstone boson it is very light, as discussed above. See Refs. [40, 42] for the singlet terms in the effective Lagrangian. The higher derivative terms in the effective Lagrangian have also been analyzed [47].

The masses of the pseudoscalar mesons which are the pseudo-Nambu-Goldstone bosons associated with chiral symmetry breaking can be obtained from $\mathcal{L}_{\text{eff}}$ of (4.5) [42]. For example,

$$m_{\pi^\pm}^2 = \frac{2c}{f_\pi^2}m_s(m_u + m_d)\,, \qquad m_{K^\pm}^2 = \frac{2c}{f_\pi^2}m_d(m_u + m_s)\,. \quad (4.6)$$

Thus, the kaon is lighter than the pion, by a factor of $m_d/(m_u + m_d)$ [42]. Note that the effective Lagrangian is quadratic in M. This arises because L_{eff} respects the Z_2 symmetry under which only the left-handed quarks change sign [12]. As we discussed in the previous section, this is almost a symmetry of the CFL phase: it would be a symmetry if instanton effects could be neglected [12]. However, instanton effects generate a nonzero, but small, ordinary $\langle\bar\psi_R\psi_L\rangle$ condensate, which breaks the Z_2 [12, 20, 31], and results in a contribution to the meson m^2 which is linear in M and which may be numerically significant [45, 58].

If we were describing pions in vacuum, or pions in nuclear matter, the only way to obtain the coefficients in the effective theory would be to measure them in an experiment or, if possible, to calculate them on the lattice. Indeed in any theory with strong interactions, the purpose of writing an effective theory for the low energy degrees of freedom is to express the predictions for many low energy processes in terms of a few parameters, which must be obtained from experiment. In the color-flavor locked phase, however, the full theory is weakly coupled at asymptotically high densities. In this regime, therefore, the coefficients f_π^2, v_π^2 and c are calculable from first principles using weak coupling methods! Up to possible logarithmic corrections, the result is[42, 43, 44, 45, 46, 48, 54, 55, 58]

$$f_\pi^2 = \frac{21 - 8\log 2}{36\pi^2}\mu^2, \quad v_\pi^2 = \frac{1}{3}, \quad c = \frac{3\Delta^2}{2\pi^2}. \tag{4.7}$$

The electromagnetic[50, 54] and nonzero temperature[54] corrections to these quantities have also been calculated. Recently, the instanton contributions to the CFL meson masses have been estimated; they may result in a significant increase [61]. Quantitatively, (see Section 3 for a discussion of estimates of Δ) we estimate that the lightest pseudoscalar meson, the kaon, has a mass in the range of 5 to 20 MeV at $\mu = 400$ MeV, and becomes lighter still at higher densities. There are two reasons why the Nambu-Goldstone bosons are so much lighter in the CFL phase than in the vacuum. First, their mass^2 is proportional to $m_{\rm quark}^2$ rather than to $m_{\rm quark}$, as at zero density. In addition, there is a further suppression by a factor of Δ/μ, which arises because the Nambu-Goldstone bosons are collective excitations of the condensates formed from particle-particle and hole-hole pairs near the Fermi surface, whereas the quark mass term connects particles with antiparticles, far from the Fermi surface [44].

In QCD with unequal quark masses, at very high densities the CFL phase is much as we have described it, except that the gaps associated with $\langle us\rangle$, $\langle ds\rangle$ and $\langle ud\rangle$ pairing will differ slightly [17, 39] and the CFL condensate may rotate in flavor space yielding a kaon condensate [56]. The formation of a kaon condensate in the CFL phase is a less dramatic effect than the formation of a kaon condensate in nuclear matter made of neutrons and protons only [62]: there, kaon condensation breaks $U(1)_S$. This symmetry is already broken in the CFL phase. Kaon condensation in the CFL phase is more akin to kaon condensation in hypernuclear matter made up of equal measures of all the octet baryons, in which $U(1)_S$ is already broken by hyperon-hyperon pairing.

At high enough baryon density and low temperature, the ground state of QCD with three flavors of quarks is the color-flavor locked (CFL) phase. In this phase, quarks of all three colors and all three flavors form Cooper pairs, meaning that all fermionic quasiparticles are gapped. The gap Δ is likely of order tens to 100 MeV at astrophysically accessible densities, with quark chemical potential $\mu \sim (350-500)$ MeV. As we shall see momentarily, the condensate is charged with respect to eight of the nine massless gauge bosons of the ordinary vacuum, meaning that eight gauge bosons get a mass. Chiral symmetry is spontaneously broken, and so is baryon number (i.e., the material is a superfluid.) The CFL phase persists for finite masses and even for unequal masses, so long as the differences are not too large [17, 39]. It is very likely the ground state for real QCD, assumed to be in equilibrium with respect to the weak interactions, over a substantial range of densities. Throughout the range of parameters over which the CFL phase exists as a bulk (and therefore electrically neutral) phase, it consists of equal numbers of u, d and s quarks and is therefore electrically neutral in the absence of any electrons [63]. The equality of the three quark number densities is enforced in the CFL phase by the fact that this equality maximizes the pairing energy associated with the formation of ud, us, and ds Cooper pairs. This equality is enforced even though the strange quark, with mass m_s, is heavier than the light quarks.

4.5. THE MODIFICATION OF ELECTROMAGNETISM

It is physically significant, and proves extremely instructive, to consider the effect of color-flavor locking on the electromagnetic properties of high-density hadronic matter. To do this, we consider coupling in the appropriate additional $U(1)_{\rm EM}$ gauge field A_μ, representing the photon. This couples to u, d, s quarks with strength $\frac{2}{3}e, -\frac{1}{3}e, -\frac{1}{3}e$, respectively. Evidently this $U(1)_{\rm EM}$ symmetry is broken by the condensate (4.1), through the terms pairing differently-charged quarks. Were this the complete story, the color-flavor locked phase would be an electromagnetic superconductor. The truth is far different, however.

The situation is analogous to what occurs in the electroweak sector of the standard model. There, the Higgs field condensate breaks both the original weak $SU(2)$ and the hypercharge $U(1)$. However, one linear combination of these symmetries leaves the condensate invariant, and remains a valid gauge symmetry of the ground state. Indeed, this is how we identify electromagnetism within the standard model. Here we must similarly consider the possibility of mixing color $SU(3)$ and electromagnetic $U(1)_{\rm EM}$ generators to find a valid residual symmetry. Indeed, we should expect this to occur, by the following argument. In QCD with three flavors, $U(1)_{\rm EM}$ is a subgroup of $SU(3)_{L+R}$. When we neglected electromagnetism, we found that in the color-flavor locked phase $SU(3)_{L+R}$ is broken but $SU(3)_{\rm color+L+R}$ is an unbroken global symmetry. We therefore expect that gauging a $U(1)$ subgroup of $SU(3)_{L+R}$ must correspond, in the CFL phase, to gauging a $U(1)$ subgroup of the unbroken $SU(3)_{\rm color+L+R}$.

Once we are alerted to this possibility, it is not difficult to identify the appropriate combination of the photon and gluons which remains unbroken [12, 64]. In the CFL phase, there is an unbroken $U(1)_{\tilde{Q}}$ gauge symmetry and a corresponding massless photon given by a linear combination of the ordinary photon and one of the gluons [12, 64]. $U(1)_{\tilde{Q}}$ is generated by

$$\tilde{Q} = Q + \eta T_8$$

with $\eta = 1/\sqrt{3}$ and where Q is the conventional electromagnetic charge generator and the color hypercharge generator T_8 is normalized such that, in the representation of the quarks, $T_8/\sqrt{3} = \mathrm{diag}(-\frac{2}{3}, \frac{1}{3}, \frac{1}{3})$ in color space. The CFL condensate is $\tilde{Q}$-neutral, the $U(1)$ symmetry generated by $\tilde{Q}$ is therefore unbroken, the associated $\tilde{Q}$-photon remains massless, and within the CFL phase the $\tilde{Q}$-electric and $\tilde{Q}$-magnetic fields satisfy Maxwell's equations. The massless combination of the photon and the eighth gluon, $A_\mu^{\tilde{Q}}$, and the orthogonal massive combination which experiences the Meissner effect, A_μ^X, are given by

$$A_\mu^{\tilde{Q}} = \cos\theta A_\mu + \sin\theta G_\mu^8 \, , \tag{4.8}$$

$$A_\mu^X = -\sin\theta A_\mu + \cos\theta G_\mu^8 \, . \tag{4.9}$$

The mixing angle θ (called α_0 in Ref. [64]) which specifies the unbroken $U(1)$ is given by

$$\cos\theta = \frac{g}{\sqrt{g^2 + \eta^2 e^2}} \, . \tag{4.10}$$

θ is the analogue of the Weinberg angle in electroweak theory. At accessible densities, the gluons are strongly coupled ($g^2/4\pi \sim 1$) and the photons are weakly coupled ($e^2/4\pi \approx 1/137$), so θ is small, perhaps of order $1/20$. The "rotated photon" consists mostly of the usual photon, with only a small admixture of the G^8 gluon.

Let us now consider the charges of all the elementary excitations which we enumerated previously. For reference, the electron couples to $A_\mu^{\tilde{Q}}$ with charge

$$\tilde{e} = \frac{eg}{\sqrt{\eta^2 e^2 + g^2}} \, . \tag{4.11}$$

which is less than e because the electron couples only to the A_μ component of $A_\mu^{\tilde{Q}}$. Now in computing the $\tilde{Q}$-charge of the quark with color and flavor indices α, a we must take the appropriate combination from

$$\frac{e(-\frac{2}{3}g, \frac{1}{3}g, \frac{1}{3}g) + g(\frac{2}{3}e, -\frac{1}{3}e, -\frac{1}{3})}{\sqrt{\eta^2 e^2 + g^2}} \, .$$

One readily perceives that the possible values are $0, \pm\tilde{e}$. Thus, in units of the electron charge, the quarks carry integer $\tilde{Q}$-charge! Quite remarkably, high-density QCD realizes a mathematically consistent gauge theory version of the old vision of Han and Nambu: the physical quark excitations have integer electric charges that depend on an internal color quantum number!

Similarly, the gluons all have $\tilde{Q}$-charges $0, \pm\tilde{e}$. Indeed, they have the $\tilde{Q}$-charges one would expect for an octet of massive vector bosons. The Nambu-Goldstone bosons arising from the breaking of chiral symmetry, of course, have the same charge assignments as the familiar π, K and η octet of pseudoscalars. The baryon superfluid mode is $\tilde{Q}$-neutral. In the color-flavor locked phase, we conclude, all the elementary excitations are integrally charged.[1] This is a classic aspect of confinement, here embodied in a controlled, weak-coupling framework.

All elementary excitations in the CFL phase are either $\tilde{Q}$-neutral or couple to $A_\mu^{\tilde{Q}}$ with charges which are integer multiples of the $\tilde{Q}$-charge of the electron $\tilde{e} = e\cos\theta$, which is less than e because the electron couples only to the A_μ component of $A_\mu^{\tilde{Q}}$. The only massless excitation (the superfluid mode) is $\tilde{Q}$-neutral. Because all charged excitations have nonzero mass and there are no electrons present [63], the CFL phase in bulk is a transparent insulator at low temperatures: $\tilde{Q}$-magnetic and $\tilde{Q}$-electric fields within it evolve simply according to Maxwell's equations, and low frequency $\tilde{Q}$-light traverses it without scattering.

It is fun to consider how color-flavor locked material would look. Imagine shining light on a chunk of dense quark matter in the CFL phase. If CFL matter occurs only within the cores of neutron stars, cloaked under kilometers of hadronic matter [65], the thought experiment we describe here in which light waves travelling in vacuum strike CFL matter can never arise in nature. If, however, the fact that quark matter features many more strange quarks than ordinary nuclear matter renders it stable even at zero pressure, then one may imagine quark stars in nature [66]. Such a quark star may be made of CFL quark matter throughout, or may have an outer layer in which a less symmetric pattern of pairing occurs. For example, quarks of only two colors and flavors may pair, yielding the 2SC phase which we shall discuss in the next section. (Some of the remaining quarks with differing Fermi momenta may also form a crystalline color superconductor, which we shall discuss below.) As in the CFL phase, the 2SC condensate leaves a (different) $\tilde{Q}$-photon massless. However, the 2SC phase is a good $\tilde{Q}$-conductor because of the presence of unpaired quarks and electrons. Thus, 2SC matter is opaque and metallic rather than transparent and insulating. Illuminating it would result in absorption and reflection, but no refraction. We shall assume that the quark matter we illuminate is in the transparent CFL phase all the way out to its surface.

Consider, then, an enormously dense, but transparent, illuminated quark star. Some light falling on its surface will reflect, and some will refract into the star in the form of $\tilde{Q}$-light. The reflection and refraction angles and the intensity of the reflected light and refracted $\tilde{Q}$-light have all been calculated [67]. The partial Meissner effect induced by a static magnetic field had been analyzed previously [64]. In Ref. [67], we analyze a time-varying electromagnetic field. As a bonus, our analysis allows us to use well understood properties of dense quark matter in the

[1] We shall see below that in two-flavor QCD, in which color-flavor locking does not occur, the color superconducting condensate which forms also leaves a $\tilde{Q}$-photon massless. The only difference relative to the CFL phase is that $\eta = -1/2\sqrt{3}$. (However, the $\tilde{Q}$-charges of the excitations are not all integral in this theory.)

CFL phase to learn about the (less well understood) QCD vacuum.

We assume that the light has ω and k both much less than the energy needed to create a charged excitation in the CFL phase. This means $\omega, k \ll \Delta$, where Δ is the fermionic gap, to avoid the breaking of pairs and the creation of quasiparticles. It also means $\omega, k \ll m_{\pi^\pm}, m_{K^\pm}$, where $\pi^\pm$ and $K^\pm$ are the charged pions and kaons of the CFL phase.

In vacuum the electromagnetic fields obey the free Maxwell's equations

$$\nabla \cdot \mathbf{D} = 0\,, \qquad \nabla \times \mathbf{E} = -\frac{\partial \mathbf{B}}{\partial t}\,, \tag{4.12}$$

$$\nabla \cdot \mathbf{B} = 0\,, \qquad \nabla \times \mathbf{H} = \frac{\partial \mathbf{D}}{\partial t}\,, \tag{4.13}$$

where $\mathbf{D} = \epsilon_0 \mathbf{E}$ and $\mathbf{B} = \mu_0 \mathbf{H}$, and ϵ_0 and μ_0 are the vacuum dielectric constant and magnetic permeability, respectively, such that the velocity of light $c = 1/\sqrt{\mu_0 \epsilon_0}$. Deep in the CFL phase, the rotated fields $\tilde{\mathbf{E}}$ and $\tilde{\mathbf{B}}$ obey the same field equations, but with dielectric constant [68]

$$\tilde{\epsilon} = \epsilon_0 \left(1 + \frac{8\alpha}{9\pi} \cos^2\theta \frac{\mu^2}{\Delta^2} \right)\,, \tag{4.14}$$

where α is the electromagnetic fine structure constant and μ is the chemical potential. This expression for $\tilde{\epsilon}$ is valid to leading order in α, and for $\omega, k \ll \Delta$. The dependence of $\tilde{\epsilon}$ on ω arises only in corrections to (4.14) which are suppressed by ω^2/Δ^2, and we therefore neglect dispersion in this letter. The magnetic permeability in the CFL phase remains unchanged to leading order, $\tilde{\mu} = \mu_0$. The index of refraction of CFL quark matter thus reduces to $\tilde{n} = \sqrt{\tilde{\mu}\tilde{\epsilon}/\mu_0\epsilon_0} = \sqrt{\tilde{\epsilon}/\epsilon_0}$. If we apply (4.14) for $\mu/\Delta \sim (4 - 10)$, we obtain $\tilde{n} \sim (1.02 - 1.1)$.

We take the surface of the CFL matter to be planar, with the CFL phase at $z > 0$ and vacuum at $z < 0$. (That is, we assume any curvature of the surface is on length scales long compared to the wavelength of the light.) For an ordinary dielectric, the analogous problem is solved in Ref. [69]. The complication here is that we must match the ordinary electric and magnetic fields in vacuum onto $\tilde{Q}$-electric and $\tilde{Q}$-magnetic fields within the CFL phase. The properties of the reflected and refracted waves will therefore depend upon both the dielectric constant $\tilde{\epsilon}$ and the mixing angle θ.

We are only interested in the reflected and refracted waves, and not in the detailed field configurations very close to the interface. This means that we can follow the strategy of Ref. [64] and encapsulate the physics of the interface into boundary conditions relating $\mathbf{E}$ and $\mathbf{B}$ on the vacuum side of the interface to $\tilde{\mathbf{E}}$ and $\tilde{\mathbf{B}}$ on the CFL side. On the CFL side, the massive X fields can be neglected as long as z is greater than some λ^{CFL}, while on the vacuum side, the confined gluon fields can be neglected as long as $|z|$ is greater than some λ^{QCD}. λ^{QCD} is a length scale characteristic of confinement. For the non-static fields of interest, and in the weak coupling regime, $\lambda^{\rm CFL}$ is of order $1/\Delta$, longer than the inverse Meissner mass $\sim g\mu$. In order to describe light whose wavelength is long compared to $\lambda^{\rm QCD}$ and $\lambda^{\rm CFL}$, we need boundary conditions relating $\mathbf{E}$ and $\mathbf{B}$ at $z = -\lambda^{\rm QCD}$ to $\tilde{\mathbf{E}}$ and $\tilde{\mathbf{B}}$ at $z = +\lambda^{\rm CFL}$.

X-magnetic fields experience a Meissner effect in the CFL phase, meaning that supercurrents in the CFL matter within $\lambda^{\rm CFL}$ of the interface screen the X-component of any ordinary magnetic field parallel to the interface on the vacuum side, yielding the boundary condition

$$\tilde{\mathbf{H}}_{\|}(t,x,y,\lambda^{\rm CFL}) = \cos\theta\, \mathbf{H}_{\|}(t,x,y,-\lambda^{\rm QCD})\,. \tag{4.15}$$

The CFL condensate is charged with respect to the X gauge boson, meaning that if there is an ordinary electric field perpendicular to the interface on the vacuum side, the X component of the electric flux will terminate in the CFL phase within $\lambda^{\rm CFL}$ of the interface, yielding

$$\tilde{\mathbf{D}}_{\perp}(t,x,y,\lambda^{\rm CFL}) = \cos\theta\, \mathbf{D}_{\perp}(t,x,y,-\lambda^{\rm QCD})\,. \tag{4.16}$$

We expect that the confined QCD vacuum should behave as if it is a condensate of color-magnetic monopoles [70]. That is, in the vacuum color magnetic field lines end: if there is a $\tilde{Q}$-magnetic field perpendicular to the interface on the CFL side, the vacuum will ensure that only the ordinary magnetic field is admitted. Thus,

$$\mathbf{B}_{\perp}(t,x,y,-\lambda^{\rm QCD}) = \cos\theta\, \tilde{\mathbf{B}}_{\perp}(t,x,y,\lambda^{\rm CFL})\,. \tag{4.17}$$

Finally, color magnetic currents on the vacuum side of the interface should exclude the color component of any $\tilde{Q}$-electric field parallel to the interface on the CFL side, ensuring that

$$\mathbf{E}_{\|}(t,x,y,-\lambda^{\rm QCD}) = \cos\theta\, \tilde{\mathbf{E}}_{\|}(t,x,y,\lambda^{\rm CFL})\,. \tag{4.18}$$

At sufficiently high density, the property of CFL matter from which (4.15) and (4.16) follow, namely the Meissner effect for X-bosons, is a weak-coupling phenomenon which can be understood analytically. The properties of the QCD vacuum used to deduce (4.17) and (4.18) follow from a reasonable and familiar description of confinement as a dual Meissner effect, but confinement is not yet understood analytically. It is therefore of interest that the analysis of refraction provides a *derivation* of (4.17) and (4.18) from (4.15) [67].

In Ref. [67], we analyze the reflection of light and refraction of $\tilde{Q}$-light when light is incident from vacuum onto CFL matter. The boundary conditions at the interface must be obeyed at all times, which immediately implies that the frequency of all the waves is the same. The boundary conditions must be obeyed at all points on the planar interface. For $1/k \gg \lambda^{\rm CFL}, \lambda^{\rm QCD}$ this implies that the incident, reflected and refracted wave vectors must lie in a plane and also implies Snell's relation between the angle of incidence i and the angle of refraction r:

$\sin i = \tilde{n} \sin r$. The kinematics of the reflection and refraction of light on CFL quark matter are unaffected by the mixing angle θ.

The boundary conditions can also be used to find the intensities of the reflected and refracted radiation [67]. Upon setting $\cos\theta = 1$, the results so obtained reproduce results for reflection and refraction off standard dielectric media (see Ref. [69]). Decreasing $\cos\theta$ decreases the A_μ component of $A_\mu^{\tilde{Q}}$, and thus favors reflection over refraction. For θ as small as in nature, the changes introduced by $\theta \neq 0$ are small. In the (unphysical) limit in which $\cos\theta = 0$, $A_\mu^{\tilde{Q}}$ would be orthogonal to A_μ making the CFL phase a superconductor with respect to ordinary electromagnetism. In this limit, we expect and find zero refraction and perfect reflection for both polarizations.

In Ref. [67], we construct the Poynting vector and check that the reflection and refraction coefficients derived from the boundary conditions respect energy conservation. We then show, however, that if we had *used* energy conservation in our derivation instead of just as a check, we could have derived all our results from the single boundary condition (4.15). That is, given only the boundary condition (4.15) which is easily derived, Snell's law which is kinematic, and energy conservation, we can derive our solutions describing the reflection and refraction of light of both polarizations and, from these electromagnetic fields, we can then derive the remaining boundary conditions (4.16), (4.17) and (4.18). This means that we have *derived* the boundary conditions motivated above by the idea that the QCD vacuum behaves like a dual superconductor filled with a condensate of color-magnetic monopoles [70]. Having analyzed the illumination of dense quark matter, we find that in addition we have illuminated our understanding of the QCD vacuum.

It is perhaps of more practical value to analyze the response of a chunk of CFL matter to a static magnetic field, as in the core of a neutron star. The effect of a chunk of color superconducting quark matter (whether in the CFL phase or in the less symmetric phase in which only up and down quarks pair) on a static magnetic field has been described in Ref. [64], and is essentially the static limit of the refraction calculation just presented. Some fraction of an externally applied ordinary magnetic field penetrates the superconductor in the form of a $\tilde{Q}$-magnetic field, while some fraction of the ordinary magnetic field is expelled by the Meissner effect. The fraction of the field which is expelled depends both on θ and on the shape of the chunk color superconducting quark matter, but it is small when θ is small, as in nature. Most of the flux is admitted, as $\tilde{Q}$-flux. This $\tilde{Q}$-magnetic field satisfies Maxwell's equations and is not restricted to flux tubes.

4.6. QUARK-HADRON CONTINUITY

The universal features of the color-flavor locked state: confinement, chiral symmetry breaking leaving a vector $SU(3)$ unbroken, and baryon number superfluidity, are exactly what one expects to find in nuclear matter in three-flavor QCD [13]. Perhaps this is not immediately obvious in the case of baryon number superfluidity, but let us recall that pairing phenomena, which would go over into neutron superfluidity and proton superconductivity in nuclear matter, are very well established in ordinary nuclei. In three-flavor QCD, there are good reasons [71] to think that the pairing interaction in the flavor singlet dibaryon channel (the so-called H-dibaryon channel) would be quite attractive in three-flavor QCD, and support a robust baryon number superfluidity. Thus, the symmetries of the color-flavor locked phase are precisely those of nuclear matter in three-flavor QCD, perhaps better referred to as hypernuclear matter [13].

Furthermore, there is an uncanny resemblance between the low-lying spectrum computed from first principles for QCD at asymptotically high density, and what one expects to find in hypernuclear matter, in a world with three degenerate quark flavors. It is hard to resist the inference that in this theory, there need be no phase transistion between nuclear density and high density [13]. There need be no sharp boundary between hypernuclear matter, where microscopic caculations are difficult but the convenient degrees of freedom are "obviously" hadrons, and the asymptotic high-density phase, where weak-coupling (but non-perturbative) calculations are possible, and the right degrees of freedom are elementary quarks and gluons, together with the collective Nambu-Goldstone modes. We call this quark-hadron continuity [13]. Perhaps the least surprising aspect of this, by now, is the continuity between the pseudoscalar mesons at nuclear density and those at asymptotically high densities, since in both regimes these are present as a result of the breaking of the same symmetry. It might seem more shocking that a quark can go over continuously into, or "be", a baryon, since baryons are supposed to contain three quarks, but remember that in the color-flavor locked phase the quarks are immersed in a diquark condensate, and so a one-quark excitation can pick two quarks up from (or lose to quarks to) the condensate at will. The difference between one and three is negotiable. What about the gluons? Within the color-flavor locked phase, similarly, they are quite literally the physical vector mesons. They are massive, as we have discussed, and have the right quantum numbers. Thus the original vision of Yang and Mills – who proposed non-abelian gauge theory as a model of ρ mesons – is here embodied.

Note that the hypothesis of continuity between hypernuclear and dense quark matter certainly does not preclude quantitative change. Far from it. The dispersion relation for a fermion — whether a quark in the CFL phase or a baryon in the hypernuclear phase — is characterized by a gap at the Fermi surface and by a gap at zero momentum, i.e. a mass. As a function of increasing density, gaps at the hyperon Fermi surfaces due to hyperon-hyperon pairing evolve continuously to become the gaps at the quark Fermi surfaces which characterize the color-flavor locked phase [17]. During this evolution, the gaps are thought to increase significantly. In contrast, the masses ("gaps at zero momentum") decrease dramatically with increasing density as they evolve from being of order the hyperon masses in hypernuclear matter to being of order the current quark masses at asymptotically high densities.

Note that in order for quark-hadron continuity to be realized, $U(1)_{EM}$ must not be broken by hyperon-hyperon pairing [17]. During the evolution of the theory as a function of increasing density, there is always an unbroken $U(1)$ and a massless $\tilde{Q}$-photon with respect to which the excitations are integer charged. As the density is increased, however, the definition of the $\tilde{Q}$-photon in terms of the vacuum photon and gluon fields rotates. The ordinary photon rotates to become the $\tilde{Q}$-photon of the CFL phase. Turning to the massive vector bosons, in hypernuclear matter there is both an octet and a singlet. The singlet must become much heavier than the octet as a function of increasing density, since in the low energy description of the color-flavor locked phase one finds the octet alone. We see that if quark-hadron continuity is realized in QCD with three degenerate quarks, it requires various quantitative (but continuous) changes. It is remarkable that it is even possible to imagine that the physical excitations of the theory evolve continuously as one dials the density up and goes from a strongly coupled hadronic world to a weak-coupling world of quarks and gluons.

If the quarks are massless, the Nambu-Goldstone bosons are massless in both hypernuclear and CFL quark matter, and in between. Once nondegenerate quark masses are introduced, however, the evolution of the Nambu-Goldstone masses as a function of increasing density becomes more intricate, as the kaon must go from being heavier than the pion to being lighter.

Finally, the story becomes further complicated once the strange quark is made as heavy as in nature [17, 39]. Although the color-flavor locked phase is certainly obtained at asymptotically densities, where quark masses are neglectable, the nuclear matter phase, made of neutrons and protons only, is not continuously connectable with the color-flavor locked phase. If quark-hadron continuity is to be realized in the phase diagram of nature, what must happen is that, as a function of increasing density, one first goes from nuclear matter to hypernuclear matter, with sufficiently high density that all the hyperons have similar Fermi surfaces. This first stage must involve phase transitions, as the symmetries of hypernuclear matter differ from those of ordinary nuclear matter. Then, as the density is increased further, the hypernuclear matter may evolve continuosly to become CFL quark matter, with pairing among hyperons becoming CFL pairing among quarks.

We now have a description of the properties of the CFL phase and its excitations, in which much can be described quantitatively if the value of the gap Δ is known. In the next two sections, we describe the less symmetric forms of color superconductivity which arise in QCD with $N_f \neq 3$. Already, however, in our idealized world (in which we either have three degenerate quarks or such high densities that the quark mass differences can be neglected) let us pause to marvel at our theoretical good fortune. The color-flavor locked phase is a concrete realization of the idea of complementarity: the same phase of a gauge theory can be described simultaneously as one in which the gauge symmetry is spontaneously broken and as one in which color is confined [11]. This means that it provides us with a weak-coupling laboratory within which we can study a confined phase from first principles at weak coupling. It is furthermore a phase of QCD wherein the physics of chiral symmetry breaking — indeed all the parameters of the chiral effective Lagrangian and all known or conjectured phenomena of the pseudoscalar meson sector, including kaon condensation — are amenable to controlled, weak-coupling calculation.

5. 2SC and Other Variants

5.1. TWO FLAVORS

In the previous section, we have described quark matter in QCD with three degenerate flavors of light quarks. In nature the strange quark is heavier than the other two, so we now go to the opposite extreme of an infinitely heavy strange quark and describe the color superconducting phase in QCD with two flavors of light quarks. The $\bar{\mathbf{3}}$ is still the most attractive color channel for quark pairing [5, 7, 8, 9], and the resulting condensate

$$\langle \psi_i^\alpha \psi_j^\beta \rangle \propto \epsilon_{\alpha\beta 3}\epsilon^{ij} \tag{5.1}$$

picks a color direction. In this case, quarks of the first two colors (red and green) participate in pairing, while the third color (blue) does not.
The ground state is invariant under an $SU(2)$ subgroup of the color rotations that mixes red and green, but the blue quarks are singled out as different. The pattern of symmetry breaking is therefore (with gauge symmetries in square brackets)

$$\begin{array}{c} [SU(3)_{\rm color}] \times [U(1)_Q] \times SU(2)_L \times SU(2)_R \\ \longrightarrow\ [SU(2)_{\rm color}] \times [U(1)_{\tilde{Q}}] \times SU(2)_L \times SU(2)_R \end{array} \tag{5.2}$$

The features of this pattern of condensation are

1. The color gauge group is broken down to $SU(2)$, so five of the gluons will become massive, with masses of order the gap (since the coupling is of order 1). Their masses have been computed in the weak-coupling theory valid at asymptotically high densities [59] and in an instanton liquid model [60]. The remaining three gluons are associated with an unbroken $SU(2)$ red-green gauge symmetry, whose confinement distance rises exponentially with density [72]. This aspect of the infrared physics of the 2SC phase is not under perturbative control, so unlike the CFL phase we cannot claim that in the 2SC phase any physical quantity can be obtained from a controlled weak-coupling calculation at sufficiently high density.

2. The red and green quark modes acquire a gap Δ, which is the mass of the physical excitations around the Fermi surface (quasiquarks). There is no gap for the blue quarks in this ansatz, and it is an interesting question whether they find some other channel in which to pair. The available attractive channels arising from the instanton interaction break rotational invariance, and are weak so the gap will be much smaller, perhaps in the keV range [8].
3. As in the CFL phase a linear combination of the photon and one gluon gives a massless gauge boson that couples to a new unbroken rotated electromagnetism $\tilde{Q}$. The two blue quarks have $\tilde{Q}$-charges 0 and 1.
4. No global symmetries are broken so there are no light scalars—the 2SC phase is not a superfluid. It has the same symmetries as the quark-gluon plasma (QGP), so there need not be any phase transition between them, but there will be a chiral phase transition between the hadronic and 2SC phases. This phase transition is first order[8, 16, 73, 19] since it involves a competition between chiral condensation and diquark condensation.[16, 19] Although the quark pair condensate appears to break baryon number, it does not. In the two flavor case baryon number is a linear combination of electric charge and isospin, $B = 2Q - 2I_3$, so baryon number is already included in the symmetry groups of (5.2). Just as an admixture of gluon and photon survives unbroken as a rotated electromagnetism, so an admixture of B and T_8 survives unbroken as a rotated baryon number.
5. Axial color is not a symmetry of the QCD action, but at asymptotically high densities where the QCD coupling g is weak, explicit axial color breaking is also weak. As a result, the pseudoscalar excitations of the condensate which would be Goldstone bosons arising from axial-$SU(3)_{\text{color}}$ to axial-$SU(2)_{\text{color}}$ breaking if g were zero may be rather light [74].

It is interesting that both the 2SC and CFL phases satisfy anomaly matching constraints, even though it is not yet completely clear whether this must be the case when Lorentz invariance is broken by a nonzero density.[75] It is not yet clear how high density QCD with larger numbers of flavors,[31] which we discuss below, satisfies anomaly matching constraints. Also, anomaly matching in the 2SC phase requires that the up and down quarks of the third color remain ungapped; this requirement must, therefore, be modified once these quarks pair to form a $J = 1$ condensate, breaking rotational invariance [8].

5.2. TWO+ONE FLAVORS

Nature chooses two light quarks and one middle-weight strange quark. If we imagine beginning with the CFL phase and increasing m_s, how do we get to the 2SC phase? This question has been answered in Refs. [17, 39]. A nonzero m_s weakens those condensates which involve pairing between light and strange quarks. The CFL phase requires nonzero $\langle us\rangle$ and $\langle ds\rangle$ condensates; because

these condensates pair quarks with differing Fermi momenta they can only exist if the resulting gaps (call them Δ_{us} and Δ_{ds}) are larger than of order $m_s^2/2\mu$, the difference between the u and s Fermi momenta in the absence of pairing. This means that as a function of increasing m_s at fixed μ (or decreasing μ at fixed m_s) there must be a first order unlocking phase transition.[17, 39] The argument can be phrased thus: the 2SC and CFL phases must be separated by a phase transition, because chiral symmetry is broken in the CFL phase but not in the 2SC phase; suppose this transition were second order; this would require Δ_{us} and Δ_{ds} to be infinitesimally small but nonzero just on the CFL side of the transition; however, these gaps must be greater than of order $m_s^2/2\mu$; a second order phase transition is therefore a logical impossibility, either in mean field theory or beyond; the transition must therefore be first order. Note that the m_s that appears in these estimates is a density dependent effective strange quark mass, somewhat greater than the current quark mass [76].
Putting in reasonable numbers for quark matter which may arise in compact stars, for $m_s = 200 - 300$ MeV and $\mu = 400 - 500$ MeV we find that the CFL phase is obtained if the interactions are strong enough to generate a gap Δ which is larger than about $40 - 110$ MeV, while the 2SC phase is obtained if Δ is smaller. $\Delta \sim 40 - 110$ MeV is within the range of current estimates and present calculational methods are therefore not precise enough to determine whether quark matter with these parameters is in the CFL or 2SC phases. At asymptotically high densities, however, the CFL phase is necessarily favored.
Note that the 2SC phase in QCD with massive strange quarks *is* a superfluid: no linear combination of baryon number and a gauge symmetry remains unbroken. Note also that in this phase, five quarks (blue up; blue down; strange quarks of all colors) are left unpaired. As we shall discuss in the next section, they may in fact pair to form a crystalline color superconductor.

5.3. FOUR OR MORE FLAVORS

We end this section with brief mention of four variants which are unphysical, but nevertheless instructive: QCD with more than three light flavors, QCD with two colors, QCD with many colors, and QCD with large isospin density and zero baryon density.
Dense quark matter in QCD with more than three flavors was studied in Ref. [31]. The main result is that the color-flavor locking phenomenon persists: condensates form which lock color rotations to flavor rotations, and the $SU(N_f)_L \times SU(N_f)_R$ group is broken down to a vector subgroup. Unlike with $N_f = 3$, however, the unbroken group is not the full $SU(N_f)_{L+R}$ which is unbroken in the vacuum. In the case of $N_f = 4$, for example, one finds $SU(4)_L \times SU(4)_R \to O(4)_{L+R}$ while in the case of $N_f = 5$, $SU(5)_L \times SU(5)_R \to SU(2)_{L+R}$.[31] For $N_f = 4, 5$ as for $N_f = 3$, chiral symmetry is broken in dense quark matter. However, because

the unbroken vector groups are smaller than $SU(N_f)_V$, there must be a phase transition between hadronic matter and dense quark matter in these theories.[31] If N_f is a multiple of three, the order parameter takes the form of multiple copies of the $N_f = 3$ order parameter, each locking a block of three flavors to color [31]. All quarks are gapped in this phase, as in the $N_f = 3$ CFL phase. For $N_f = 6$, the resulting symmetry breaking pattern is $SU(6)_L \times SU(6)_R \to SU(3)_{L+R} \times U(1)_{L+R} \times U(1)_{L-R}$ [31]. The unbroken $SU(3)_{L+R}$ is a simultaneous rotation of both three flavor blocks for L and R and a global color rotation. Note that the unbroken $U(1)$'s are subgroups of the original $SU(6)$ groups: they correspond to vector and axial flavor rotations which rotate one three flavor block relative to the other. Note that for $N_f = 6$, unlike for $N_f = 3, 4, 5$, chiral symmetry is not completely broken at high density: an axial $U(1)$ subgroup remains unbroken. As the primary condensate we have just described leaves no quarks ungapped, there is no reason to expect the formation of any subdominant condensate which could break the unbroken chiral symmetry. Both because of this unbroken chiral symmetry and because the unbroken vector symmetry differs from that of the vacuum, there must be a phase transition between hadronic matter and dense quark matter in QCD with $N_f = 6$ [31].

5.4. TWO COLORS

The simplest case of all to analyze is QCD with two colors and two flavors. The condensate is antisymmetric in color and flavor, and is therefore a singlet in both color and flavor. Because it is a singlet in color, dense quark matter in this theory is not a color superconductor. Although the condensate is a singlet under the ordinary $SU(2)_L \times SU(2)_R$ flavor group, it nevertheless does break symmetries because the symmetry of the vacuum in QCD with $N_f = N_c = 2$ is enhanced to $SU(4)$. One reason why $N_c = 2$ QCD is interesting to study at nonzero density is that it provides an example where quark pairing can be studied on the lattice.[77] The $N_c = 2$ case has also been studied analytically in Refs. [9, 78]; pairing in this theory is simpler to analyze because quark Cooper pairs are color singlets.

5.5. MANY COLORS

The $N_c \to \infty$ limit of QCD is often one in which hard problems become tractable. However, the ground state of $N_c = \infty$ QCD is a chiral density wave, not a color superconductor [79]. At asymptotically high densities, color superconductivity persists up to N_c's of order thousands [80, 81] before being supplanted by the phase described in Ref. [79]. At any finite N_c, color superconductivity occurs at arbitrarily weak coupling whereas the chiral density wave does not. For $N_c = 3$, color superconductivity is still favored over the chiral density wave (although not by much) even if the interaction is so strong that the color superconductivity gap is $\sim \mu/2$ [82].

5.6. QCD AT LARGE ISOSPIN DENSITY

The phase of $N_c = 3$ QCD with nonzero isospin density ($\mu_I \neq 0$) and zero baryon density ($\mu = 0$) *can* be simulated on the lattice [83]. The sign problems that plague simulations at $\mu \neq 0$ do not arise for $\mu_I \neq 0$. Although not physically realizable, physics with $\mu_I \neq 0$ and $\mu = 0$ is very interesting to consider because phenomena arise which are similar to those occurring at large μ and, in this context, these phenomena are accessible to numerical "experiments". Such lattice simulations can be used to test calculational methods which have also been applied at large μ, where lattice simulation is unavailable. At low isospin density, this theory describes a dilute gas of Bose-condensed pions. Large μ_I physics features large Fermi surfaces for down quarks and anti-up quarks, Cooper pairing of down and anti-up quarks, and a gap whose g-dependence is as in (3.5), albeit with a different coefficient of $1/g$ in the exponent [83]. This condensate has the same quantum numbers as the pion condensate expected at much lower μ_I, which means that a hypothesis of continuity between hadronic — in this case pionic — and quark matter as a function of μ_I. Both the dilute pion gas limit and the asymptotically large μ_I limit can be treated analytically; the possibility of continuity between these two limits can be tested on the lattice [83]. The transition from a weak coupling superconductor with condensed Cooper pairs to a gas of tightly bound bosons which form a Bose condensate can be studied in a completely controlled fashion.

6. Crystalline Color Superconductivity

At asymptotic densities, the ground state of QCD with quarks of three flavors (u, d and s) with equal masses is expected to be the color-flavor locked (CFL) phase. This phase features a condensate of Cooper pairs of quarks which includes ud, us, and ds pairs. Quarks of all colors and all flavors participate in the pairing, and all excitations with quark quantum numbers are gapped. As in any BCS state, the Cooper pairing in the CFL state pairs quarks whose momenta are equal in magnitude and opposite in direction, and pairing is strongest between pairs of quarks whose momenta are both near their respective Fermi surfaces. Pairing persists even in the face of a stress (such as a chemical potential difference or a mass difference) that seeks to push the Fermi surfaces apart, although a stress that is too strong will ultimately disrupt Cooper pairing [17, 39]. Thus, the CFL phase persists for unequal quark masses, so long as the differences are not too large [17, 39]. This means that the CFL phase is the ground state for real QCD, assumed to be in equilibrium with respect to the weak interactions, as long as the density is high enough.

Imagine decreasing the quark number chemical potential μ from asymptotically large values. The quark matter at first remains color-flavor locked, although the CFL condensate may rotate in flavor space as terms of order m_s^4 in the free energy become important [56]. Color-flavor locking is maintained until a transition to a state in which some quarks become ungapped. This "unlocking transition", which must be first order [17, 39], occurs when [17, 39, 63, 65]

$$\mu \approx m_s^2/4\Delta_0 \, . \tag{6.1}$$

In this expression and throughout this section, we write the gap in the BCS state as Δ_0. As we have seen, estimates in both models and asymptotic analyses suggest that it is of order tens to 100 MeV. m_s is the strange quark mass parameter, which includes the contribution from any $\langle \bar{s}s \rangle$ condensate induced by the nonzero current strange quark mass, making it a density-dependent effective mass, decreasing as density increases and equaling the current strange quark mass only at asymptotically high densities. At densities which may arise at the center of compact stars, corresponding to $\mu \sim 400 - 500$ MeV, m_s is certainly significantly larger than the current quark mass, and its value is not well-known. In fact, m_s decreases discontinuously at the unlocking transition [76]. Thus, the criterion (6.1) can only be used as a rough guide to the location of the unlocking transition in nature [76]. Given this quantitative uncertainty, there remain two logical possibilities for what happens as a function of decreasing μ. One possibility is a first order phase transition directly from color-flavor locked quark matter to hadronic matter, as explored in Ref. [65]. The second possibility is an unlocking transition [17, 39] to quark matter in which not all quarks participate in the dominant pairing, followed only at a lower μ by a transition to hadronic matter. We assume the second possibility here, and explore its consequences.

Once CFL is disrupted, leaving some species of quarks with differing Fermi momenta and therefore unable to participate in BCS pairing, it is natural to ask whether there is some generalization of the ansatz in which pairing between two species of quarks persists even once their Fermi momenta differ. Crystalline color superconductivity is the answer to this question. The idea is that it may be favorable for quarks with differing Fermi momenta to form pairs whose momenta are *not* equal in magnitude and opposite in sign [84, 85]. This generalization of the pairing ansatz (beyond BCS ansätze in which only quarks with momenta which add to zero pair) is favored because it gives rise to a region of phase space where *both* of the quarks in a pair are close to their respective Fermi surfaces, and such pairs can be created at low cost in free energy. Condensates of this sort spontaneously break translational and rotational invariance, leading to gaps which vary in a crystalline pattern. As a function of increasing depth in a compact star, μ increases, m_s decreases, and Δ_0 changes also. If in some shell within the quark matter core of a neutron star (or within a strange quark star) the quark number densities are such that crystalline color superconductivity arises, rotational vortices may be pinned in this shell, making it a locus for glitch phenomena [85].

An analysis of these ideas in the context of the disruption of CFL pairing is complicated by the fact that in quark matter in which CFL pairing does not occur, up and down quarks may nevertheless continue to pair in the usual BCS fashion. In this 2SC phase, the attractive channel involves the formation of Cooper pairs which are antisymmetric in both color and flavor, yielding a condensate with color (Greek indices) and flavor (Latin indices) structure $\langle q_a^\alpha q_b^\beta \rangle \sim \epsilon_{ab}\epsilon^{\alpha\beta 3}$. This condensate leaves five quarks unpaired: up and down quarks of the third color, and strange quarks of all three colors. Because the BCS pairing scheme leaves ungapped quarks with differing Fermi momenta, crystalline color superconductivity may result.

Most analyses of crystalline color superconductivity have been done in the simplified model context with pairing between two quark species whose Fermi momenta are pushed apart by turning on a chemical potential difference [85, 86, 87, 88], rather than considering CFL pairing in the presence of quark mass differences. In Ref. [89], we investigate the ways in which the response of the system to mass differences is similar to or different from the response to chemical potential differences. We can address this question within the two-flavor model by generalizing it to describe pairing between massless up quarks and strange quarks with mass m_s. For completeness, we introduce

$$\begin{aligned} \mu_u &= \mu - \delta\mu \\ \mu_s &= \mu + \delta\mu \,, \end{aligned} \tag{6.2}$$

allowing us to consider the effects of m_s and $\delta\mu$ simultaneously. We shall use this two-flavor toy model throughout, deferring an analysis of crystalline color superconductivity induced by the effects of m_s on three-flavor quark matter to future work.

6.1. CONSEQUENCES OF $\delta\mu \neq 0$, WITH M_S=0

Before describing the consequences of $m_s \neq 0$, let us review the salient facts known about the consequences of $\delta\mu \neq 0$, upon taking $m_s = 0$. If $|\delta\mu|$ is nonzero but less than some $\delta\mu_1$, the ground state in the two-flavor toy-model is precisely that obtained for $\delta\mu = 0$ [90, 91, 85].[2] In this 2SC state, red and green up and strange quarks pair, yielding four quasiparticles with superconducting gap Δ_0. Furthermore, the number density of red and green up quarks is the same as that of red and green strange quarks. As long as $|\delta\mu|$ is not too large, this BCS state remains unchanged (and favored) because maintaining equal number densities, and thus coincident Fermi surfaces, maximizes the pairing and hence the interaction energy. As $|\delta\mu|$ is increased, the BCS state remains the ground state of the system only as long as its negative interaction energy offsets the large positive free energy

[2] In this two-flavor toy-model the diquark condensate is a flavor singlet. As the condensate breaks no flavor symmetries, there is no analogue of the rotations of the condensate in flavor space which occur within the CFL phase with nonzero $\delta\mu$ [56].

cost associated with forcing the Fermi seas to remain coincident. In the weak coupling limit, in which $\Delta_0/\mu \ll 1$, the BCS state persists for $|\delta\mu| < \delta\mu_1 = \Delta_0/\sqrt{2}$ [90, 85]. For larger Δ_0, the $1/\sqrt{2}$ coefficient changes in value. These conclusions are the same whether the interaction between quarks is modeled as a point-like four-fermion interaction or is approximated by single-gluon exchange. The loss of BCS pairing at $|\delta\mu| = \delta\mu_1$ is the analogue in this toy model of the unlocking transition.

If $|\delta\mu| > \delta\mu_1$, BCS pairing between u and s is not possible. However, in a range $\delta\mu_1 < |\delta\mu| < \delta\mu_2$ near the unpairing transition, it is favorable to form a crystalline color superconducting state in which the Cooper pairs have nonzero momentum. This phenomenon was first analyzed by Larkin and Ovchinnikov and Fulde and Ferrell [84] (LOFF) in the context of pairing between electrons in which spin-up and spin-down Fermi momenta differ. It has proven difficult to find a condensed matter physics system which is well described simply as BCS pairing in the presence of a Zeeman effect: any magnetic perturbation that may induce a Zeeman effect tends to have much larger effects on the motion of the electrons, as in the Meissner effect. The QCD context of interest to us, in which the Fermi momenta being split are those of different flavors rather than of different spins, therefore turns out to be the natural arena for the phenomenon first analyzed by LOFF.

The crystalline color superconducting phase (also called the LOFF phase) has been described in Ref. [85] (following Refs. [84]) upon making the simplifying assumption that quarks interact via a four-fermion interaction with the quantum numbers of single gluon exchange. In the LOFF state, each Cooper pair carries momentum $2\mathbf{q}$ with $|\mathbf{q}| \approx 1.2\delta\mu$. The condensate and gap parameter vary in space with wavelength $\pi/|\mathbf{q}|$. Although the magnitude $|\mathbf{q}|$ is determined energetically, the direction $\hat{\mathbf{q}}$ is chosen spontaneously. The LOFF state is characterized by a gap parameter Δ and a diquark condensate, but not by an energy gap in the dispersion relation: the quasiparticle dispersion relations vary with the direction of the momentum, yielding gaps that vary from zero up to a maximum of Δ. The condensate is dominated by those regions in momentum space in which a quark pair with total momentum $2\mathbf{q}$ has both members of the pair within $\sim \Delta$ of their respective Fermi surfaces. These regions form circular bands on the two Fermi surfaces. Making the ansatz that all Cooper pairs make the same choice of direction $\hat{\mathbf{q}}$ corresponds to choosing a single circular band on each Fermi surface. It corresponds to a condensate which varies in position space like

$$\langle\psi(\mathbf{x})\psi(\mathbf{x})\rangle \propto \Delta e^{2i\mathbf{q}\cdot\mathbf{x}} . \tag{6.3}$$

This ansatz is certainly *not* the best choice. If a single plane wave is favored, why not two? That is, if one choice of $\hat{\mathbf{q}}$ is favored, why not add a second $\mathbf{q}$, with the same $|\mathbf{q}|$ but a different $\hat{\mathbf{q}}$? If two are favored, why not three? This question, namely the determination of the favored crystal structure of the crystalline color superconductor, is unresolved but is under investigation. Note, however, that if we find a region $\delta\mu_1 < |\delta\mu| < \delta\mu_2$ in which the simple LOFF ansatz with a single $\hat{\mathrm{q}}$ is favored over the BCS state and over no pairing, then the LOFF state with whatever crystal structure turns out to be optimal must be favored in *at least* this region. Note also that even the single $\hat{\mathrm{q}}$ ansatz, which we use henceforth, breaks translational and rotational invariance spontaneously. The resulting phonon has been analyzed in Ref. [87].

Crystalline color superconductivity is favored within a window $\delta\mu_1 < |\delta\mu| < \delta\mu_2$. As $|\delta\mu|$ increases from 0, one finds a first order phase transition from the ordinary BCS phase to the crystalline color superconducting phase at $|\delta\mu| = \delta\mu_1$ and then a second order phase transition at $|\delta\mu| = \delta\mu_2$ at which Δ decreases to zero. Because the condensation energy in the LOFF phase is much smaller than that of the BCS condensate at $\delta\mu = 0$, the value of $\delta\mu_1$ is almost identical to that at which the naive unpairing transition from the BCS state to the state with no pairing would occur if one ignored the possibility of a LOFF phase, namely $\delta\mu_1 = \Delta_0/\sqrt{2}$. For all practical purposes, therefore, the LOFF gap equation is not required in order to determine $\delta\mu_1$. The LOFF gap equation is used to determine $\delta\mu_2$ and the properties of the crystalline color superconducting phase [85]. In the limit of a weak four-fermion interaction, the crystalline color superconductivity window is bounded by $\delta\mu_1 = \Delta_0/\sqrt{2}$ and $\delta\mu_2 = 0.754\Delta_0$, as first demonstrated in Refs. [84]. These results have been extended beyond the weak four-fermion interaction limit in Ref. [85].

We now know that the use of the simplified point-like interaction significantly underestimates the width of the LOFF window: assuming instead that quarks interact by exchanging medium-modified gluons yields a much larger value of $\delta\mu_2$ [88]. This can be understood upon noting that quark-quark interaction by gluon exchange is dominated by forward scattering. In most scatterings, the angular positions on their respective Fermi surfaces do not change much. In the LOFF state, small-angle scattering is advantageous because it cannot scatter a pair of quarks out of the region of momentum space in which both members of the pair are in their respective circular bands, where pairing is favored. It is therefore natural that a forward-scattering dominated interaction like single-gluon exchange is more favorable for crystalline color superconductivity that a point-like interaction, which yields s-wave scattering. Thus, although for the present we use the point-like interaction in our analysis of m_s-induced crystalline color superconductivity, it is worth remembering that this is very conservative.

6.2. CONSEQUENCES OF $M_S \neq 0$

In the absence of any interaction, and thus in the absence of pairing, the effect of a strange quark mass is to shift the Fermi momenta to

$$p_F^u = \mu - \delta\mu$$

$$p_F^s = \sqrt{(\mu + \delta\mu)^2 - m_s^2} \,. \qquad (6.4)$$

Assuming both $|\delta\mu|/\mu$ and m_s/μ are small, the separation between the two Fermi momenta is $\approx |2\delta\mu - m_s^2/2\mu|$. This suggests the conjecture that even when $m_s \neq 0$ the description given in the previous subsection continues to be valid upon replacing $|\delta\mu|$ by $|\delta\mu - m_s^2/4\mu|$. We show in Ref. [89] that this conjecture is *incorrect* in one key respect: whereas if $m_s = 0$ a $|\delta\mu|$ which is nonzero but smaller than $\delta\mu_1$ has no effect on the BCS state, the BCS gap Δ_0 decreases with increasing m_s^2. We show that for small m_s^2, $\Delta_0(m_s)/\Delta_0(0)$ decreases linearly with m_s^2. Because Δ_0 occurs in the free energy in a term of order $\Delta_0^2\mu^2$, the m_s-dependence of Δ_0 corrects the free energy by of order $\Delta_0(0)^2 m_s^2$. As $\delta\mu$ has no analogous effect, we conclude that $m_s^2/4\mu$ and $\delta\mu$ have qualitatively different effects on the paired state.
At another level, however, the story *is* quite similar to that for $m_s = 0$: if $|\delta\mu - m_s^2/4\mu|$ is small enough, we find the BCS state; if $|\delta\mu - m_s^2/4\mu|$ lies within an intermediate window, we find LOFF pairing; if $|\delta\mu - m_s^2/4\mu|$ is large enough, no pairing is possible. The boundaries between the phases, however, are related to a $\Delta_0(m_s)$, rather than simply to a constant Δ_0. That is, the definitions of "small enough" and "large enough" are m_s-dependent. We map the $(m_s, \delta\mu)$ plane in Ref. [89]. We find that the appropriate variable to use to describe the width of the crystalline color superconductivity window is $\delta\mu/\Delta_0(m_s)$, as opposed to $\delta\mu/\Delta_0(0)$. When described with this variable, m_s-induced and $\delta\mu$-induced crystalline color superconductivity are approximately equally robust. At all but the weakest of couplings, the width of the crystalline color superconductivity window increases with m_s, meaning that crystalline color superconductivity is somewhat more robust if it is m_s-induced than if it is $\delta\mu$-induced. Indeed, we find that at the moderate coupling corresponding to $\Delta_0(0) = 100$ MeV, m_s-induced crystalline color superconductivity occurs whereas $\delta\mu$-induced crystalline color superconductivity does not.

6.3. OPENING THE CRYSTALLINE COLOR SUPERCONDUCTIVITY WINDOW

In Ref. [88], we analyze the crystalline color superconducting phase upon assuming that quarks interact by the exchange of a medium-modified gluon, as is quantitatively valid at asymptotically high densities. We obtain $\delta\mu_2$, the upper boundary of the crystalline color superconductivity window. This analysis is controlled at asymptotically high densities where the coupling g is weak.
At weak coupling, quark-quark scattering by single-gluon exchange is dominated by forward scattering. In most scatterings, the angular positions of the quarks on their respective Fermi surfaces do not change much. As a consequence, the weaker the coupling the more the physics can be thought of as a sum of many (1+1)-dimensional theories, with only rare large-angle scatterings able to connect

one direction in momentum space with others [26]. In the LOFF state, small-angle scattering is advantageous because it cannot scatter a pair of quarks out of the region of momentum space in which both members of the pair are in their respective rings, where pairing is favored. It is therefore natural to expect that a forward-scattering-dominated interaction like single-gluon exchange is more favorable for crystalline color superconductivity than a point-like interaction, which yields s-wave scattering.

Suppose for a moment that we were analyzing a truly $(1+1)$-dimensional theory. The momentum-space geometry of the LOFF state in one spatial dimension is qualitatively different from that in three. Instead of Fermi surfaces, we would have only "Fermi points" at $\pm\mu_u$ and $\pm\mu_d$. The only choice of $|\mathbf{q}|$ which allows pairing between u and d quarks at their respective Fermi points is $|\mathbf{q}| = \delta\mu$. In $(3+1)$ dimensions, in contrast, $|\mathbf{q}| > \delta\mu$ is favored because it allows LOFF pairing in ring-shaped regions of the Fermi surface, rather than just at antipodal points [84, 85]. Also, it has long been known that in a true $(1+1)$-dimensional theory with a point-like interaction between fermions, $\delta\mu_2/\Delta_0 \to \infty$ in the weak-interaction limit [92].

We expect that in $(3+1)$-dimensional QCD with the interaction given by single-gluon exchange, as $\mu \to \infty$ and $g(\mu) \to 0$ the $(1+1)$-dimensional results should be approached: the energetically favored value of $|\mathbf{q}|$ should become closer and closer to $\delta\mu$, and $\delta\mu_2/\Delta_0$ should diverge. We derive both these effects in Ref. [88] and furthermore show that both are clearly in evidence already at the rather large coupling $g = 3.43$, corresponding to $\mu = 400$ MeV using the conventions of Refs. [28, 34]. At this coupling, $\delta\mu_2/\Delta_0 \approx 1.2$, meaning that $(\delta\mu_2 - \delta\mu_1) \approx (1.2 - 1/\sqrt{2})\Delta_0$, which is much larger than $(0.754 - 1/\sqrt{2})\Delta_0$. If we go to much higher densities, where the calculation is under quantitative control, we find an even more striking enhancement: when $g = 0.79$ we find $\delta\mu_2/\Delta_0 > 1000$! We see that (relative to expectations based on experience with point-like interactions) the crystalline color superconductivity window is wider by more than four orders of magnitude at this weak coupling, and is about one order of magnitude wider at accessible densities if weak-coupling results are applied there.[3]

We have found that $\delta\mu_2/\Delta_0$ diverges in QCD as the weak-coupling, high-density limit is taken. Applying results valid at asymptotically high densities to those of interest in compact stars, we find that even here the crystalline color supercon-

[3] LOFF condensates have also recently been considered in two other contexts. In QCD with $\mu_u < 0$, $\mu_d > 0$ and $\mu_u = -\mu_d$, one has equal Fermi momenta for $\bar{u}$ antiquarks and d quarks, BCS pairing occurs, and consequently a $\langle\bar{u}d\rangle$ condensate forms [83, 93]. If $-\mu_u$ and μ_d differ, and if the difference lies in the appropriate range, a LOFF phase with a spatially varying $\langle\bar{u}d\rangle$ condensate results [83, 93]. The result of Ref. [88] that the LOFF window is much wider than previously thought applies in this context also. Suitably isospin asymmetric nuclear matter may also admit LOFF pairing, as discussed recently in Ref. [94]. Here, the interaction is not forward-scattering dominated.

ductivity window is an order of magnitude wider than that obtained previously upon approximating the interaction between quarks as point-like. The crystalline color superconductivity window in parameter space may therefore be much wider than previously thought, making this phase a *generic* feature of the phase diagram for cold dense quark matter. The reason for this qualitative increase in $\delta\mu_2$ can be traced back to the fact that gluon exchange at weaker and weaker coupling is more and more dominated by forward-scattering, while point-like interactions describe s-wave scattering. What is perhaps surprising is that even at quite *large* values of g, gluon exchange yields an order of magnitude increase in $\delta\mu_2 - \delta\mu_1$.

This discovery has significant implications for the QCD phase diagram and may have significant implications for compact stars. At high enough baryon density the CFL phase in which all quarks pair to form a spatially uniform BCS condensate is favored. Suppose that as the density is lowered the nonzero strange quark mass induces the formation of some less symmetrically paired quark matter before the density is lowered so much that baryonic matter is obtained. In this less symmetric quark matter, some quarks may yet form a BCS condensate. Those which do not, however, will have differing Fermi momenta. These will form a crystalline color superconducting phase if the differences between their Fermi momenta lie within the appropriate window. In QCD, the interaction between quarks is forward-scattering dominated and the crystalline color superconductivity window is consequently wide open. This phase is therefore generic, occurring almost anywhere there are some quarks which cannot form BCS pairs. Evaluating the critical temperature T_c above which the crystalline condensate melts requires solving the nonzero temperature gap equation obtained as in Ref. [86] for the case of a point-like interaction. As in that case, we expect that all compact stars which are minutes old or older are much colder than T_c. This suggests that wherever quark matter which is not in the CFL phase occurs within a compact star, crystalline color superconductivity is to be found. As we discuss in the next section, wherever crystalline color superconductivity is found rotational vortices may be pinned resulting in the generation of glitches as the star spins down.

7. Color Superconductivity in Compact Stars

Our current understanding of the color superconducting state of quark matter leads us to believe that it may occur naturally in compact stars. These are the only places in the universe where we expect very high densities and low temperatures. They typically have masses close to 1.4 solar masses, and are believed to have radii of order 10 km. Their density ranges from around nuclear density near the surface to higher values further in, although uncertainty about the equation of state leaves us unsure of the value in the core.

Much of the work on the consequences of quark matter within a compact star has focussed on the effects of quark matter on the equation of state, and hence on the radius of the star. As a Fermi surface phenomenon, color superconductivity has little effect on the equation of state: the pressure is an integral over the whole Fermi volume. Color superconductivity modifies the equation of state at the $\sim (\Delta/\mu)^2$ level, typically by a few percent [8]. Such small effects can be neglected in present calculations, and for this reason we will not attempt to survey the many ways in which observations of neutron stars are being used to constrain the equation of state [95]. Color superconductivity gives mass to excitations around the ground state: it opens up a gap at the quark Fermi surface, and makes the gluons massive. One would therefore expect its main consequences to relate to transport properties, such as mean free paths, conductivities and viscosities.

The critical temperature T_c below which quark matter is a color superconductor is high enough that any quark matter which occurs within neutron stars that are more than a few seconds old is in a color superconducting state. In the absence of lattice simulations, present theoretical methods are not accurate enough to determine whether neutron star cores are made of hadronic matter or quark matter. They also cannot determine whether any quark matter which arises will be in the CFL or 2SC phase, and if the latter whether the quarks that do not participate in 2SC pairing form a crystalline color superconductor. Just as the higher temperature regions of the QCD phase diagram are being mapped out in heavy ion collisions, we need to learn how to use neutron star phenomena to determine whether they feature cores made of 2SC (possibly crystalline) quark matter, CFL quark matter or hadronic matter, thus teaching us about the high density region of the QCD phase diagram. It is therefore important to look for astrophysical consequences of color superconductivity.

7.1. THE TRANSITION REGION

There are two possibilities for the transition from nuclear matter to quark matter in a neutron star: a mixed phase, or a sharp interface. The surface tension of the interface determines which is favored.

Here, we consider the case where the strange quark is light enough so that quark pairing is always of the CFL type. That is, we assume a direct transition from hadronic matter to CFL quark matter, with no intervening window of 2SC and/or crystalline color superconductivity. Figure 2 shows the μ_B-μ_e^{eff} phase diagram, ignoring electromagnetism. The lightly shaded region is where nuclear matter (NM) has higher pressure. The darker region is where quark matter (QM) has higher pressure. Where they meet is the coexistence line. The medium solid lines labelled by values of the pressure are isobars. Below the coexistence line they are given by the NM equation of state, above it by the QM equation of state.

The thick (red) lines are the neutrality lines. Each phase is negatively charged above its neutrality line and positively charged below it. Dotted lines show extensions onto the unfavored sheet (NM above the coexistence line, QM below it).

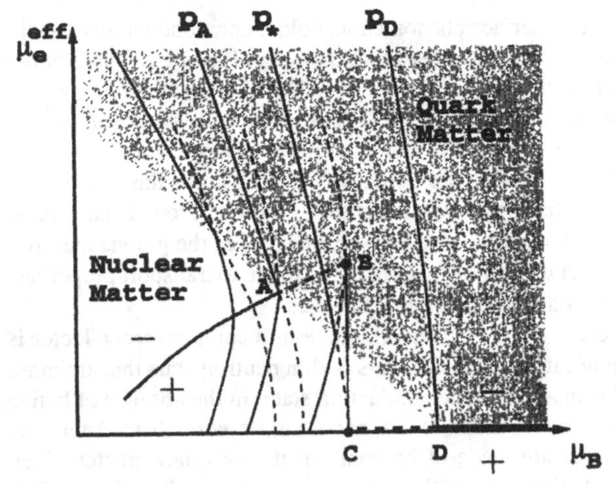

Figure 2. A schematic form of the μ_B-μ_e phase diagram for nuclear matter and CFL quark matter, ignoring electromagnetism. For an explanation see the text.

The electric charge density is

$$Q = -\left.\frac{\partial p}{\partial \mu_e}\right|_{\mu_B} \tag{7.1}$$

so the neutrality line goes through the right-most extremum of each isobar, since there the derivative of pressure with respect to μ_e is zero. For the CFL phase, the neutrality line is $\mu_e = 0$ [63].

Two possible paths from nuclear to CFL matter as a function of increasing μ are shown. In the absence of electromagnetism and surface tension, the favored option is to progress along the coexistence line from A to D, giving an overall neutral phase made of appropriate relative volumes of negatively charged CFL matter and positively charged nuclear matter.

If, on the other hand, Coulomb and surface energies are large, then the system remains on the nuclear neutrality line up to B, where there is a single interface between nuclear matter at B and CFL matter at C. The effective chemical potential $\mu_e^{\rm eff}$ changes across the interface, meaning that there must be an electric field at the interface [65]. As a result, charged boundary layers develop with negative charge leaking into the CFL phase in the form of electrons and perhaps CFL mesons, leaving a net positive charge (in the form of both protons and a depletion of electrons) on the nuclear side of the interface. These charged boundary layers, the analogues of inversion layers in semiconductor physics, are of order tens of fermi thick [65]. This minimal interface, with its attendant charged boundary layers occurs between phases with the same μ_e, $\mu = \mu_B = \mu_C$, and pressure

P_*. The effective chemical potential $\mu_e^{\rm eff}$ changes across the interface, though, as a result of the presence of the electric field.

The single interface creates a dramatic density discontinuity in the star: CFL quark matter at about four times nuclear density floats on nuclear matter at about twice nuclear density. This may affect the mass vs. radius relationship for neutron stars with quark matter cores. It may also have qualitative effects on the gravitational wave profile emitted during the inspiral and merger of two compact stars of this type.

If the surface tension σ_{QCD} and the electrostatic forces are ignored, then a mixed phase is favored over a sharp interface. [96, 97, 98]. If we treat $\sigma_{\rm QCD}$ as an independent parameter, we can estimate the surface and Coulomb energy cost of the mixed phase [98, 65]. We find that if $\sigma_{\rm QCD} \gtrsim 40~{\rm MeV/fm}^2$, as seems likely, the single sharp interface with its attendant charged boundary layers is free-energetically favored over the mixed phase [65].

7.2. COOLING BY NEUTRINO EMISSION

We turn now to neutron star phenomena which *are* affected by Fermi surface physics. For the first 10^{5-6} years of its life, the cooling of a neutron star is governed by the balance between heat capacity and the loss of heat by neutrino emission. How are these quantities affected by the presence of a quark matter core? This has been addressed recently in Refs. [99, 100], following earlier work in Ref. [101]. Both the specific heat C_V and the neutrino emission rate L_ν are dominated by physics within T of the Fermi surface. If, as in the CFL phase, all quarks have a gap $\Delta \gg T$ then the contribution of quark quasiparticles to C_V and L_ν is suppressed by $\sim \exp(-\Delta/T)$. There may be other contributions to L_ν [99], but these are also very small. In the CFL phase, the specific heat is dominated by that of the superfluid mode — i.e. the Goldstone boson associated with the spontaneous breaking of $U(1)_B$ — and there may also be small contributions from the light but not massless pseudo-Goldstone bosons associated with chiral symmetry breaking. Although further work is required, it is already clear that both C_V and L_ν are much smaller than in the nuclear matter outside the quark matter core. This means that the total heat capacity and the total neutrino emission rate (and hence the cooling rate) of a neutron star with a CFL core will be determined completely by the nuclear matter outside the core. The quark matter core is "inert": with its small heat capacity and emission rate it has little influence on the temperature of the star as a whole. As the rest of the star emits neutrinos and cools, the core cools by conduction, because the electrons keep it in good thermal contact with the rest of the star. These qualitative expectations are nicely borne out in the calculations presented in Ref. [100].

The analysis of the cooling history of a neutron star with a quark matter core in the 2SC phase is more complicated. The red and green up and down quarks pair with a gap many orders of magnitude larger than the temperature, which is of order 10 keV, and are therefore inert as described above. The remaining quarks may form a crystalline color superconductor. In addition, strange quarks may form an $\langle ss \rangle$ condensate with angular momentum $J = 1$ which locks to color in such a way that rotational invariance is not broken [102]. The resulting gap has been estimated to be of order hundreds of keV,[102] although applying results of Ref. [8] suggests a somewhat smaller gap, around 10 keV. The critical temperature T_c above which no condensate forms is of order the zero-temperature gap Δ. ($T_c = 0.57\Delta$ for $J = 0$ condensates.[25]) Therefore, if there are quarks for which $\Delta \sim T$ or smaller, these quarks do not pair at temperature T. Such quark quasiparticles will radiate neutrinos rapidly (via direct URCA reactions like $d \rightarrow u + e + \bar{\nu}$, $u \rightarrow d + e^+ + \nu$, etc.) and the quark matter core will cool rapidly and determine the cooling history of the star as a whole [101, 100]. The star will cool rapidly until its interior temperature is $T < T_c \sim \Delta$, at which time the quark matter core will become inert and the further cooling history will be dominated by neutrino emission from the nuclear matter fraction of the star. If future data were to show that neutron stars first cool rapidly (direct URCA) and then cool more slowly, such data would allow an estimate of the smallest quark matter gap. We are unlikely to be so lucky. The simple observation of rapid cooling would *not* be an unambiguous discovery of quark matter with small gaps; there are other circumstances in which the direct URCA processes occur. However, if as data on neutron star temperatures improves in coming years the standard cooling scenario proves correct, indicating the absence of the direct URCA processes, this *would* rule out the presence of quark matter with gaps in the 10 keV range or smaller. The presence of a quark matter core in which *all* gaps are $\gg T$ can never be revealed by an analysis of the cooling history.

7.3. SUPERNOVA NEUTRINOS

We now turn from neutrino emission from a neutron star which is many years old to that from the protoneutron star during the first seconds of a supernova. Carter and Reddy [103] have pointed out that when this protoneutron star is at its maximum temperature of order 30-50 MeV, it may have a quark matter core which is too hot for color superconductivity. As such a protoneutron star core cools over the next few seconds, this quark matter will cool through T_c, entering the color superconducting regime of the QCD phase diagram. For $T \sim T_c$, the specific heat rises and the cooling slows. Then, as T drops further and Δ increases to become greater than T, the specific heat drops rapidly. Furthermore, as the number density of quark quasiparticles becomes suppressed by $\exp(-\Delta/T)$, the neutrino transport mean free path rapidly becomes very long [103]. This means that all the neutrinos previously trapped in the now color superconducting core are able to escape in a sudden burst. If a terrestrial neutrino detector sees thousands of neutrinos from a future supernova, Carter and Reddy's results suggest that there may be a signature of the transition to color superconductivity present in the time distribution of these neutrinos. Neutrinos from the core of the protoneutron star will lose energy as they scatter on their way out, but because they will be the last to reach the surface of last scattering, they will be the final neutrinos received at the earth. If they are released from the quark matter core in a sudden burst, they may therefore result in a bump at late times in the temporal distribution of the detected neutrinos. More detailed study remains to be done in order to understand how Carter and Reddy's signature, dramatic when the neutrinos escape from the core, is processed as the neutrinos traverse the rest of the protoneutron star and reach their surface of last scattering.

7.4. MAGNETIC FIELD EVOLUTION

Next, we turn to the physics of magnetic fields within color superconducting neutron star cores [104, 64]. The interior of a conventional neutron star is a superfluid (because of neutron-neutron pairing) and is an electromagnetic superconductor (because of proton-proton pairing). Ordinary magnetic fields penetrate it only in the cores of magnetic flux tubes. A color superconductor behaves differently. At first glance, it seems that because a diquark Cooper pair has nonzero electric charge, a diquark condensate must exhibit the standard Meissner effect, expelling ordinary magnetic fields or restricting them to flux tubes within whose cores the condensate vanishes. This is not the case, as we have seen: a linear combination of the $U(1)$ gauge transformation of ordinary electromagnetism and one (the eighth) color gauge transformation remains unbroken even in the presence of the condensate. This means that the ordinary photon A_μ and the eighth gluon G^8_μ are replaced by the new linear combinations of (4.8) and (4.9), where $A^{\tilde{Q}}_\mu$ is massless and A^X_μ is massive. That is, $B_{\tilde{Q}}$ satisfies the ordinary Maxwell equations while B_X experiences a Meissner effect. If a color superconducting neutron star core is subjected to an ordinary magnetic field, it will either expel the X component of the flux or restrict it to flux tubes, but it can (and does[64]) admit the great majority of the flux in the form of a $B_{\tilde{Q}}$ magnetic field satisfying Maxwell's equations. The decay in time of this "free field" (i.e. not in flux tubes) is limited by the $\tilde{Q}$-conductivity of the quark matter. A color superconductor is not a $\tilde{Q}$-superconductor — that is the whole point — but it may turn out to be a very good $\tilde{Q}$-conductor due to the presence of electrons: if a nonzero density of electrons is required in order to maintain charge neutrality, the $B_{\tilde{Q}}$ magnetic field likely decays only on a time scale which is much longer than the age of the universe [64]. This means that a quark matter core within a neutron star can serve as an "anchor" for the magnetic field: whereas in ordinary nuclear matter the magnetic flux tubes can be dragged outward by the neutron superfluid vortices as the star spins down [105], the magnetic flux within the color superconducting core simply cannot decay. Even though this distinction is a qualitative one, it will be difficult to confront it with data since what is observed is the total dipole moment of the neutron star. A color superconducting core anchors those magnetic flux lines which pass through the core, while in a neutron star with no quark matter core the entire internal magnetic field can decay over time. In both cases, however, the total dipole moment can change since the magnetic flux lines which do not pass through the core can move.

7.5. CRYSTALLINE COLOR SUPERCONDUCTIVITY AND GLITCHES IN QUARK MATTER

The final consequence of color superconductivity we wish to discuss is the possibility that (some) glitches may originate within quark matter regions of a compact star.

We do not yet know whether compact stars feature quark matter cores. And, we do not yet know whether, if they contain quark matter, that quark matter is color-flavor locked, meaning that quarks of all colors and flavors participate in BCS pairing, or whether the BCS condensate leaves some quarks unpaired. The lesson we take from the toy model analysis is that because the interaction between quarks in QCD is dominated by forward scattering, rather than being an s-wave point-like interaction, the difference in Fermi momenta between the unpaired quarks need not fall within a narrow window in order for them to form a crystalline color superconductor.

We wish now to ask whether the presence of a shell of crystalline color superconducting quark matter in a compact star (between the hadronic "mantle" and the CFL "inner core") has observable consequences. A quantitative formulation of this question would allow one either to discover crystalline color superconductivity, or to rule out its presence. (The latter would imply either no quark matter at all, or a single CFL-nuclear interface [65].)

Many pulsars have been observed to glitch. Glitches are sudden jumps in rotation frequency Ω which may be as large as $\Delta\Omega/\Omega \sim 10^{-6}$, but may also be several orders of magnitude smaller. The frequency of observed glitches is statistically consistent with the hypothesis that all radio pulsars experience glitches [106]. Glitches are thought to originate from interactions between the rigid neutron star crust, typically somewhat more than a kilometer thick, and rotational vortices in a neutron superfluid. The inner kilometer of crust consists of a crystal lattice of nuclei immersed in a neutron superfluid [107]. Because the pulsar is spinning, the neutron superfluid (both within the inner crust and deeper inside the star) is threaded with a regular array of rotational vortices. As the pulsar's spin gradually slows, these vortices must gradually move outwards since the rotation frequency of a superfluid is proportional to the density of vortices. Deep within the star, the vortices are free to move outwards. In the crust, however, the vortices are pinned by their interaction with the nuclear lattice. Models [108] differ in important respects as to how the stress associated with pinned vortices is released in a glitch: for example, the vortices may break and rearrange the crust, or a cluster of vortices may suddenly overcome the pinning force and move macroscopically outward, with the sudden decrease in the angular momentum of the superfluid within the crust resulting in a sudden increase in angular momentum of the rigid crust itself and hence a glitch. All the models agree that the fundamental requirements are the presence of rotational vortices in a superfluid and the presence of a rigid structure which impedes the motion of vortices and which encompasses enough of the volume of the pulsar to contribute significantly to the total moment of inertia. Although it is premature to draw quantitative conclusions, it is interesting to speculate that some glitches may originate deep within a pulsar which features a quark matter core, in a region of that core which is in the crystalline color superconductor phase. If this phase occurs within a pulsar it will be threaded by an array of rotational vortices. It is reasonable to expect that these vortices will be pinned in a LOFF crystal, in which the diquark condensate varies periodically in space. The diquark condensate vanishes at the core of a rotational vortex, and for this reason the vortices will prefer to be located with their cores pinned to the nodes of the LOFF crystal.

A real calculation of the pinning force experienced by a vortex in a crystalline color superconductor must await the determination of the crystal structure of the LOFF phase. We can, however, attempt an order of magnitude estimate along the same lines as that done by Anderson and Itoh [109] for neutron vortices in the inner crust of a neutron star. In that context, this estimate has since been made quantitative [110, 111, 108]. For one specific choice of parameters [85], the LOFF phase is favored over the normal phase by a free energy $F_{\rm LOFF} \sim 5 \times (10\ {\rm MeV})^4$ and the spacing between nodes in the LOFF crystal is $b = \pi/(2|{\bf q}|) \sim 9$ fm. The thickness of a rotational vortex is given by the correlation length $\xi \sim 1/\Delta \sim 25$ fm. The pinning energy is the difference between the energy of a section of vortex of length b which is centered on a node of the LOFF crystal vs. one which is centered on a maximum of the LOFF crystal. It is of order $E_p \sim F_{\rm LOFF}\, b^3 \sim 4\ {\rm MeV}$. The resulting pinning force per unit length of vortex is of order $f_p \sim E_p/b^2 \sim (4\ {\rm MeV})/(80\ {\rm fm}^2)$. A complete calculation will be challenging because $b < \xi$, and is likely to yield an f_p which is somewhat less than that we have obtained by dimensional analysis. Note that our estimate of f_p is quite uncertain both because it is only based on dimensional analysis and because the values of Δ, b and $F_{\rm LOFF}$ are uncertain. (We have a reasonable understanding of all the ratios Δ/Δ_0, $\delta\mu/\Delta_0$, q/Δ_0 and consequently $b\Delta_0$ in the LOFF phase. It is of course the value of the BCS gap Δ_0 which is uncertain.) It is premature to compare our crude result to the results of serious calculations of the pinning of crustal neutron vortices as in Refs. [110, 111, 108]. It is nevertheless remarkable that they prove to be similar: the pinning energy of neutron vortices in the inner crust is $E_p \approx 1 - 3$ MeV and the pinning force per unit length is $f_p \approx (1 - 3\ {\rm MeV})/(200 - 400\ {\rm fm}^2)$.

The reader may be concerned that a glitch deep within the quark matter core of a neutron star may not be observable: the vortices within the crystalline color superconductor region suddenly unpin and leap outward; this loss of angular momentum is compensated by a gain in angular momentum of the layer outside the LOFF region; how quickly, then, does this increase in angular momentum manifest itself at the *surface* of the star as a glitch? The important point here is that the rotation of any superfluid region within which the vortices are able to move freely is coupled to the rotation of the outer crust on very short time scales [112]. This rapid coupling, due to electron scattering off vortices and the fact that the electron fluid penetrates throughout the star, is usually invoked to explain that the core nucleon superfluid speeds up quickly after a crustal glitch: the only long relaxation time is that of the vortices within the inner crust [112]. Here, we invoke it to explain that the outer crust speeds up rapidly after a LOFF glitch has accelerated the quark matter at the base of the nucleon superfluid. After a glitch in the LOFF region, the only long relaxation times are those of the vortices in the LOFF region and in the inner crust.

A quantitative theory of glitches originating within quark matter in a LOFF phase must await further calculations, in particular a three flavor analysis and the determination of the crystal structure of the QCD LOFF phase. However, our rough estimate of the pinning force on rotational vortices in a LOFF region suggests that this force may be comparable to that on vortices in the inner crust of a conventional neutron star. Perhaps, therefore, glitches occurring in a region of crystalline color superconducting quark matter may yield similar phenomenology to those occurring in the inner crust. This is surely strong motivation for further investigation.

There has been much recent progress in our understanding of how the presence of color superconducting quark matter in a compact star would affect five different phenomena: cooling by neutrino emission, the pattern of the arrival times of supernova neutrinos, the evolution of neutron star magnetic fields, r-mode instabilities and glitches. Nevertheless, much theoretical work remains to be done before we can make sharp proposals for which astrophysical observations can teach us whether compact stars contain quark matter, and if so whether it is in the 2SC or CFL phase and whether it is a crystalline color superconductor.

I am very grateful to the collaborators with whom I have been exploring the condensed matter physics of QCD: Mark Alford, Juergen Berges, Jeff Bowers, Joydip Kundu, Adam Leibovich, Cristina Manuel, Sanjay Reddy, Eugene Shuster and Frank Wilczek. I am also very grateful to Jean-Paul Blaizot and Edmond Iancu for having organized a wonderful school. They created an environment which was stimulating for students and lecturers alike while at the same time the magic of Cargèse left at least this lecturer tanned and relaxed. Research supported in part by the DOE under cooperative research agreement DE-FC02-94ER40818.

References

1. M. Alford, hep-ph/0102047; K. Rajagopal, F. Wilczek, hep-ph/0011333; T. Schäfer, E. Shuryak, nucl-th/0010049; K. Rajagopal, hep-ph/0009058; D. Rischke, R. Pisarski, hep-ph/0004016.
2. D. J. Gross and F. Wilczek, Phys. Rev. Lett. **30**, 1343 (1973); H. D. Politzer, Phys. Rev. Lett. **30**, 1346 (1973).
3. J. C. Collins and M. J. Perry, Phys. Rev. Lett. **34**, 1353 (1975).
4. J. Bardeen, L. N. Cooper and J. R. Schrieffer, Phys. Rev. **106**, 162 (1957); **108**, 1175 (1957).
5. B. Barrois, Nucl. Phys. **B129**, 390 (1977); S. Frautschi: "Asymptotic Freedom And Color Superconductivity In Dense Quark Matter". In: *Proceedings of 1978 Erice workshop "Hadronic matter at extreme density", Erice, Italy, Oct 13-21, 1978*, ed. by N. Cabibbo and L. Sertorio, (Plenum, New York, 1980) pp. 19-27
6. B. Barrois, "Nonperturbative effects in dense quark matter", Cal Tech PhD thesis, UMI 79-04847-mc (1979).
7. D. Bailin and A. Love, Phys. Rept. **107**, 325 (1984), and references therein.
8. M. Alford, K. Rajagopal and F. Wilczek, Phys. Lett. **B422**, 247 (1998) [hep-ph/9711395].
9. R. Rapp, T. Schäfer, E. V. Shuryak and M. Velkovsky, Phys. Rev. Lett. **81**, 53 (1998) [hep-ph/9711396].
10. S. Elitzur, Phys. Rev. **D12**, 3978 (1975).
11. E. Fradkin and S. Shenker, Phys. Rev. **D19**, 3682 (1979); T. Banks and E. Rabinovici, Nucl. Phys. **B160**, 349 (1979).
12. M. Alford, K. Rajagopal and F. Wilczek, Nucl. Phys. **B537**, 443 (1999) [hep-ph/9804403].
13. T. Schäfer and F. Wilczek, Phys. Rev. Lett. **82**, 3956 (1999) [hep-ph/9811473].
14. For a recent review and references, see O. Philipsen, hep-lat/0011019.
15. S. Chandrasekharan and U. Wiese, Phys. Rev. Lett. **83**, 3116 (1999) [cond-mat/9902128].
16. J. Berges and K. Rajagopal, Nucl. Phys. **B538**, 215 (1999) [hep-ph/9804233].
17. M. Alford, J. Berges and K. Rajagopal, Nucl. Phys. **B558**, 219 (1999) [hep-ph/9903502].
18. B. Vanderheyden and A. D. Jackson, Phys. Rev. **D62**, 094010 (2000) [hep-ph/0003150]; S. Pepin, A. Schäfer, hep-ph/0010225
19. G. W. Carter and D. Diakonov, Phys. Rev. **D60**, 016004 (1999) [hep-ph/9812445].
20. R. Rapp, T. Schäfer, E. V. Shuryak and M. Velkovsky, Annals Phys. **280**, 35 (2000) [hep-ph/9904353].
21. N. Evans, S. D. H. Hsu and M. Schwetz, Nucl. Phys. **B551**, 275 (1999) [hep-ph/9808444]; Phys. Lett. **B449**, 281 (1999) [hep-ph/9810514].
22. T. Schäfer and F. Wilczek, Phys. Lett. **B450**, 325 (1999) [hep-ph/9810509].
23. D. T. Son, Phys. Rev. **D59**, 094019 (1999) [hep-ph/9812287].
24. T. Schäfer and F. Wilczek, Phys. Rev. **D60**, 114033 (1999) [hep-ph/9906512].
25. R. D. Pisarski and D. H. Rischke, Phys. Rev. **D60**, 094013 (1999) [nucl-th/9903023]; Phys. Rev. **D61**, 051501 (2000) [nucl-th/9907041]; Phys. Rev. **D61**, 074017 (2000) [nucl-th/9910056];
26. D. K. Hong, Phys. Lett. **B473**, 118 (2000) [hep-ph/9812510]; Nucl. Phys. **B582**, 451 (2000) [hep-ph/9905523].
27. D. K. Hong, V. A. Miransky, I. A. Shovkovy and L. C. Wijewardhana, Phys. Rev. **D61**, 056001 (2000), erratum *ibid.* **D62**, 059903 (2000) [hep-ph/9906478].
28. T. Schäfer and F. Wilczek, Phys. Rev. **D60**, 114033 (1999) [hep-ph/9906512].
29. W. E. Brown, J. T. Liu and H. Ren, Phys. Rev. **D61**, 114012 (2000) [hep-ph/9908248]; Phys. Rev. **D62**, 054016 (2000) [hep-ph/9912409]; Phys. Rev. **D62**, 054013 (2000) [hep-ph/0003199].
30. S. D. Hsu and M. Schwetz, Nucl. Phys. **B572**, 211 (2000) [hep-ph/9908310].

31. T. Schäfer, Nucl. Phys. **B575**, 269 (2000) [hep-ph/9909574].
32. I. A. Shovkovy and L. C. Wijewardhana, Phys. Lett. **B470**, 189 (1999) [hep-ph/9910225].
33. N. Evans, J. Hormuzdiar, S. D. Hsu and M. Schwetz, Nucl. Phys. **B581**, 391 (2000) [hep-ph/9910313].
34. K. Rajagopal and E. Shuster, Phys. Rev. **D62**, 085007 (2000) [hep-ph/0004074].
35. C. Manuel, Phys. Rev. **D62**, 114008 (2000) [hep-ph/0006106].
36. Q. Wang and D. H. Rischke, Phys. Rev. D **65**, 054005 (2002) [nucl-th/0110016].
37. M. Iwasaki, T. Iwado: Phys. Lett. **B350**, 163 (1995); M. Iwasaki: Prog. Theor. Phys. Suppl. **120**, 187 (1995)
38. M. Srednicki and L. Susskind, Nucl. Phys. **B187**, 93 (1981).
39. T. Schäfer and F. Wilczek, Phys. Rev. **D60**, 074014 (1999) [hep-ph/9903503].
40. R. Casalbuoni and R. Gatto, Phys. Lett. **B464**, 111 (1999) [hep-ph/9908227].
41. M. Alford, J. Berges and K. Rajagopal, Phys. Rev. Lett. **84**, 598 (2000) [hep-ph/9908235].
42. D. T. Son and M. A. Stephanov, Phys. Rev. **D61**, 074012 (2000) [hep-ph/9910491]; erratum, *ibid.* **D62**, 059902 (2000) [hep-ph/0004095].
43. M. Rho, A. Wirzba and I. Zahed, Phys. Lett. **B473**, 126 (2000) [hep-ph/9910550].
44. D. K. Hong, T. Lee and D. Min, Phys. Lett. **B477**, 137 (2000) [hep-ph/9912531].
45. C. Manuel and M. H. Tytgat, Phys. Lett. **B479**, 190 (2000) [hep-ph/0001095].
46. M. Rho, E. Shuryak, A. Wirzba and I. Zahed, Nucl. Phys. **A676**, 273 (2000) [hep-ph/0001104].
47. K. Zarembo, Phys. Rev. **D62**, 054003 (2000) [hep-ph/0002123].
48. S. R. Beane, P. F. Bedaque and M. J. Savage, Phys. Lett. **B483**, 131 (2000) [hep-ph/0002209].
49. D. H. Rischke, Phys. Rev. **D62**, 054017 (2000) [nucl-th/0003063].
50. D. K. Hong, Phys. Rev. **D62**, 091501 (2000) [hep-ph/0006105].
51. T. Schäfer, nucl-th/0007021.
52. M. A. Nowak, M. Rho, A. Wirzba and I. Zahed, hep-ph/0007034.
53. V. A. Miransky, I. A. Shovkovy and L. C. Wijewardhana, hep-ph/0009173.
54. C. Manuel and M. Tytgat, hep-ph/0010274.
55. R. Casalbuoni, R. Gatto and G. Nardulli, hep-ph/0010321.
56. P. F. Bedaque and T. Schäfer, hep-ph/0105150. See also D. B. Kaplan and S. Reddy, hep-ph/0107265.
57. V. A. Miransky and I. A. Shovkovy, hep-ph/0108178; T. Schafer, D. T. Son, M. A. Stephanov, D. Toublan and J. J. Verbaarschot, hep-ph/0108210; D. T. Son, hep-ph/0108260.
58. T. Schaefer, hep-ph/0109052.
59. D. H. Rischke, Phys. Rev. **D62**, 034007 (2000) [nucl-th/0001040].
60. G. Carter and D. Diakonov, Nucl. Phys. **B582**, 571 (2000) [hep-ph/0001318].
61. T. Schaefer, hep-ph/0201189.
62. D. B. Kaplan and A. E. Nelson, Phys. Lett. **B175**, 57 (1986).
63. K. Rajagopal and F. Wilczek, Phys. Rev. Lett. **86**, 3492 (2001) [hep-ph/0012039].
64. M. Alford, J. Berges and K. Rajagopal, Nucl. Phys. **B571**, 269 (2000) [hep-ph/9910254].
65. M. G. Alford, K. Rajagopal, S. Reddy and F. Wilczek, hep-ph/0105009.
66. E. Witten, Phys. Rev. **D30**, 272 (1984); E. Farhi and R. L. Jaffe, Phys. Rev. **D30**, 2379 (1984); P. Haensel, J. L. Zdunik and R. Schaeffer, Astron. Astrophys. **160**, 121 (1986); C. Alcock, E. Farhi and A. Olinto, Phys. Rev. Lett. **57**, 2088 (1986); Astrophys. J. **310**, 261 (1986).
67. C. Manuel and K. Rajagopal, Phys. Rev. Lett.
68. D. F. Litim and C. Manuel, hep-ph/0105165.
69. J. D. Jackson, "Classical Electrodynamics", 3rd Ed., (Wiley, New York, 1998).
70. G. 't Hooft, in "High Energy Physics," ed. A. Zichichi (Editrice Compositori, Bologna, 1976); S. Mandelstam, Phys. Rept. **23**, 245 (1976).
71. R. L. Jaffe, Phys. Rev. Lett. **38**, 195, 617(E) (1977).

72. D. Rischke, D. Son, M. Stephanov: hep-ph/0011379
73. R. D. Pisarski and D. H. Rischke, Phys. Rev. Lett. **83**, 37 (1999) [nucl-th/9811104].
74. V. A. Miransky, I. A. Shovkovy and L. C. Wijewardhana, Phys. Rev. **D62**, 085025 (2000) [hep-ph/0009129].
75. F. Sannino, Phys. Lett. **B480**, 280 (2000) [hep-ph/0002277]; R. Casalbuoni, Z. Duan and F. Sannino, hep-ph/0004207; S. D. Hsu, F. Sannino and M. Schwetz, hep-ph/0006059.
76. For a recent exploration of the μ-dependence of m_s, see M. Buballa and M. Oertel, hep-ph/0109095.
77. UKQCD Collaboration, Phys. Rev. **D59** (1999) 116002; S. Hands, J. B. Kogut, M. Lombardo and S. E. Morrison, Nucl. Phys. **B558**, 327 (1999) [hep-lat/9902034]; S. Hands, I. Montvay, S. Morrison, M. Oevers, L. Scorzato and J. Skullerud, hep-lat/0006018.
78. J. B. Kogut, M. A. Stephanov and D. Toublan, Phys. Lett. **B464**, 183 (1999) [hep-ph/9906346]; J. B. Kogut, M. A. Stephanov, D. Toublan, J. J. Verbaarschot and A. Zhitnitsky, Nucl. Phys. **B582**, 477 (2000) [hep-ph/0001171].
79. D. V. Deryagin, D. Yu. Grigoriev and V. A. Rubakov, Int. J. Mod. Phys. **A7**, 659 (1992).
80. E. Shuster and D. T. Son, Nucl. Phys. **B573**, 434 (2000) [hep-ph/9905448].
81. B. Park, M. Rho, A. Wirzba and I. Zahed, Phys. Rev. **D62**, 034015 (2000) [hep-ph/9910347].
82. R. Rapp, E. Shuryak and I. Zahed, hep-ph/0008207.
83. D. T. Son and M. A. Stephanov, hep-ph/0005225.
84. A. I. Larkin and Yu. N. Ovchinnikov, Zh. Eksp. Teor. Fiz. **47**, 1136 (1964) [Sov. Phys. JETP **20**, 762 (1965)]; P. Fulde and R. A. Ferrell, Phys. Rev. **135**, A550 (1964).
85. M. Alford, J. Bowers and K. Rajagopal, Phys. Rev. D **63**, 074016 (2001) [hep-ph/0008208].
86. J. A. Bowers, J. Kundu, K. Rajagopal and E. Shuster, Phys. Rev. D **64**, 014024 (2001) [hep-ph/0101067].
87. R. Casalbuoni, R. Gatto, M. Mannarelli and G. Nardulli, Phys. Lett. B **511**, 218 (2001) [hep-ph/0101326]; R. Casalbuoni, these proceedings, hep-th/0108195.
88. A. K. Leibovich, K. Rajagopal and E. Shuster, hep-ph/0104073.
89. J. Kundu and K. Rajagopal, hep-ph/0112206.
90. A. M. Clogston, Phys. Rev. Lett. **9**, 266 (1962); B. S. Chandrasekhar, App. Phys. Lett. **1**, 7 (1962).
91. P. F. Bedaque, hep-ph/9910247.
92. A. I. Buzdin and V. V. Tugushev Zh. Eksp. Teor. Fiz. **85**, 735 (1983) [Sov. Phys. JETP **58**, 428 (1983)]; A. I. Buzdin and S. V. Polonskii, Zh. Eksp. Teor. Fiz. **93**, 747 (1987) [Sov. Phys. JETP **66**, 422 (1987)].
93. K. Splittorff, D. T. Son and M. A. Stephanov, hep-ph/0012274.
94. A. Sedrakian, nucl-th/0008052. The related unpairing transition was discussed in the absence of LOFF pairing in A. Sedrakian and U. Lombardo, Phys. Rev. Lett. **84**, 602 (2000).
95. For a review, see H. Heiselberg and M. Hjorth-Jensen, Phys. Rept. **328**, 237 (2000).
96. N. K. Glendenning, Phys. Rev. D **46**, 1274 (1992).
97. M. Prakash, J. R. Cooke and J. M. Lattimer, Phys. Rev. D **52**, 661 (1995).
98. N. K. Glendenning and S. Pei, Phys. Rev. C **52**, 2250 (1995).
99. D. Blaschke, T. Klahn and D. N. Voskresensky, Astrophys. J. **533**, 406 (2000) [astro-ph/9908334]; D. Blaschke, H. Grigorian and D. N. Voskresensky, astro-ph/0009120.
100. D. Page, M. Prakash, J. M. Lattimer and A. Steiner, Phys. Rev. Lett. **85**, 2048 (2000) [hep-ph/0005094].
101. C. Schaab *et al*, Astrophys. J. Lett **480** (1997) L111 and references therein.
102. T. Schäfer, Phys. Rev. **D62**, 094007 (2000) [hep-ph/0006034].
103. G. W. Carter and S. Reddy, Phys. Rev. **D62**, 103002 (2000) [hep-ph/0005228].
104. D. Blaschke, D. M. Sedrakian and K. M. Shahabasian, Astron. and Astrophys. **350**, L47 (1999) [astro-ph/9904395].

105. For reviews, see J. Sauls, in Timing Neutron Stars, J. Ögleman and E. P. J. van den Heuvel, eds., (Kluwer, Dordrecht: 1989) 457; and D. Bhattacharya and G. Srinivasan, in X-Ray Binaries, W. H. G. Lewin, J. van Paradijs, and E. P. J. van den Heuvel eds., (Cambridge University Press, 1995) 495.
106. M. A. Alpar and C. Ho, Mon. Not. R. Astron. Soc. **204**, 655 (1983). For a recent review, see A.G. Lyne in *Pulsars: Problems and Progress*, S. Johnston, M. A. Walker and M. Bailes, eds., 73 (ASP, 1996).
107. J. Negele and D. Vautherin, Nucl. Phys. **A207**, 298 (1973).
108. For reviews, see D. Pines and A. Alpar, Nature **316**, 27 (1985); D. Pines, in *Neutron Stars: Theory and Observation*, J. Ventura and D. Pines, eds., 57 (Kluwer, 1991); M. A. Alpar, in *The Lives of Neutron Stars*, M. A. Alpar et al., eds., 185 (Kluwer, 1995). For more recent developments and references to further work, see M. Ruderman, Astrophys. J. **382**, 587 (1991); R. I. Epstein and G. Baym, Astrophys. J. **387**, 276 (1992); M. A. Alpar, H. F. Chau, K. S. Cheng and D. Pines, Astrophys. J. **409**, 345 (1993); B. Link and R. I. Epstein, Astrophys. J. **457**, 844 (1996); M. Ruderman, T. Zhu, and K. Chen, Astrophys. J. **492**, 267 (1998); A. Sedrakian and J. M. Cordes, Mon. Not. R. Astron. Soc. **307**, 365 (1999).
109. P. W. Anderson and N. Itoh, Nature **256**, 25 (1975).
110. M. A. Alpar, Astrophys. J. **213**, 527 (1977).
111. M. A. Alpar, P. W. Anderson, D. Pines and J. Shaham, Astrophys. J. **278**, 791 (1984).
112. M. A. Alpar, S. A. Langer and J. A. Sauls, Astrophys. J. **282**, 533 (1984).